CELL MEMBRANES

Biochemistry, Cell Biology & Pathology

Adapted from
Hospital Practice

Illustrated by
Bunji Tagawa
and
Albert Miller
(*charts and graphs*)
and other contributing artists.
For specific illustration and data source credits, see page 283.

Designed by
Robert S. Herald

CELL MEMBRANES
Biochemistry, Cell Biology & Pathology

Edited by

Gerald Weissmann, M.D.

Professor of Medicine,
Director, Division of Rheumatology, Department of Medicine,
New York University Medical Center School of Medicine, New York

and

Robert Claiborne

Senior Editor, Hospital Practice

HP Publishing Co., Inc. • Publishers • New York, N.Y.

CELL MEMBRANES: Biochemistry, Cell Biology & Pathology
The material in this book has been updated through April 1975.

Printed in the United States of America

HP Publishing Co., Inc.
485 Madison Avenue
New York, N.Y. 10022

ISBN 0-913800-06-6

Library of Congress Catalog Card Number 75-7642

Table of Contents

Contributing Authors

ALEC D. BANGHAM	Head, Biophysics Unit, Agricultural Research Council, Institute of Animal Physiology, Babraham, Cambridge, England.
MAX M. BURGER	Professor and Acting Chairman, Department of Biochemistry, Chairman of the Biocenter of the University of Basel, Switzerland.
DENNIS CHAPMAN	Professor, Department of Chemistry, University of Sheffield, England.
PEDRO CUATRECASAS	Professor, Department of Pharmacology and Experimental Therapeutics and Associate Professor of Medicine, Johns Hopkins University School of Medicine, Baltimore.
JAMES F. DANIELLI	Director, Center for Theoretical Biology, State University of New York at Buffalo.
NELSON D. GOLDBERG	Professor of Pharmacology, Laboratory Medicine and Pathology, University of Minnesota Medical School, Minneapolis.
JOSEPH F. HOFFMAN	Eugene Higgins Professor and Chairman, Department of Physiology, Yale University School of Medicine, New Haven.
ERIC HOLTZMAN	Professor, Department of Biological Sciences, Columbia University, New York.
HARRY S. JACOB	Chief, Section of Hematology and Professor of Medicine, University of Minnesota Medical School, Minneapolis.
JAMES D. JAMIESON	Associate Professor, Section of Cell Biology, Yale University School of Medicine, New Haven.
STEPHEN C. KINSKY	Professor of Pharmacology and Biological Chemistry, Department of Pharmacology, Washington University School of Medicine, St. Louis.
WERNER R. LOEWENSTEIN	Professor and Chairman, Department of Physiology and Biophysics, University of Miami School of Medicine.
JACK A. LUCY	Professor of Biochemistry, University of London; Head, Department of Biochemistry and Chemistry, Royal Free Hospital School of Medicine, London.
VINCENT T. MARCHESI	Professor and Chairman, Department of Pathology, Yale University School of Medicine, New Haven.
GEORGE D. PAPPAS	Professor of Neuroscience and Anatomy, Rose F. Kennedy Center for Research in Mental Retardation and Human Development, Albert Einstein College of Medicine, New York.
EFRAIM RACKER	Albert Einstein Professor of Biochemistry, Molecular and Cell Biology, Cornell University Division of Biological Sciences, Ithaca.
HOWARD RASMUSSEN	Benjamin Rush Professor of Biochemistry, University of Pennsylvania School of Medicine, Philadelphia.

SAUL ROSEMAN — Professor, Department of Biology, McCollum-Pratt Institute, Johns Hopkins University, Baltimore.

PHILIP SEEMAN — Professor of Pharmacology, Faculty of Medicine, University of Toronto, Canada.

PHILIP SIEKEVITZ — Professor of Cell Biology, Rockefeller University, New York.

S. J. SINGER — Professor of Biology, University of California at San Diego, La Jolla.

BENJAMIN F. TRUMP — Professor and Chairman, Department of Pathology, University of Maryland School of Medicine, Baltimore.

JONATHAN W. UHR — Professor and Chairman, Department of Microbiology, and Professor of Medicine, University of Texas Southwestern Medical School, Dallas.

LEONARD WARREN — American Cancer Society Professor of Therapeutic Research, University of Pennsylvania School of Medicine, Philadelphia.

GERALD WEISSMANN — Professor of Medicine and Director, Division of Rheumatology, Department of Medicine, New York University Medical Center.

VICTOR P. WHITTAKER — Director and Chairman, Department of Neurochemistry, Max-Planck-Institute for Biophysical Chemistry, Göttingen, Federal Republic of Germany.

Acknowledgments

Because "CELL MEMBRANES: Biochemistry, Cell Biology & Pathology" is a compendium of articles originally published in HOSPITAL PRACTICE (and subsequently updated for book publication), many individuals made significant contributions toward its preparation in both of its incarnations. This introductory note is an effort, by no means all-inclusive, to give credit where it is due.

During the course of publication of the various articles in HOSPITAL PRACTICE, Gertrude Halpern served as managing editor and later as associate editorial director of the publication. In these roles, she made invaluable contributions to assuring the clarity of verbal presentation, as well as to the integration of illustrative material with the text.

The design of the articles, as well as of the book, was the responsibility of Robert S. Herald, our art director, with the able assistance of the members of the HOSPITAL PRACTICE art staff. Adele Spiegler and Herb Yavel were most skillful in preparing the actual pages for production and printing.

We were fortunate in obtaining the services of Ruth M. Kimmerer in providing the index.

Finally, special appreciation must be expressed to Herb Cornell, Director of the Book Division, who coordinated the entire publishing effort with tireless help from Angel Kuchinski on the administrative side, and from Esther Goldman and Katherine Bloch, editorially.

DAVID W. FISHER
President
HP Publishing Company, Inc.

Foreword

Suddenly it is all membranes. Activities, speculations, and enthusiasms that were devoted to informational macromolecules in the decade between 1955 and 1965 have now swung towards problems of biologic membranes. Thus, although the primary structures and functions of DNA, RNA, and proteins have by no means been entirely solved, the major outlines of this area have been sketched out, and much of what is being done seems to be a kind of filling in. This is not true for the field of membrane research. The unknown is still there, seemingly just around the corner of the nearest electron microscope or behind the next mass spectrometer. Scientists and clinicians, trained in a broad range of disciplines and outfitted from a rich arsenal of techniques, are beginning to sort out the structure and function of biologic membranes. The present volume has been designed to summarize what we know in this field and to chart in diagrammatic–though by no means unduly oversimplified–fashion those features of the territory of biomembranes that have caused this new landscape to capture the imagination of experimentalists.

This volume deals with mammalian cell membranes (both the external plasma membrane and intracellular membranes) and has naturally divided itself into three distinct groupings: 1) biochemistry, 2) cell biology, and 3) pathology. The first section presents a description of the major topographic features common to biologic membranes—their architecture, their arrangement in space, and their configuration—and provides an analysis of their building blocks. We will therefore examine the lipids, proteins, and sugars of cell membranes and determine how these may arrange themselves into *ideal* (or nonexistent!) membranes, such as those of model systems. We will then describe how they are actually disposed in the real membranes of animal cells from which they have been isolated.

In the second section, dealing with cell biology and physiology of biologic membranes, we learn how cells join together, how they communicate, how they transport ions in order to maintain microenvironments for the capture of energy, and how they function in the transmission of neural or humoral signals. Then, beginning with the outer plasma membrane, the authors explain how the various intracellular membranes—those of the mitochondria, the endoplasmic reticulum, and the lysosomes—relate to cell economy in general, and how the important metabolic transformations are regulated.

In the third section, a number of chapters deal with specific areas of experimental pathology where membrane research has already borne fruit with respect to analysis of human disease. These include discussions of membrane changes in neoplastic cells, genetic diseases, immunologic reactions, and tissue injury, as well as a discussion of membranes as receptors for pharmacologically-active substances. Throughout the book, emphasis has been placed not only upon an analysis of the membranes themselves, but also upon how properties of membranes can explain the vital functions of cells and organelles. To this end, extensive discussions have been included on how ions, cyclic nucleotides, and hormones influence processes that could not exist save for the compartmentalization of molecules in space: the major function of membranes in animal cells.

How would one define a membrane? I think most biologists would agree that a biologic membrane is a structure that separates cells or organelles into compartments, that this structure is composed of various proportions of lipids and proteins (together with appropriate prosthetic groups of sugars), and that it possesses a characteristic trilaminar structure (approximately 75 Å) when viewed by electron microscopy. If one considers the cell as an individual, its membranes may be thought of as an essential organ that serves the cell 1) by performing a variety of functions, such as osmotic regulation, the provision of localized bursts of energy, and the communication of signals to the outside; and 2) by maintaining discrete internal environments within which important biochemical reactions can proceed in optimum fashion. However, I think a more poetic description of biomembranes has been given by Lewis Thomas who, in his recent book, "Lives of a Cell," suggests:

> It takes a membrane to make sense out of disorder in biology. You have to be able to catch energy and hold it, storing precisely the needed amount and releasing it in measured shares. A cell does this, and so do the organelles inside. Each assemblage is poised in the flow of solar energy, tapping off energy from metabolic surrogates of the sun. To stay alive, you have to be able to hold out against equilibrium, maintain imbalance, bank against entropy, and you can only transact this business with membranes in our kind of world.

Evidence for this poetic view of membrane function must surely come from a consideration of biologic evolution. In the beginning, there must have been a membrane! Whatever flash of lightning there was that organized purines, pyrimidines, and amino acids into macromolecules capable of reproducing themselves, it would not have yielded cells but for the organizational trick afforded by the design of a membrane wrapping. Macromolecules are unlikely to have found each other in the big reaction vessel of the pre-Cambrian sea except in a limited, discrete environment. Even if the first DNA were to have persuaded the first RNA to have synthesized the first protein within the limitless ocean tides, both enzyme and product would have been diffused away. Consequently, one really has to postulate the presence of a kind of lipid bubble enclosing the first, primitive products of these tentative synthetic reactions. To put it another way, whatever macromolecules were first engendered in the unfriendly ocean buffer, these would readily have been dissipated or dissolved in the high salt of the early sea, were they not in some way originally sequestered by what we must now recognize as a membrane. One of the striking experiences that has evolved from studies with model lipid structures (such as those described by A.D. Bangham in Chapter 3) has been that simple lipids in salt solutions tend to form closed, liquid-crystalline, bilayer structures that make it possible to capture substances originally present in the fluid phase. These structures, which we have termed liposomes, have been used experimentally to capture such things as enzymes, macromolecules, and dyes. Were lipids, of whatever origin, to have found themselves in the vicinity of primordial reactions involving DNA, RNA, and protein synthesis, it would not be too difficult to imagine them forming self-assembled bubbles within which to segregate this new thing—life, as it were—from the hostile sea.

Whatever be the implications for modern cell biology of this ontologic hypothesis, it is clear that the quest for insight into the structure and function of biologic membranes did not begin with speculation. The present revolution in our state of knowledge regarding biologic membranes has not been due to advances in theoretical biology. It might be argued that most of the solid knowledge of biomembranes that we do possess has resulted directly from advances in scientific technology and instrumentation, advances that usually precede biologic innovation in any area. Thus, it would be impossible for us to present these new views in any enlightened fashion whatsoever without the newer techniques of preparative biochemistry, protein isolation, high resolution electron microscopy, x-ray crystallography, freeze-fracture and freeze-etch techniques, and, perhaps above all, newer spectroscopic techniques, such as circular dichroism, electron spin resonance, and nuclear magnetic resonance.

Indeed, our paradigm of the generalized biologic membrane (the Singer - Nicolson model shown on the jacket and on page 37 of this volume) grew out of the spectroscopic and ultrastructural evidence that proteins, both extrinsic and intrinsic to biologic membranes, seem to float in a sea of lipids. The forces governing the assemblage and fluid motion of these constituents are inherent in the chemical structures of the individual molecules. This formulation would have been only theoretical and not susceptible to experimental validation or refutation were it not for general availability of the techniques outlined above.

To a large degree the chapters in this volume will demonstrate not only that this paradigm structure bears some resemblance to many fundamental features of all biologic membranes, but that inherent in the model is a considerable explanatory potential. It is, for example, clear that many important biologic communications, such as those between cells and their neighbors, or between cells and their environment, are transduced by proteins intimately associated with the membrane. Plasma and intracellular membranes are called upon not only to segregate specialized functions within cells but also to regulate the foreign affairs of cells and organelles; and well defined receptors (some protein, some lipid) serve this embassy. Interactions of some membrane proteins with other proteins or with membrane lipids generate, by launching biochemical reactions, a series of messengers (e.g., cyclic nucleotides) to other membrane-bound structures within the cell in order to evoke what we usually call a *physiologic response*. Not only technologic but also conceptual advances coming from many disciplines (immunology, genetics, pharmacology, etc.) have made it possible to probe in detail the means whereby membranes translate external messages into an internal stimulus-response. Therefore, a general topography of membranes includes well-defined structural and functional sites, such as ion pumps, transplantation antigens, receptors for neurohumors, intercellular junctions, immunoglobulins, etc., to the point where the sea of lipid is as well charted as the Bermuda Triangle, and perhaps as treacherous!

In one sense, therefore, this volume can be considered an introduction to modern biochemistry and cell biology viewed from the outside in, as it were, beginning with the external plasma membrane and continuing through the cytosol to the organelles and the lurking assembly of the nucleus. Unlike more traditional outlines, which begin with analyses of the chemistry of informational macromolecules, our introduction will be as much morphologic as it will be biochemical or functional. That's necessary because it is in the Lucretian nature of membranes that they appear to divide cells and organs into units of space so organized as to form the specialized reaction flasks required for biochemical processes. For it is no longer possible to view biochemical reactions or physiologic responses as proceeding in two dimensions, once it has been grasped that each of the relevant events occurs not only in time, but in space. Consequently, a volume of this sort, filled with a large number of diagrams, electron photomicrographs, and pictures of lipid assemblies, should prove instructional for students in all branches of biologic science. It is, I think, our conception that membranes define the basic spatial lattice of living things that has now grasped our imagination and that promises to unravel the tangled biologic problems of the future. If this volume of elegant pictures and solid evidence of experimental achievement helps the reader to share our interest in the field, it will have realized its purpose.

GERALD WEISSMANN, M.D.

New York
May, 1975

Section One

Biochemistry

1

The Bilayer Hypothesis Of Membrane Structure

JAMES F. DANIELLI

State University of New York at Buffalo

It is unlikely that there are living organisms, no matter how simple, that are not physically separated from the environment.

This principle – which was emphasized by the Soviet physiologist A. I. Oparin in his classic work, *The Origin of Life* – is one of those fundamental truths that are completely obvious once they have been stated. Since a living organism, by definition, differs from its environment in both its physical properties and its chemical composition, it must – also by definition – possess some structure that can maintain the existence of these differences by walling off the organism from its surroundings. The principle, which Oparin developed in relation to the hypothetical evolution of the first single-celled organisms, applies equally to the individual cells of any multicelled organism; in both cases, the "wall" is the cell membrane.

At this point, a moment's thought will tell us that separating cell and environment is only part of the role of the membrane. No living organism, or cell, is an island; the very fact that it lives means that it interacts dynamically with its environment, taking in needed substances and emitting its own characteristic products and byproducts. Thus the cell membrane, in addition to separating the interior of the cell from its surroundings, must also make possible the selective interaction of these two regions, between which substances and information are continually exchanged.

When I first began studying cell membranes during the 1930's, about the time Oparin was writing his book, these evolutionary and philosophic considerations were admittedly not much in my mind. As a physical chemist, I was much less interested in cell function than in the chemical and physical properties of membranes and the molecules that compose them, i.e., molecular biology. Nonetheless, from these essentially physicochemical considerations I was able, in discussions with E. Newton Harvey and Hugh Davson, to develop a model of membrane structure that fulfilled the membrane's functional properties as defined above. The basic features of this model, I think it fair to say, have survived all efforts to demolish them.

In what might be called the folklore of cytology, the so-called Danielli-Davson bilayer model of membrane structure is primarily associated with the behavior of lipids in membranes. This strikes me as something of an irony. While the theory did indeed involve certain postulates about membrane lipid molecules, these cannot, in my judgment, be considered as the most important part even of my original contribution, or the most difficult part to make, since they flowed almost automatically from consideration of the basic properties of these molecules and would have been obvious to any competent physical chemist. Indeed, as I later discovered, they had been anticipated some 10 years earlier by two Dutch investigators, Gorter and Grendel. Much more important, in my view, were those aspects of the model that concerned the behavior of proteins in the membrane – problems widely considered today to have been first tackled by other and later researchers. Nonetheless, even though the lipid part of the model was very easily conceived in the light of available theory, an understanding of it is essential to an understanding of the whole theory, so it is a reasonable point at which to begin the discussion.

Simple hydrocarbons, such as those that make up gasoline and kerosene, and the larger part of lipids, such as the fatty acids found in animal and vegetable fats, are nonpolar; that is, they carry no net electrical charge or dipoles. Water, by contrast, is a strongly polar molecule.

Polar Lipid in Water

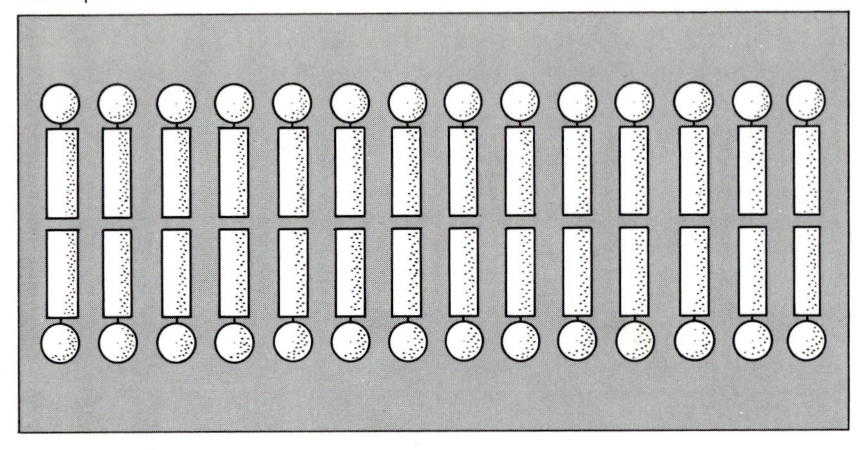

Polar Lipid in Air

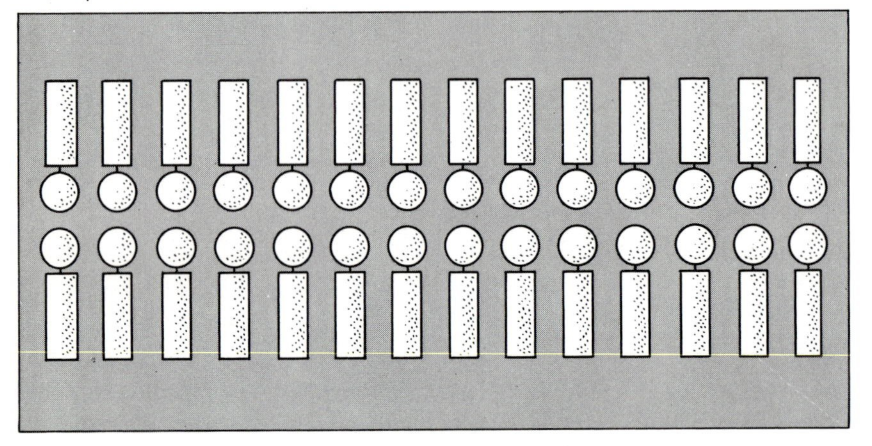

Bilayer of polar lipid in water (top) has hydrophilic "heads" outside in contact with water molecules, with hydrophobic "tails" inside facing one another. In air, bilayer shows inverse structure, with tails facing outward, heads inward, as in soap bubbles.

Water and other polar molecules attract one another strongly. Similarly, a polar group on an otherwise nonpolar molecule interacts strongly with water. Hydrocarbons, on the other hand, interact with water much less strongly. Consequently, if nonpolar hydrocarbons and polar molecules, such as water, are mixed together, the nonpolar molecules and the polar molecules tend to separate from one another, primarily because the polar molecules attract one another so strongly and interact with nonpolar molecules to a much lesser extent. Thus from a mixture of polar and nonpolar molecules, the nonpolar components tend to be squeezed out into a nonpolar phase. Where a molecule contains both polar and nonpolar parts, it tends to arrange itself so that the polar part of the molecule can interact with polar molecules and the nonpolar part of the molecule interacts with nonpolar molecules, e.g., at an interface between hydrocarbon and water.

The lipids found in membranes are more complex structures than the simple hydrocarbons; these molecules consist of nonpolar hydrocarbon combined with various other groups – phosphate groups in particular – that can take on an electrical charge. As a result, while the hydrocarbon end of the molecule is still nonpolar, or hydrophobic, the other end is polar, or hydrophilic; in effect, one end will mix with water, the other will not. This places these substances in the larger class of "surface active" substances, which includes the soaps and detergents.

When a membrane lipid (or other surface active substance) is placed in an aqueous solution, its molecules will tend to arrange themselves so that the hydrophilic "heads" will remain in contact with the water molecules, while the hydrophobic "tails" will orient themselves toward nonpolar space, e.g., the air or the container – or other tails. Under certain circumstances, the molecules may form tiny, spheroidal micelles, with water-facing exteriors consisting of heads and interiors of tails. More commonly, the lipid molecules will tend to arrange themselves in a double layer, both of whose surfaces will consist of heads, with their tails facing one another inside the bilayer membrane. This lipid layer is like a liquid crystal.

As I have already noted, this concept was, as we now say, no big deal. The basic physics had already been worked out by physicists such as Irving Langmuir, and, given the theory, the bilayer was simply the most efficient (and therefore the most probable) way for lipid molecules to arrange themselves consistent with minimization of free energy. Indeed, a comparable concept had been put forward years before to explain the structure of soap bubbles. Here, of course, the structure is the inverse of the cell-membrane bilayer: the hydrophobic tails of the soap molecules face outward, toward the air inside and outside the bubbles, while their hydrophilic heads face inward.

A cell membrane is, of course, much more durable than a soap bubble. The physical properties of cell membranes are different in many ways from those of simple lipid membranes. Evidently, then, the membranes, though they certainly contained lipid, must also contain something else. It seemed very probable that the "something else" was protein. We had already experimented with adding proteins such as hemoglobin and ovalbumin to solutions in contact with lipid-water interfaces, and found that they adsorbed on the interfaces. Indeed, given the thermodynamics of the situation, there was, so to speak, no way in which the proteins could *not* be adsorbed, barring some special mechanism for keeping them away (for which there was no evidence).

At first we believed that the proteins were just loosely attached to the two surfaces of the membrane by polar forces. As our original model, then, we visualized a lipid bilayer more or less covered on both sides with molecules of globular protein, their polar areas facing outward and their nonpolar areas enclosed in the interior of the globule, well away from the surrounding water and the polar surface of the lipid bilayer.

Within a couple of years, however, it became apparent that this model was inadequate to explain the actual behavior of proteins vis-à-vis lipids. Experiments showed that they were by no means loosely attached to the lipids, but rather very tightly bound. For example, when we took a solution of hemoglobin in water and brought it into contact with a water-lipid interface, the adsorbed protein completely lost its solubility; calculations indicated that the free energy of adsorption was of the order of 100,000 calories per mole, arising from interaction between the nonpolar moieties of the protein and lipid molecules. This energy was normally available for stabilizing the globular protein, but in "unrolled" proteins it was available for adsorption to nonpolar surfaces.

As our second model, then, we visualized a situation in which the proteins had somehow become unrolled from their normal globular structure, with portions of them intimately bound to the liquid lipid bilayer. Exactly how this could occur was unclear; it must be remembered that at this time, more than 30 years ago, far less was known about protein structures than at present. But it seemed plausible that the cell membrane was

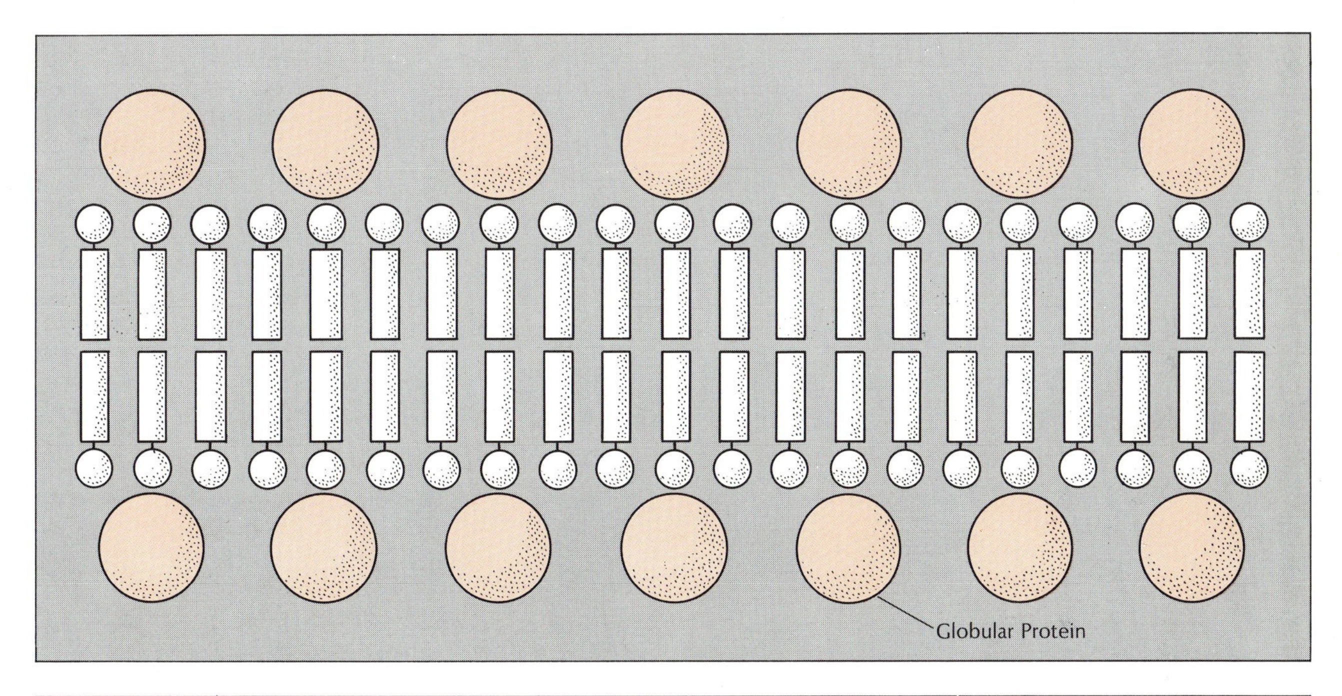

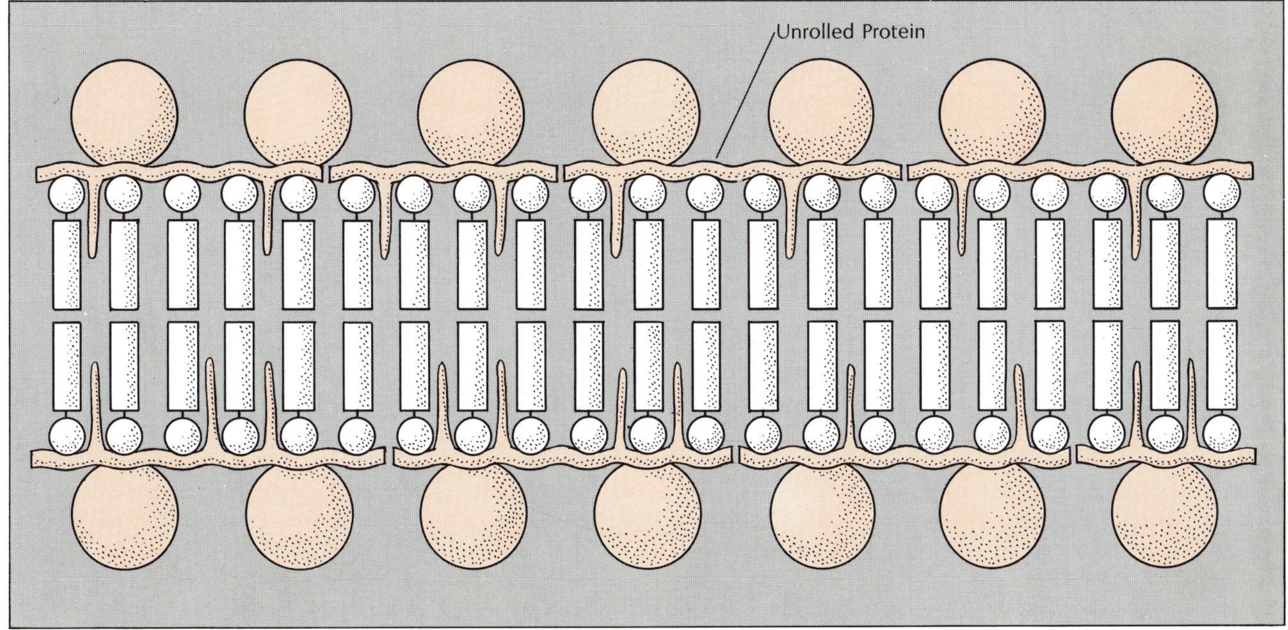

Original (1934) schematic model of cell membrane (top) included lipid bilayer sandwiched between layers of adsorbed globular protein. When experiments showed that protein was actually bound tightly to the lipids, this model was revised in 1937 (lower drawing) to include two additional layers of "unrolled" protein in close contact with the lipid layers, with globular proteins outside as before. The unrolled protein interacts with the lipid through both polar and nonpolar forces.

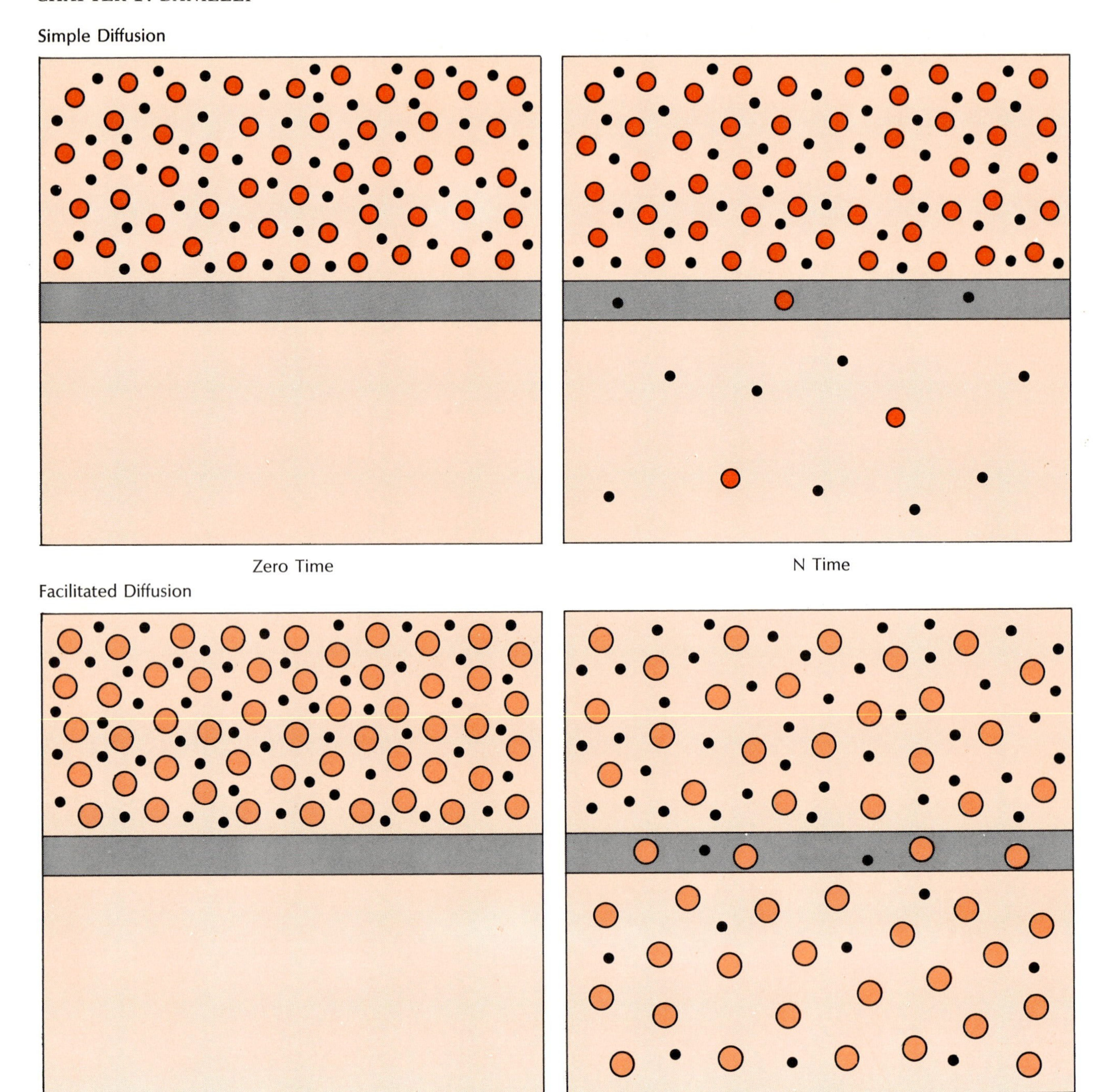

In simple diffusion (top), molecules pass through a membrane at rates determined by their molecular size, the smaller molecules tending to diffuse more rapidly. Facilitated diffusion (bottom) permits far greater quantities of certain molecules (pink) to pass the membrane in a given time – but only to the point where concentrations on both sides of the membrane are equalized.

a sort of sandwich – two slices of lipid between two slices of flattened-out protein – and probably with a "garnish" of globular protein over all, so that the basic membrane consisted of six layers of molecules. Thus, my first model was considerably revised by 1937 to take account of nonpolar forces binding proteins to membranes.

So far, we had pretty well limited ourselves to drawing pictures based on atom models and on general considerations of thermodynamics – and anybody can draw pictures. The problem then was to discover whether our revised picture could account for the permeability properties of membranes.

The most obvious properties of membranes are precisely those suggested by the basic functions noted at the beginning of this chapter: most substances pass through them very slowly (i.e., they separate the cell from its environment) but some very rapidly (i.e., they permit the selective interchange of substances between cell and environment). Through studies extending over the period 1937-41, I was able to demonstrate that for most substances the membrane behaves, *as a first approximation*, like a simple lipid bilayer. That is, most substances pass through the membrane, by simple diffusion, in proportion to their solubility in lipids and also their molecular size. This diffusion rate is far less (by a factor of 10^6 to 10^9) than the "free" diffusion rate that would obtain in the absence of the membrane. Evidently the membrane acts as a remarkably efficient "wall," especially considering that its lipid bilayer is of the order of only 50 angstroms (Å) thick. But a few molecular species move across the membrane much faster than would be predicted from their lipid solubility or size; these include urea, glycerol, and glucose with many cells, as well as the sodium and potassium ions for most cells.

The abnormally rapid movement of all these "preferred" substances is sometimes lumped together under the heading of active transport, but in fact the phenomena involved are of two distinct kinds. With such molecules as urea, glycerol, and glucose, and also the amino acids, the movement is often abnormally rapid, but will not occur against a concentration gradient, as is the case with simple diffusion. With glucose, movement may occur at up to 100,000 times the estimated rate for a simple lipid membrane, based on its molecular size and interaction with lipids. But if the concentration of glucose is greater outside the cell than inside (or vice versa), the sugar will rapidly pass through the membrane until the concentrations are equalized, and thereafter no further changes will take place. The process, moreover, requires no energy, i.e., it is not hampered when cellular respiration is blocked with poisons such as cyanide. Thus in all respects it seems to operate much like ordinary diffusion, but much faster, for which reason I called it "facilitated diffusion."

In the case of the sodium and potassium ions, on the other hand, often some mechanism actually "pumps" them across the membrane against a gradient, enabling the cell to maintain an ionic concentration greater or less than that of the surrounding medium; as one would expect, this active transport process requires the expenditure of energy, supplied either by respiration or (in some cases) glycolysis. It is only an energized process that mer-

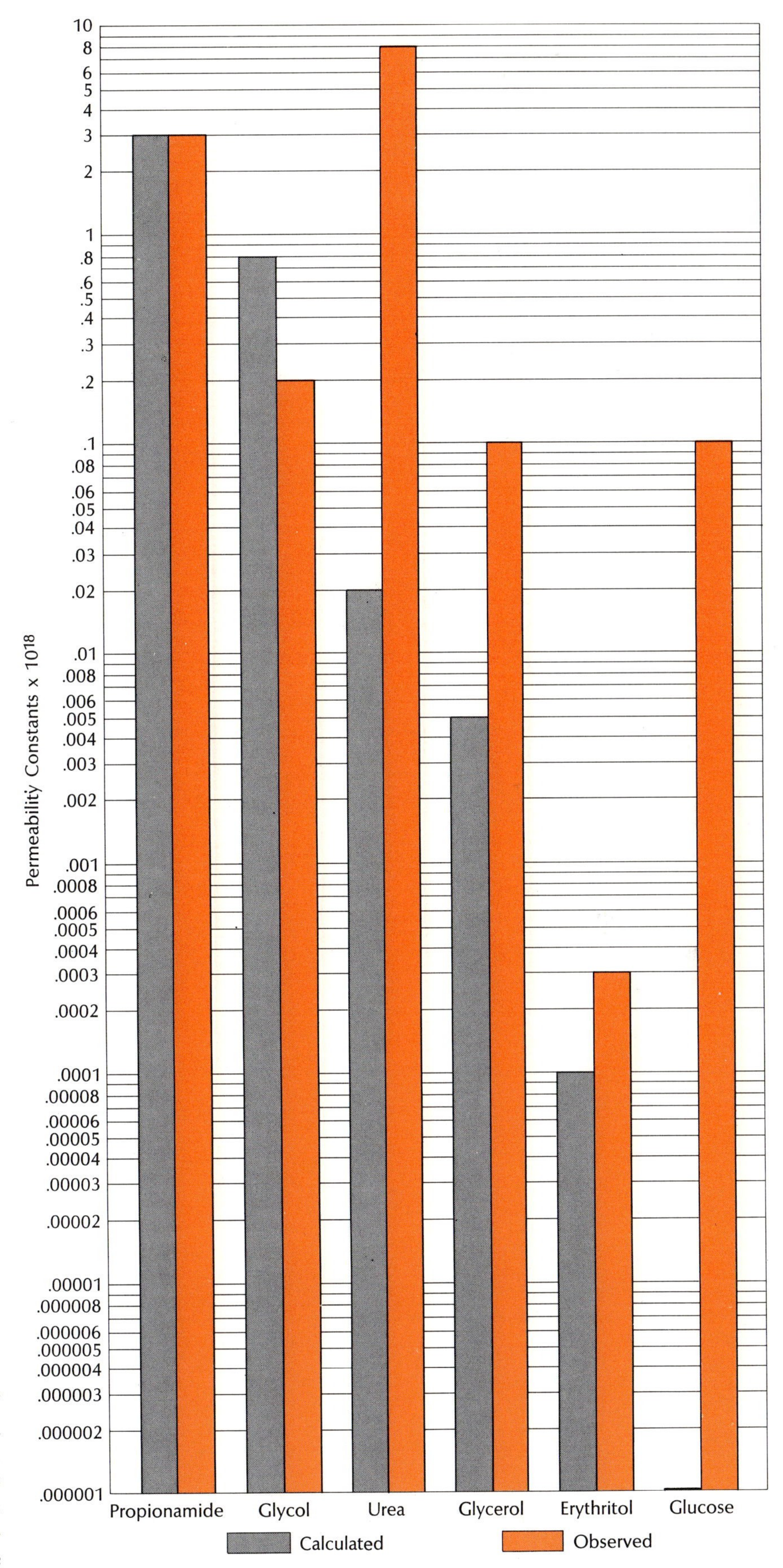

Facilitated diffusion is demonstrated by comparing permeability constants calculated for a simple membrane model with actual constants observed in studies with erythrocyte membrane. Urea, glycerol, and glucose all diffuse far faster than predicted.

its the name of active transport.

Leaving aside the detailed molecular mechanics of active transport – about which very little is known now and even less was known then – it appeared that the membrane "sandwich" must contain "pores" or "active regions" of some sort, resembling enzymes, impermeable to most types of molecules, but through which selected molecules could diffuse at their characteristically rapid rates without passing through the lipid barrier. It was even possible to calculate how much of the membrane surface was involved in these active regions. Since it had been known for some time that some active transports and facilitated diffusions could be blocked by enzyme poisons such as copper, we were able to determine, by measuring the quantity of copper necessary completely to block facilitated diffusion of glycerol, that not more than 1% of the membrane surface was involved. In the case of glucose, this meant that its passage through the active regions must be more than a million times the rate through the rest of the membrane.

Though it seemed clear that both facilitated diffusion and active transport must involve enzyme-like "patches" of this sort on – and through – the membrane surface, it was some time before we were able to propose a plausible mechanism whereby such a mosaic of lipid and protein could exist. The "patches" or "pores" were particularly hard to visualize. Clearly they must be hydrophilic in nature, since they permitted the passage of such hydrophilic molecules as glucose, urea, and so on, which would be blocked by a hydrophobic structure. Yet a hydrophilic opening in the hydrophobic lipid bilayer would normally be expected to enlarge, because of surface tension, to the point where the membrane would be destroyed – and this was obviously not happening.

By the early 1950's, however, advances in the understanding of protein structure suggested a physically plausible structure that could explain both the tight binding of protein to the lipid bilayer and the active hydrophilic patches extending through it. Proteins, of course, are formed from amino acids that, in addition to the amino (NH_2) and acid (COOH) groups, contain various side chains, some polar, some not. It became apparent that hemoglobin (and probably many other globular proteins) could assume a flattened-out or lamellar structure, with the polar chains on one side and the nonpolar chains on the other.

It seemed to me, therefore, that one could predicate an arrangement whereby the hydrophobic side chains on one surface of a protein lamella would work their way through interstices between the lipid molecules into the interior of the lipid bilayers, where they would become closely bound to the lipid hydrophobic tails; the hydrophilic side chains would remain outside but in close contact with the lipids' hydrophilic heads. So far as the pores were concerned, these could consist of gaps in the lipid bilayer "lined" all around with protein lamellae. Again, the hydrophobic surfaces of the lamellae would face the tails of the lipid molecules, while their hydrophilic surfaces would face one another in the center of the pore. And the attraction between these polar surfaces would counteract the surface tension forces that would otherwise enlarge the pore and tear the membrane asunder.

In substance, then, the "final" model of the cell membrane, circa 1955, consisted of not two but six layers. At the center were the two lipid layers, back to back – or tail to

Active Transport

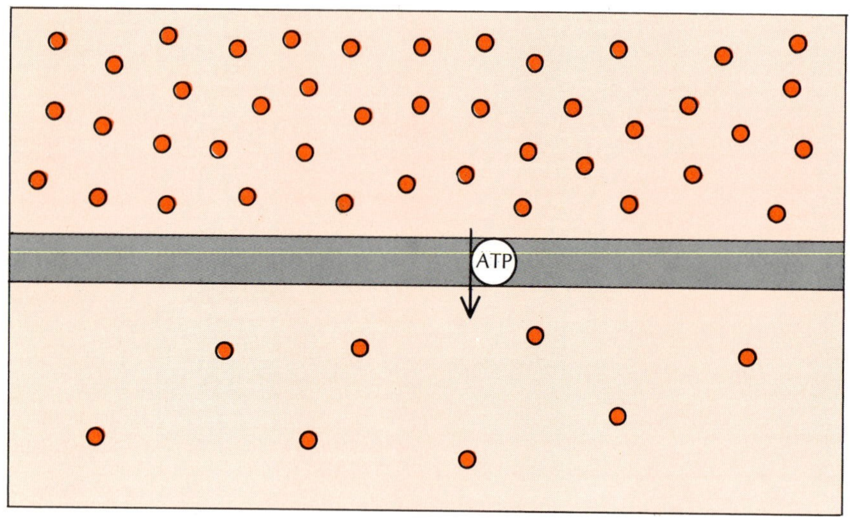

Zero Time

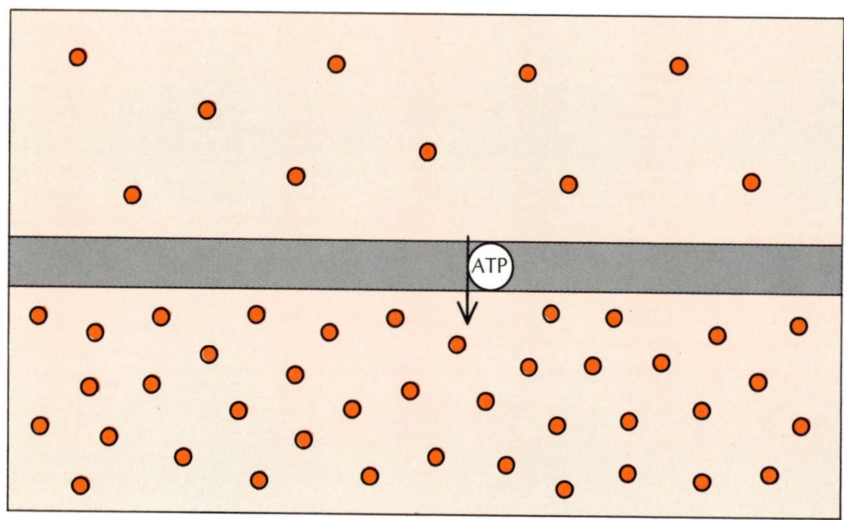

N Time

Active transport differs from diffusion, either simple or facilitated, in that particles (usually ions) can be "pumped" across a membrane against a concentration gradient. This process requires energy, symbolized by adenosine triphosphate (ATP).

tail; surrounding them were two layers of "bound" protein more or less interwoven with the lipids by both nonpolar and polar bonding and in some places (the active patches) penetrating through them. The outermost portions were composed of two layers of loosely attached, adsorbed proteins. The latter, it appeared, served to buffer the membrane against external stresses and also, perhaps, to seize on certain specific molecules in the environment for which the cell had a special need, which could then be "passed on" to the bound protein for transport into the cell through the active patches.

Whether and to what extent the details of this 1950's model of the cell membrane have been confirmed by subsequent research will no doubt be discussed extensively by other contributors to this volume. Personally, I would just as soon leave it at that – the more so in that my own work during the past 15 or more years has been mainly concerned with other biologic problems. What seems incontestable, however, is that the cell membrane does consist of a lipid-protein "mosaic" of the sort originally suggested – the lipids serving to isolate the cell from its environment, with the liquid lipid bilayer acting as a two-dimensional solvent for the macromolecules. The proteins serve both to reinforce the lipid bilayer, insulating it from external physical and chemical stresses, promoting the selective interchange of substances with the environment on which cellular life depends, acting as receptors, etc., and to provide polar pathways through the thickness of the membrane.

It is worth noting that there are, in retrospect, strong genetic-evolutionary reasons for postulating a mosaic of this type. It is quite possible, indeed, to imagine other types of cell membrane – for instance, one composed entirely of protein molecules. From a strictly physicochemical standpoint such a membrane is perfectly plausible, but from an evolutionary standpoint it is essentially impossible.

The basic reason is that the cell membrane, if it is to serve as an effective barrier between the cell and its surroundings, must present an essentially continuous impermeable surface. And considering the number and

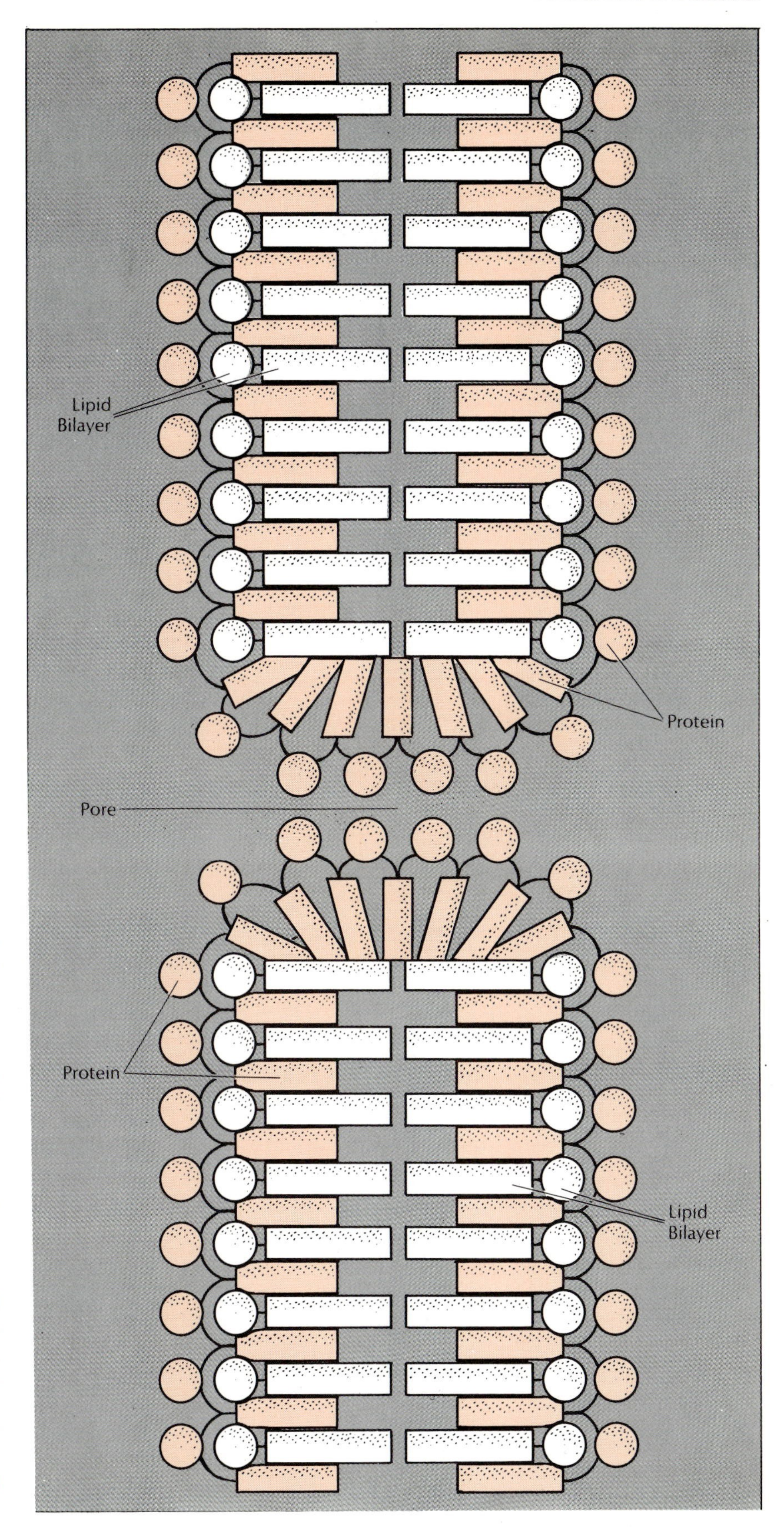

Final Danielli-Davson model of cell membrane included pores or active patches "lined" with unrolled protein; attraction between polar areas of protein keeps pore from expanding and thereby destroying membrane. Pores were postulated to explain such phenomena as facilitated diffusion and active transport, which earlier models could not.

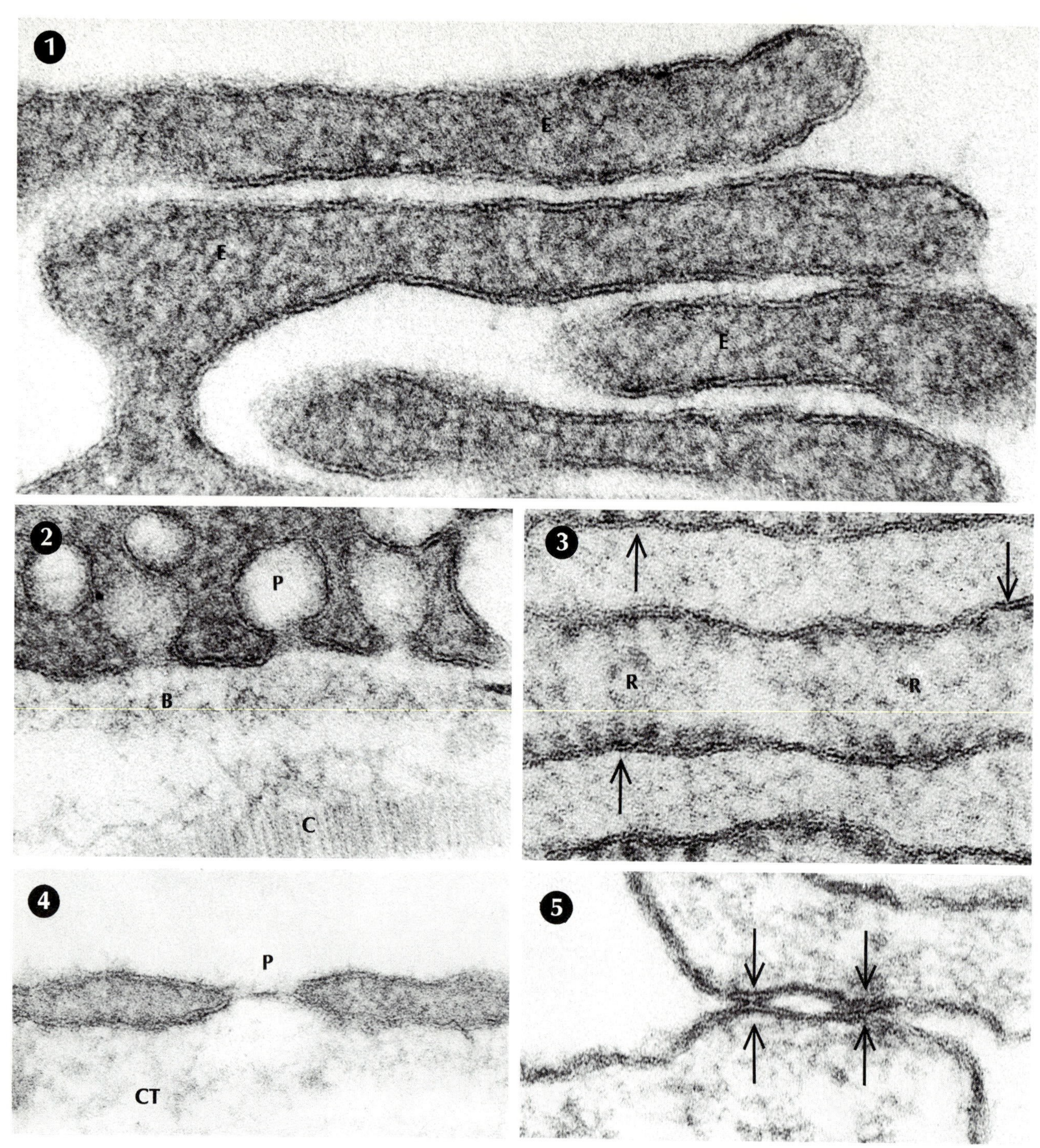

Electron micrographs of thin sections, by Drs. George Pappas and Joe Keeter of Albert Einstein College of Medicine, reveal the bilayer nature of the cell membrane and show it to have an overall thickness of 80 to 95 Å. In figure 1, a portion of interdigitated endothelial cell processes (E) taken from a rat pancreas is reproduced (magnification x222,000). The unit membrane structure of the plasma (or cell) membrane is shown; i.e., two dense lines separating a clear area. Overall thickness measures about 80 Å. Figure 2 (magnification x200,000) shows a portion of an endothelial cell in the rat pancreas in which micropinocytotic vesicles (P) have been formed by invagination of the plasma (cell) membrane. Again, unit membrane structure is apparent. In this thin section, a condensate of material of intermediate density making up the basal lamina (B) is visible. Periodicity can be observed in a nearby collagen fibril (C). A portion of a fibroblast in amphibian connective tissue is shown in figure 3 (magnification x200,000); the unit membrane structure of the endoplasmic reticulum is seen to good advantage at the arrows. Ribosomes (R), not well defined, are attached on the cytoplasmic surface of the membranes of the endoplasmic reticulum. Part of an attenuated endothelial cell of a thin capillary in the rat pancreas is shown in figure 4 (x185,000). A pore (P) seems to be covered with a "diaphragm," which is apparently a nonmembranous structure. (CT denotes connective tissue space.) Figure 5 (x190,000) shows a portion of the junction between two endothelial cells of a capillary in amphibian muscle. Unit membrane structure is evident. Note the closeness of the plasma membranes of the two apposing cells at the arrowed points.

variety of functions carried out by the membrane, it surely must include a large number of *different* proteins. Given these two requirements, for variety and for continuity, it would seem evident that if only proteins were present, all of them (at a minimum, more than a dozen) would have to fit together like the pieces of a jigsaw puzzle.

All this is not beyond the bounds of possibility. Though nearly all protein molecules are large and complex structures, the problem of "designing" a dozen, or even a hundred, of them that can be fitted together into a watertight membrane may be no more difficult than many other problems evolution has solved successfully. Where the difficulty arises is in "improving" such a membrane once it has evolved. When a mutation occurs affecting one protein – which must happen repeatedly in the course of evolution as cells acquire new ways of interacting with their environments – mutations must occur simultaneously, affecting at least some of the adjacent proteins, in order to preserve their fit with the original mutant. Moreover, these secondary mutations must not only be structurally complementary to the original mutation but also functionally neutral; that is, the transport activities of the proteins must be left unaltered despite their structural change. Otherwise cellular function would be disrupted.

The probability of even two such compatible mutations occurring simultaneously is obviously very small; in the case of half a dozen simultaneous mutations, the probability drops to the vanishing point. Thus, if a cell with an all-protein membrane did appear at some point in the course of evolution – which it may have done – its subsequent rate of evolution would have been virtually nil; neither the kind nor the efficiency of its metabolic functions could have changed appreciably, so that sooner or later it would have been displaced by cells whose membranes were not subject to these stringent limitations.

Lipid molecules, being far smaller and simpler than protein molecules, can fit together far more easily; indeed, it has been shown that in a number of organisms the specific lipids present in the membrane can be changed considerably by diet without impairing membrane integrity. Equally, the small size of the lipid molecules vis-à-vis the proteins means that the former can fit equally well with a wide variety of the latter. To take a three-dimensional analogy, an apple, a banana, and a carrot will none of them fit at all neatly with one another, or even with other members of their species – yet any of them can be wedged firmly into a bowl of BB shot. The lipid layer, in other words, serves not only as protective barrier between cell and environment but also as a fluid matrix that can receive almost any protein molecule that evolution, as it were, chooses to insert in it. It is this plastic, protean quality, I suspect, that chiefly accounts for the fact that of the multitude of cells, with their endless differences in structure and metabolic activities, all, so far as we know, have membranes of the same basic type.

I should like to conclude by mentioning one of the central and still unresolved problems in elucidating membrane structure and function: the nature of the membrane proteins. These molecules are obviously essential to the special functions of the membranes, including transport and diffusion activities, and since the membranes are of many types, involving many different compounds, one would expect that the proteins that mediate them would be similarly diversified; my own guess would be that the "average" cell has something like 100 different membrane proteins. It is no less clear that to understand the molecular mechanics of these transport activities we shall need much more information on the structures of the individual proteins; that is, we shall require pure protein preparations. And there's the rub. As noted earlier, a large proportion of membrane protein is tightly bound to the lipid bilayer, to the point where it is extremely difficult to extract the protein without denaturing it. Even to the extent that this can be done, one ends up not with a mixture of discrete protein molecules (which could be separated fairly easily into their various components by such techniques as electrophoresis) but rather with a collection of particles of varied sizes, containing two, three, or more different protein species and often bits of lipid as well. With present techniques, there is no way one can make any biochemical sense out of a heterogeneous collection of this sort. I might just add, moreover, that even if we were able to obtain pure preparations of individual protein species, it is entirely possible that their properties in purified form would be quite different from their properties in their "natural environment" – i.e., embedded in the lipid bilayer.

It seems evident, therefore, that some fundamental advances will be required in techniques for segregating and analyzing proteins – and that these advances will take effort and cost money. A really thorough comprehension of the nature and activities of membrane proteins could well represent the most fundamental advance in biochemistry since the discovery of the Watson-Crick model of the DNA molecule – a breakthrough in the full sense of that perennially abused word.

2

Lipid Dynamics In Cell Membranes

DENNIS CHAPMAN
University of Sheffield (England)

Cell membranes can be viewed in several different ways: in terms of their anatomy or gross structure as seen in an electron microscope; in terms of their physiology or function – how they admit and exclude substances to and from the cell or how they organize cellular enzymes, the visual pigments, and perform many other important physiologic tasks; and viewed in terms of their chemistry, i.e., their composition and how the constituent molecules of membranes are organized to form a three-dimensional structure. These different approaches to cell membranes are presently converging. The last, their chemical composition, will constitute the main focus of this chapter.

As is known, any field of scientific knowledge goes through three general stages: collection, classification, and explanation. Having gathered together as many examples as possible of the phenomenon under consideration, the scientist arranges them in some sort of rational system, taking account of both their similarities and their differences, and, finally, devises a rationale that explains, or at least elucidates, the arrangement.

So far as the chemistry of cell membranes is concerned, it must be admitted that we are not far out of the collection stage. We have identified a host of substances present in one or another membrane. We have classified some of them, after a fashion, in terms of their chemical structures, but the classification suffers from the fact that we still have much to learn about which structural features are functionally significant and which are not. And when it comes to the "why" of the classification – why membranes consist of these types of substances and not others, and why this or that substance is present in a particular type of membrane – we have as yet little more than a few plausible hypotheses and conjectures. The one thing that can be said categorically about membrane chemistry is that it is a complex business, and is made no simpler by the emerging realization that membranes are dynamic, not static, structures, whose chemical composition and molecular arrangement can and do change with changing conditions.

Before elaborating on these generalities, I should perhaps make one overall point about cell membranes: though most of the early studies dealt only with the exterior plasma membrane, which separates the cell from its environment, membranes of various sorts are also found inside the cell. Thus the mitochondria – the cell's "power plants"– consist largely of intricately folded membranes; so does the endoplasmic reticulum, in which protein synthesis takes place. These and other intracellular membranes, such as those that define the Golgi apparatus and lysosomes, though they naturally differ from the plasma membrane in many respects, also possess clear structural and chemical similarities to it, and some of the data to be cited come from analyses of these interior structures.

Membranes are composed primarily of lipids and proteins; other constituents include: water, cholesterol, sugar groups, and metal ions. Cholesterol is not found in all types of membrane. A very significant chemical fact about membranes is that the proportion of protein varies greatly from one to another. Thus in the myelin membranes sheathing the nerve fibers, the proportion of lipid to protein is something like nine to one, while at the other extreme we have the mitochondrial membrane, in which the ratio is about one to one. The most probable explanation of this particular difference is that the myelin membrane is primarily an insulator, its role being to exclude substances that would interfere with the transmission of nerve impulses; as we know from the demyelinating diseases, damage to these membranes produces severe impairment of nerve function. The mitochondrial membranes, on the other hand, are thought to serve the function of organizing – in space and perhaps in time – the action of the many enzymes associated with the mitochondrial system, and it is likely that much or all of its "extra" protein consists of these same enzymes. Beyond this, we may also note that most membranes appear to contain not one but several species of protein, which, considering the variety of metabolic jobs performed by most membrane types, is not surprising. Another possibly significant fact is that the proteins in general seem characterized by a relatively high proportion of amino acid residues with long nonpolar side chains, such as leucine and isoleucine. These could help to associate the proteins to the lipids through their affinity for the latter's fatty acid chains.

Very few proteins have been fully characterized in terms of their chemical composition or molecular structure, one important reason being the difficulty of extracting intact proteins from their associated lipids. Recently a glycoprotein from the erythrocyte membrane has been studied in some detail (and these sorts of proteins will be discussed in later chapters of this book). This discussion will deal chiefly with the lipid components of cell membranes.

The Lipids of Cell Membranes

Most (though not all) membrane lipids are based on the glycerol molecule – a simple three-carbon structure with a hydroxyl (-OH) group attached to each of the carbons – and thus resemble the triglycerides, in which most of the body's fat is stored. In the latter, however, all three hydroxyl groups are replaced by long-chain fatty acids (palmitic, stearic, oleic, etc.), while in membrane lipids only two fatty-acid chains are present, the third position being occupied instead by a polar group. Often this is a group that carries, or can accept, an electric charge and is therefore soluble in water. Since the two fatty-acid chains are insoluble in water, they are thought to project from the glycerol chain in a direction opposite to that taken by the polar group. In structure, they are sometimes compared to a tuning fork, with the fatty-acid chains forming the "prongs" and the polar group the "handle."

The membrane lipids differ among themselves in the identity of the fatty acids or of the polar group, or both. Thus in the phospholipids – the most numerous class – the molecule may contain any two of more than half a dozen fatty acids. These can be concisely characterized by two numbers, the first giving the number of carbon atoms in the chain and the second, the number of double bonds, or their degree of unsaturation. Thus stearic acid, for example, is 18:0, with a fully saturated 18-carbon chain; oleic acid is 18:1, possessing one double bond. Other possibilities are myristic (14:0), palmitic (16:0), linoleic (18:2), linolenic (18:3), and arachidonic (20:4) acids.

Usually a phospholipid from a biologic membrane contains a saturated and an unsaturated fatty acid, the latter generally lying in the middle position on the glycerol chain, between the saturated acid at one end and the polar group at the other. As yet we have no good explanation for this particular arrangement, or indeed why the two fatty acids should almost always be of different types. One of the exceptions to this arrangement of two different fatty acids is the alveolar membrane, which contains quantities of a dipalmitate lipid – i.e., one with both prongs containing saturated fatty acids. It is conceivable that this "preference" for completely saturated chains in the lung membrane is due to the fact that these fatty acids are less easily oxidized than unsaturated chains – but this is sheer conjecture at present.

The polar groups of the phospholipids are no less variegated than are the fatty acids. All of them, as the name implies, contain a phosphate group, but additional groups are also present and these can often be fairly complicated. Thus in lecithin, for example, the phosphate has a choline group attached. Other possible addenda include amines of similar structure to choline; the amino acids serine and threonine; various modifications of the glycerol molecule, sometimes

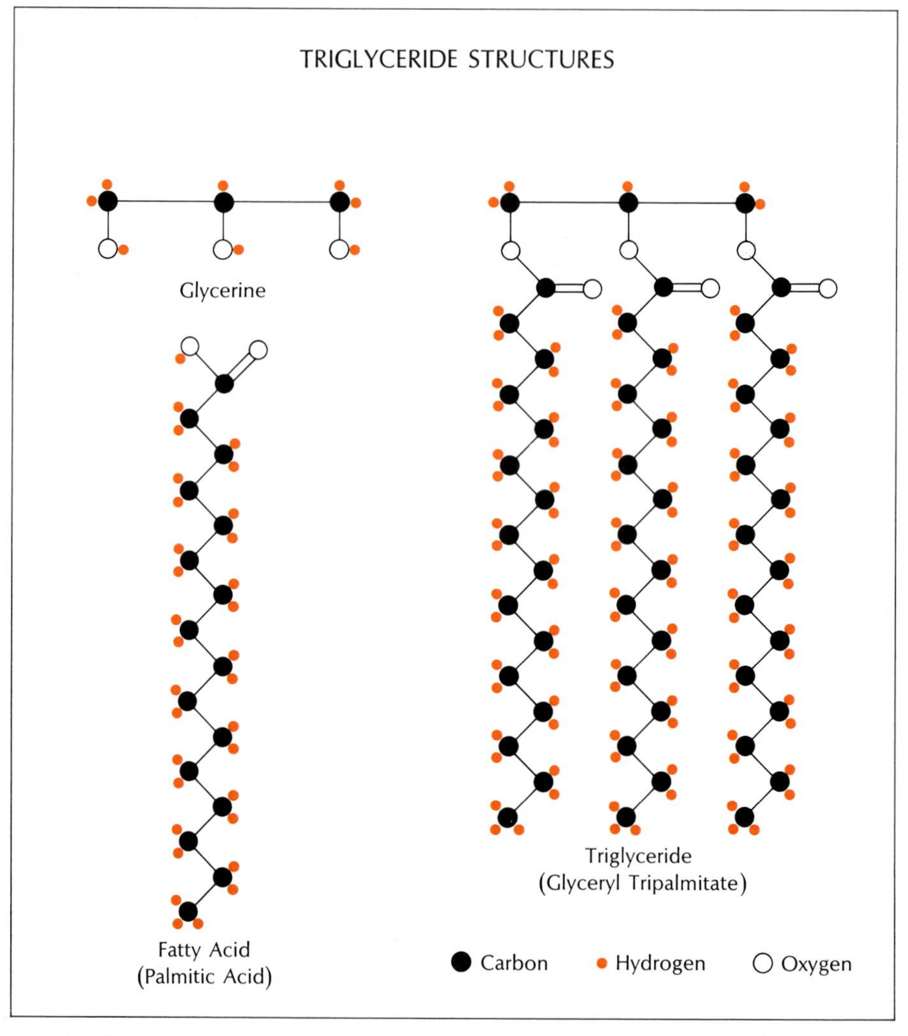

Nearly all body lipids are built around the three-carbon glyceryl framework. In the neutral lipids of body fats, a fatty-acid chain is attached to each of these carbons, as in the example above. In phospholipids, the most numerous class of membrane lipids,

with an additional phosphate attached; the sugar inositol, sometimes with one or even two extra phosphates attached to it; and an elaborate chain of three sugars. In other cases the phosphate has a metallic ion (often calcium) attached.

The second major class of membrane lipids is that of the sphingolipids; these are based on the molecule sphingosine. Though they possess the basic tuning-fork design of the phospholipids they differ from them in several ways. The first long-chain component is always a 15:1 hydrocarbon, which, moreover, is linked to the base by a simple carbon-carbon bond rather than the ester bond (-COO-) found in the phosphoglycerides. In addition, a hydroxyl group is retained. The second prong, as in the phosphoglycerides, may be one of several fatty acids but is linked to a second carbon by an amide (-CONH-) rather than an ester bond; this carbon, too, may retain its hydroxyl on occasion. The polar group may be a phosphate-plus-choline structure (sphingomyelin), a simple hydroxyl (the ceramides), a simple sugar (the cerebrosides), or a complex polysaccharide (the gangliosides).

Nearly all membranes contain a variety of lipids. Thus myelin membrane lipid, for example, has as major constituents cerebroside, phosphatidylethanolamine, and the nonpolar molecule cholesterol, plus smaller quantities of lecithin, sphingomyelin, phosphatidylserine, and other substances. Erythrocyte membrane lipid contains many of the same constituents, though in somewhat different proportions – notably, much more sphingomyelin – but no cerebroside whatever. Mitochondrial membrane lipids, finally, consist almost entirely of lecithin, phosphatidylethanolamine, and cardiolipin, with little cholesterol and no phosphatidylserine, sphingomyelin, or cerebroside. These differences in lipid composition clearly reflect differences in the physiologic functions of the membranes in question, but we still cannot define the relationship between composition and function exactly, although we are beginning to understand this in a general way.

Thus cell membranes contain different lipid classes and each of the lipid classes has not one but a variety of discrete fatty acids associated with it. Lecithin, for example, is actually not *a* compound but is rather *a group* of compounds, all of which con-

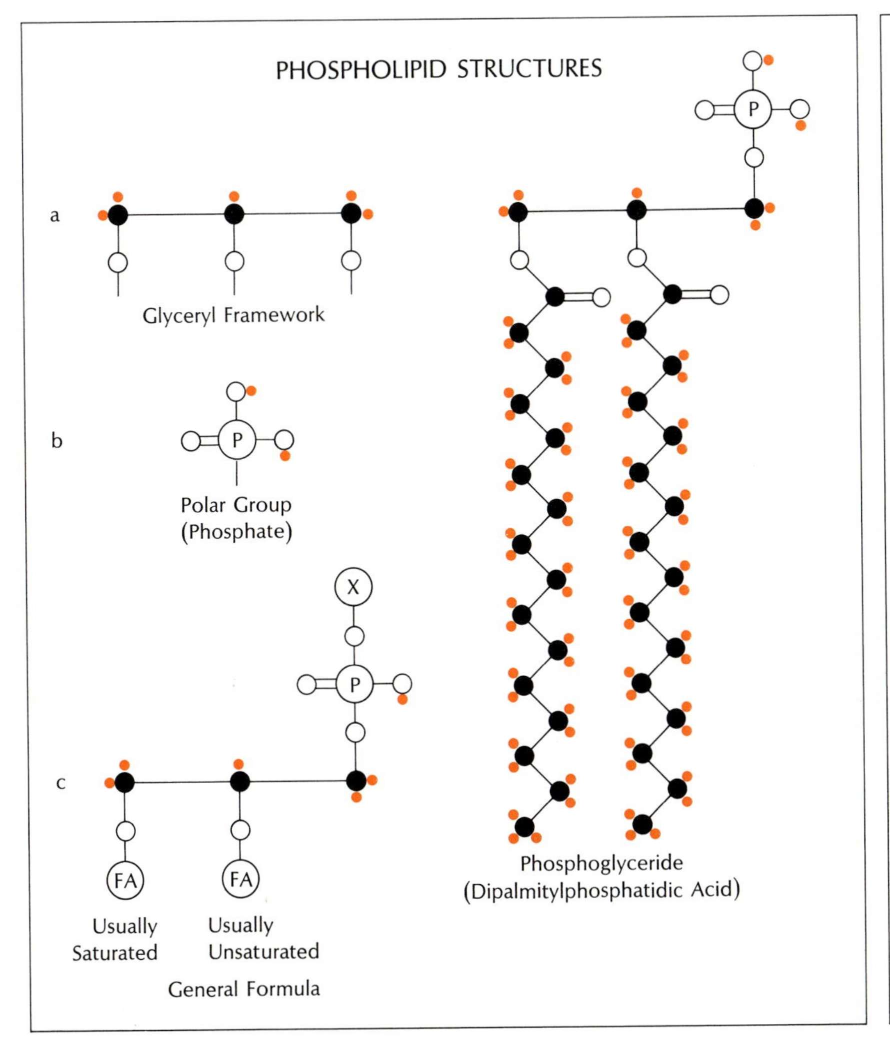

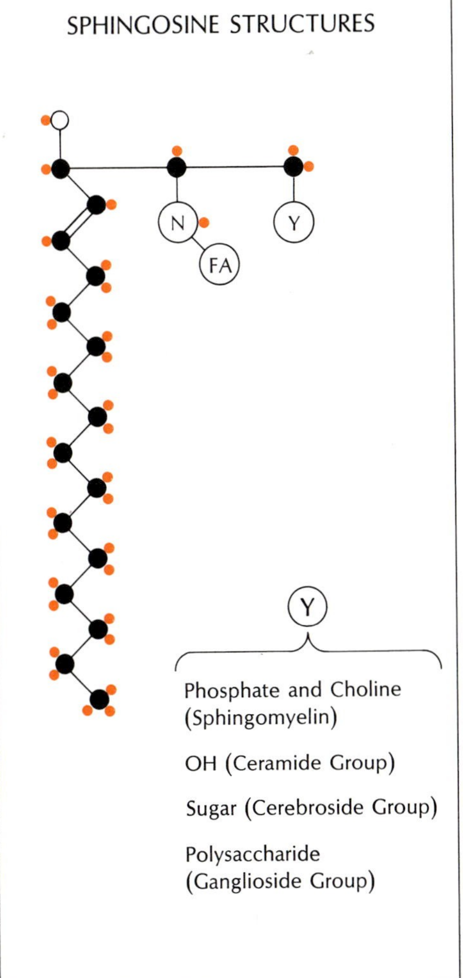

a phosphate, usually with an added group (X), replaces one fatty acid; variations in these groups, and in the fatty acids, make possible an enormous variety of phospholipids, as suggested on the next two pages. In sphingolipids, also found in membranes, the basic structure is further altered, and only one of the fatty-acid chains is subject to variation.

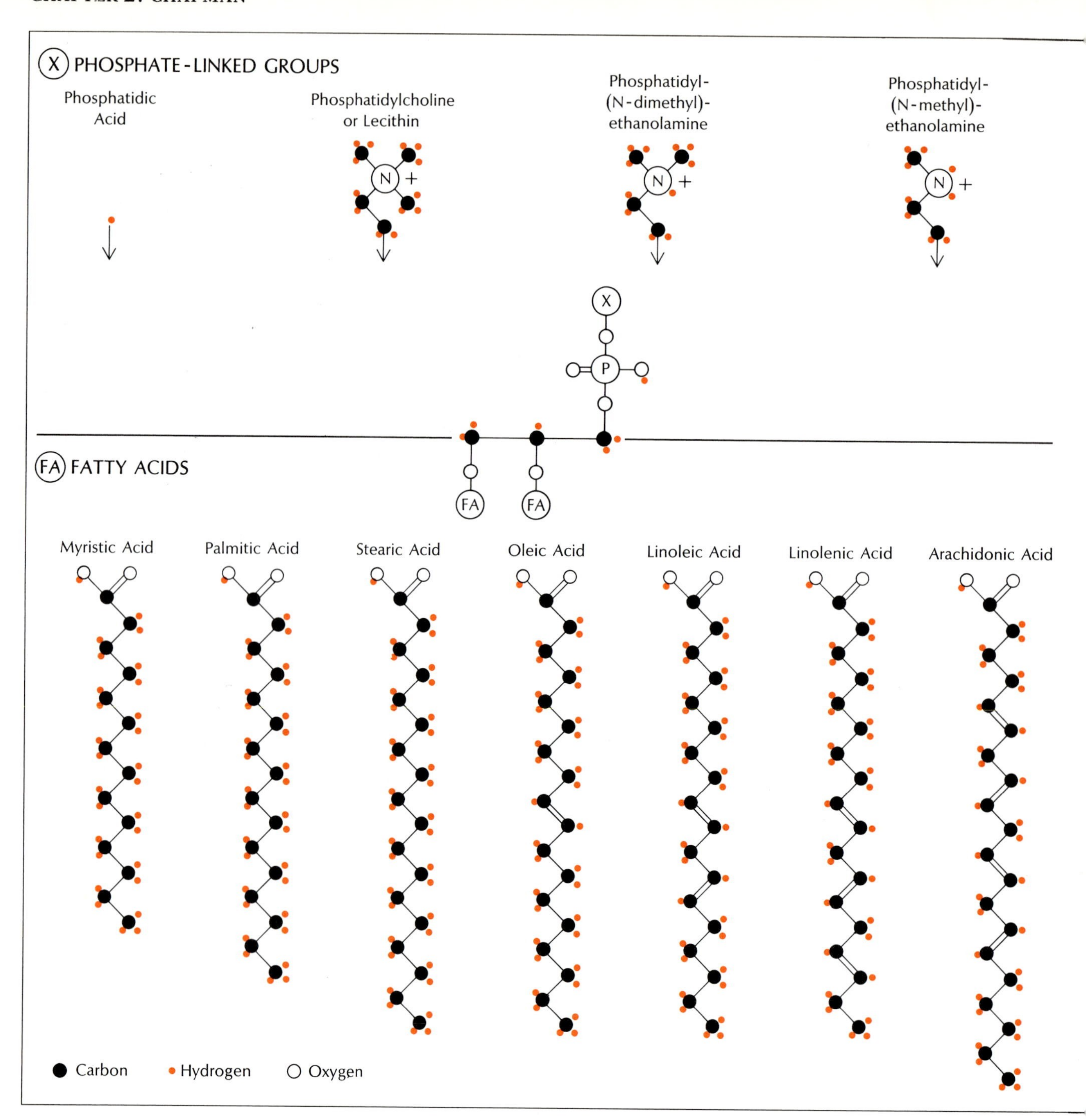

Almost infinite variety of phospholipids is possible through variations in the fatty acids (above, bottom) attached to two of the glyceryl carbons, and in the groups attached to the phosphate (above, top, and at right). Most phospholipids include two different fatty acids; variations in the chain-length and degree of unsaturation (i.e., the number of double bonds) are known to

tain the (polar) phosphatidylcholine group but different fatty-acid chains. The latter, however, are always linked to the glycerol framework by ester bonds.

The Fatty Acids of Membranes

Some interesting points emerge from interspecific comparisons of fatty acids – both those found in the polar, membrane lipids and those in the neutral, nonmembrane triglycerides. The latter show noticeable differences from one species to another but much similarity in different tissues from the same species. Thus in the mouse, rat, and rabbit the predominant fatty acids in neutral lipids from any tissue are in all cases 16:0 (palmitic) and 18:1 (oleic), but whereas the mouse has much more of the latter, the rabbit has considerably more of the former, with the rat occupying an intermediate position. The intraspecific consistency from one type of tissue to another is no more than one would expect for neutral lipids, since these substances serve mainly as a nutritional reserve and pass fairly freely

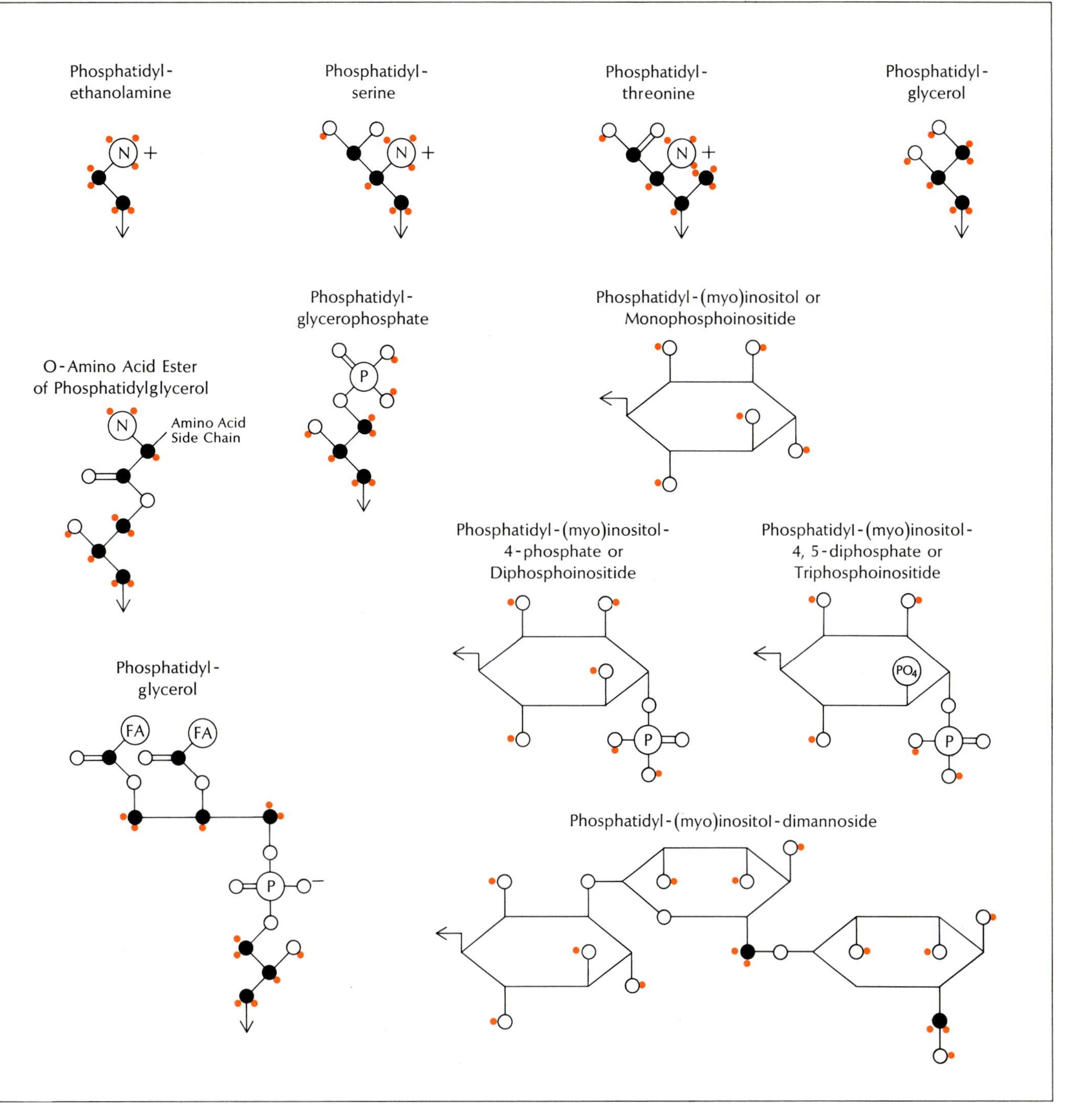

affect membrane rigidity and permeability. Functions of the phosphate-attached groups are less well understood, but it is believed that among other things they determine the "response" of particular portions of the membrane to certain molecules impinging on it, thereby determining whether or not the molecule will pass through the membrane.

from one physiologic "compartment" to another. The interspecific differences, too, are expectable; they doubtless reflect species differences in physiology and also, quite possibly, in diet (it is known, for instance, that the composition of human fat can vary considerably, depending on what one has eaten).

When we turn to the phospholipids, we find a quite contrasting situation. Their fatty-acid composition differs markedly from one *tissue* to another within the same species, but turns out to be remarkably similar in a particular type of tissue, regardless of species. Thus brain phospholipids from seven different animals (the three above, plus pig, ox, horse, and sheep) all show consistently more 18:1 than 16:0, a sizable amount of 18:0 about equal to that of 16:0 (it is only a minor component of neutral lipids), and no or almost no 18:2 (a significant, sometimes major, component of neutral lipids). Phospholipids from lung, on the other hand, contain (with only one exception) less 18:1 than 16:0, and invariably less 18:0 than either, plus significant quantities of

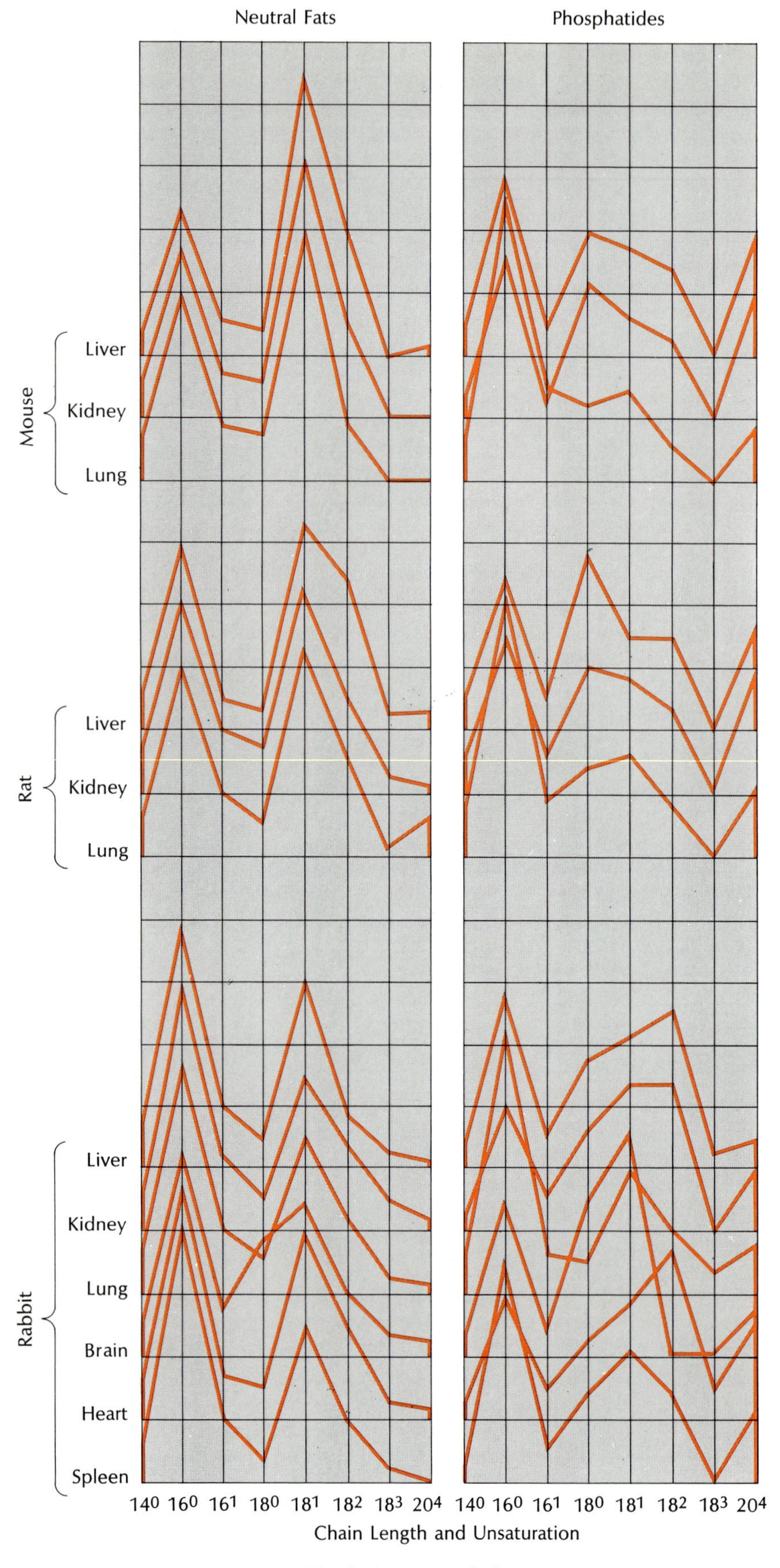

Fatty-acid composition of neutral lipids (left) varies little in a given species, no matter from what organ obtained. By contrast, phospholipids, obtained from membranes, show considerable variation in composition from one organ to another (right).

18:2. These tissue-specific rather than species-specific similarities are again unsurprising. If we take it that the natural "home" of phospholipids is in cell membrane, we must assume that a major proportion of it is bound up in these membranes, rather than free to circulate in a body pool. Differences in fatty-acid proportions between brain and lung would then reflect differences in the physiologic functions of membrane in these organs, but since lung physiology or brain physiology does not differ very markedly from one (mammalian) species to another, we would expect that membrane function – and therefore the fatty-acid composition of membrane phospholipids – would show a comparable limited range of variation.

The Fluidity of Lipids and Cell Membranes

Some clues to the significance of these variations in membrane fatty acids have come from model experiments with purified phospholipid preparations. If, for example, we take pure dipalmitoyl lecithin, which plays such a predominant role in alveolar membranes, we find that in the anhydrous condition it melts at around 100° C. This transition temperature, I should note, does not involve the change of state from solid crystal to "normal" lipid usually implied by the term "melting," but rather a shift from a crystalline gel to a "liquid crystal." In this state the glycerol and polar groups retain a fairly regular organization although the polar group does have considerable mobility, but the fatty-acid chains melt and acquire considerably more mobility, with the methyl end of the chain having greatest motion. (In a bilayer, this would mean that the interior is more fluid than the two faces exposed to the aqueous environment.)

If we now add water to the anhydrous phospholipid, to a degree approximating the physiologic condition, we find that the transition temperature drops sharply, to about 42°. Thus below 42° the lipid is in a gel condition, while above 42° it is in a fluid condition. (Another observation is that some of the water molecules appear to strongly bind themselves to the polar groups in such a way that they cannot be frozen, i.e., they are not as free in their mobility as ordinary water. These water molecules also provide a sort of cement for linking one polar group to another and keep the lipids together to form the membrane structure.)

Experiments with a variety of phospholipids that differ only in the nature of their fatty-acid chains have established that the transition temperature is dependent on chain length and degree of saturation. Thus the transition temperature of hydrated lecithin, for instance, can be raised as high as 60° by increasing the length of its fatty-acid chains, or it can be lowered to temperatures below 0° by shortening the chains or by incorporating unsaturated chains.

The same phenomena have also been observed in natural membranes from bacteria (e.g., *Escherichia coli* membranes and *Achole plasma laidlawii* membranes). The transition temperature can be most easily observed and measured in a calorimeter, which gives sensitive readings on heat absorption by a preparation of natural membranes (or by one of artificial phospholipids). Since the shift from gel to liquid crystal is endothermic – i.e., relatively large quantities of heat must be absorbed in order to break the bonds that hold the molecules in the more-or-less rigid gel structure – a sharp peak in heat absorption at a given temperature signals that the change of state has in fact occurred. By growing certain bacteria on media containing different sorts of fatty acids, one can vary the composition of the membrane and observe the effect of chain length and saturation on the transition temperature. The same experiments, incidentally, also give evidence that the membrane lipids are at least partially arranged in a bilayer structure, since otherwise one would not expect them to behave in this way.

Similar experiments, with both artificial systems and natural membranes, have identified yet another variable affecting the transition temperature: the presence of cholesterol. Addition of this substance appears to abolish the transition entirely, in the sense that the lipids retain the loose, liquid-crystal organization even when cooled well below the physiologic range of temperatures. In essence, the cholesterol molecules interpose themselves between the lipid chains and prevent them from assuming the orderly crystalline-gel configuration.

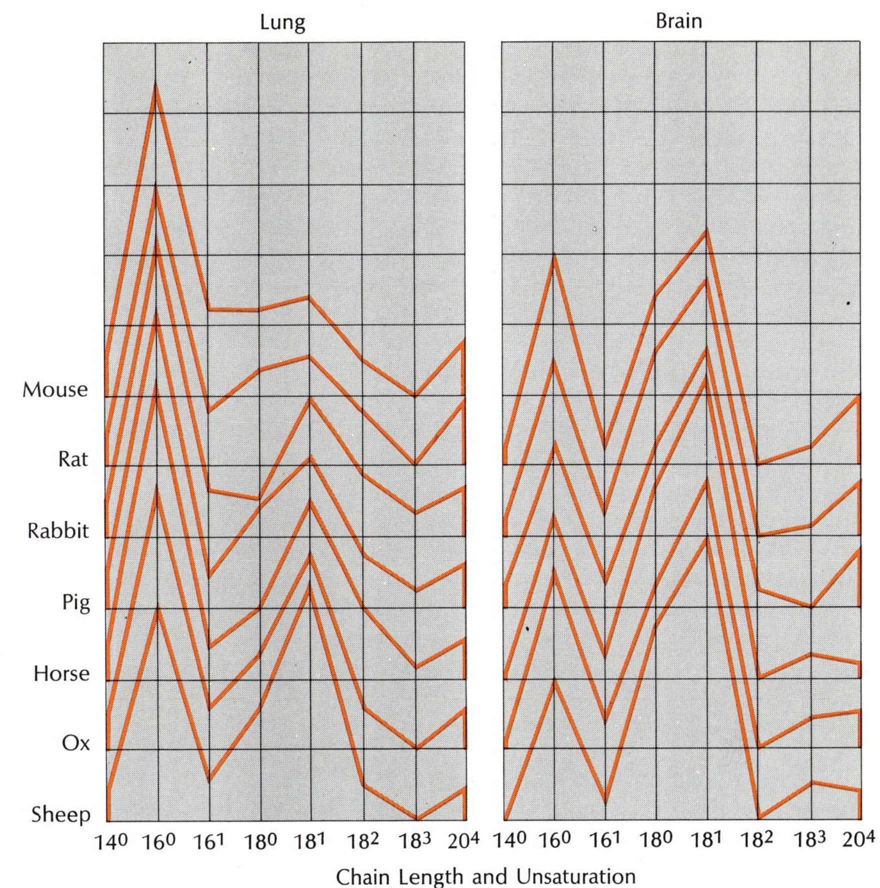

Fatty-acid composition of membrane phospholipids shows marked similarities in a given organ, no matter from what species, but marked differences between, e.g., lung and brain. These are presumed to reflect functional differences in the cells.

To summarize the foregoing, then, a given membrane – or portion of a membrane – is able to exist in the liquid-crystal state at physiologic temperatures, depending on 1) the length of its fatty-acid chains, 2) the degree of saturation of those chains, and 3) the presence – and perhaps the proportion – of cholesterol. One obvious fact is that in the liquid-crystal state the membrane will interpose far less of a barrier to the passage across it of organic molecules, which can more readily "dissolve" and penetrate the membrane. Furthermore, small polar molecules – notably, for instance, the water molecules – can also more readily cross the membrane, but this cannot be the whole story. If, for example, we compare the transition behavior of mitochondrial and myelin membranes, we find that both are in the liquid-crystal state at physiologic temperatures (37°), but for quite different reasons: in the mitochondria, this is because the fatty acids are relatively unsaturated, whereas in myelin, it is because of the high cholesterol content (around 25%). The effect of cholesterol is not only to keep the lipids in a fluid condition. We can see this when we consider the organization of a number of lipid molecules arranged side by side, as they are in bilayer membranes. The motion in one fatty-acid chain is to some extent transmitted to the chains next to it, and so on down the line in a cooperative manner. In this manner, changes – information – can be transmitted from one part of a membrane to the remainder. Molecules of cholesterol, however, would be expected to block this transmitting system, so that the cooperative motion of the chains will be quickly damped out, ensuring that changes in the membrane remain relatively local instead of involving the membrane as a whole.

We have, in fact, some direct evidence that even more pronounced local differentiation exists in some membranes. In certain bacterial membranes – and probably in others – probing by highly sophisticated physicochemical techniques has established that some of the membrane is in the crystalline-gel configuration, with portions of it in the looser, liquid-crystal structure. We do not know why this should be so; conceivably, the relative "stiffness" of these particular membranes may serve some structural function in these cell systems.

Whatever the structural role of the physical changes in the membrane, there can be no doubt that the permeability characteristics of particular membranes at physiologic temperatures is basic to their functioning; in fact, some animals can adjust the composition of their membrane lipids to take account of temperature changes. Fish, for example, are often called "cold-blooded" animals; more accurately, they are poikilothermic. They adjust the fatty acids of lipids to suit environmental temperature.

Bacteria also modify their membrane lipids to suit the growth temperature. Lowering the growth temperature of *E. coli* results in an increase in the level of unsaturated fatty acids in the lipids, indicating that the organism attempts to counterbalance the effects of low temperature by maintaining higher levels of unsaturated fatty acids in their membrane lipids to keep the same fluidity.

Another aspect of membrane fluidity characteristics is thought to involve not only the transport of water and other simple molecules in and out of the cell but also the relationship of membrane proteins to the lipids. Several researchers have hypothesized that the proteins float like icebergs in a "sea" of liquid-crystalline lipid. I do not want to say much about this theory as it will be discussed at length by Dr. Singer in Chapter 4, so I will

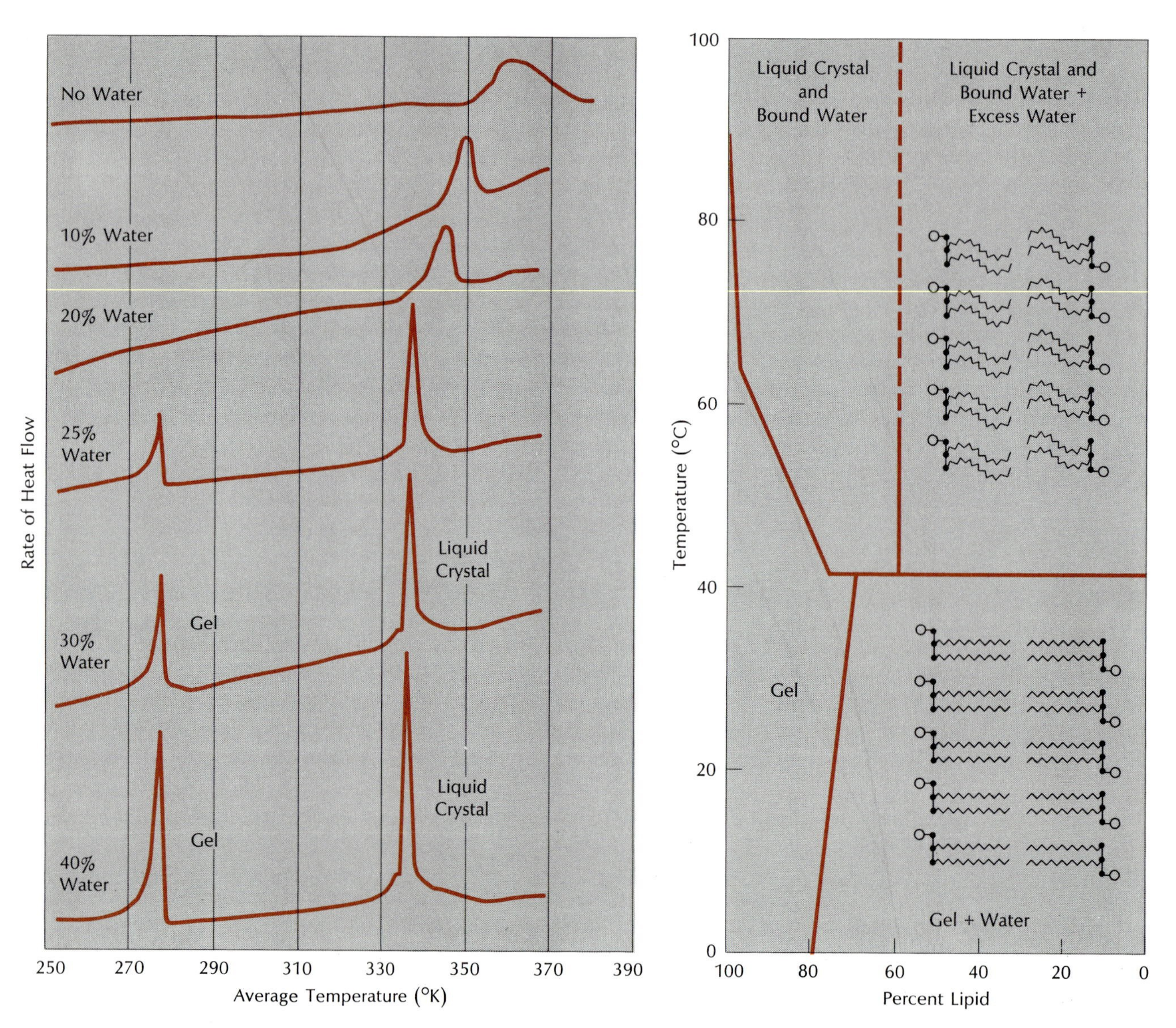

Graphs of heat flow in phospholipid preparations (left) show melting peaks at right marking the transition from gel to liquid crystal. The transition temperature drops as more water is added. Appearance of the peak at the left shows freezing of water; with less than 25% water, no ice-melting peak occurs, i.e., all water molecules are bound to the lipid and cannot freeze (distearoyl lecithin). Phase diagram (right) graphs transitions between gel, in which the fatty-acid chains are relatively rigid, and liquid crystal, in which the chains acquire some flexibility (dimyristoyl lecithin).

merely note that one basis for it is the belief that some membrane proteins, in order to perform their function of transporting substances into and out of the cell, must be free to rotate within their lipid matrix, as well as to move in the plane of the membrane. This is very probably true in certain cases, and here it is obvious that changes of fluidity in the lipids – i.e., from the liquid crystal to the gel state and vice versa – could facilitate or inhibit the activity of the proteins. In fact, recent experiments with *E. coli* membranes show that the activity of the enzyme succinic dehydrogenase has a temperature transition that can be related to the transition of the phospholipids forming the membrane. The activity shows a discontinuity at 19° when oleate lipids are used and at 28° when elaidate lipids are used. On the other hand, however, it has also been demonstrated that in artificial membrane systems certain polypeptides, such as gramicidin A, engage in ion transport, and continue to do so even when the lipid around them is "frozen."

Trigger Mechanisms

Thus far, I have been writing as though the transition temperatures of the lipids in various membranes or membrane regions were fixed, static properties deriving from the fatty-acid chains of the lipids. In fact, it is not as simple as that. Transition temperatures and lipid fluidity can be shifted up or down in temperature by adding ions such as sodium potassium or calcium, which bind to the polar groups of the membrane. The interaction of protein with the lipid polar group can also affect membrane permeability. More recently, we have found that certain antidepressant drugs can also exert similar shifts on the lipid transition temperature – and thereby, evidently, on membrane permeability. The key to understanding the action of a number of drug molecules may be in their interaction with membranes. It seems clear that temporary alterations of membrane characteristics by exogenous and endogenous substances are potentially a most important area of research. This area is already being studied by some of the most sophisticated techniques for probing molecular structures with-

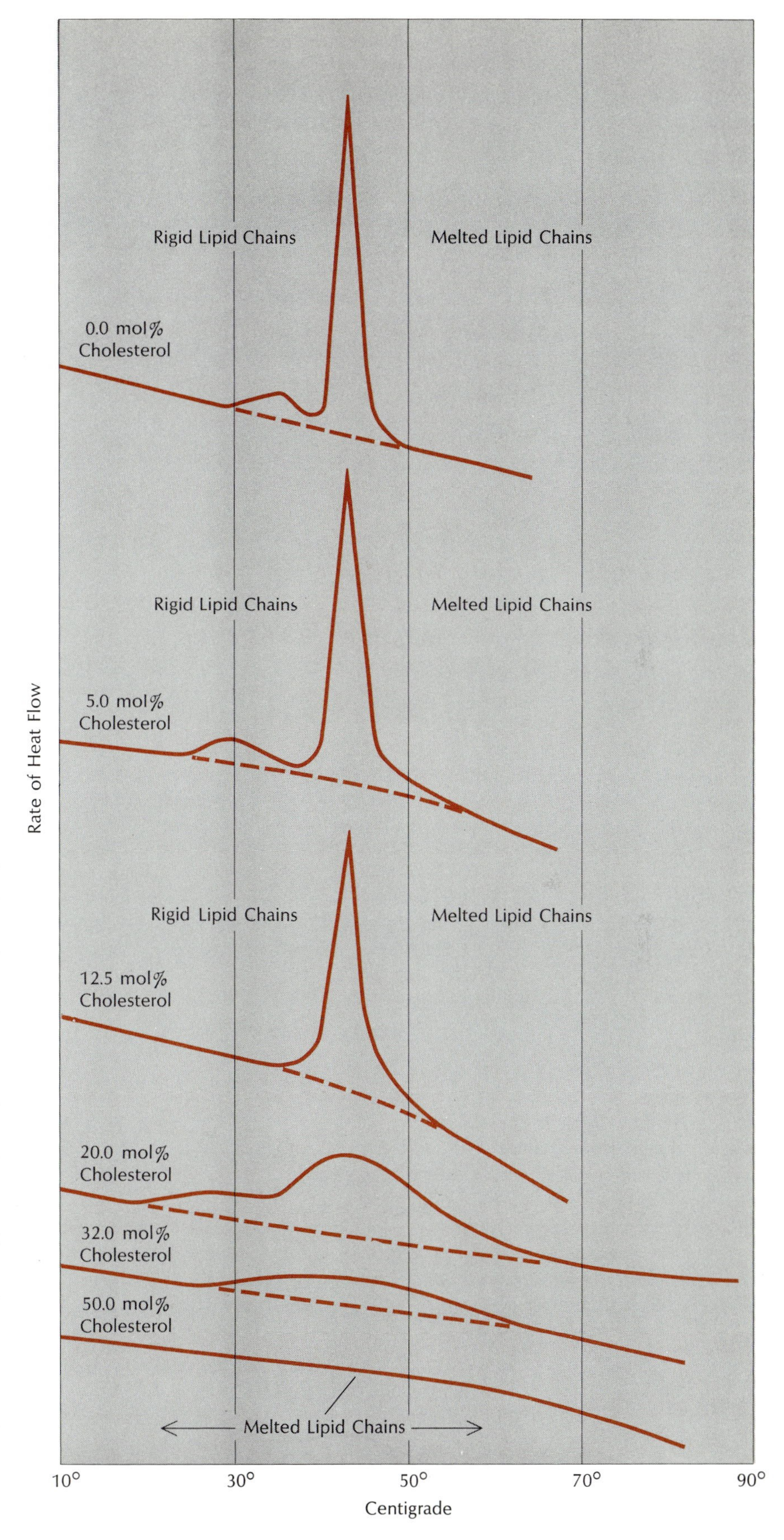

Adding cholesterol to phospholipid preparation blurs transition between liquid crystal and gel states. As relative amount of cholesterol approaches 50%, transition vanishes and lipid remains a liquid crystal, with fluid lipid chains, even at low temperature.

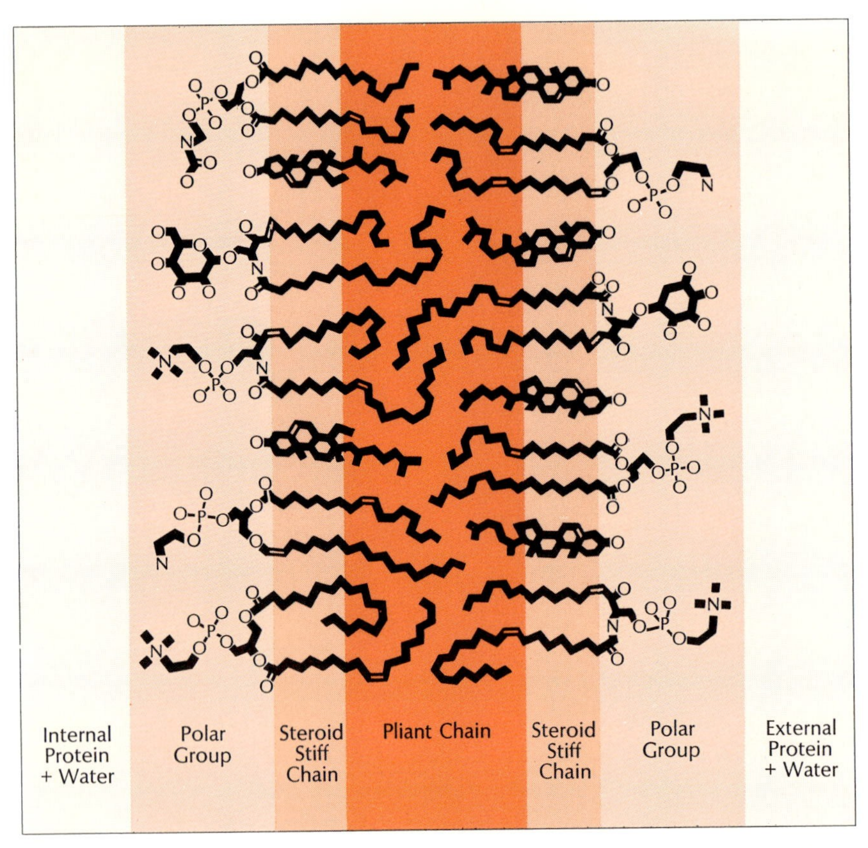

Actual complexity of membrane structures, as opposed to the highly schematized diagrams hitherto depicted, is suggested by this drawing – itself considerably simplified – showing a portion of myelin membrane. This tiny section, only about 30 x 30 Å., contains six cholesterol molecules, five phospholipid molecules of three different types, and four sphingolipid molecules of two different types.

out disrupting them, such as infrared spectroscopy, Raman spectroscopy measurements of nuclear magnetic resonance, electron spin resonance, etc.

Already such techniques have established that the concept of membrane lipids as being fixed in place is only relatively true. Individual lipid molecules can and do migrate or diffuse from one part of the membrane to another, and can even (albeit rarely) "flip over" from the outside to the inside of the bilayer and vice versa.

Thus we are beginning to picture membrane lipids – and, by extension, proteins and other membrane constituents – as in a state of constant dynamic activity. The fatty-acid chains, so long as the lipid is above its transition temperature, are in motion, flexing and twisting either "cooperatively" or – in the presence of cholesterol – less so. The lipid molecules are also in motion, and can move laterally from one part of the membrane to another. And, finally, the molecules are engaging in dynamic interactions with ions and molecules from both within and without the cell. The precise nature of these actions will clearly depend on a number of things, especially the nature of the triggering substance and its capacity to bind to particular membrane proteins or to the polar groups of membrane lipids. It seems fairly certain, for instance, that one important distinction among the diverse polar groups enumerated earlier will turn out to be whether they possess a net negative charge (as does the phosphatidylserine group, for instance) that enables them to bind metal ions (which of course have a positive charge), or a balanced charge – one positive and one negative. This interaction between the trigger molecule and specific proteins and/or polar groups, in turn, can set off interactions among lipid molecules and/or their fatty-acid chains, which can raise or lower membrane permeability.

Many researchers, having in mind the classic "lock and key" model of enzyme action, have applied a similar concept to membrane action, visualizing the membrane as containing a great variety of "holes" or other structures, each type keyed to the transport of a particular substance. This is very probably the case in some instances; for example, in some artificial systems certain polypeptides can discriminate between sodium and potassium ions, allowing the former to pass and excluding the latter, apparently just on the basis of ionic size. But as we learn more about membrane dynamics, it may well turn out that some substances may carry their own key to the membrane's locked "doors," i.e., their molecules may possess the property of triggering local changes in the membrane that enable them to pass through it – a sort of chemical "Open, Sesame!" Similar dynamic interactions may well help explain recognition of one cell by another, immunologic reactions, and many of the other obscure yet fascinating properties that make membranes and cell surfaces the objects of so much current scientific effort.

3

Models of Cell Membranes

ALEC D. BANGHAM

Institute of Animal Physiology, Babraham, Cambridge (England)

Workers in nearly all fields of science are constantly devising conceptual models. Biologists and physiologists in particular have, for some generations, been drawing simplified pictures of the structures they were investigating as one way of visualizing how and why these structures behave as they do. Considerably less common, I would say, are actual physical models of biologic structures. I do not here refer to the static molecular models (e.g., that of the DNA double helix) familiar to every chemistry student, but rather to artificial, simplified versions of a cell or cell organelle in which the particular structure's *dynamic* properties can be simulated and studied.

Model-making of this sort, as it happens, has become particularly common in cell membrane research – and has proved remarkably fruitful. Not only have the models provided valuable information on the general properties of biomembranes, they have also furnished clues to the "fine structure" of specific pathologic processes and drug effects. And quite apart from their heuristic value as tools of investigation, the model structures themselves show considerable promise of being put to practical use as tools of therapy.

The simplest (and, as one would expect, the oldest) model of the cell membrane is a lipid monolayer. This is commonly produced on the surface of water confined within a glass or Teflon trough to which polar lipids, either natural or synthetic, are added. Because molecules of these substances are amphiphilic – hydrophilic at one end, hydrophobic at the other – they tend to line up neatly on the surface of the water, the polar, hydrophilic ends downward, in contact with the water molecules, and the hydrophobic ends sticking up into the air.

As noted previously by Danielli, just such a model was used by the Dutch workers Gorter and Grendel in the early 1920's for their earliest bilayer version; they used phospholipids extracted from the membranes of human erythrocytes. By using the now classic methods of Langmuir and Adam, they were able to determine how much lipid was required to cover a surface of a given area completely with a single lipid layer; at the same time, they were able to estimate, from the size, shape, and number of the erythrocytes required to produce this quantity of lipid, how large an area of cell membrane the trough monolayer represented. Had the two areas been approximately equal, they would have been forced to conclude that the membrane, like the trough surface, contained only a single layer of lipids. In fact, however, the trough area was roughly double that of the calculated membrane area, indicating that in the membrane the lipids must be "doubled up" into a bilayer.

More recently, monolayers have been used to study a variety of other membrane phenomena – for example, the interaction of metal ions and lipids. It has been determined, for instance, that lipids in monolayers show little or no "preference" in binding Li^+, Na^+, and K^+ ions – though they react somewhat differently with such divalent ions as C^{++} and Mg^{++}. The inference is that differences in the transport of, say, Na^+ and K^+ across a membrane must involve "special" – i.e., nonlipid – mechanisms; as we shall see later, other membrane models have provided clues as to what those mechanisms may be.

Other monolayer studies have established relationships between the fatty-acid composition of lipids and their "packing" characteristics – that is, the force required to produce the most tightly packed monolayer configuration. It turns out, for example, that packing is densest at a given force when both of the two fatty-acid chains in the lipid are long and fully saturated; a looser configuration is obtained when one of the chains (or both) is either shorter or unsaturated. The finding meshes neatly with studies of the bulk behavior of the various types of lipids described previously in this text (see Chapter 2, "Lipid Dynamics in Cell Membranes," by Chapman) in which unsaturated or short chains preserve the looser, liquid-crystal configuration at physiologic temperatures.

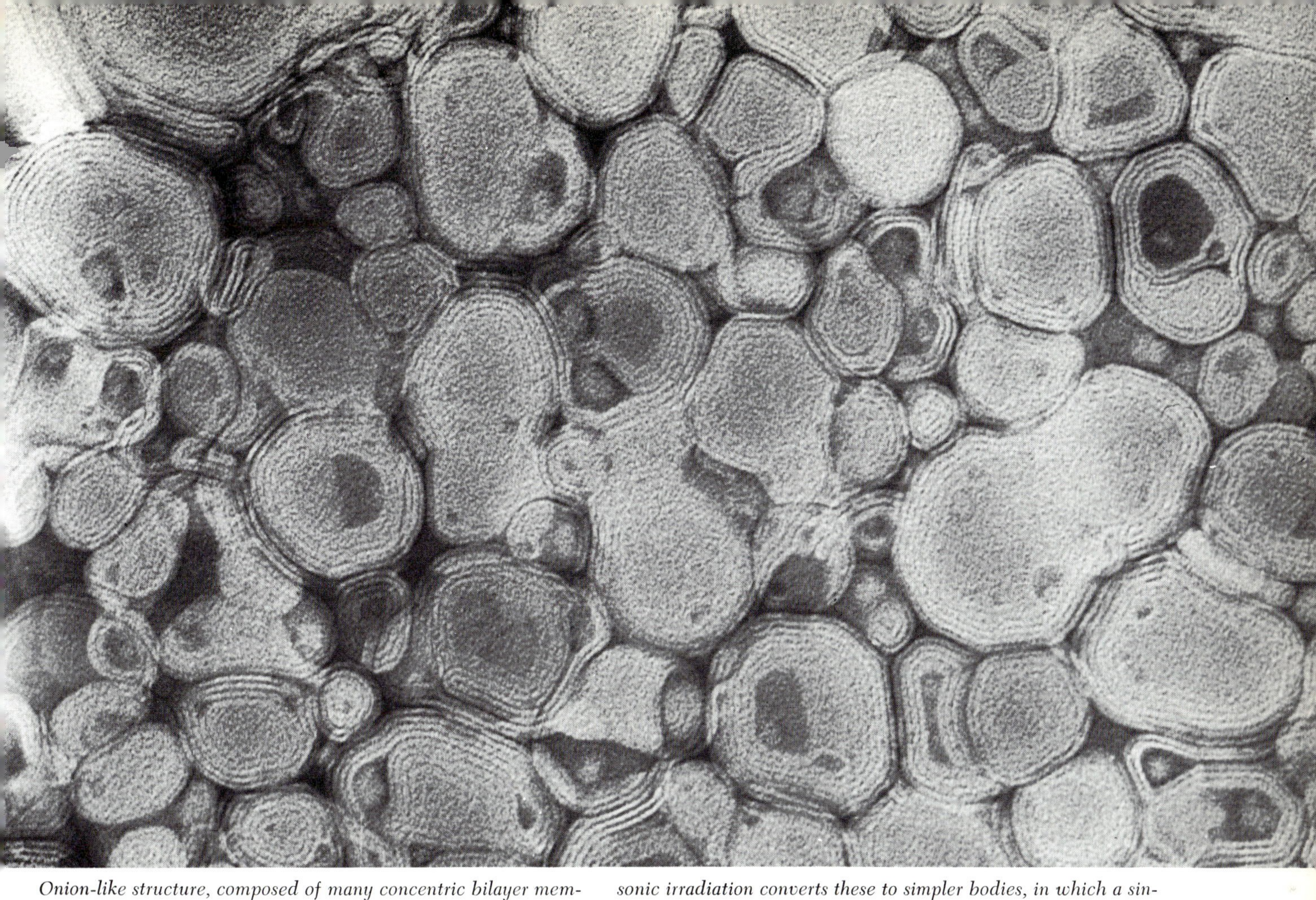

Onion-like structure, composed of many concentric bilayer membranes, typifies early version of "liposome" model (top). Ultrasonic irradiation converts these to simpler bodies, in which a single membrane surrounds a single aqueous compartment (bottom).

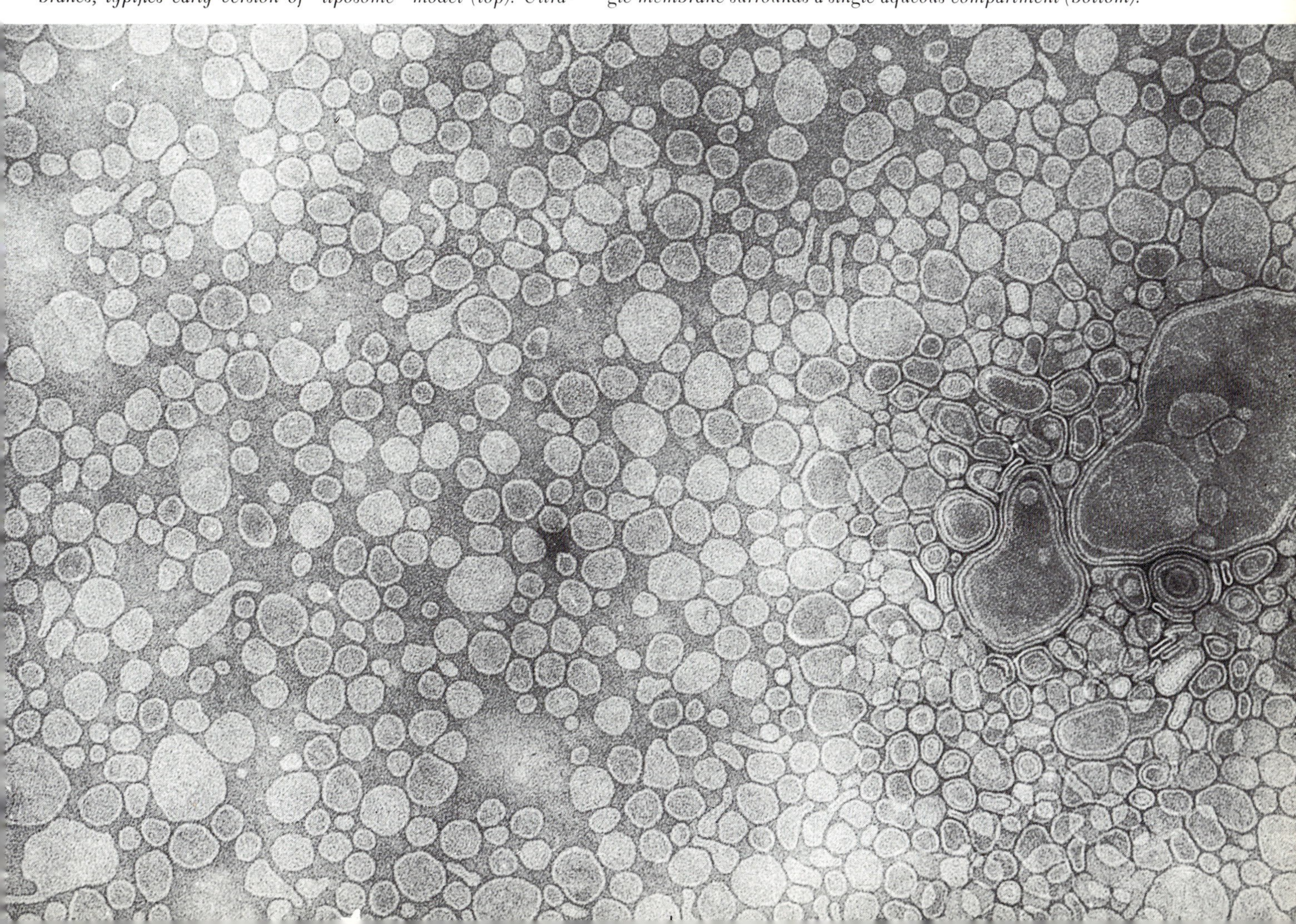

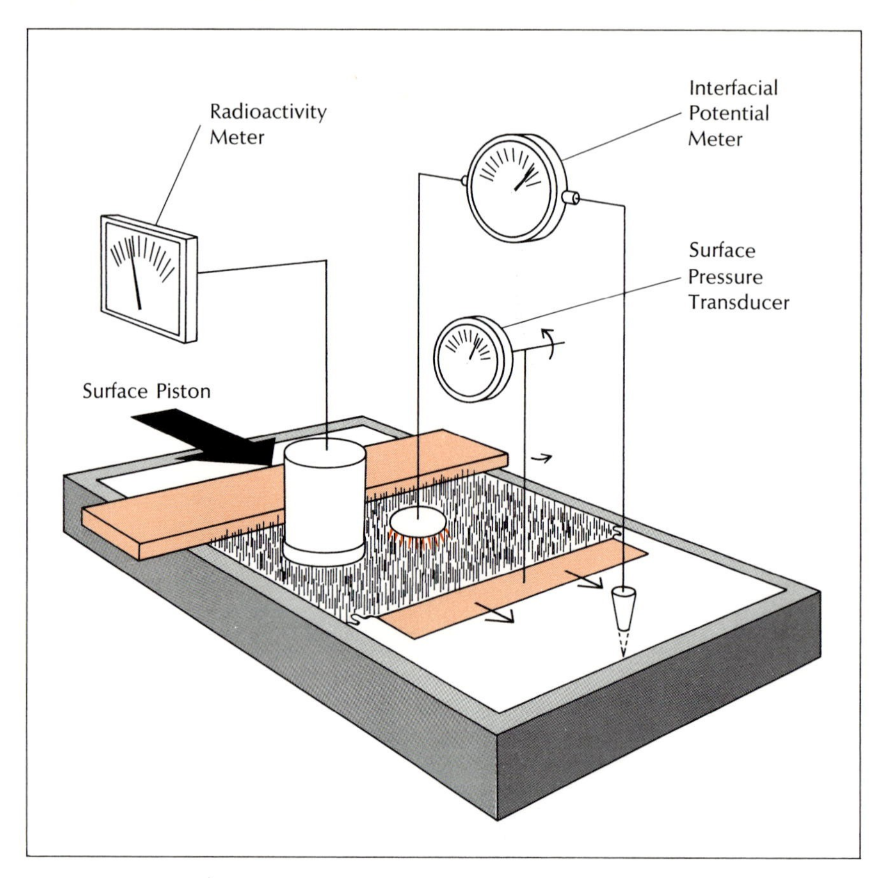

Monolayer membrane model is formed by floating lipid molecules that have an affinity for water in a "trough." Movable piston and pressure transducer measure strength of attractive and repulsive forces between membrane's molecules, potentiometer connected to top of membrane and water beneath it measures electrical forces across it.

The monolayer model, however, has two major limitations. First, it *is* a monolayer, meaning that one must be cautious in projecting its properties as applying to the bilayers that make up natural membranes. Second, it has water on only one side, meaning that it is unsuited to study most diffusion and transport phenomena, which in nature proceed from an aqueous phase on one side of the membrane to an aqueous phase on the other.

About 14 years ago, this limitation was overcome by a more realistic membrane model, which was both a bilayer and had water on both sides. This model can be produced quite easily; the apparatus required is a small pot made of a hydrophobic material, e.g., polyethylene or Teflon, open at the top and with a small hole in the side. The pot is immersed in water, which also enters it through the hole until the levels are equal. A watercolor brush is then loaded with lipid dissolved in a solvent that is immiscible in water, such as straight-chain decane (a constituent of gasoline), and drawn across the hole. The result is something like a soap bubble, which like such a bubble progressively thins, giving rise to a series of iridescent color changes due to light interference, ultimately appearing "black" when it reaches its lowest energy state, the "pure" bilayer configuration. For this reason such bilayers are often called black, or bilayer, lipid membranes (BLM); depending on their size and environment, they remain stable for minutes, hours, or even days.

With such a system, a variety of experiments can be performed, it being particularly suitable for studying transmembrane electrical events; one need only insert an electrode into the pot fluid and another into the surrounding fluid. Since BLM's of pure lipid are essentially nonconducting, the system acts, in fact, as a capacitator whose value will depend on the area of the BLM (known), the dielectric constant of the lipid in question (which can be closely estimated), and the thickness of the membrane. This latter parameter, which is of some interest, can be accurately estimated by such electric technique without in any way disturbing the synthetic bilayer itself. The actual thickness turns out to be of the order of 50 Å. A parallel experiment measures not the capacitance of the membrane but its conductance when ionizing salts are introduced into one of the compartments. Since, as already noted, the BLM itself is nonconducting, the system's conductance under these conditions measures the relative mobility with which cations and anions, including protons, diffuse across the membrane, a matter obviously relevant to the behavior of natural membranes (diagram, page 28).

With a similar model system, one can also measure water diffusion. Here any impermeable solute placed in one compartment creates an osmotic gradient that "attracts" water molecules from the other. As they pass from one compartment to the other, the membrane will be bowed outward by the increasing volume on one side; measurements of the volume displaced, by various methods, will give the speed with which the water molecules pass through the membrane. As it turns out, this is encouragingly close to the speed inferred for various biologic systems (diagram page 29).

Thus far we have been speaking of models made from pure lipids, which are obviously only very crude models of actual membranes, the latter being anything but pure in their chemical composition. In fact, it has proved possible to "dope" the models with various other types of molecules and thereby endow them with properties more closely simulating those of natural membranes. An early experiment along these lines involved a substance obtained from rotted egg whites, which its discoverers christened excitability inducing material (EIM). When such material was introduced into the fluid on one side of the membrane, it endowed the latter with electrical characteristics resembling those of neuron membranes – specifically, a sudden jump in electrical resistance in response to an applied current above a certain threshold voltage. (Oddly enough, a similar phenomenon is found in the electronic device called a tunnel diode.) Subsequent inclusion of a second additive – one of the simple proteins called protamines – produced electrical "spikes" and rhythmic pulses closely resembling the action potentials observed in neurons. Unfortunately, it has proved impossible thus far chemically to characterize EIM, or even to manufacture it with consistency; egg whites do not rot reliably to order (diagram, page 30).

Other additives can radically increase the model membrane's permeability to ions; even more interestingly, they show very great selectivity among various species of ions. Their properties have been studied both with BLM's and with another type of membrane model, which I shall describe in detail in a moment. Ionic permeability is measured either by conductance – the movement of ions across the membrane being equivalent to the movement of electric charge – or by using tracer isotopes. In the presence of the antibiotic valinomycin, K^+ ions have been shown to pass through the membrane at several hundreds of times the normal rate, i.e., x 10^7 or more, but the passage of Na^+ ions is almost unaffected. Similar results have been obtained with other substances, including nonactin, gramicidin, and enniatin.

Most of these substances were originally isolated from the bodies or byproducts of various microorganisms; all are antibiotic in some sense, and most of them are rather simple polypeptides. Perhaps their most important resemblance, however, is that they are without exception cyclic, forming amphiphilic rings or doughnuts, the outside being hydrophobic and the inside, or "hole," hydrophilic. The ions, it appears, bond to the hydrophilic groups, if, that is, they fit the holes; K^+, for example, does, but Na^+ does not, hence it is not facilitated in passing through membranes.

Exactly how the facilitation works is still to be determined. For compounds such as gramicidin and alamethicin it is tempting to visualize the rings as tiny pores or portholes embedded in the bilayer membrane through which the ions, which would find difficulty in passing the hydrophobic interior of the bilayer, can move by simple diffusion. For valinomycin and a number of other antibiotics it seems more useful to think of the rings as "carriers" that pass through the membrane along with the

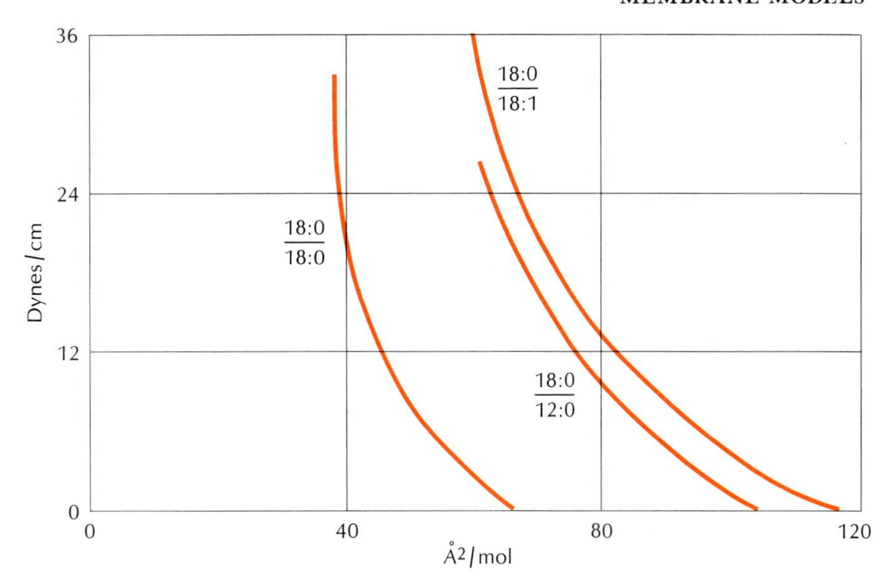

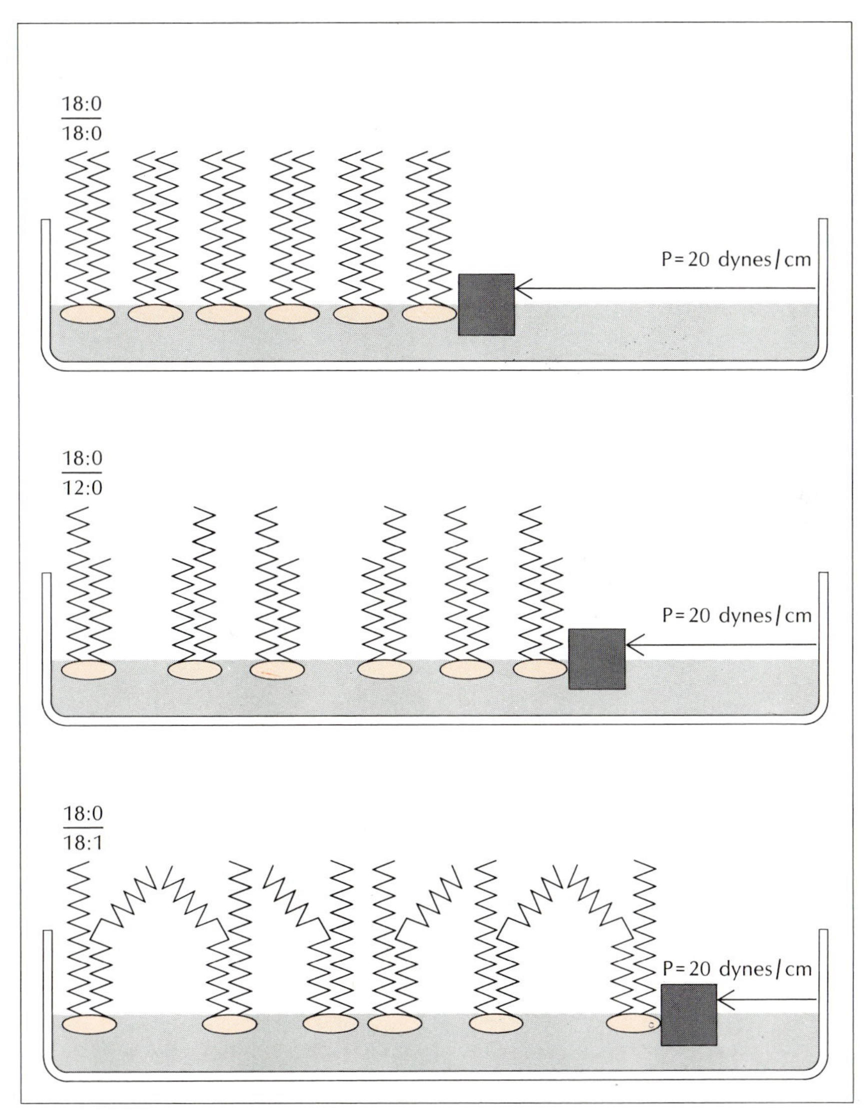

Measurements on monolayer models show that the surface area occupied by a lipid molecule varies with the type of lipid (graph at top); schematic drawings below show why. Molecules with two long, saturated chains (18:0/18:0) occupy minimum space at a given pressure; with one short chain (18:0/12:0), space required is greater because of lessened attraction between short chains. When one chain is unsaturated (18:0/18:1), the "kink" in the unsaturated chain produces maximum separation.

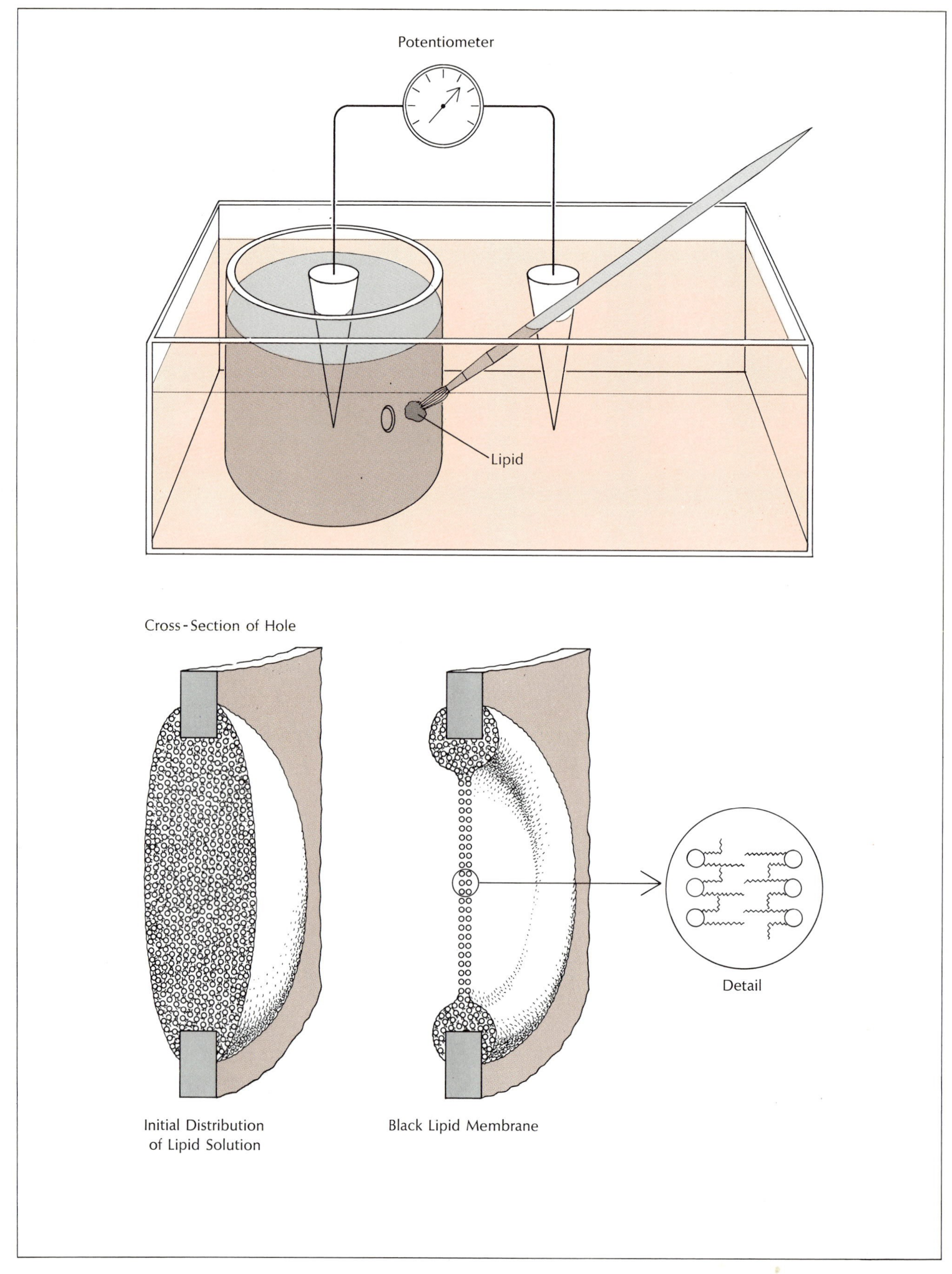

Black lipid membrane, a bilayer with aqueous phase on both sides, is formed by brushing a lipid solution across a hole in a "pot" of hydrophobic material (top); initial thick layer of lipid (bottom left) rapidly thins to a single "black" bilayer (bottom right). Electrodes in pot and its container permit measurements of membrane conductance and capacitance.

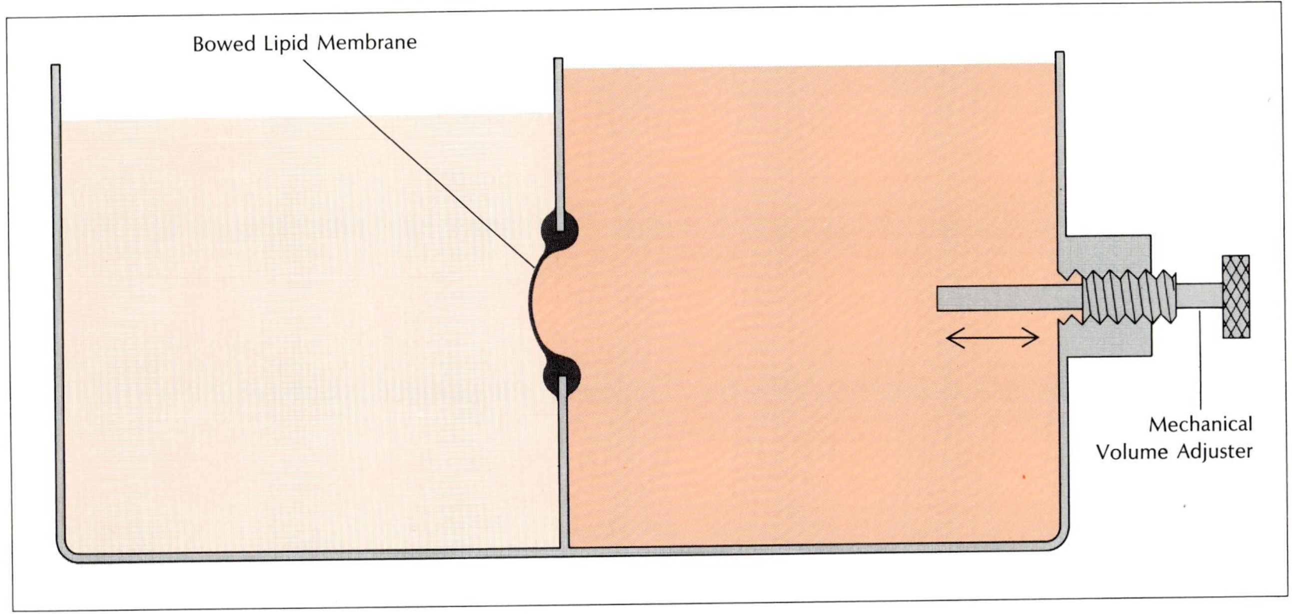

The diffusion of water molecules across the membrane, produced by the osmotic gradient between the two compartments, results in the "bowing" of the membrane. The net volume of water moving into the right-hand compartment in a given time can be accurately measured by mechanically adjusting the volume to eliminate the bow.

ions they enclose. Certainly there is abundant evidence that, here, the rings do not take up permanent residence in the membrane, though they do seem to have a marked affinity for it (diagram, page 31, bottom).

Nor can we say at present whether ion transport – including selective transport – in natural membranes is mediated by similar cyclic molecules. Nonetheless, I find it impressive that researchers have been able to modify their models to simulate an increasing number of natural membrane properties. In one case, indeed, they have far outdone any known physiologic process. In the presence of the cyclic molecule alamethicin, membrane conductivity – the rate of ion passage – actually increases as the sixth power of the alamethicin concentration! Whatever alamethicin does and however it does it, it seems to be a remarkably busy molecule (diagram, page 31 top).

A third model is the liposome, prepared from a wide spectrum of synthetic as well as natural components (e.g., phospholipids) extracted from cell membranes (e.g., erythrocyte "ghosts") with, say, chloroform methanol. When the extracting fluid is evaporated to dryness, one is left with a waxy or soapy deposit of phospholipid on the slide or test tube. If to this is added a small quantity of water, it can be shown by electron microscopy or by x-ray diffraction that the phospholipids rearrange themselves so that the (limited) amount of water is shared by and located close to the head-group region of the phospholipids. Characteristically, and of great importance in our understanding of the liposome model, this arrangement requires that the mass of lipid is penetrated throughout by a series of long interconnecting tubes down which solute may freely diffuse. With certain very pure preparations, the phospholipid does not form tubes containing water but sheets or micelles separated by layers of water, which are, again, intercommunicating.

If, however, the proportion of water to phospholipid goes much over 50%, the tubular or micellar structure becomes unstable. The reason is that in these configurations the hydrophobic phospholipid "tails" are unable effectively to "escape" from the water. The phospholipid re-forms into liposomes: minute onion-like structures consisting of a series of concentric, roughly spherical, closed membranes, each separated from the next by a layer of water and surrounded by water on the outside (see EM, page 25).

At one time some investigators questioned whether the liposome membrane layers were in fact closed structures – though dynamic considerations indicated that only in this way could the hydrophobic tails avoid contact with water. An alternative suggestion was that the liposomes were constructed rather like what the English call a Swiss roll, or the Americans a jelly roll, with the water – corresponding to the "jelly" – being enfolded in the intricately rolled sheets of lipid – the "cake" – but still maintaining direct contact, however circuitous, with the surrounding water. Data from light-scattering x-ray crystallography and other techniques were consistent with this possibility.

Measurements of ion permeability, however, demonstrated that the jelly-roll model was incorrect. By adding a labeled salt solution, such as potassium chloride, to the original phospholipid forming the liposomes, as described above, then filtering them off from the "mother" liquid and placing them in clean water, one could measure the rate at which the radiopotassium in the liposomes leaked into the surrounding solution. As it turned out, this rate was far too slow to be consistent with any system of continuous clefts as in the jelly-roll model; the liposomes were clearly made of a series of closed sacs. The clinching evidence came when we performed a similar experiment using both sodium and potassium isotopes, added valinomycin to the surrounding medium, and found that *only* the potassium could escape from the liposomes. (This, by the way, represented the original finding on the preferential

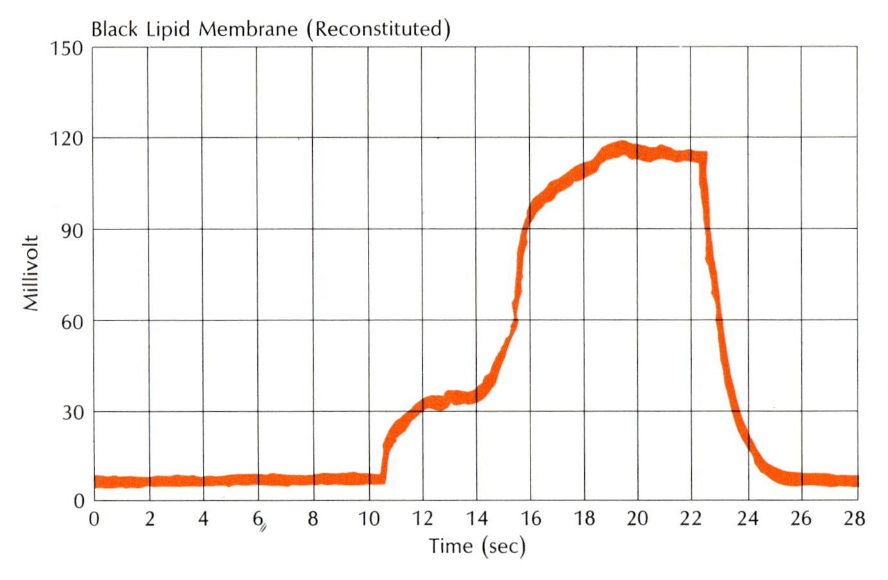

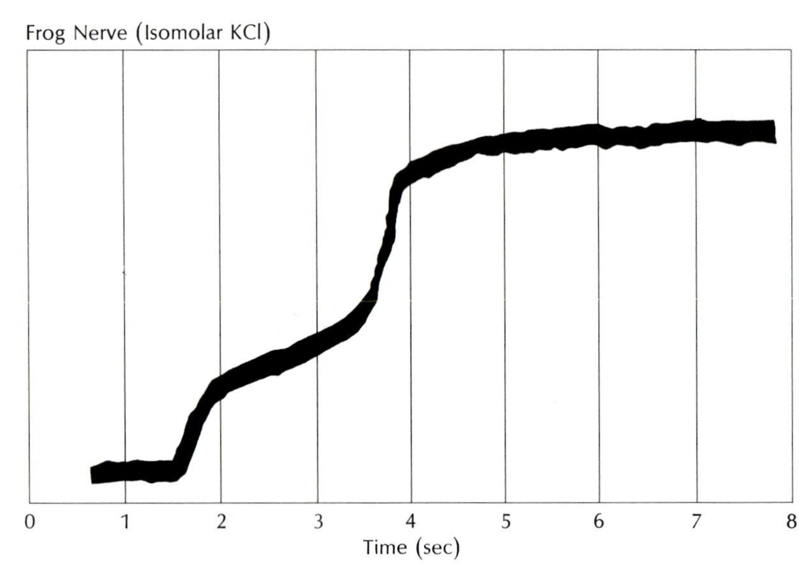

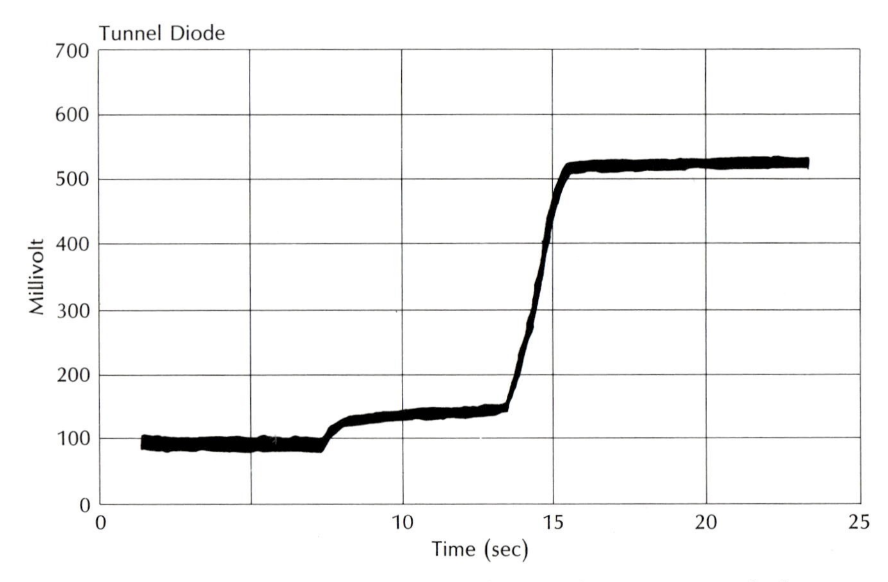

Black lipid membrane "doped" with various substances shows a jump to a higher state of electrical resistance in response to initial low-voltage pulse (top). Very similar resistance changes are seen in the natural membrane of frog nerves (center) and also in the electronic semiconductor known as a tunnel diode (bottom).

transport of potassium by valinomycin in a model system, though subsequently much related work was done on the BLM models.)

The liposome model has helped elucidate a number of problems in membrane physiology. To me, as a once-practicing physician, one of the most interesting has been the role of membranes in anesthesia. As is well known, surgical anesthesia can be achieved with a great variety of compounds – chloroform, ether, cyclopropane, xenon, perhaps even ethanol if one drinks enough of it. It is also evident that many or most of these substances do not produce anesthesia by chemical means, since they are excreted unchanged by the body. (Xenon is chemically inert except under special, nonphysiologic conditions.) A long-held theory relates anesthesia to the anesthetic's solubility in lipids particularly to what is called its partition coefficient between lipid and water. Molecules that dissolve preferentially in lipid are likely to be anesthetic. (For a full discussion of this, see Chapter 24, by Seeman.)

The apparent role of lipid solubility naturally suggested that anesthetics might work by altering the cell membrane, since membranes are one of the main loci of cellular lipids. It seemed to my associates and me that the liposome model would be a good means of exploring this possibility. We could not, of course, inquire of a liposome whether it was anesthetized, but we very quickly determined that anesthetics made the liposome membranes leaky: when we labeled them with radiopotassium, the isotope moved into the surrounding medium with a speed proportionate to the concentration of anesthetic in the medium. Further experiments at different temperatures, pressures, and concentrations of anesthetics established that the leak was due to a partial loosening of the orderly, bilayer membrane structure. I do not mean to suggest that a patient becomes anesthetized because his cell membranes have been made leaky: the leakiness is simply what we measure in the model. Rather, I think, the operative mechanism of anesthesia is membrane disorder, of which the leak is simply one expression. The partial disruption of the membrane structure would be expected to affect all cells to some extent, and in fact this has

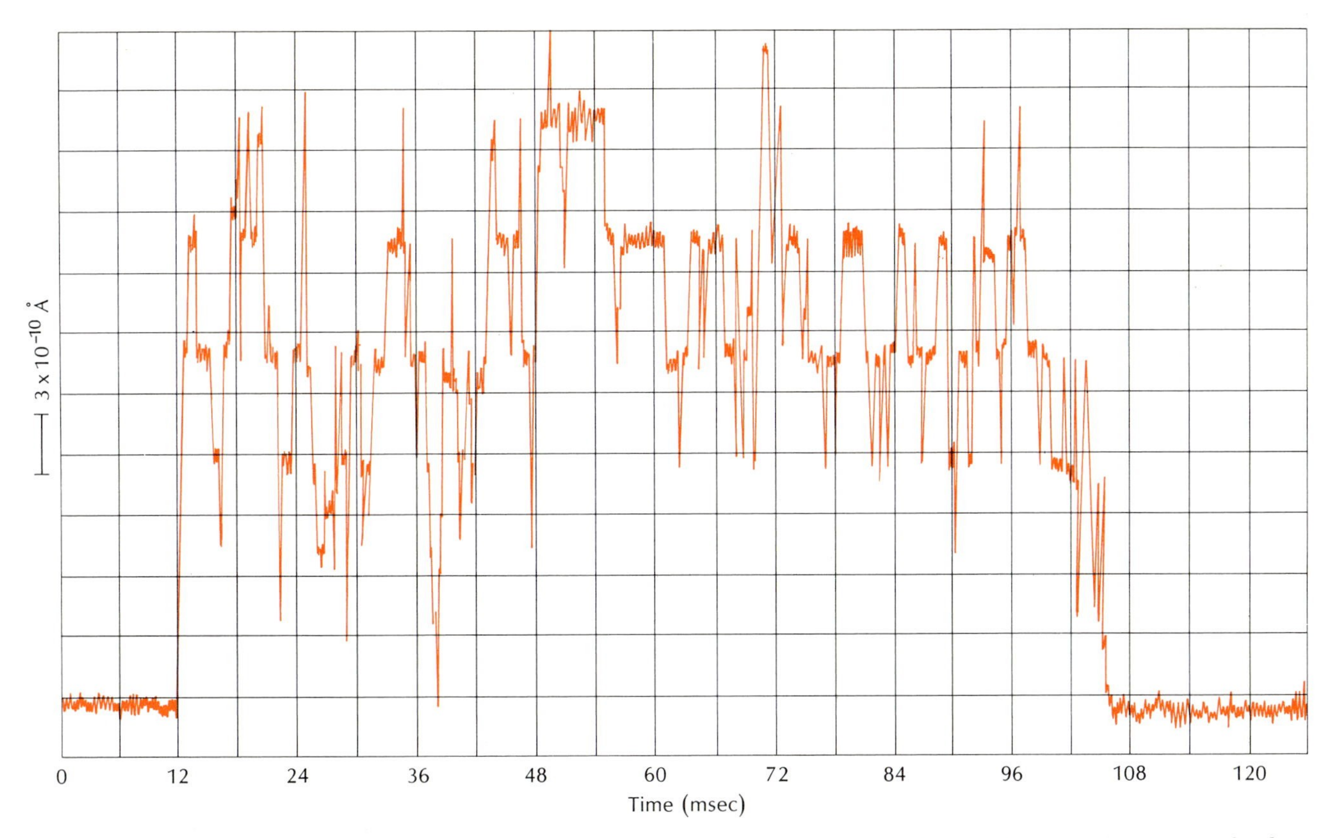

Tracing above represents a group of short-lived fluctuations of electrical current through a black lipid membrane induced by the presence of alamethicin molecules in the membrane and an applied voltage of 210 mV. The discrete and fluctuating levels of conductance are not observed when valinomycin is present in a BLM as an ion carrier (after Haydon and Hladky).

long been known to be the case. Anesthetics must, however, affect nerve cells preferentially – chiefly, I believe, because the integrity of the neuronal membrane is absolutely essential if the cell is to perform its job of transmitting nerve impulses. (It is not unlikely, also, that some anesthetics act preferentially on neurons partly because their molecules have a greater affinity for neuronal membranes, because of specific chemical properties of the latter.) Some earlier experiments, at the University of California School of Medicine, San Francisco, using monolayers rather than bilayers as models, indicated that the addition of only 2% chloroform or ether molecules to a membrane is enough to disrupt its functioning to the point of producing surgical anesthesia.

While we are on the subject of anesthesia, a colleague of mine has been able to carry out an amusing but revealing experiment whereby liposomes are subjected to a bizarre sequence of events. The experiment was originally carried out with newts or tadpoles, anesthetized with chloroform or ether, to the point at which they sank to the

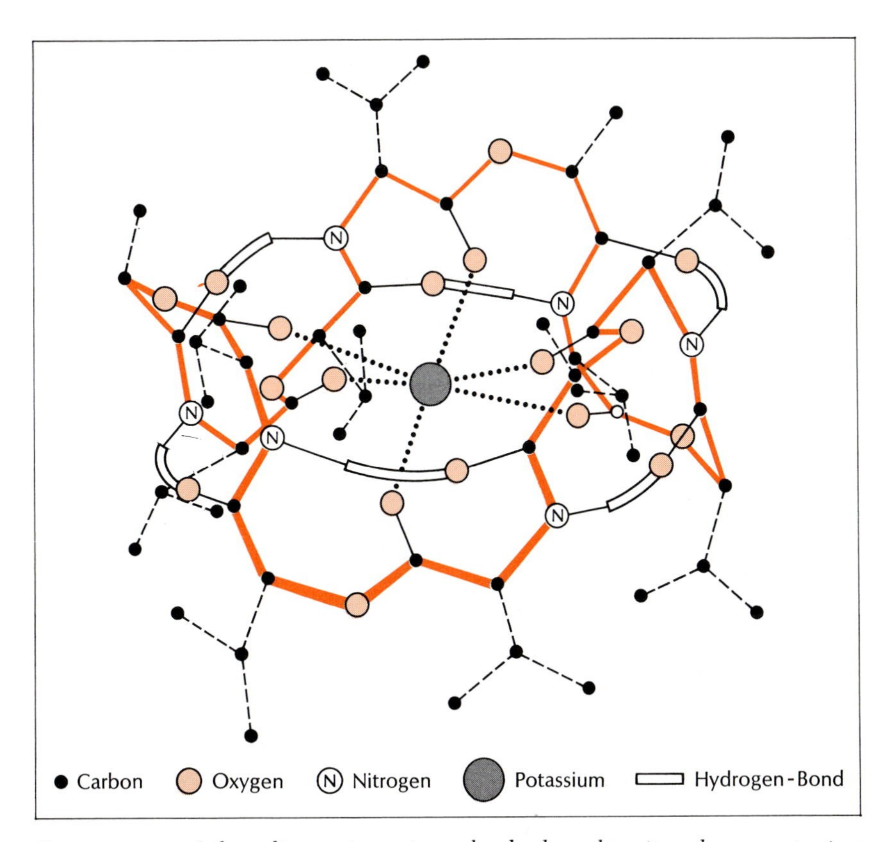

Ring structure of the valinomycin carrier molecule shows how it encloses a potassium ion in a favorable, water-like oxygen environment while at the same time it is presenting a lipophilic periphery to the lipid membrane (after Shemyakin).

Formation of liposomes containing labeled ions (colored dots) begins with almost total separation of water from lipid (top left). As more water is added, it forms tubes within the lipid, which are lined with the molecules' hydrophilic heads (top right). With still more water, lipid first forms lamellar bilayers (center left), then enclosed liposomes (center right). Dialysis removes labeled medium (bottom left) after which liposomes tend to retain their enclosed label (bottom right).

bottom of a tank; upon application of a pressure of some 90 atmospheres, the animals were observed to resume swimming!

Evidently the increased pressure was forcing the membrane molecules back into their normal orderly configuration. Interestingly, liposome models showed the same effect. Labeled liposomes developed an abnormally high leak rate at the same anesthetic concentrations that immobilized the newts. And when we put both liposomes and newts into the pressure chamber, we found that the pressures that reawakened the animals also returned the liposome leak rate to normal. In theory, of course, these results might have been due not to a forcible reordering of the "contaminated" membranes but rather to a squeezing out of the anesthetic molecules. A control experiment, however, pretty well disposed of this possibility. Here we used "unanesthetized" liposomes, and found that at high pressures their leakage rate dropped to *subnormal* levels; the only explanation seemed to be that the pressure was forcing the membrane lipids into an abnormally orderly and "tight" structure.

It is perhaps worth noting that this reversal of anesthesia by pressure is the apparent opposite of what has been observed in the case of gaseous rather than liquid anesthetics. Divers, for example, must limit the depths at which they operate lest they succumb to nitrogen narcosis, a form of anesthesia that manifests at high pressures, or must breathe an artificial mixture of oxygen and helium, which is nonanesthetic. Here, however, we are dealing with a specifically gaseous pressure effect. Nitrogen, like all gases, becomes more soluble in water as pressure increases (compare, for example, the carbon dioxide in a closed and an open bottle of soda water). Thus the effect of increasing depth is to increase the concentration of nitrogen in the diver's body fluids and therefore its anesthetic effect. Theoretically, if the diver were to go deep enough, he would reach a point at which the continuing increase in pressure would reverse the nitrogen narcosis by reordering the molecules in his neuronal membranes – but I doubt that any actual diver would be rash enough to undertake the experiment.

Another series of experiments with

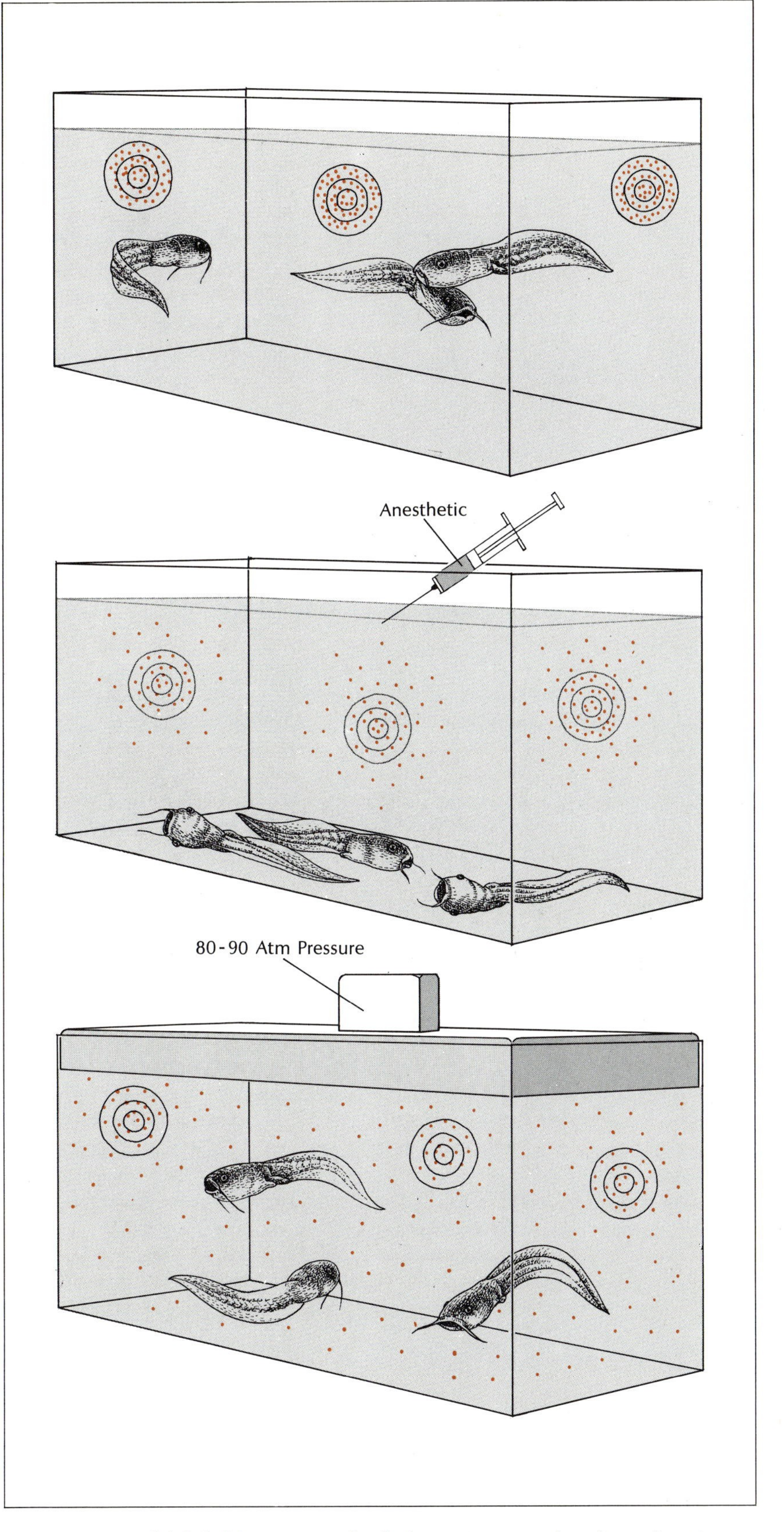

Experiments with labeled liposomes and tadpoles (top) suggest the relationship of membrane integrity to the mechanisms of anesthesia. The addition of anesthetic renders tadpoles unconscious and disorders liposome membranes, releasing radioactive label to medium (center); when high pressure is applied, the tadpoles awaken and order is also restored to the membranes, blocking release of label (bottom).

liposomes was carried out during the past few years by a group at New York University headed by Gerald Weissmann, in connection with studies in the physiology of gout. From a variety of evidence obtained by other investigators, chiefly D. J. McCarthy, these investigators had postulated that acute inflammation in gout results from phagocytosis of monosodium urate (MSU) crystals by leukocytes. The crystal-containing vacuole merges with one of the cell's lysosomes and the crystal interacts with the lysosomal membrane, rupturing it and releasing lysosomal enzymes into the cell's cytoplasm. The result is cell death and lysis, with release of inflammatory cellular enzymes into the gouty joint (see Chapter 26, "The Molecular Basis of Acute Gout," by Weissmann).

As a check on their reasoning, the NYU researchers used liposomes to simulate lysosomes and found that MSU crystals did indeed break down the liposomes, releasing the labeled ions incorporated into them. They even produced "boy" and "girl" liposomes, by incorporating traces of testosterone and 17-beta-estradiol respectively, and found that the former were far more susceptible to MSU-induced lysis. These observations chime neatly with the ancient clinical observation that gout occurs only in men and postmenopausal women. Interestingly, they also found that lysis occurred only if the liposome contained cholesterol. Just possibly, this might provide a biochemical rationale for the old theory that gout attacks are brought on by a "rich," i.e., high-cholesterol, diet.

Yet another application of the liposome model has been in relating chemical differences among membrane lipids to functional differences in the membranes themselves. Since this has already been explored in the previous chapter and will be dealt with again later (see Chapter 8, by Lucy), I need say little more about it, other than to observe that the liposome studies have yielded results fully consistent with those obtained by other methods. Liposomes, modified in various ways, have also been used to study antigen-antibody reactions at the cell surface, as well as the transmission of nerve impulses at neuromuscular junctions, but these experiments will also be discussed later at length (see Chapter 23, "Immune Reactions of Model Membranes," by Kinsky; Chapter 17, "Membranes in Synaptic Function," by Whittaker; and Chapter 9, "Junctions Between Cells," by Pappas).

Meanwhile, on the "practical" side, there is reason to suspect that liposomes, in addition to serving as models of cell membranes, may find clinical application as tools for the precise or timed administration of therapeutic substances. In the enzymatic approach to treating genetic disease, for example, it has been found that only a very limited number of enzymes can be injected into the circulation with any hope of actually reaching the interior of the cells where they are needed. Preliminary studies suggest that enclosing the enzyme in liposomes may be one way around this problem, the membrane models providing both a protective "capsule" for the enzyme – preventing its breakdown in the bloodstream – and a carrier facilitating its cellular absorption.

Another possibility was foreshadowed by a 1935 patent application by I. G. Farben, which came to my attention by chance some time ago; it dealt with a number of techniques to slow the absorption of parenterally administered drugs. One of the methods described reads, with hindsight, like a recipe for manufacturing liposomes – though I do not think the I. G. Farben chemists realized what they had got hold of. Certainly it seems quite feasible to "package" a drug in liposomes. Since their permeability characteristics can be varied almost at will, they could be tailored to allow their contents to diffuse into the circulation at any desired rate, thereby paralleling the action of "spansules" in orally administered preparations.

One can speculate further. In principle, it should be perfectly possible to incorporate into the liposome membrane specific "target" molecules that would ensure release of the drug only when or where needed. For example, liposomes containing a tranquilizer could be equipped with target molecules that only respond to supernormal levels of adrenalin. Likewise, liposomes loaded with antitumor agents or with powerful antibiotics could be keyed to the specific antigenic properties of tumor cells or of some pathogenic microorganism respectively, so that the drug would be released only when the liposome was actually in contact with its therapeutic target. By this method, it would be possible to employ drugs far too toxic for administration through the general circulation..

Whether or not membrane models ever achieve therapeutic applications of this sort, I think it safe to say that we can expect these fascinating structures to continue to provide us with basic information of the properties of natural membranes. Considering the number of natural membrane properties that have already been built into models of various sorts, there seems no reason to doubt that increasingly sophisticated models will simulate natural membranes ever more closely year by year.

4

Architecture and Topography Of Biologic Membranes

S. J. SINGER
University of California, San Diego

Up to this point, the discussions on biomembranes have been concerned primarily with membrane lipids. This is in part because their physical and chemical properties, and especially their structural interrelations, are relatively well understood. However, in order to discuss the overall structure and conformation of membranes, we must necessarily involve ourselves in the more difficult problem of membrane proteins. These substances are the other major chemical component of biomembranes, in fact, generally the predominant one. Moreover, their structural relationship to the lipids is one of the most discussed (and disputed) aspects of membrane studies.

Insofar as it concerns itself with membrane proteins, this chapter will inevitably overlap to some extent with those immediately to follow. The conceptual barriers that separate the fields of protein function, protein structure, and protein chemistry can be fairly described as highly permeable, and discussion of any one of these aspects is bound to reflect that fact. In particular, as we shall see, experiments contrived to answer questions in one field have often proved to be relevant to either or both of the others.

Proteins, to begin with, are the major component of nearly all membranes – from 50% to 70% by weight, the balance being mostly lipids with some admixture of cholesterol. The single known exception is myelin, in which lipid plus cholesterol accounts for nearly 90% of the weight. The reason for this exception seems to be that most membranes are physiologically very active, transporting molecules into and out of the cell, reacting to various extraneous substances, and so on – functions that are thought to be chiefly if not exclusively the province of their predominant protein moieties. With myelin, on the other hand, its chief physiologic function is precisely to remain *in*active, i.e., to insulate the neuron it surrounds from external, perturbing influences, so that large quantities of protein would not only be unnecessary, they would be undesirable.

A question that arises immediately is why are some proteins soluble, found in the cytoplasm or extracellular fluids, while others are specifically associated with membranes? Any successful model or theory of membrane structure must provide an explanation for the difference between molecules of soluble proteins, such as hemoglobin or serum albumin, on the one hand, and membrane-bound enzymes and receptor proteins on the other. The membrane model I discuss in this chapter provides such an explanation.

I have suggested, as a first approach to an understanding of membrane proteins, that they can be classified into two categories, peripheral and integral. The former, which may constitute up to 25% of total protein, are defined operationally as only loosely attached to the membrane, as indicated by the fact that they can be isolated by such "mild" techniques as changing the ionic concentration of the medium or adding a chelating agent such as EDTA; examples are the cytochrome C of mitochondrial membrane and the protein complex called the spectrin of erythrocyte membrane. Integral proteins, on the other hand, can be isolated only by quite drastic treatment – with detergents, bile acids, organic solvents, and so on. Even then they often remain associated with some lipids and can be freed only at the price of being denatured. Though I shall have something more to say about the peripheral proteins later on, they are essentially irrelevant to our central problem, the basic lipid-protein structure of the membrane, since they are literally peripheral to that structure.

Starting about 1960, the work of protein chemists and

especially of protein x-ray crystallographers has told us a great deal about the factors that are involved in folding up a protein molecule into its three-dimensional active form. Every protein molecule has ionic and highly polar amino acid residues along with other residues that are nonpolar. To fold up in a stable form, the ionic and highly polar, or hydrophilic, residues must be largely exposed to water, while the nonpolar, or hydrophobic, portions should be kept away from contact with water. The reader may recall that the "classic" Danielli-Davson model of membrane structure visualized a lipid bilayer covered on both sides by a layer of unfolded protein, with an additional layer of globular protein over all (see Chapter 1 by Danielli). In order to account for the various types of observed forms of accelerated molecular transport across the membrane, which could not occur with an intact lipid bilayer, Danielli eventually postulated "pores" lined with protein and extending through the lipid bilayer. One would conclude that this model is not satisfactory on thermodynamic grounds. A continuous unfolded layer of protein would of necessity have many nonpolar residues exposed to water. Furthermore, the protein layer would also cover up the hydrophilic, "water-seeking," heads of the lipid molecules. The few pores were introduced into the model in an ad hoc fashion, without an explanation of which proteins they involved or what made them stable.

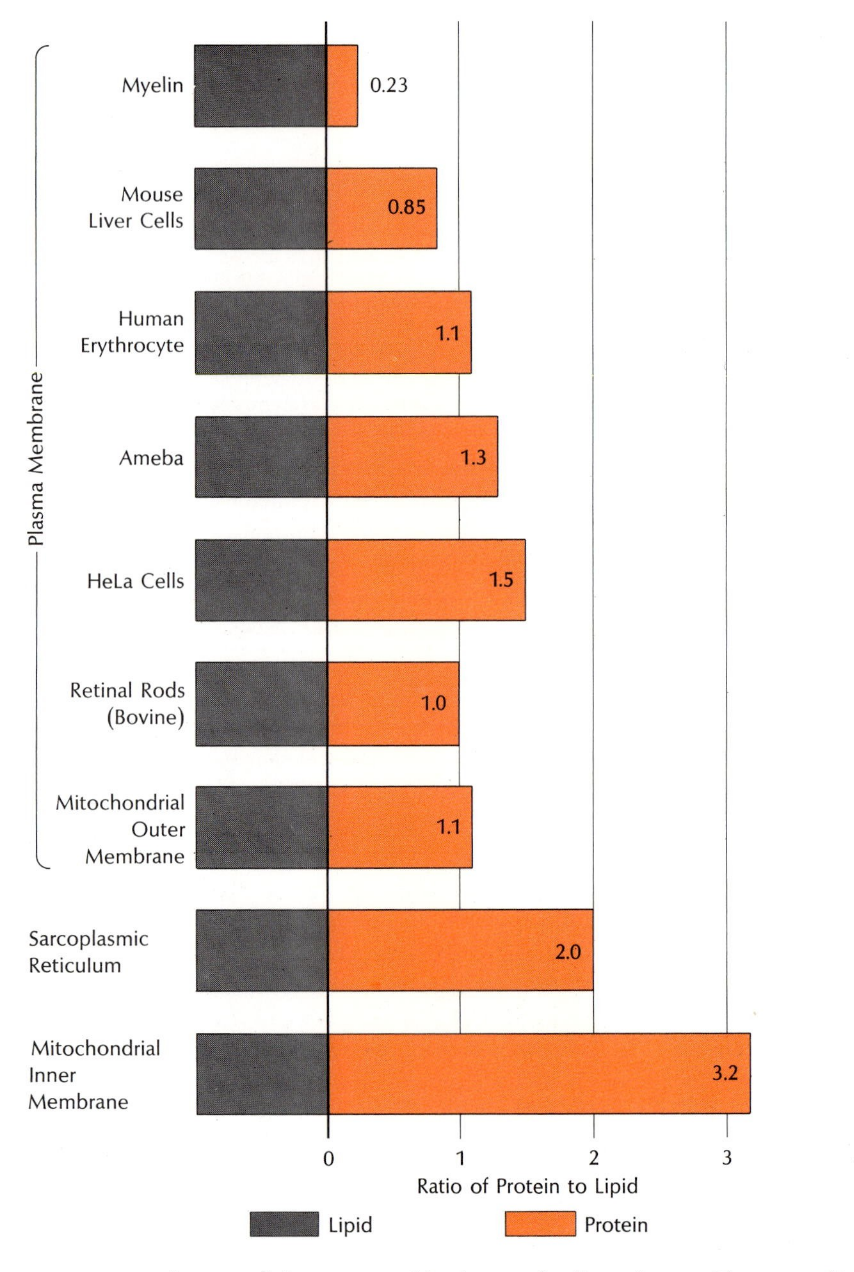

Proteins equal or exceed the quantity of lipid in nearly all membranes. The outstanding exception is myelin, thought to act as an insulator of the neuron – a function that would be incompatible with a high proportion of active (protein) molecules.

In addition to these theoretical objections, however, there is now much experimental evidence against the "unfolded protein" concept. For example, optical measurements of integral membrane proteins in intact membranes reveal that some 40% of their length is definitely not unfolded but rather is in the α-helix configuration; the proportion of α-helix is, in fact, greater than that found in most soluble proteins of known globular structure. But if we then visualize the integral proteins as a layer of globular (rather than unfolded) protein attached to the exterior of the lipid bilayer (which is thermodynamically possible), the membranes would be much thicker than those actually observed (70 to 90 Å).

In the light of these and other considerations, Don Wallach, then at Harvard University, and I, around 1966, independently postulated a quite different model of protein-lipid architecture. We visualized the proteins as globular and folded up so as to be amphipathic – possessing one hydrophobic and one hydrophilic end, just as do the membrane lipid molecules, though of course the proteins would be much bigger. The hydrophobic end would be embedded in the interior of the lipid bilayer, in contact with the hydrophobic lipid "tails," while the hydrophilic end would be ringed with hydrophilic lipid "heads" and would also project out into the aqueous medium surrounding the membrane. The fundamental structure has been compared to that of icebergs (protein) floating in the sea (lipid). If the proteins were large enough to span the entire thickness of the membrane, I suggested that they might have two protruding hydrophilic ends and a hydrophobic embedded interior.

The suggestion that membrane proteins were amphipathic was made at a time when no such proteins had as yet been investigated in sufficient structural details. But it was attractive because it could explain why membrane proteins were special compared with soluble proteins, and yet were generally not unusual in their amino acid compositions. The x-ray crystallographers have learned that soluble proteins have their ionic residues more or less uniformly spread over the surface of the molecule; they are therefore soluble because they interact strongly with water over their whole surface. Amphipathic proteins, on the other hand, would tend to be insoluble in water because of their water-repelling hydrophobic ends, the same ends that would be especially adapted to bind to the hydrophobic lipid interior of the membrane. However, their amino acid *compositions* need not be unusual; only the way the amino acid residues were unsymmetrically arranged would be different for membrane-bound compared with soluble proteins.

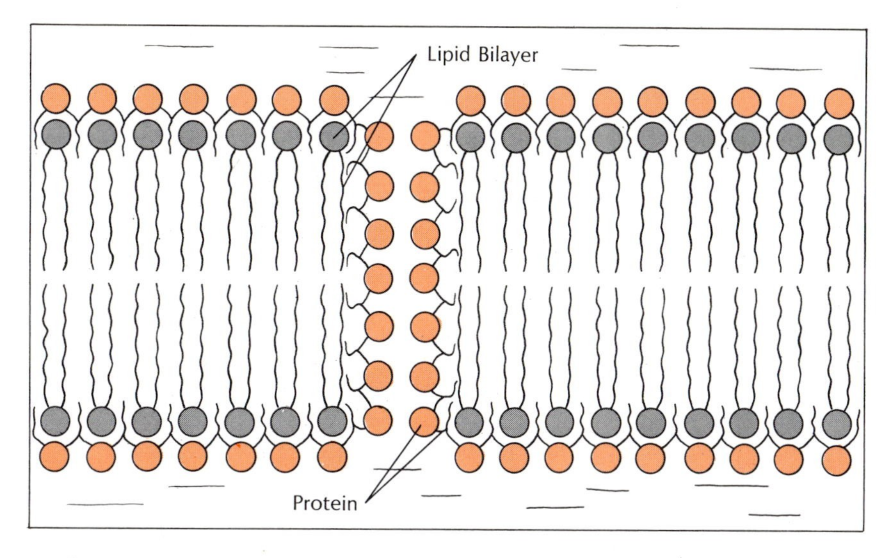

Early (Danielli) model of membrane structure visualized "unrolled" protein covering both sides of lipid bilayer, with hydrophobic amino acid residues interacting with similar lipid chains; hydrophilic residues were thought to form "pores" for molecular transport.

In the years since this model was introduced, a series of experiments has given strong support to it and much extended it. Some of them concern the actual structures of membrane proteins, a few of which have now been isolated and studied. Cytochrome B_5 of microsomal membranes, for example, is a demonstrably amphipathic protein. Here the hydrophilic portion of the protein, containing about 100 amino acid residues, the heme group, and the enzyme active site, can be "clipped" from the intact membrane with proteolytic enzymes, showing that it protrudes from the membrane. The remainder of the protein, consisting of about 40 amino acid residues that are disproportionately hydrophobic, is clearly embedded in the lipid. The whole molecule, when isolated and then added back to microsomal membranes at 37°C, will become attached spontaneously to the membrane, but not if the hydrophobic end is first clipped off. Another fairly well-studied integral protein is glycophorin, recently isolated from erythro-

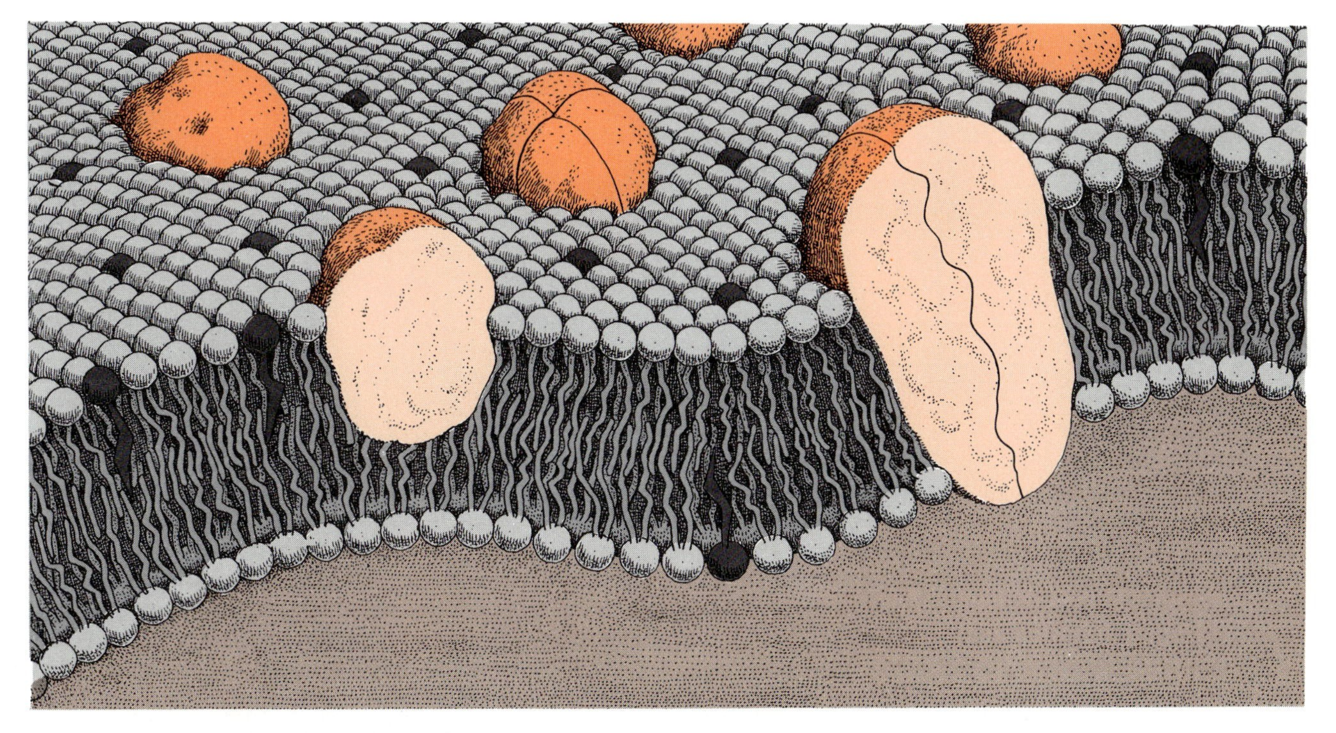

Current (Singer) membrane model sees proteins as predominantly globular (red) and amphipathic, with their hydrophilic ends protruding from the membrane and their hydrophobic ends embedded in the bilayer of lipids (gray) and cholesterol (black). The proteins make up the membrane's "active sites"; some of them are simply embedded on one or the other side, while others pass entirely through the bilayer. Some of the latter presumably contain transport pores.

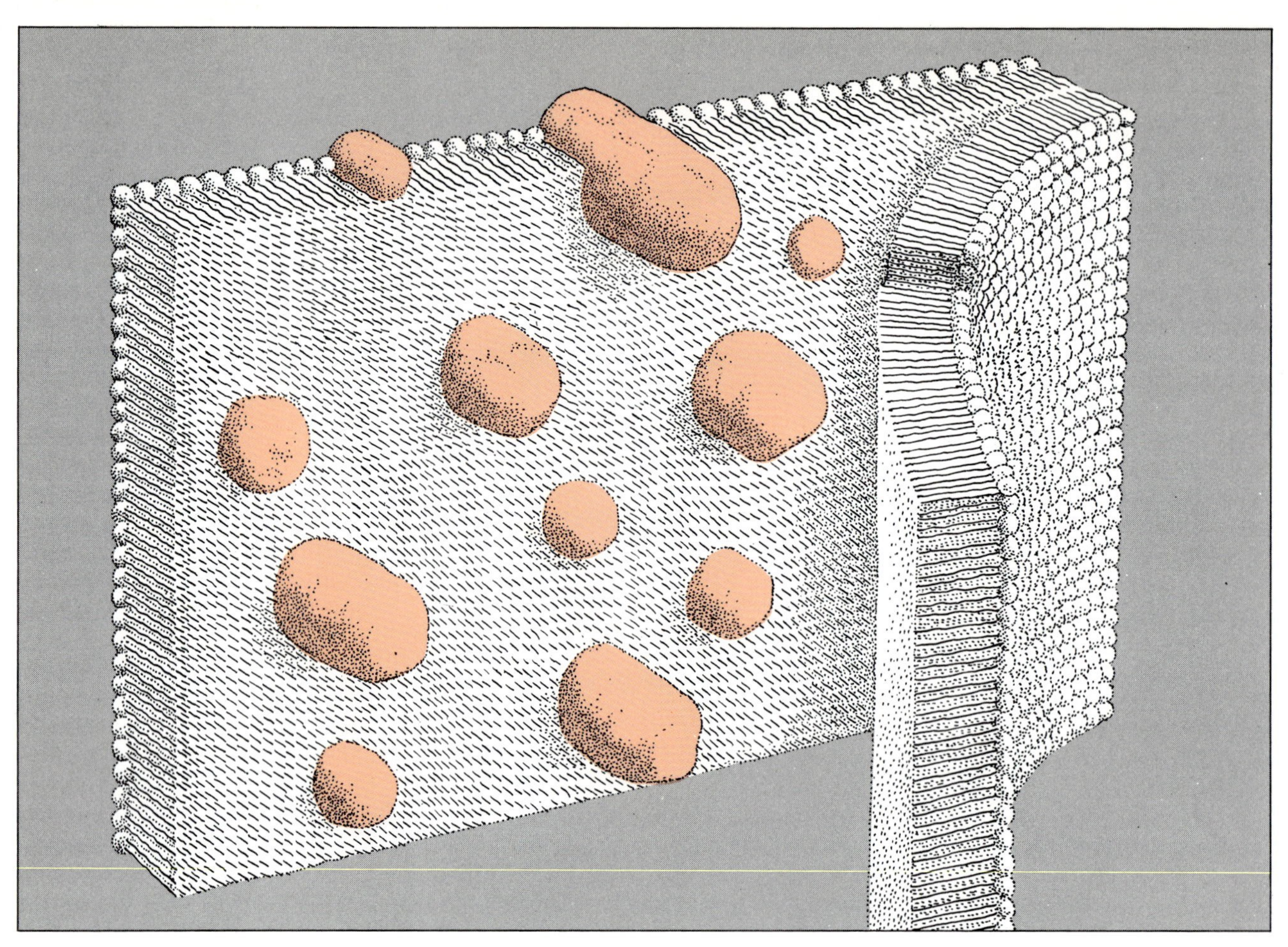

Splitting of membrane shows interior of model at bottom of preceding page; schematized above, it is seen below in freeze-etched electron micrograph by da Silva and Branton. Fracture plane through preparation reveals the rough fracture surfaces (F), the bumps on which are thought to represent individual protein molecules or subunit aggregates.

cyte membranes and analyzed by Vincent Marchesi and his colleagues at the National Institutes of Health. It consists of a protein chain of about 200 amino acid residues to which are attached a large number of oligosaccharide chains. Most remarkably, these sugar residues – which are, of course, highly polar – are all found on half the chain (about the first hundred residues from the amino end). This region is followed by a highly hydrophobic stretch of about 25 amino acid residues and finally by a hydrophilic region (but with no sugar residues) near the carboxyl end of the chain. This structure is consistent with other evidence that the molecule of glycophorin spans the entire erythrocyte membrane, with hydrophilic portions protruding from either surface of the membrane and an intervening hydrophobic portion embedded in the membrane interior. Both cytochrome B_5 and glycophorin are even more literally amphipathic than Wallach and I had imagined, since we had visualized the *three-dimensional* folded structure as demarcated into distinct hydrophilic and hydrophobic regions, whereas these proteins are *linearly* amphipathic. This probably means that each linear portion of the chain can fold up independently. (For a full discussion of this, see Chapter 5 by Marchesi.)

Even more direct visual evidence can be obtained by freeze etching. Cells, or isolated membranes, are frozen into a block of ice, which is then mechanically cleaved. It has been demonstrated by Daniel Branton and his collaborators at the University of California at Berkeley that certain portions of the cleavage plane split the membrane through its interior – between the two layers of lipid. And on these interior areas the electron microscope reveals quantities of particles protruding above the lipid layer very much in the way one would expect if one were looking at the "icebergs" from below. Every membrane that has been looked at this way has shown the presence of similar particles, although their number and size vary with different membranes. In the case of the erythrocyte, it has been established that the particles contain glycoproteins and, in particular, glycophorin. It seems likely that the intramembranous particles in other mem-

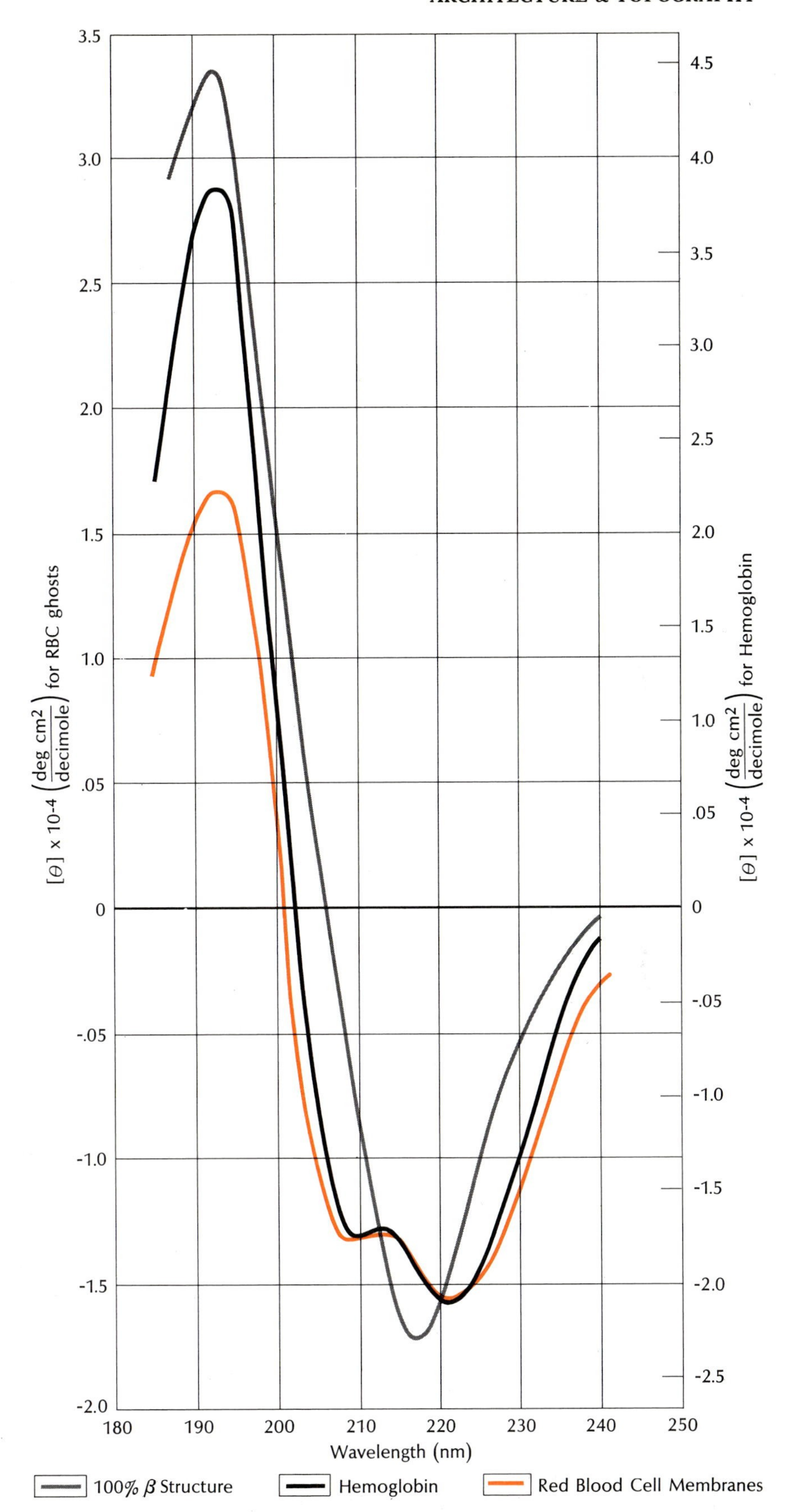

Circular dichroism spectra of intact erythrocyte membrane and of soluble hemoglobin both show characteristic double minimum at 208 and 222 nm, indicating that much of protein is in alpha helix form, about 75% for hemoglobin and 40% for the average of the proteins of the membrane. Hypothetical protein entirely in beta ("unrolled") form would show only a single minimum at 218 nm.

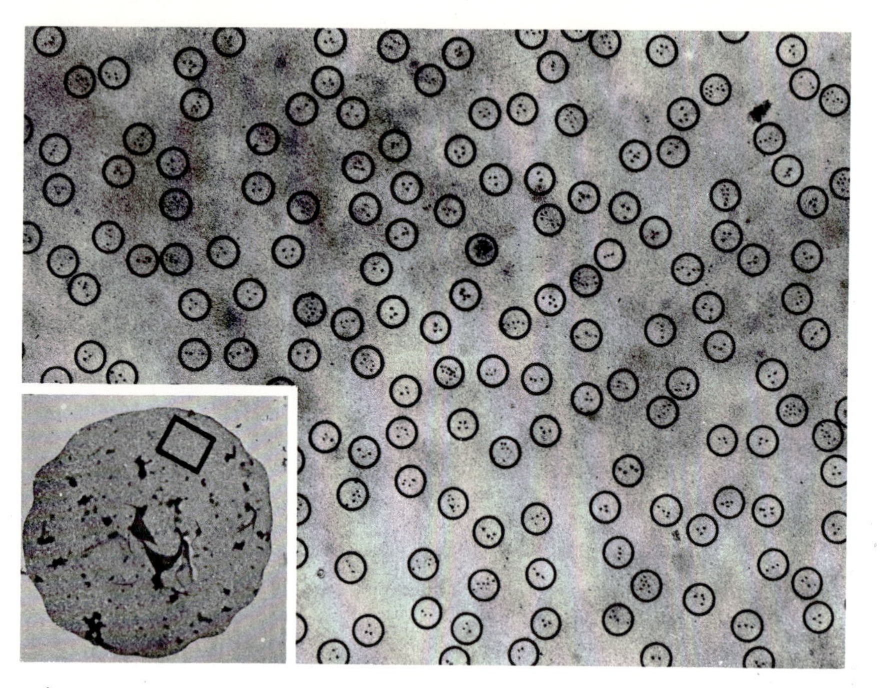

Protein antigen (the Rho(D) antigen) sites on human erythrocyte membrane are visualized by indirect staining with ferritin-conjugated antibody. First, an unlabeled specific human antibody is bound to the Rho(D) antigen sites on the membrane surface. Then a ferritin-conjugated goat antibody to human γ-globulin is reacted with the membrane. A single Rho(D) site is marked by a cluster of 2 to 8 ferritin particles (circled). Inset shows erythrocyte ghost at low magnification (Science *175:725, 1972).*

branes are also protein in nature.

An important point about these particles is that in the erythrocyte membrane they account for something like half the total integral membrane protein, yet the observed particles represent *only* those protein lumps whose bottoms project beyond the midline of the lipid bilayer. Since there is no reason to think that all membrane proteins are that deeply embedded – and some reason to suspect that many are not – it seems probable that a large fraction, if not all, of the integral protein exists in the embedded, iceberg form.

Other experiments concern a postulate implied by the iceberg model: the protein bergs should be free to float about in the lipid sea, so long as the latter is in a liquid or liquid-crystal condition, which, as noted earlier, is typically the case at physiologic temperatures (see Chapter 2, "Lipid Dynamics in Cell Membranes," by Chapman). For the unfolded, continuous-layer model, such mobility would be impossible; rather, one would expect a relatively rigid and, in all likelihood, orderly arrangement of proteins. Studies of erythrocyte membranes in which antigenic proteins have been tagged with labeled antibodies show that their arrangement is not orderly but random. Even more significant is an experiment by C. D. Frye and M. Edidin of Johns Hopkins University, using human and mouse cells fused with Sendai virus. Immediately after fusion, the human and mouse antigens, as revealed by immunofluorescence, are largely segregated in the two halves of the fused membrane, but after 40 minutes at 37° C the antigens become thoroughly intermixed throughout the entire membrane. Addition of substances that inhibit protein synthesis, ATP formation, or other key metabolic processes does not slow the rate of mixture – but lowering the temperature below 15° C does. This indicates the process is not energy dependent, nor the result of synthesis and insertion of *new* protein molecules, but rather a matter of simple diffusion of the existing molecules within the membrane.

Within the last two years, the field of cellular immunology has been electrified by the demonstration from several laboratories that the binding of an antigen to its specific antibody receptor molecules on the surface of lymphocytes can cross-link the receptors in the fluid membrane and drag them to one pole of the cell. It is thought by many immunologists that this process is in some way critical to turning the cell on to make and secrete its specific antibody.

A further postulate of the iceberg model is that the proteins, while free to move laterally in the lipid sea, would not be free to pass from one side of the membrane to the other. Such a process, however envisaged, would involve "submerging" the hydrophilic end of the protein among the hydrophobic lipid tails – a thermodynamically most unlikely process. This amounts to saying that proteins projecting from the exterior of the membrane, having presumably been placed in that position during membrane biosynthesis, will not be found on the interior face of the membrane. This has in fact been demonstrated experimentally. As noted above, erythrocyte membrane contains quantities of sugars attached to some of its proteins, whose location can be visualized by certain plant proteins called agglutinins; these bind to specific sugars much as antibodies bind to antigens. By the use of labeled agglutinins, it can be shown that the sugars in question are found *only* on the exterior of the erythrocyte and other mammalian cell membranes; they do not in any way "flip over" to the membrane interior surface. Similarly, it has been demonstrated that certain enzymatic activities of erythrocyte membranes are restricted to the inner (cytoplasmic) membrane surface.

This is all consistent with what we know about membrane function, which suggests that the outside and inside of the membrane must do quite different things. Membrane-bound sugars, for example, seem to be much involved in the processes of cell recognition and in many immunologic reactions, all of which concern primarily the "outside world" rather than the cell interior. One would expect, then, that maintenance of structural and chemical asymmetry between the inside and outside surfaces would be essential to cell function – and in fact it appears to be a thermodynamically necessary consequence of the iceberg model.

Before considering some other implications of the iceberg model, it might be well to say a word about another model that has been put forward by some investigators. This is the inverse of the iceberg model – not protein bergs floating in a lipid sea but rather pools of lipid scattered across a protein "ice field"–i.e., a mosaic of protein molecules in contact with one another and interacting through hydrogen bonding and similar forces. Thermodynamically this is as plausible as the iceberg model, but structurally it seems most unlikely, given the great variety of proteins in any given membrane – on the order of 50 species or even more. And even assuming that such an elaborate jigsaw could be constructed, it would necessarily be a rigid structure, i.e., the demonstrated mobility of membrane proteins previously described would be inexplicable. (As noted in an earlier chapter, there are also evolutionary considerations weighing against this model: in order to preserve the continuity of the membrane, a mutational change in one membrane protein would have to be accompanied by complementary structural mutations in the adjacent pieces of the jigsaw, a highly unlikely eventuality.)

Interestingly, some recent studies indicate that localized regions in certain membranes may consist of large numbers of protein molecules in tight contact with one another, producing a relatively rigid sort of structure. An example is the synaptosome, the region of contact between a nerve and a muscle cell that "recognizes" the chemical signal from a neuron. Significantly, however, such regions contain only one or a few different proteins, thereby greatly simplifying the jigsaw puzzle problem noted above. It should be pointed out, moreover, that these more-or-less rigid protein structures still account for only a small proportion of the cells' plasma membrane. What we have, apparently, are "super-icebergs" floating in the lipid sea that are perhaps a hundred times larger than usual types.

Given the basic iceberg-sea concept, an obvious question is whether the lipid is merely an essentially passive matrix supporting the proteins, or whether instead some sort of lipid-protein interaction is essential to protein function. Here the data are conflicting. On the one hand, the demonstrated mobility of proteins, with no apparent loss of physiologic function, would suggest that their location with respect to specific varieties of lipid is unimportant. On the other, we possess data concerning the influence of lipids on the activity of a number of membrane enzymes suggesting that at least some kinds of lipid must play a "meaningful" role in protein function, possibly by influencing the proteins' tertiary, or three-dimensional, structure. The most plausible conjecture seems to be that while the proteins do not interact with the bulk of the lipid component, in some cases they may interact with specific lipids, presumably those adjacent to them. Jost and Griffith of the University of Oregon have recently found evidence of such tightly bound lipid in the protein complex cytochrome oxidase of mitochondrial membranes.

A no less obvious question is whether the iceberg-sea model is compatible with the various types of known membrane function – specifically, information transfer (e.g., cell recognition, immune reactions, hormone responses, etc.), enzymatic activity, and transport activity. In regard to information transfer, in which certain proteins or glycoproteins in the plasma membrane respond to substances in the environment by inducing changes in the cell's interior, there seems to be no incompatibility with the model insofar as the protein response is concerned. On how the response is transmitted through the membrane to the cytoplasm we are still pretty much in the dark, but the model may suggest some educated guesses to explain this phenomenon. With respect to enzymatic activity, the model also seems compatible,

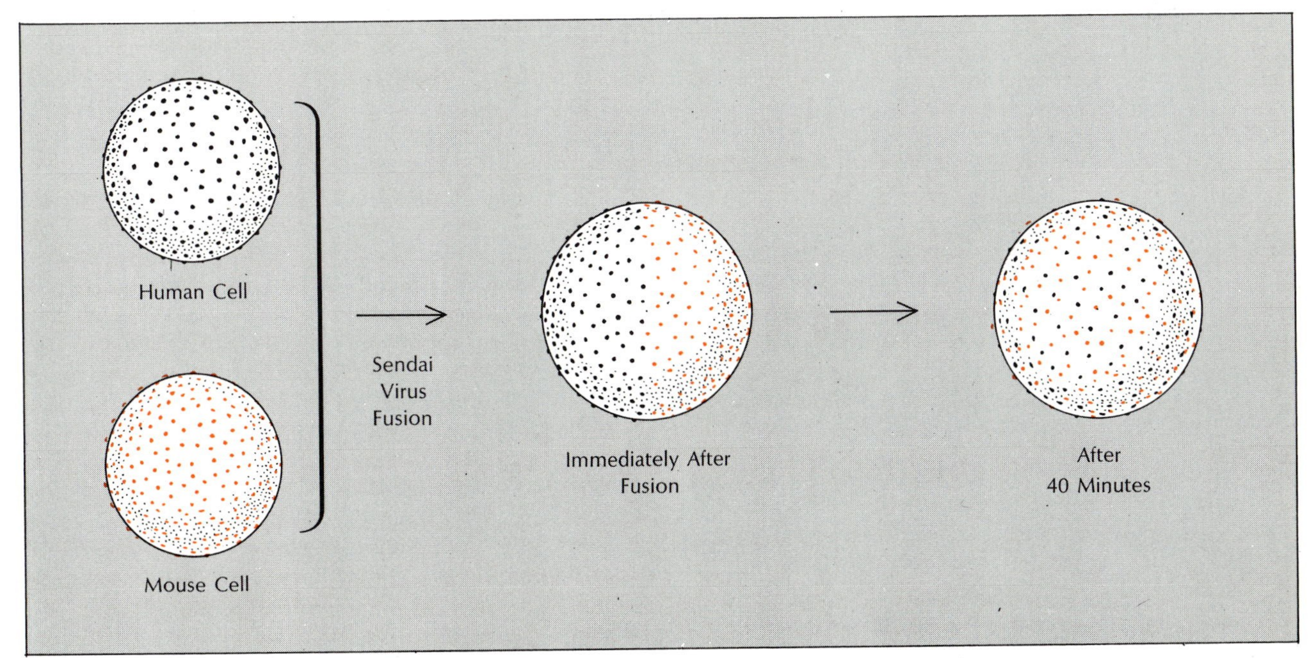

Mobility of proteins in lipid matrix of membrane has been demonstrated experimentally by fusion of human and mouse cells. Immediately after fusion, characteristic human and murine antibodies (which can be visualized by immunofluorescence) are found in discrete areas of fused membrane, but less than an hour later they have intermixed over entire surface (Frye and Edidin).

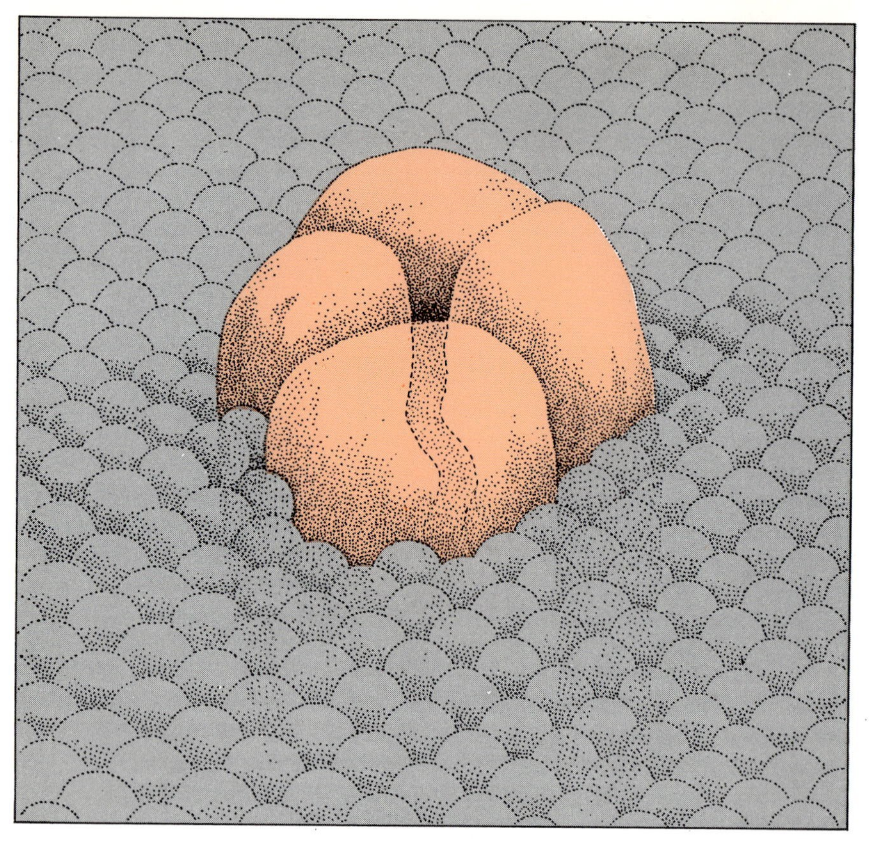

Water-filled channels for transport of specific ions and hydrophilic molecules through membrane may be formed by groupings of four (or more) protein subunits, as schematized above. Such a channel, about 10 Å across, is known to exist between the subunits of the hemoglobin molecule, though the protein is of course not found in membranes.

since there seems to be no reason why an appropriately amphipathic protein molecule could not have its active site region exposed while the hydrophobic end was partly embedded in the lipid matrix, as in the case of cytochrome B_5.

The phenomenon of molecular and ionic transport across the membrane requires a somewhat more extended discussion. One early theory had envisaged transport as involving either free "carrier" proteins that could bind to the transported ions and molecules and move with them through the membrane, or embedded proteins that by rotating in the plane of the membrane could move bound molecules or ions from the outside to the inside or vice versa. (The latter hypothesis is sometimes called the "revolving door" model.) It will be apparent that neither of these mechanisms is compatible with the thermodynamic considerations cited earlier. A "free" transport protein floating in the aqueous medium will, according to our views, have a hydrophilic exterior, meaning that it will require large inputs of energy to diffuse through the hydrophobic interior of the membrane; the same consideration applies to the hydrophilic end of an embedded, rotating protein. Experiments with synthetic phospholipid vesicles have shown that the amphipathic lipid molecules move from one side of the bilayer to the other very slowly – in a matter of days, if at all – and the same would surely be true of the much larger and more hydrophilic protein molecules.

A much more thermodynamically plausible hypothesis is that the transport proteins extend all the way through the membrane. That is, they are not bipartite – hydrophobic at the embedded end, hydrophilic at the other, but tripartite – hydrophobic in the embedded middle and hydrophilic at both protruding ends, as appears to be the case with glycophorin of erythrocyte membranes. It should be noted that of the presumed protein particles observed by freeze etching, at least some project far enough from the plane of cleavage (the membrane midline) to pass all the way through the other half of the membrane if it were present.

But how would such tripartite proteins, which for thermodynamic reasons would have to remain essentially fixed in their position vis-à-vis the two surfaces of the bilayer matrix, transport substances through their own structure? Here one must postulate some sort of water-filled pore, as has been proposed by many investigators. The general knowledge of protein structures that has been acquired in the last 10 years or so suggests how such pores could exist. We now know that most soluble proteins do not consist of single chains but rather of several chain subunits – two, four, or more; hemoglobin, with four subunits, is a good example. And the hemoglobin molecule has a water-filled channel, circuitous but continuous (its diameter is about 10 Å), running down its center. In fact, such a central channel is geometrically almost inevitable whenever four quasi-cylindrical bodies are combined.

One can thus visualize the transport proteins as consisting of several embedded subunits forming such a channel, which would be lined with ionic sidegroups. An ion or polar molecule could then either pass through the channel by diffusion or, by binding to its interior surface, could induce a change in the spatial relationship of the subunits (with or without the assistance of energy in the form of ATP). This conformational change would in effect force the ion or molecule through the channel and eject it on the other side. It is worth noting that the work of Max Perutz and others on the hemoglobin molecule has demonstrated that oxygen binding can induce only minimal conformational changes in individual hemoglobin chains, but that by interactions among these subunits the changes are "amplified" so as to induce substantial rearrangements of the subunits relative to one another. It is, of course, precisely such changes that account for hemoglobin's remarkable capacity to bind and release oxygen.

It must be admitted that we do not yet have much evidence to confirm (or, for that matter, disprove) this transport hypothesis (see Chapter 10, "Ionic Transport Across the Plasma

Membrane," by Hoffman). It may be worth noting that freeze-etch studies of myelin membrane have revealed none of the (presumed) protein particles found in other membranes. If we assume that at least some of the particles in the latter are transmembrane transport proteins, their absence in myelin would be consistent with the essentially insulating function of that substance. Really convincing evidence on the nature of membrane transport, however, will almost certainly have to wait until someone isolates an intact transport protein and determines its structure. All we can say now is that the subunit "channel" theory is consistent with the facts as we know them as well as with the laws of thermodynamics.

Having established as well as we can the basic architecture of lipids and integral proteins in biomembranes, what can we say about the peripheral proteins? There can be little doubt that these substances have quite diverse functions, many of which can only be guessed at today. Two examples will suggest some possibilities.

Cytochrome C is found in association with mitochondrial membrane, where it functions enzymatically in electron transport as part of the energy-yielding respiratory chain. As a peripheral protein, it is easily dissociated from the membrane and when dissociated is perfectly soluble in water, indicating that it is not amphipathic but essentially hydrophilic. Its three-dimensional structure, recently worked out by researchers at the California Institute of Technology, confirms this supposition; the surface of the molecule includes only a few small hydrophobic patches. These are presumed to be sites of attachment to other substances in the membrane, possibly including cytochrome oxidase. Cytochrome oxidase is known to be functionally related to cytochrome C and is also believed to be an integral protein of the mitochondrial membrane. Thus cytochrome C would appear to be an ordinary enzyme whose activity requires some sort of loose physical link with its associated enzyme.

Rather more remarkable is the peripheral protein called spectrin, found on the inner surface (and only the inner surface) of erythrocyte membranes (its name comes from the

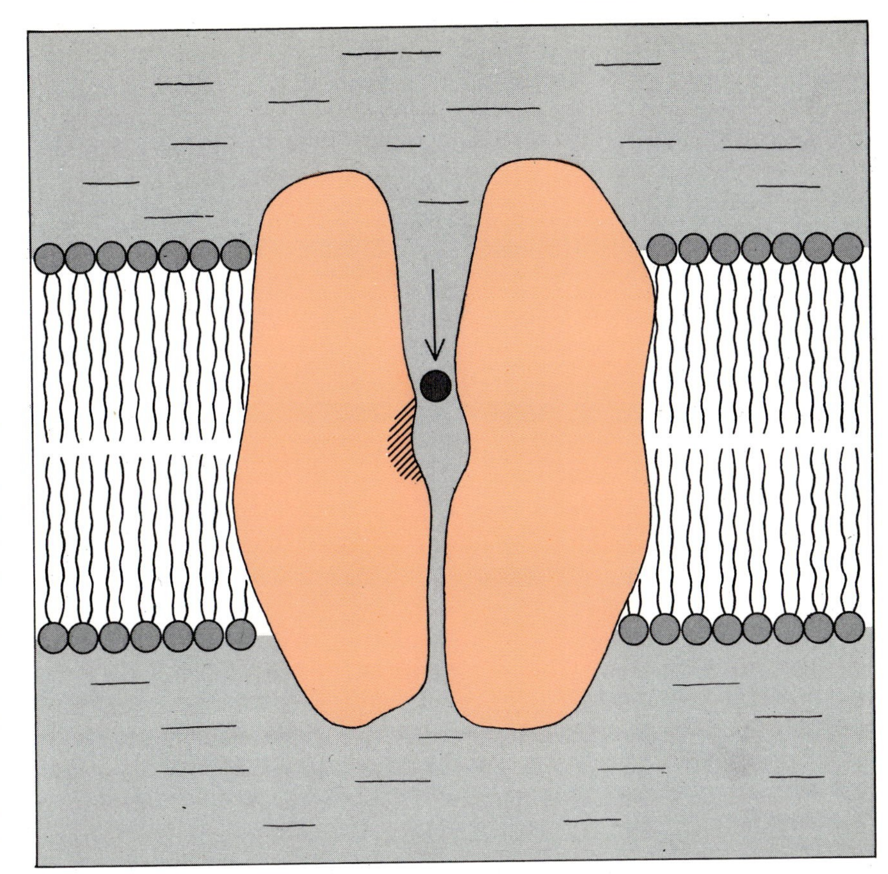

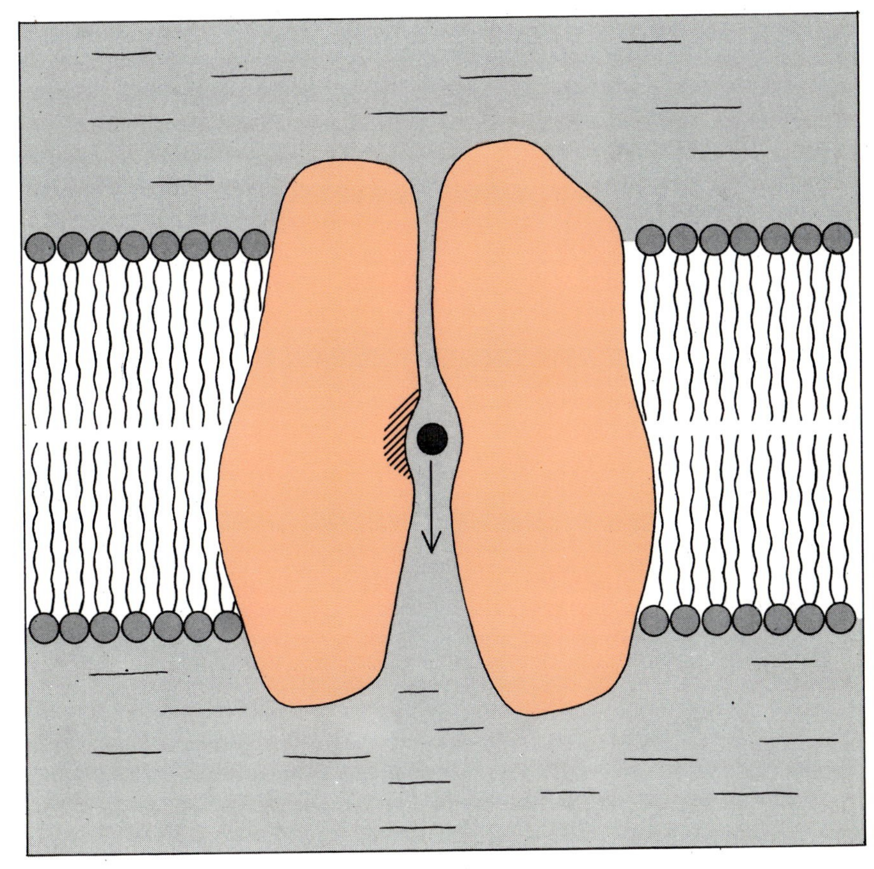

Active transport of molecule through membrane protein channel is visualized two-dimensionally. Molecule impinges (top) on active site (shaded) of protein, following which some energy-yielding enzyme reaction triggers shift in subunit configuration (bottom) that "squeezes" the molecule through the membrane.

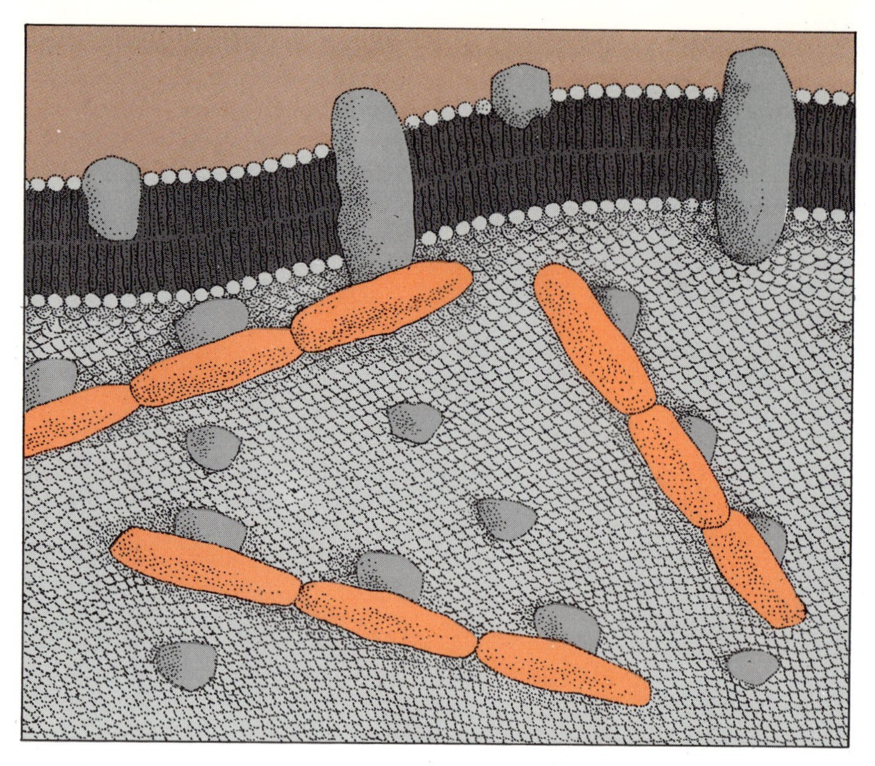

Hypothetical bracing action of the peripheral protein spectrin is visualized in this view of the inner face of an erythrocyte membrane. Rodlike assemblages of spectrin molecules (color) are thought to attach to, and form bridges between, several integral protein molecules in the membrane. It is believed they thereby stiffen the membrane and aid the cell to maintain its characteristic shape.

erythrocyte "ghosts" from which it is obtained). It accounts for some 25% of *total* protein in the erythrocyte membrane, and so it must be important, but it is not yet clear what its function is. Recently, Dr. Michael Sheetz and I have found that rabbit antibodies induced by the muscle protein myosin, isolated from the human uterus, will also bind (although weakly) to human erythrocyte spectrin. This shows that spectrin belongs to the family of proteins responsible for muscle activity and suggests that spectrin's function in the erythrocyte membrane is a mechanical one. This is further supported by our finding that one of the other proteins in the erythrocyte membrane closely resembles muscle actin. From these and other studies, we have suggested that spectrin (probably in conjunction with actin) serves as a sort of reinforcing scaffolding for the erythrocyte membrane. Spectrin molecules in solution are rodlike and tend to aggregate. One can envisage an aggregate of spectrin molecules lined up end-to-end, with an active site on each spectrin molecule binding to a specific site on an integral protein protruding from the membrane. By tying together these integral proteins in this manner, considerable mechanical strength would be imparted to the membrane without making it rigid.

There are a number of reasons why the erythrocyte would be expected to need special reinforcement of this sort. First, unlike most other cells it floats free, receiving no mechanical support from neighboring cells. Second, it is subject to a good deal of mechanical deformation as it is forced through the finer capillaries. Third, its peculiar biconcave structure–quite different from the approximately spherical shape that would be produced by surface tension alone (and which is found in immature erythrocytes) – is essential to its functioning. By maximizing the cell's surface area relative to its volume, the biconcave shape speeds the processes of oxygen uptake and release. Thus it seems not unlikely that a spectrin scaffolding would help the erythrocyte retain, or rapidly regain, its characteristic conformation in the face of the special stresses to which it is subjected.

It has been suggested that spectrin or some similar protein may play a mechanicochemical role in other cells by deforming the cell membrane in response to certain chemical stimuli. Such a process could account, for example, for formation of the annulus between the two halves of a dividing cell. One would expect such a protein to be peripheral, in the sense that it would attach itself to the membrane only at certain places and at certain times in the division cycle. Similar processes might account for the membrane deformations seen in such processes as phagocytosis and cell locomotion.

The model of membrane structure discussed in this chapter, after a few years' gestation in the scientific literature, has emerged and been widely accepted in the last year or so, a time in which many different kinds of experimental results were obtained that could be explained by the model. In the next few years, many membrane proteins will be isolated and their structures analyzed, and various predictions of the model will be tested. At present, however, it seems to be a good working model from which to begin to understand the complex and fascinating problems of membrane function.

5

The Structure and Orientation Of a Membrane Protein

VINCENT T. MARCHESI
Yale University

Our understanding of the lipid moiety of cell membranes has progressed considerably faster than our knowledge of membrane proteins. This has occurred not primarily because protein chemistry is more complicated than lipid chemistry but rather because until quite recently no one had managed to isolate membrane proteins in anything approaching their intact form. To free them from their lipid matrix meant at the same time partially to disrupt their molecular structures; the resulting heterogeneity and impurity of the preparations inevitably vitiated, or at least sharply limited, the conclusions that could be drawn from analytic studies. (This limitation applied primarily to the so-called integral membrane proteins; the peripheral proteins, as their name implies, are less intimately bound to membrane lipids and hence pose a less intractable problem; see Chapter 4, "Architecture and Topography of Biologic Membranes," by Singer.)

Now, however, a major integral protein of the human erythrocyte membrane has been isolated through studies my associates and I conducted at the National Institute of Arthritis, Metabolism, and Digestive Diseases and also through the work of other laboratories. From what appear to be pure preparations of this substance, we have been able to determine a number of important facts about its structure and properties. These, taken together with further data obtained by electron microscopy of intact membranes, have enabled us to reach certain conclusions as to the orientation of these molecules in the membrane; both the structural and orientational findings turn out to be quite consistent with what is known about the functions of this substance. As would be expected, our discoveries concerning this specific protein appear to have implications for the more general problem of protein-lipid interrelations in membranes.

The substance in question is one of the class often called glycoproteins – that is, proteins with carbohydrate molecules attached at various points along the amino acid chain. The term is actually somewhat misleading, since most proteins contain small quantities of carbohydrate, on the order of 1% by weight. The compounds we are concerned with, however, are far richer in carbohydrate – up to 50% or more; these typically occur in the form of oligosaccharides, containing from 4 to as many as 15 sugars. Glycoproteins of this sort are found in a number of types of plasma membrane, notably that of the human erythrocyte, where glycoprotein accounts for some 10% of total protein. If we consider only integral protein, the proportion of glycoprotein rises to 20%, making it a major integral membrane constituent.

As noted earlier in this book, erythrocyte membranes were among the first to be isolated from their cytoplasmic contents, in the form of "ghosts"; lipids extracted from these membranes have been intensively studied for years and have by now been characterized chemically in much detail. Efforts to isolate the proteins, however, were less successful. Preparations obtained by treating ghosts with such reagents as phenol, butanol, pyridine, and formic acid yielded heterogeneous mixtures of glycoproteins whose reported molecular weights ranged from 31,000 to 160,000. However, these and other experiments did establish a number of basic facts about the glycoproteins. First, they contain about 60% carbohydrates by weight. Second, though they are integrally bound to the membrane lipids, a portion of their molecular structure projects beyond the lipid layer on the exterior of the membrane, whence it can be "snipped" off by treatment with proteolytic agents such as trypsin. Third, the carbohydrate moieties of these molecules are antigenic, carrying deter-

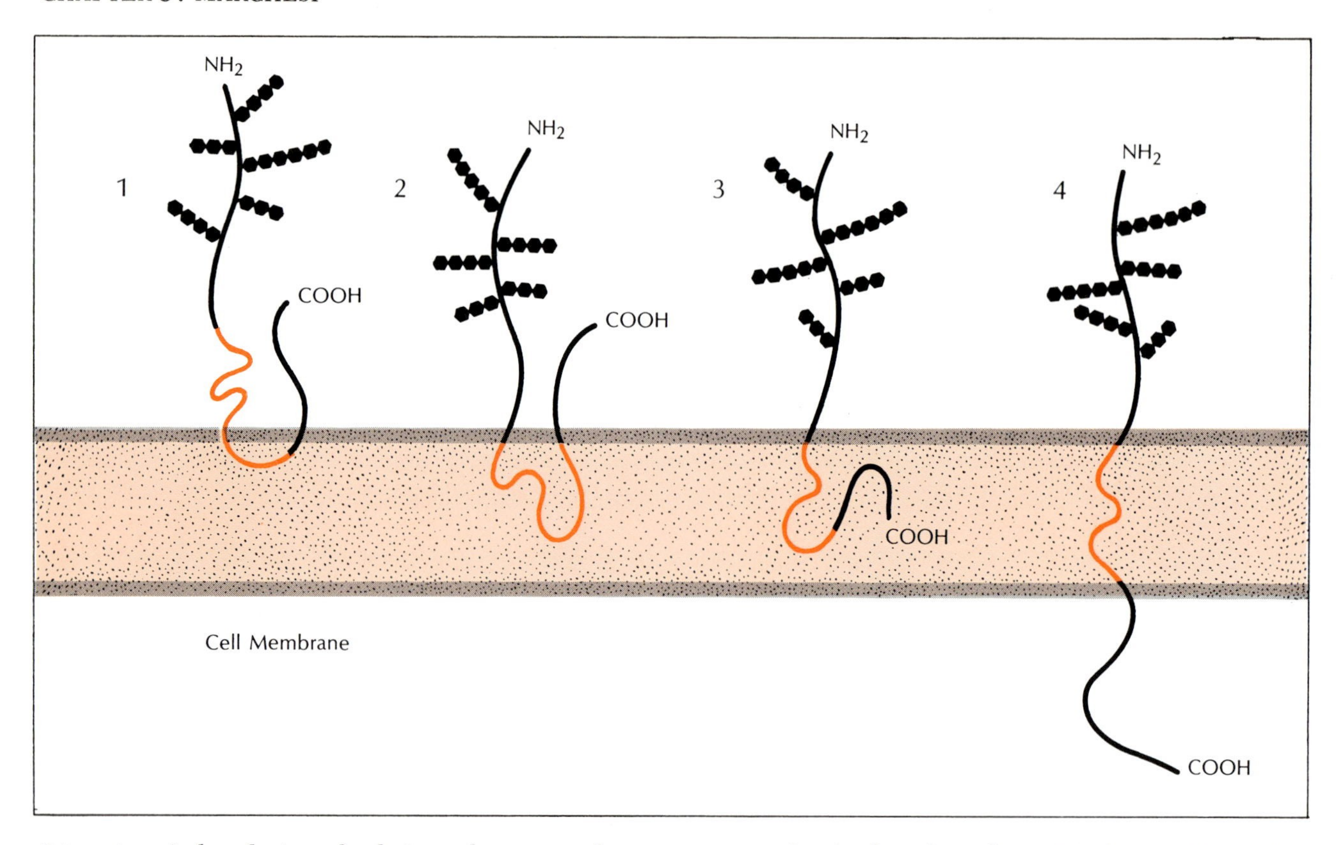

Orientation of glycophorin molecule in erythrocyte membrane can be deduced by considering four possible models. The first and third models are excluded because they are either inconsistent with the solubility properties of glycophorin (1) or involve incompatible relationships between hydrophilic and hydrophobic portions of molecule and membrane (3); the second possibility is excluded by labeling experiments showing that only the amino (NH_2) end of the molecule is accessible from the membrane exterior. The only remaining possibility is that the molecule passes entirely through the membrane (4).

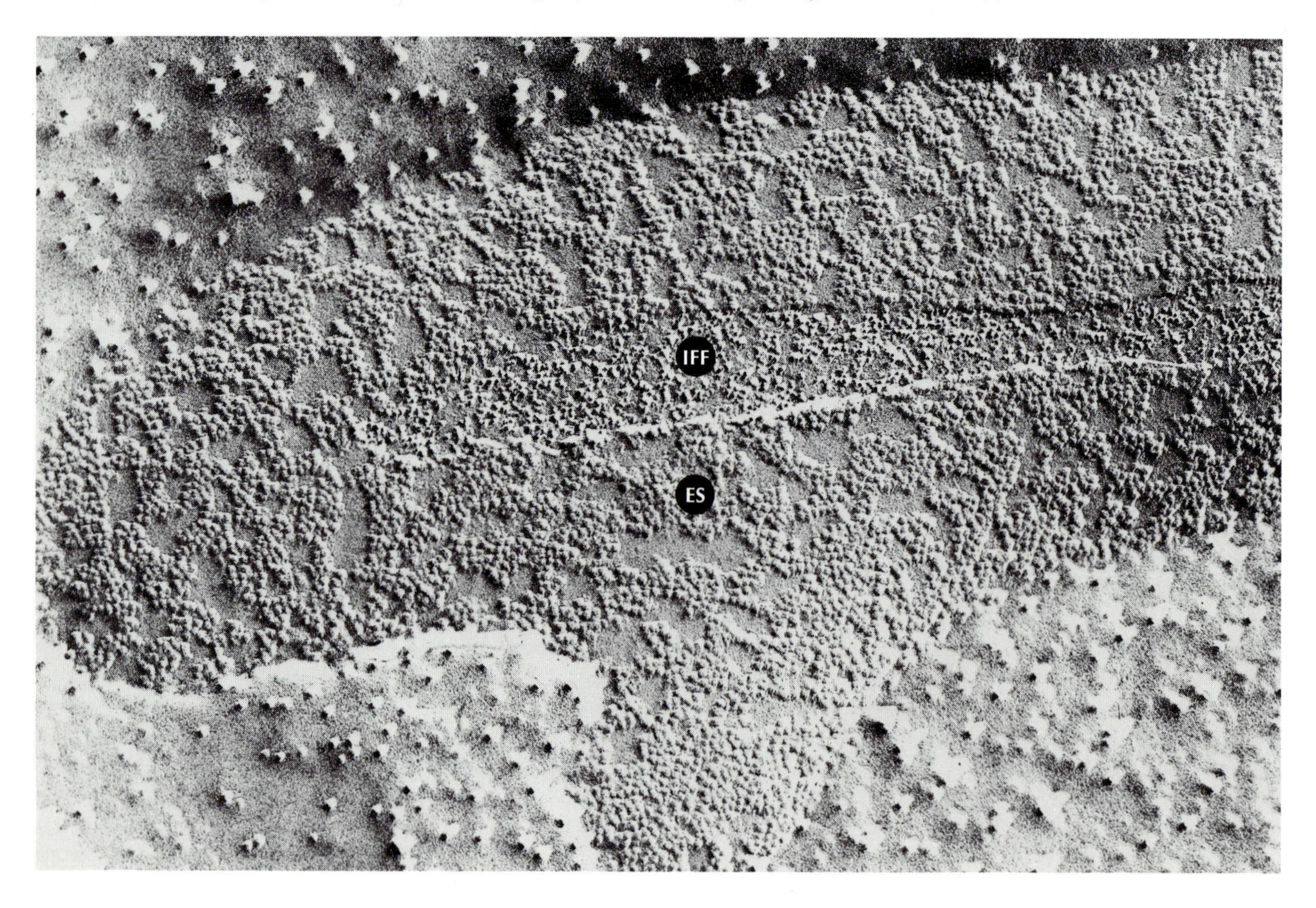

minants for both the ABO and MN blood groups, and also serve as receptors for viruses and plant agglutinins. Finally, residues of sialic acid—a nine-carbon amino sugar that is an important constituent of the carbohydrate moiety – seem to be responsible for most of the negative charge at the erythrocyte surface, which among other things is believed to prevent (through the repulsion of like charges) the cells from clumping together in the blood vessels.

Our own treatment of erythrocyte ghosts employed a new reagent, lithium diiodosalicylate (LIS), discovered by other investigators, which for reasons still uncertain appears uniquely capable of detaching membrane proteins from lipids. With the aid of this reagent, we were able to obtain purified preparations of glycoprotein having a molecular weight ranging from 50,000 to 55,000. The extracted substance migrated as a single component when subjected to chromatography or electrophoresis. Our apparently pure preparations proved, on assay, to possess most of the antigenic and receptor characteristics already assigned to erythrocyte glycoprotein that have been listed above. Quantitative studies, in fact, indicate that this substance accounts for a large part of the erythrocyte membrane carbohydrate and almost all of the sialic acid, for which reason we have given it the specific name of erythrocyte glycophorin—"glycophorin" meaning "sugar-bearing substance" and "erythrocyte" in recognition of the likelihood that other glycophorins, of somewhat different structure and composition, can be expected to turn up in membranes of other types of cells.

Further work with erythrocyte glycophorin has progressed along two lines: efforts to determine its molecular structure and attempts to characterize its spatial and functional relationship to the membrane as a whole. Discrete molecules can, indeed, be compared to discrete organs: both function typically as parts of larger organ systems, to which they are related in both structure and function.

Electron micrograph of freeze-etched erythrocyte membrane (left) shows "bumps" on both exterior surface (ES) and inner fracture face (IFF). ES bumps are ferritin-labeled agglutinin attached to protein-bound sugars; the "bumps" on the IFF are globules of protein within the membrane. Close examination reveals that pattern of "bumps" is apparently continuous across the two regions, suggesting that the two proteins are actually one.

Glycophorin appears to be a single-chain polypeptide of some 200 amino acid residues. Attached to this chain are 20 to 30 oligosaccharides, the points of attachment being asparagine, threonine, and serine residues. About half the sugar chains are tetrasaccharides, composed mainly of *N*-acetyl galactosamine, galactose, and sialic acid in various permutations and combinations; the other half, some attached to asparagine, are longer, with 8 to 12 sugar residues.

It is worth emphasizing that although, as noted above, the polysaccharides serve a number of different antigenic and receptor functions, there are far more polysaccharide chains than there are functions, even making a generous allowance for functions now unknown. Putting it another way, each glycophorin molecule evidently includes multiple copies of each species of receptor type. This has been demonstrated directly by removing sections of the "protruding" part of the molecule, by treating cells or ghosts with trypsin, and then testing the remainder for antigen or receptor activity. It turns out, for example, that the membrane will continue to bind the plant lectin phytohemagglutinin even after a substantial amount of the polysaccharide has been removed, and the same is true of other receptor activities.

Perhaps the most interesting fact about the glycophorin molecule is that, like Gaul, it is divided into three parts. This has been determined by the usual technique of breaking up the molecule into pieces with such reagents as trypsin and cyanogen bromide and then analyzing the composition of the individual fragments and determining the order in which they are linked together. (Eventually, of course, we hope to determine the precise sequence of amino acids for the entire chain, as well as the precise location and structure of the oligosaccharide side chains, but this is a good way off as yet.)

Meanwhile, it has become apparent that the oligosaccharides are attached only to about half the chain – the first 90 to 100 amino acids, figuring from the amino end of the polypeptide. Of course, not all of these have saccharide attachments, since there are approximately 30 of the latter, but of the remaining 100+ amino acids, few, if any have detectable saccharides attached. However, this portion of the polypeptide is not without interest in its own right. One segment (see diagram page 50) appears to consist entirely of amino acids without charge groups (e.g., serine, proline, leucine, isoleucine), while the remainder – the carboxyl end of the chain – is rich in charged amino acid residues such as aspartic acid, glutamic acid, and arginine.

Those who have read earlier chapters will get the point without difficulty. What we have is a polypeptide, one end of which is rich in sugars – i.e., is strongly hydrophilic, while the other end is rich in charged groups – i.e., is equally hydrophilic. The middle portion, by contrast, is devoid of charged groups – i.e., is strongly hydrophobic. The molecule, in other words, might have been expressly designed to mesh with a lipid bilayer, the hydrophobic middle portion interacting with the hydrophobic lipid "tails" in the interior of the bilayer, while the hydrophilic ends interact with the hydrophilic outer surfaces of the bilayer and with the aqueous medium that surrounds it on both sides. And as we shall see, there can be little doubt that, from an evolutionary standpoint, the molecule was indeed "designed" for that purpose.

In theory, this tripartite molecule could be disposed within the bilayer in several different ways, all equally consistent with thermodynamic principles: the hydrophobic middle portion would still necessarily be embedded in the hydrophobic interior of the bilayer, but *both* hydrophilic ends could project on one side (the exterior) of the bilayer instead of one on each side. However, this possibility has been pretty well disposed of through labeling experiments by us and others. When intact cells or intact ghosts are labeled with radioiodine and an enzyme, lactoperoxidase, which attaches it to the tyrosine residues in the polypeptide, *only* the end containing the oligosaccharide (i.e., the presumed exterior) is labeled. If, however, the ghosts are disrupted, so that the interior surface becomes accessible, *both* ends are labeled, strongly suggesting that the other (carboxyl) end protrudes into, or at any rate is in contact with, the aqueous medium inside the cell.

Further evidence on the orientation of glycophorin in the membrane has come through freeze-etching electron microscopic (EM) studies. Frozen preparations of isolated membrane can be cleaved so that the bilayer is split through the middle. On these exposed surfaces the electron microscope reveals numbers of intramembranous particles – protruding lumps thought to represent proteins embedded in or passing through the membrane. If the cleavage plane is shifted to pass close to the exterior surface of the membrane, either "outside" or "inside," and not through the middle, it becomes possible to visualize the exterior of the membrane. In untreated cells no structures are visible on these surfaces, for reasons to be discussed later. However, if the membranes have been previously treated with phytohemagglutinin (PHA) chemically conjugated with ferritin, electron micrographs show a "pebbled" texture. Moreover, if the cleavage plane is located so as to expose simultaneously both the bilayer interior and the exterior membrane surface, the exterior and interior bumps appear to form part of a continuous pattern.

This effect becomes much more marked if the membranes are first treated with trypsin. This, as already noted, detaches a substantial amount of the polysaccharide portion of the molecule, but does not eliminate its capacity to bind PHA. In addition, for reasons perhaps having to do with alterations of charge, the particles tend to clump together. This shows up under the electron microscope as a cell surface in which a reticulated pattern of clumped PHA-ferritin sites alternates with extensive "bald" patches. And in the partially split "exterior-interior" cell preparations, the pattern clearly continues across the line separating the etched exterior from the split-open interior.

Still further confirmation comes from examination of cell ghosts treated with dilute preparations of influenza virus. The virus particles embed themselves in the membrane, so that when it is cleaved the interior surface shows viral "footprints" – virus-sized dents – and fragments of the virus capsule. Since the particles are far bigger than the receptor sites or intramembranous particles, it is impossible to determine in normal preparations whether the particles are or are not associated with the smaller bodies. However, in trypsinized preparations the footprints show up *only* on the reticulated areas representing clumped intramembranous particles, never on the bald patches.

The conclusion seems clear. Since PHA is known to bind to glycophorin, it follows that each of the exterior sites visualized by the PHA-ferritin treatment must represent one or more glycophorin molecules. And since the distribution and patterning of the intramembranous particles are essentially identical with that of the binding sites, and remain so under various conditions (as does the distribution of virus footprints), these too must represent one or more glycophorin molecules. Finally, since the intramembranous particles project well past the midline of the membrane (along which it splits), it follows that they must at least be deeply embedded in it, and could equally well project all the way through it.

What one would like to be able to do, of course, would be to actually observe the polysaccharide portion of the molecule on the cell exterior, without the need for adding PHA-ferritin, and likewise the hydrophilic "tail" of the molecule projecting from the interior membrane surface. So far as the polysaccharide section is concerned, this should be possible in theory, assuming this part of the molecule does in fact protrude more or less straight out from the surface. Since we do not observe any such structures, it seems likely that this section of the molecule does not protrude in this manner. The polysaccharide chains are strongly hydrophilic and also, like most sugars, flat (planar) in molecular configuration; consequently, there seems no reason why they might not lie relatively flat along the membrane surface, interacting with the hydrophilic heads of the lipids. The effect would be rather like seaweed lying limply over a rock at low tide; from a distance one sees the shape of the rock, not the individual strands of weed.

Electron micrograph (right) of freeze-etched, trypsinized membrane shows clear continuity of pattern between aggregations of ferritin-agglutinin particles on exterior surface and those of intramembranous particles on inner fracture face. There can be little doubt the same molecules are being visualized in the two regions.

Direct visualization of the interior "tail" of the glycophorin molecule presents no less intractable problems. Freeze-etched EM's of the membrane's inner surface do reveal a surface texture, but it is fibrous rather than like the pebbly or lumpy surfaces described above. The fibers are believed to be the (peripheral) protein spectrin, which (as Singer notes in the previous chapter) is suspected of acting as a sort of scaffolding within the membrane that aids the cell in maintaining its characteristic shape. But even if the spectrin were removed, the "tails" would remain invisible. They consist at most of some 65 amino acid residues with a total molecular weight of no more than 6,000, making them unresolvable by present techniques. Eventually we hope to visualize them indirectly, by labeling them with specific antibodies or other substances. However, finding the proper reagent is likely to take some time, since the tail, unlike the other end of the molecule, is not designed to act as a binding site. (Exactly what it *is* designed to do is not known, though we can guess in a general way.)

Let us conclude by considering what we have learned about glycophorin in terms of its probable physiologic functions – insofar as we understand them. Concerning sialic acid, which forms the terminal residue on most of the oligosaccharide chains, one important function, as already indicated, seems to be to maintain the negative charge on the surface of the cell, so that it repels other erythrocytes instead of clumping with them. We know, for example, that as red cells approach their normal life-span limit of about 120 days, their membranes contain increasingly smaller amounts of sialic acid – and the probability of their being removed from the circulation by the spleen increases. In-

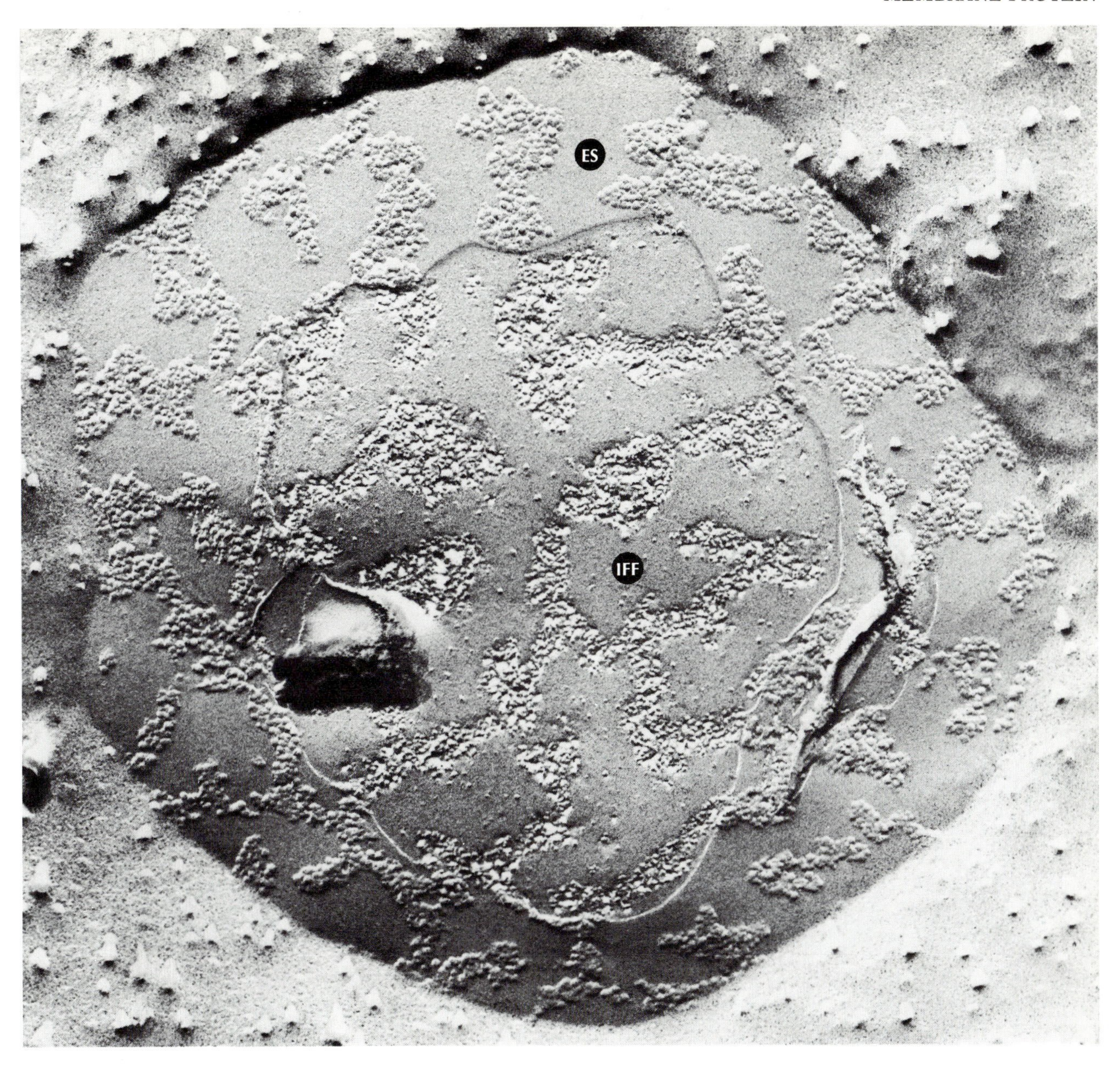

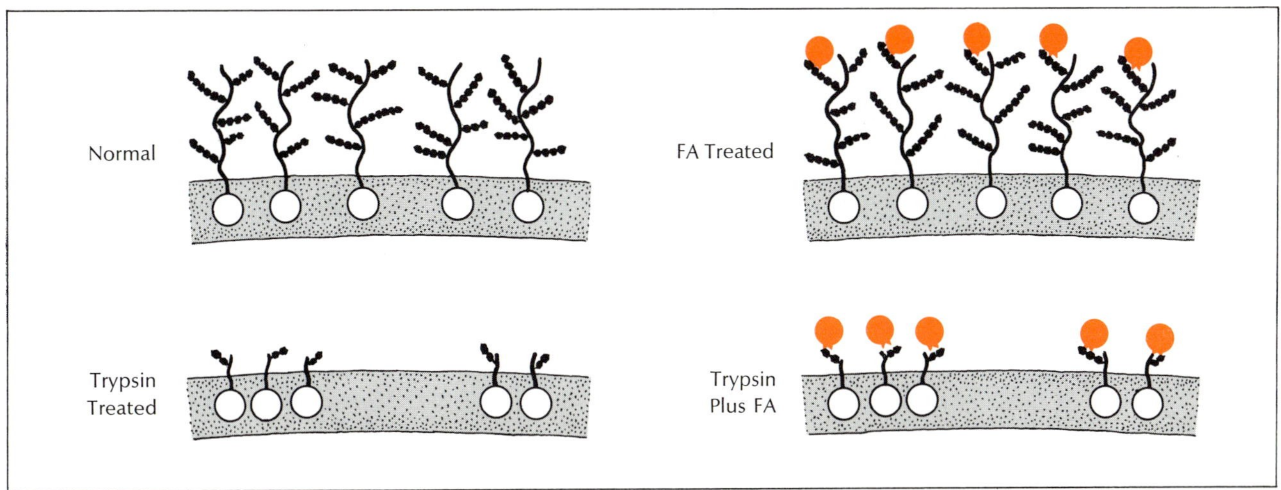

Ferritin-agglutinin complex (FA) binds both to normal glycophorin molecules and to trypsinized preparations in which most of the sugar-containing protein has been removed, indicating that the molecule contains two or more saccharide receptors for the agglutinin. Trypsin treatment, for unknown reasons, also causes glycophorin molecules to group into tight agglomerations.

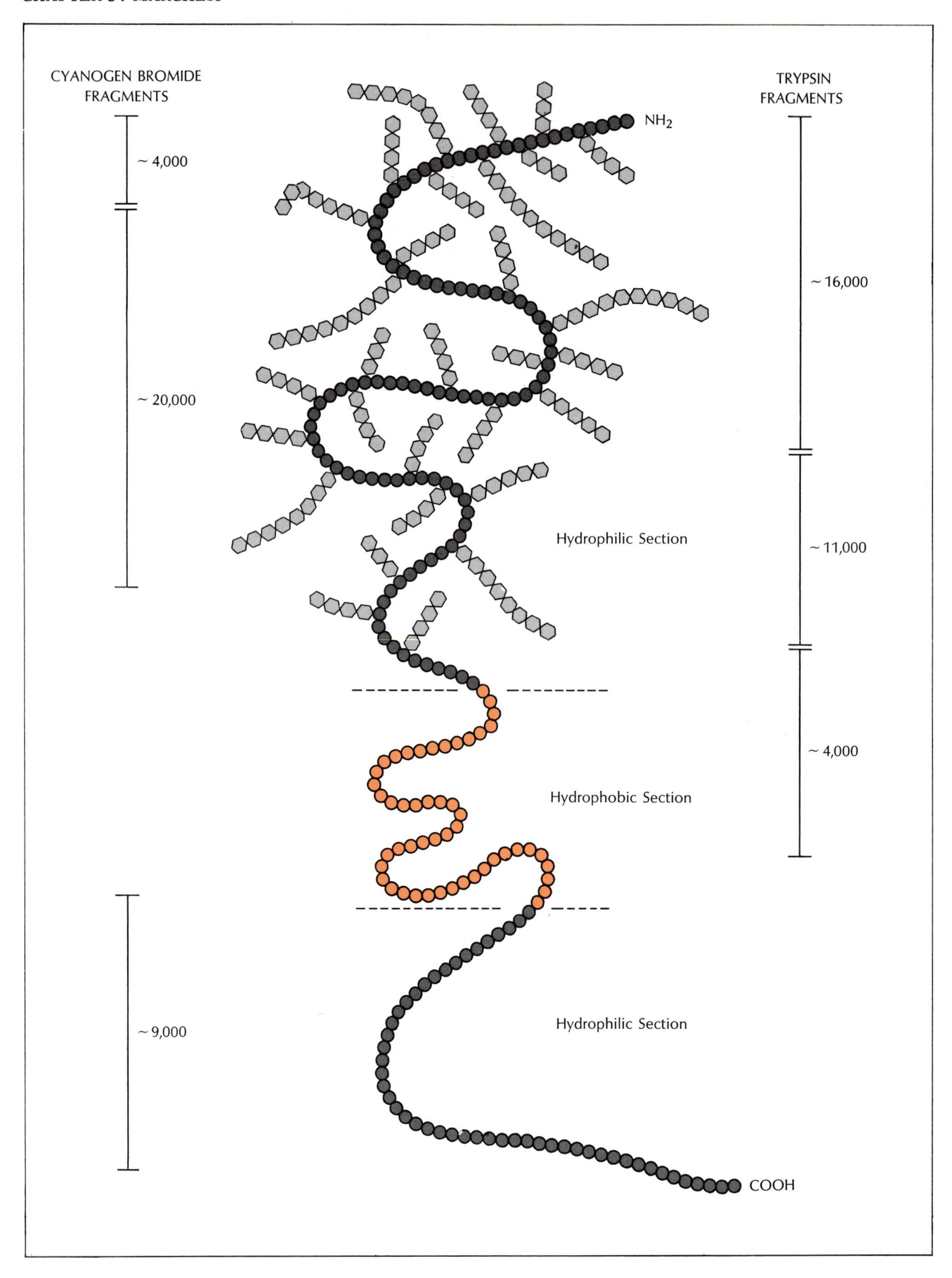

Glycophorin molecule consists of three main parts. Hydrophilic portion (top) carries many oligosaccharides (hexagons) attached to protein chain; hydrophobic portion (color), containing only uncharged amino acids, is thought to be embedded in lipid matrix of membrane. Hydrophilic carboxyl "tail" (bottom) includes no sugars but numerous charged amino acids.

deed, it has been shown experimentally that if animal erythrocytes are treated so as to remove most or all of their sialic acid, labeled, and then returned to the circulation, the labeled cells localize in the spleen in large numbers.

But how does the spleen "know" that these aged erythrocytes, and no others, are to be retired from the circulation? Not, it appears, because the spleen cells "recognize" such red cells simply through their loss of charge, but rather through something approaching an immune reaction, in which the spleen cells recognize the altered erythrocytes as "foreign bodies" because of the alterations in the sialic acid side chains. Normally, the penultimate sugar link in these chains is galactose, which is exposed when the acid is removed or drops off. And it has been found that if, after chemically removing the sialic acid from animal erythrocytes, the galactose is also removed from the membrane, the cells are no longer recognized by the spleen as scheduled for removal, or at any rate such recognition is greatly diminished.

Exactly what the relationship is between glycophorin's loss of sialic acid and the general process of erythrocyte aging remains unclear. Aging processes in the cell's interior may trigger the loss of sialic acid, providing a signal that these cells are due for retirement. But alternatively, or in addition, the loss of sialic acid may itself be one important aspect of erythrocyte aging. Certainly if we are correct in thinking that its presence is essentially what maintains the cells' surface charge, thereby preventing aggregation, cells that have lost their sialic acid will for that reason alone need to be removed from the circulation in fairly short order, since a build-up of such "decharged" cells could lead to clot formation. Just to complicate matters a bit further, we may note that other types of cells, when chemically stripped of the external sections of their glycoprotein by trypsin treatment, can regenerate the missing portions quite rapidly. Erythrocytes evidently cannot do this, at least as regards sialic acid, probably because they possess no nuclei to direct the necessary chemical synthesis.

From these findings on erythrocytes and the spleen, one might guess that the physiologic function – or at least a function – of erythrocyte glycophorin (and, by extension, of other glycophorins in other types of cells) is to serve as the cell's communication system with its environment, the latter term including both other cells (e.g., those of the spleen) and free, biologically active molecules (e.g., hormones). And in fact there is considerable reason to believe that this is the case. There is now much evidence that glycoproteins bind hormones such as insulin, and some reason to suspect that they may be involved in such intercellular interactions as contact inhibition. The blood-group antigens found on glycophorin are suspected of playing some role in intercellular interactions, though we do not yet know what. Certainly these antigens can hardly have evolved simply for the purpose of making life difficult for transplant surgeons – though some surgeons might perhaps disagree!

Another type of communication with the environment, albeit an "unnatural" one, is represented by the action of PHA. As we have seen, this substance binds to erythrocyte glyco-

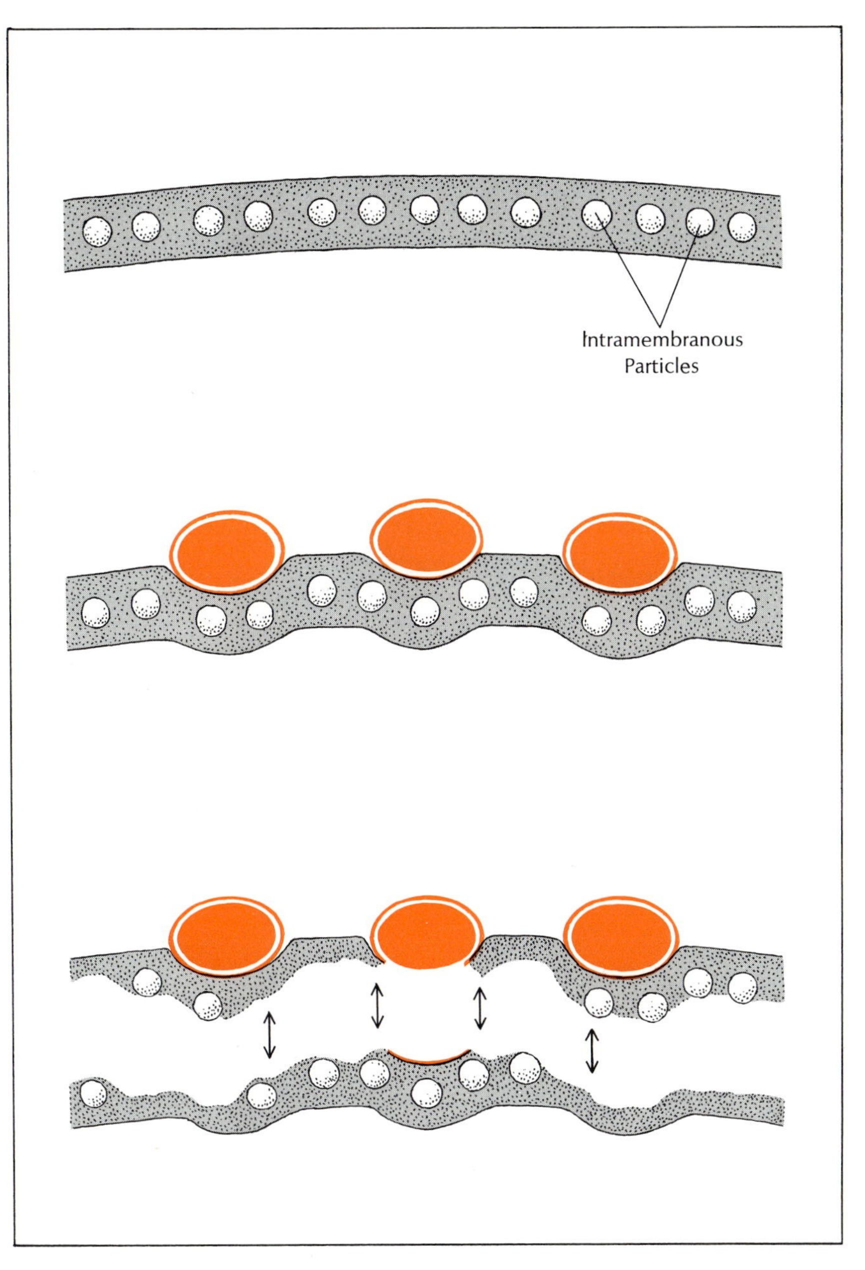

Influenza virus particles (color) attached to erythrocyte membrane leave "footprints" in membrane interior (center), sometimes including portions of virus capsule; these are revealed when membrane is cleaved to expose the inner fracture face (bottom).

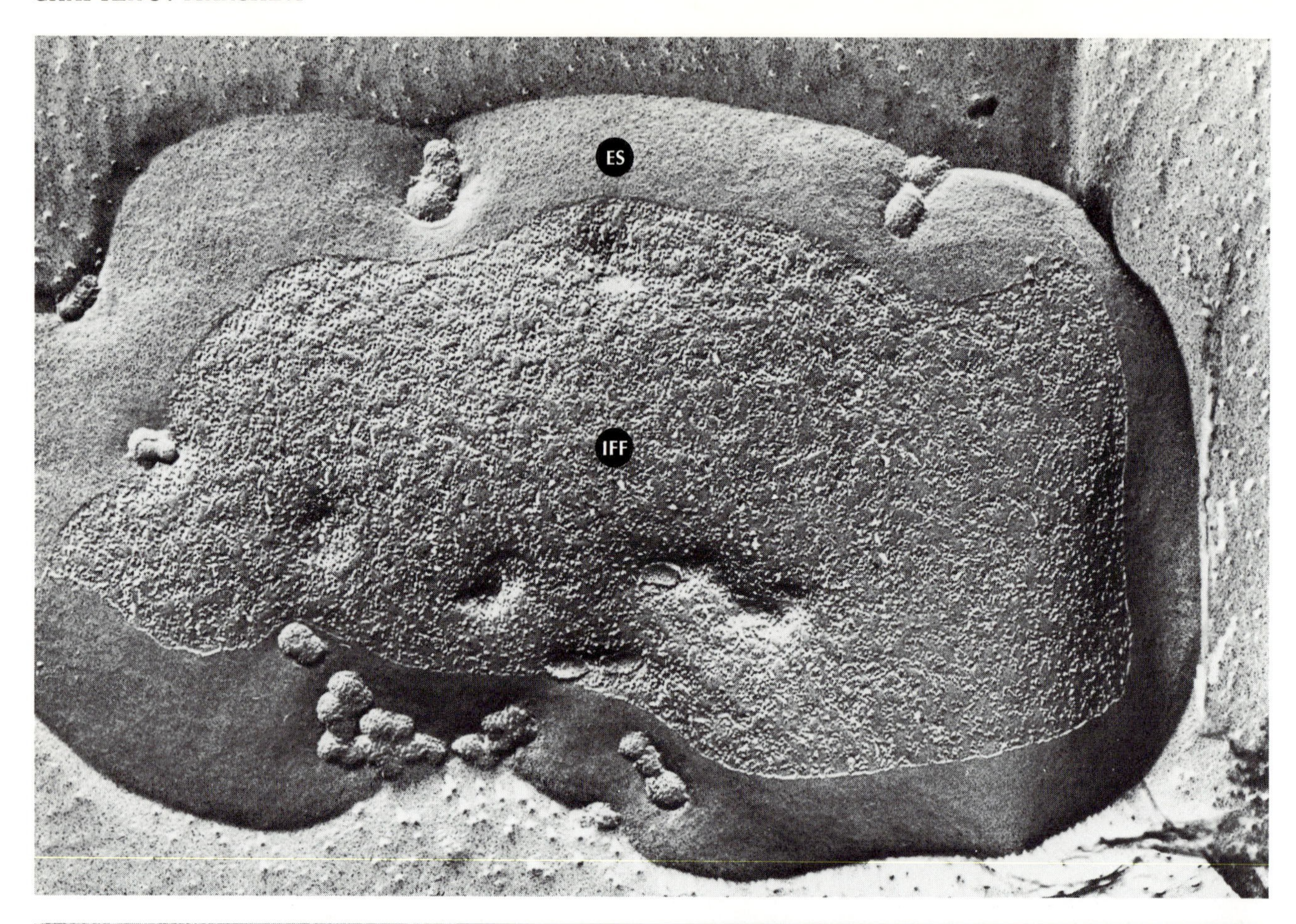

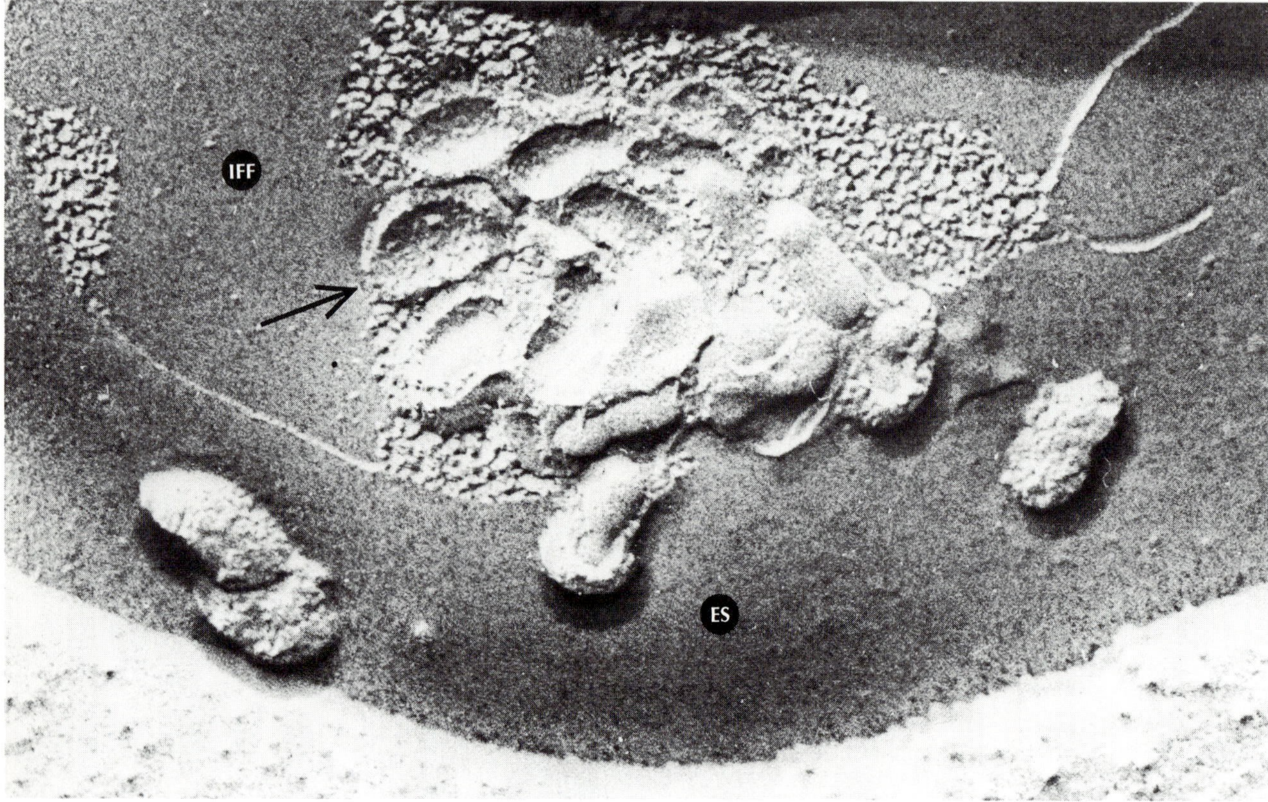

In normal membrane (top), virus "footprints" cannot, because of their size, be associated with particular intramembranous particles. However, in trypsinized membrane (bottom), "footprints" appear only on aggregates of the particles (arrow), showing that these molecules – presumed to be glycophorin – include receptors for the virus as well as for substances such as agglutinin.

phorin and also, as is well known, stimulates nucleated cells to divide (being widely used for this purpose in the preparation of cell karyotypes for chromosome studies). And if we can assume that it binds to glycophorin in these other cells, as it does in erythrocytes (which seems very likely), then it would appear that the binding somehow triggers the process of cell division. There is, at any rate, some evidence that once PHA has given the "signal," it can be removed from the cell exterior without interfering with the division it has set off.

The likelihood that PHA can transmit a message across the cell membrane *without itself crossing the membrane* – which is much more certainly the case with such hormones as insulin – brings us back to our tripartite model of the glycophorin molecule. For if PHA (or insulin) does not itself carry the message into the cell interior, something else must – and that "something" would seem to be the glycophorin molecule, which, as we have seen, extends all the way through the membrane. One can hypothesize that the binding of PHA to glycophorin sets off some structural change in the molecule that reaches as far as the carboxyl tail lying within, or in contact with, the cytoplasm; this change, in turn, would activate or release some substance (most likely an enzyme) that would induce or permit the division process to go forward. But whatever the exact nature of the process, it would seem that only a tripartite molecule such as we have described, with a "receiving terminal" outside the membrane and a "sending terminal" in contact with the cytoplasm inside, could serve to transmit messages across the membrane. According to this model, the transmitting system would be "insulated" by the lipid membrane matrix.

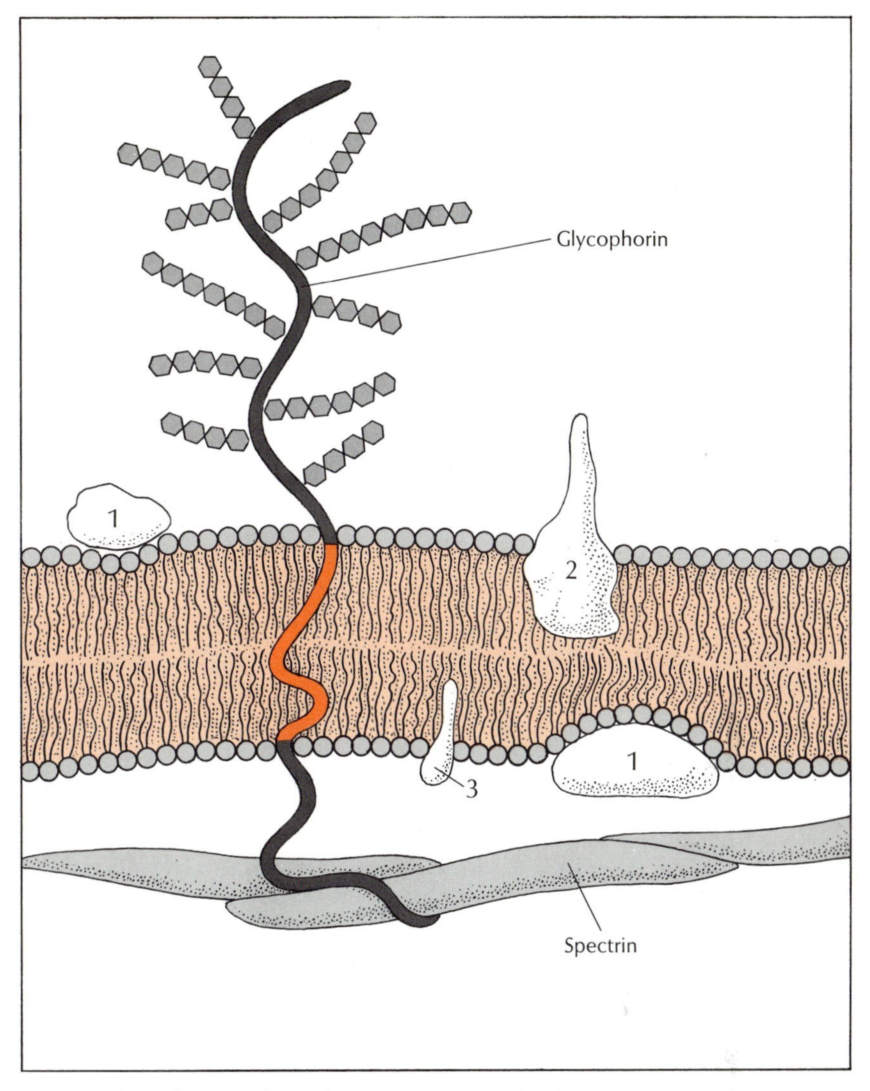

Proteins show diverse relationships to membrane's lipid matrix. Some are adsorbed or otherwise attached to the membrane surface (1 and spectrin), others are embedded in membrane lipid, but with portions projecting into the surrounding medium (2) or the cell interior (3). Still others, such as glycophorin, pass entirely through the membrane. The association shown here between the C-terminal segment of the glycophorin molecule and spectrin is largely speculative.

How glycophorin actually transmits its messages – if it does – and what other functions, if any, it serves, remain to be determined. But we can at any rate say that its structure as thus far determined seems appropriate to its functions – also as thus far determined. More broadly, it seems fair to say that the isolation and characterization of this membrane protein promise a new and exciting era in the study of these fascinating and biologically important substances.

6

Sugars of the Cell Membrane

SAUL ROSEMAN
Johns Hopkins University

In the preceding chapters, the chemical composition of cell membranes has been discussed almost entirely in terms of the lipids and proteins that make up a major part of their structure. However, virtually all plasma membranes (and perhaps some intracellular membranes as well) contain another class of substances, the sugars; nearly all cells, it appears, are more or less "sugar-coated."

Membrane sugars almost never occur in the form of simple sugars such as glucose, but rather as oligosaccharides – linked groupings of from two to 15 simple sugar residues, or as longer polymers, i.e., polysaccharides. In these substances, monosaccharides are linked to form the polymers in much the same way (but with different types of bonds) that amino acids are polymerized to form proteins. As compared with protein structures, many of which are known in detail, the structures of membrane oligosaccharides are still poorly defined (for reasons I shall discuss in a moment), while their biologic functions are, if anything, even less understood. Nonetheless, enough is already known to enable us to say with some confidence that these substances play important roles in cell biology, being implicated in such processes as cell recognition and intercellular adhesion.

Before delving further into the subject, I should note that virtually everything we know about membrane sugars refers only to those associated with the plasma membrane. Many investigators have tried to discover, by methods of subcellular fractionation, whether sugars are present on such intracellular membranes as those of the mitochondria and endoplasmic reticulum, but their findings have been severely limited by the thus-far intractable problem of cross-contamination between fractions. About all that can be said at this point, therefore, is that if sugars are present at all on intracellular membranes, their concentrations must be low, as opposed to their very significant concentrations on plasma membrane.

The sugars of the plasma membrane fall into three broad categories. First are the "free" polysaccharides, which are only loosely attached to the membrane exterior. In organisms such as pneumococci, these substances, in the form of gel, act as a protective coating or capsule around the cell. Similar compounds, such as the extracellular matrix in connective tissue, are secreted by, for example, mammalian fibroblasts. The other two categories of membrane sugars are not free but, rather, are firmly linked by covalent bonds to membrane proteins (glycoproteins) and membrane lipids (glycolipids) respectively. The glycolipids are by all odds the best understood category.

So far as the glycolipids are concerned, most – or at any rate the most abundant – of them have fatty acid linked to sphingosine as their lipid moiety, with from one to eight sugars linked to carbon 1 of the sphingosine group (see Chapter 2, "Lipid Dynamics in Cell Membranes," by Chapman). Thus far, no glycolipid with more than eight sugars has been completely characterized, but there is reason to suspect that some of them may have many more.

The glycoproteins are a far more diverse class. For one thing – and rather surprisingly – they appear to include nearly all proteins of serum. Among them are such substances as enzymes (e.g., ribonuclease B, glucose oxidase), hormones (e.g., human chorionic gonadotropin), the blood-group proteins of the erythrocyte membrane, collagen, interferon, and ovalbumin. Only a handful of sugar-free serum proteins are known, though they include such well-known substances as insulin and albumin. The proportion of sugar in glycoproteins, moreover, varies enormously from one to another, from about 1% for ovalbumin to as much as 85% for the blood-group substances. One source of this variability, which is far greater than that found among the glycolipids, is

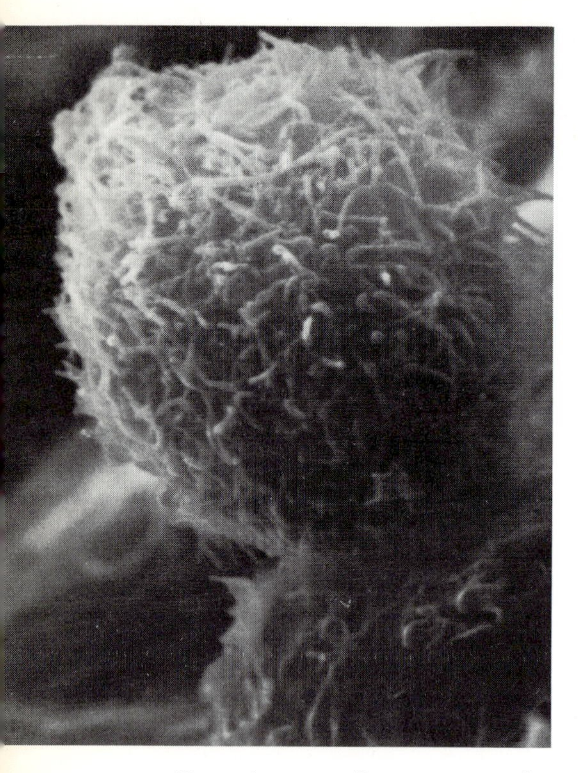

Adhesion between cells appears to involve interactions between spaghetti-like protrusions of plasma membrane, as seen in scanning EM of Balb/c 3T3 fibroblasts.

the fact that while the latter, so far as we know, have only one available point of attachment for a sugar or oligosaccharide (the carbon atom cited above), sugars can be joined to proteins at one or more sites of the polypeptide chain.

Some of the structural complexities of membrane oligosaccharides can best be understood by comparing them with a more familiar group of complex substances, proteins in general. The latter, as everyone knows, are formed from a limited group of simple components, the 20-odd amino acids; these combine into long chains of amino acid residues, linked by peptide bonds, in which the amino group of one link combines with the carboxyl (acidic) group of the next, with the loss of a molecule of water. As is also well known, these chains can assume an enormous variety of structures and biologic properties, depending on the number, identity, and order of the component residues. They are, however, subject to one fundamental limitation: the chains cannot branch. Since amino acids as a rule contain only one amino and one carboxyl group, each of them can be attached only to two others. Even in the case of the few amino acids containing two aminos or two carboxyls, the limitation still holds; at any rate, there are few examples of pure proteins with a branched structure.

The oligosaccharides attached to proteins are, like the proteins themselves, composed of a few simple components – only nine, in fact. Three of them are D-glucose, D-galactose, and D-mannose, simple sugars of the same composition ($C_6H_{12}O_6$) but differing slightly in structure (stereoisomers). Another, L-fucose, is similar to the first three, but has one less oxygen atom. Two other simple sugars are L-arabinose and D-xylose, stereoisomers with the composition $C_5H_{10}O_5$. Of the remaining three components, two, N-acetyl-D-glucosamine and N-acetyl-D-galactosamine, are also as their names imply stereoisomers: simple 6-carbon sugars with an aminoacetyl (-$NHCOCH_3$) group attached. The last and most complex, N-acetylneuraminic acid (sometimes called sialic acid) includes both an aminoacetyl and a carboxyl group, along with several extra carbons and hydroxyl groups.

But though the number of simple components of oligosaccharides is markedly less than that of the amino acids that form proteins, the ways in which they can combine are far greater, since each one can link to another at half a dozen different points – meaning that a chain of only two of these sugars could exist in some 36 (6×6) different structurally isomeric forms. Moreover, this opens up the possibility that multiple linkages can occur, with one sugar joined not to two others (as is the case with amino acids in a polypeptide) but to three or (in theory, anyhow) more; the oligosaccharides, that is, need not be limited to a simple chain structure, but can and do form branched chains.

All these multiple possibilities make determination of the precise structure of any given oligosaccharide a formidable problem. John Clamp of the University of Bristol has calculated the possible permutations of a 13-residue oligosaccharide consisting of three each of mannose, N-acetylglucosamine, galactose, and neuraminic acid, plus another N-acetylglucosamine that serves to connect the oligosaccharide with a protein. The number of possibilities is on the order of 10^{24}! And even if the exact sequence of residues were known, the number of possible isomers, based on variations in bonding arrangements, would be on the order of six million.

In fact, as we shall see in a moment, the situation is not quite this bad, since analyses of specific glycoproteins, though they have by no means provided exact structural information, have at any rate revealed certain constraints on possible structures. Perhaps the most productive means of attack has been the sequential degradation of purified glycoproteins. This begins with degradation by

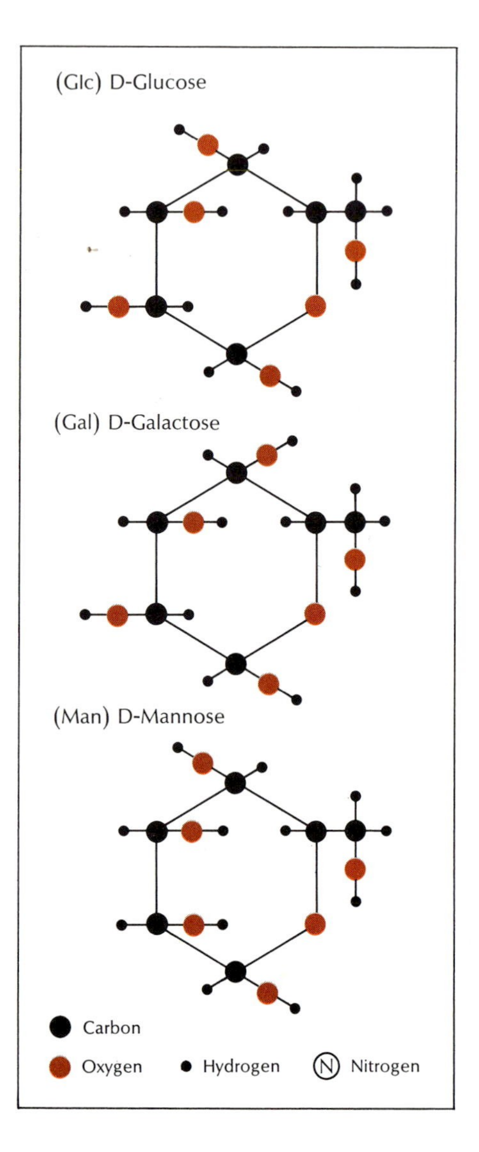

proteolytic enzymes, which break the substance into a number of single amino acids and oligopeptides, to some of which the oligosaccharide side chains remain attached. These, in turn, are treated successively with various glycosidases – enzymes that snip off a particular sugar residue from the ends of the side chains. Through this technique, along with various other analytic methods, the sequence of residues, and most or sometimes all of their fine structure, has been determined for the side chains of several glycoproteins, including the blood-group proteins, ovalbumin, and collagen.

One common characteristic of the oligosaccharides thus far analyzed is that though all of them, apart from the shortest and simplest, show branching structures, in no case does a given residue result in more than two branches. A chain may form two branches at a given point, each of which may then divide into two sub-branches – but never more than two. Moreover, so far as we now know, the process of proliferation never proceeds beyond the sub-branch stage. To take the analogy of a tree, the "trunk" of an oligosaccharide may divide into two "limbs," each of which may throw off one or more "branches" – but the branches do not then split into "twigs."

These generalities are admittedly provisional, since they may reflect merely our still-limited knowledge of oligosaccharide structures. Rather less provisional are findings concerning the "roots" of the oligosaccharide "trees," that is, the junctions between the sugar side chains and the proteins to which they are attached. It appears that these junctions involve only five of the nine sugars, and only five different amino acids, on what is almost a one-to-one basis. N-acetylglucosamine can join up only with asparagine, galactose only with hydroxylysine, xylose with serine, and arabinose with hydroxyproline. The only exception to the one-to-one rule is N-acetylgalactosamine, which can link up with either threonine or (like xylose) with serine.

In all cases, the bond between sugar and amino acid is of the glycosidic type, in which a hydroxyl from the sugar and a hydrogen from the amino acid combine to form a molecule of water, leaving a bond between the two molecules. In the N-acetylglucosamine-asparagine bond, typically found in serum glycopro-

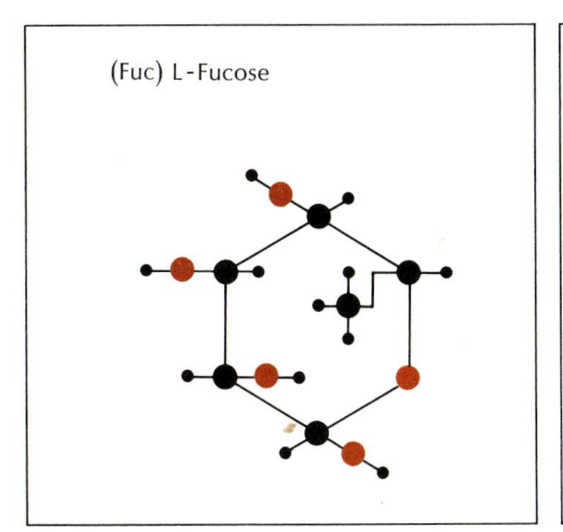

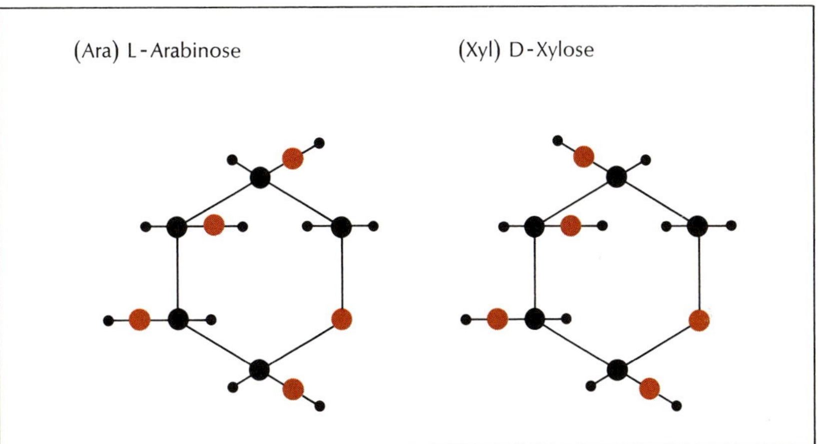

Membrane oligosaccharides are constructed from only nine different components. Six are simple sugars–glucose and two of its stereoisomers, galactose and mannose; fucose, also six-carbon, and arabinose and xylose, both five-carbon sugars (above). Glucosamine and galactosamine are simple sugars with an acetylamino group added; more complex is acetylneuraminic acid, also called sialic acid. Properties of particular oligosaccharides are thought to depend not merely on identity and sequence of their components but also on points at which these are joined (each can link up at several different places).

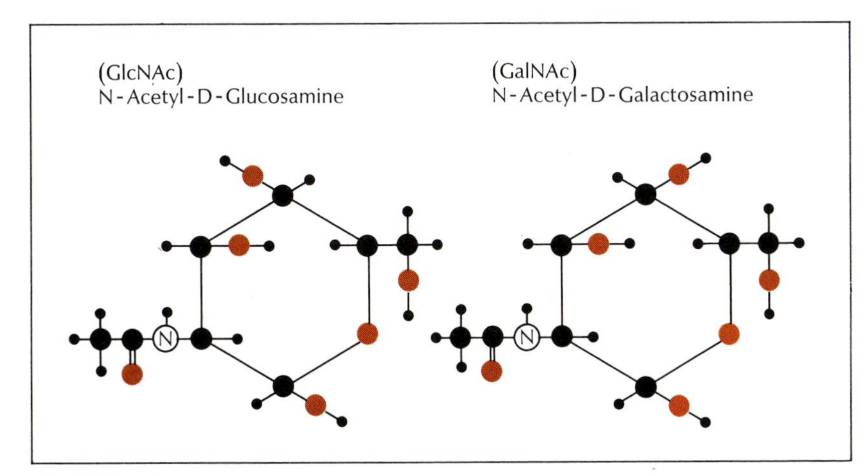

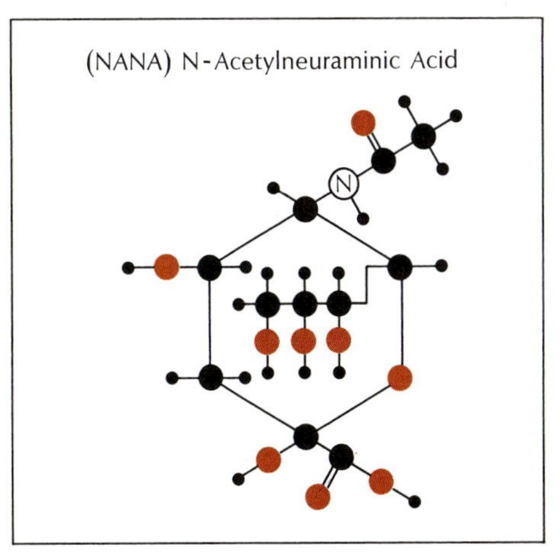

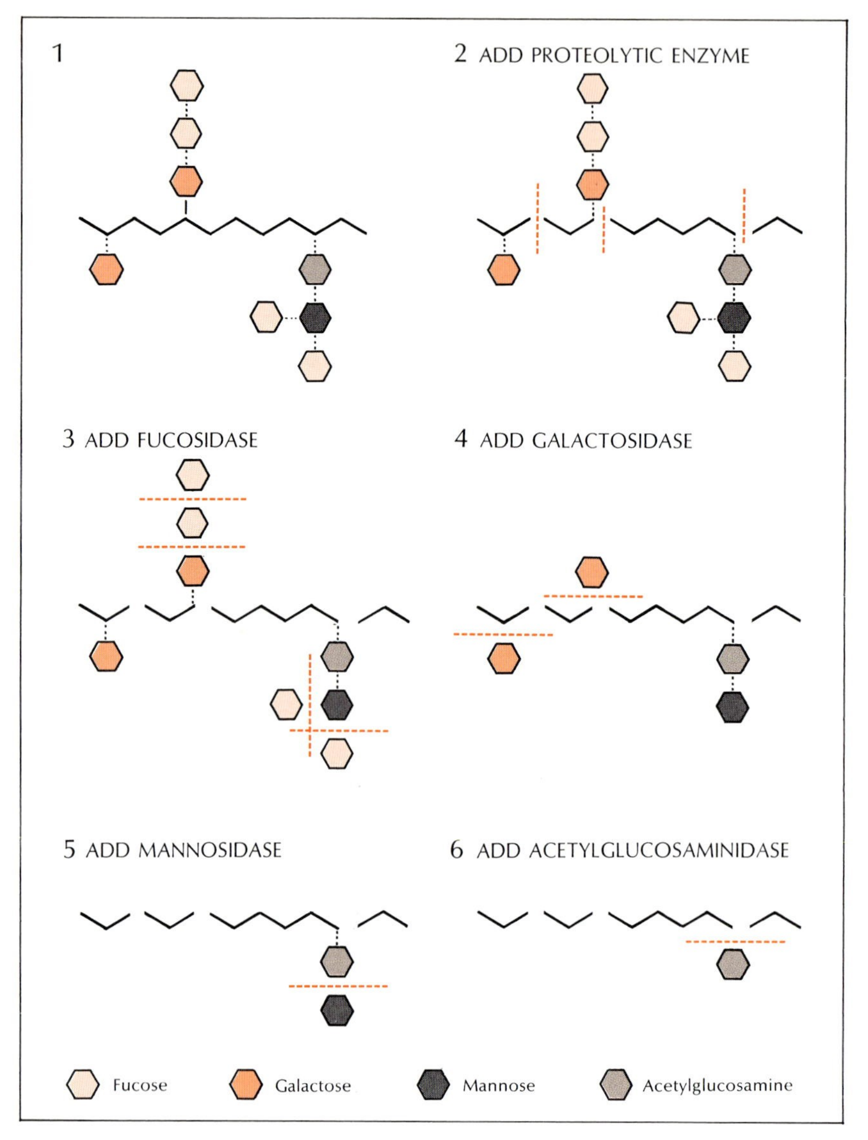

Sequential degradation of glycoproteins has provided valuable clues to the structures of their oligosaccharide components. First, the protein itself is broken up into small fragments, some with sugar chains attached. Then the successive addition of various sugar-detaching enzymes "snips off" specific terminal sugars where they are present, aiding the experimenter in deducing the original oligosaccharide structures.

teins, the linkage is between a carbon in the sugar and a nitrogen in the amino acid, while in the four other sugar-amino acid junctions the link is between a carbon and an oxygen. The N-acetylgalactosamine bond to the oxygen of threonine or serine is typically found in the mucins – glycoproteins prominent in salivary secretions – while the other three carbon-oxygen-type bonds are found in the mucopolysaccharides (of collagen, etc.). As implied by the "polysaccharide" in the name, the mucopolysaccharides differ from the other glycoproteins in having much longer sugar chains.

With the membrane glycoproteins we are on much more shaky ground, since the difficulty of obtaining really pure preparations has severely limited our knowledge of their structures. There is reason to believe, however, that at least some of these substances include bonds of both the "serum" (carbon-nitrogen) and "mucin" (carbon-oxygen) type.

Concerning the number of oligosaccharide chains per protein molecule, matters are even more complicated; the serum glycoproteins may have from one to 30, while the mucins may have up to 800. Again, matters are much less clear in the case of membrane polysaccharides, though it seems likely that the number is at the low end of the scale.

Amid all these complexities there is one note of relative simplicity: in most (though not all) glycoproteins, all the sugar chains within a given molecule are similar if not identical. Basically they appear to be a single structure, but in various stages of completion. To offer an analogy, one might find attached to a given protein chains of the type ABCDEF (the F linked to the protein), or BCDEF, or CDEF, or DEF, or even F; what one would not find would be such structures as ABCD, or ADCF, or DEACBF, etc.

To understand the probable reason for this curiously limited diversity, we must again contrast oligosaccharides with proteins, this time from the standpoint of their biosynthesis. Proteins are, of course, synthesized by the ribosomes, where their structures – specific sequences of amino-acid residues – are determined by specific sequences of nucleotide bases, with each "triplet" of bases representing the code for a particular amino acid. Ribosomal enzymes, though they undoubtedly play a role in protein synthesis, have no influence on protein structure; if the protein is synthesized at all, it must follow the RNA blueprint.

The protein moiety of a glycoprotein is, like all proteins, synthesized in the ribosomes. The oligosaccharides, however, are added elsewhere in the cell; exactly where is not known, though the Golgi apparatus seems a plausible site. They are put together by enzymes known collectively as glycosyltransferases, each of which is highly specific to both the donor sugar and the receptor, which may be either a sugar or an amino acid. For example, the enzyme that attaches N-acetylgalactosamine to galactose will not attach it to anything else, nor will it attach any other sugar to galactose. The specificity goes even further: for example, the enzyme that binds glucose to galactose in the formation of collagen glycoprotein will do so only if the galactose is already attached to the protein with free galactose being "ignored."

What this means is that the structure of a particular oligosaccharide is determined, not by any blueprint comparable to RNA but by the enzymes that are present at that locus in the cell, by what substrates are available for the enzymes to act upon, and by other factors not well understood. Moreover, oligosaccharide assembly evidently proceeds in a stepwise fashion, beginning with the attachment of the "root" sugar to the protein, with each enzyme able to do its job only when the preceding steps in the synthesis have taken place. For a particular variety of chain, it appears, the process may stop at any point; though the complete chain may have six different links, some of its companions may have only five, or four, or even fewer. The sequence of sugars in these incomplete versions must, however, be identical with those in the complete chain and must begin with the "root" sugar next to the protein. Thus the composition of a particular glycoprotein oligosaccharide, unlike that of a protein (or of most other body constituents), is not rigidly fixed, but rather represents a sort of average among the complete and incomplete chains.

On the face of it, this would appear to be an intrinsically inaccurate and hence inefficient process. It should be emphasized, however, that one cannot rationally judge the efficiency (or otherwise) of glycoprotein synthesis without knowing the physiologic purpose of the synthesis—the specific function the substance is to subserve. And this, in almost all cases, is still unknown. In some cases we have a fairly good notion of the structure of the oligosaccharides, at least as regards the sequence of their sugars, but have no notion of what they or the protein as a whole is doing. In others, we may know the function of the protein but cannot determine how the oligosaccharides help it perform that function.

An example of the first type is the blood-group glycoproteins, whose oligosaccharides (the "complete" chains, that is) are in the form of multiple branched chains containing 11 or 13 monosaccharide units. In type A blood-group substance, two of the branches terminate in N-acetylgalactosamine residues, while in type B these are replaced by galactose (in type O, neither is present). The difference is due to the presence of two different enzymes, either of which can be inherited in simple mendelian fashion (as blood types indeed are). But as to why the erythrocyte and other cell membranes possess these proteins at all, we are wholly ignorant.

In the case of a glycoprotein such as ribonuclease B, by contrast, we are in no doubt whatever as to its enzymatic function—but its oligosaccharide contributes nothing to that function. The situation is even more puzzling when we note that this enzyme possesses a nonidentical twin, ribonuclease A, whose protein structure and enzymatic function appear to be identical to those of ribonuclease B but which contains no sugars whatever.

Given the multiple ambiguities of oligosaccharide structures, it seems unlikely that any attempt to relate a known structure to a particular function will be successful in the near future. Partly for this reason, my associates and I have tackled the problem from the opposite direction, starting with a known function of the cell surface and attempting to determine what role, if any, membrane sugars play in it.

The particular function we have focused on is cellular adhesion. It is, in the first place, a basic property of virtually all cell types of higher organisms, serving to preserve the structural and, ultimately, the functional integrity of tissues against mechanical stress. At the same time, the inability – or at least the failure – of certain cells to adhere to certain others helps impart discreteness to different organs, which otherwise might tend to fuse together into a single lump.

These complementary processes of adhesion and nonadhesion are deeply implicated in the processes of embryonic growth and differentiation. The growth of the embryo from a single zygote to billions of differentiated cells, arranged in elaborately pat-

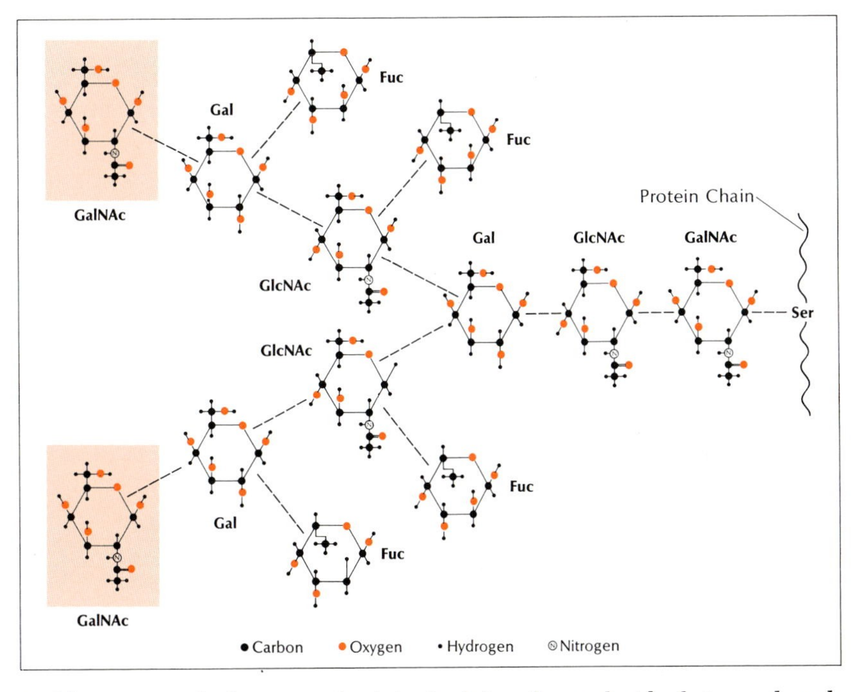

Unlike proteins, which consist only of simple chains, oligosaccharide chains can branch. Typical are those of the blood-group glycoproteins, shown here in their type A version. In type B, the shaded galactosamine residues are replaced by galactose; in type O, neither terminal residue is present. The broken lines between the sugars indicate that their exact points of connection are not known, though their sequence is.

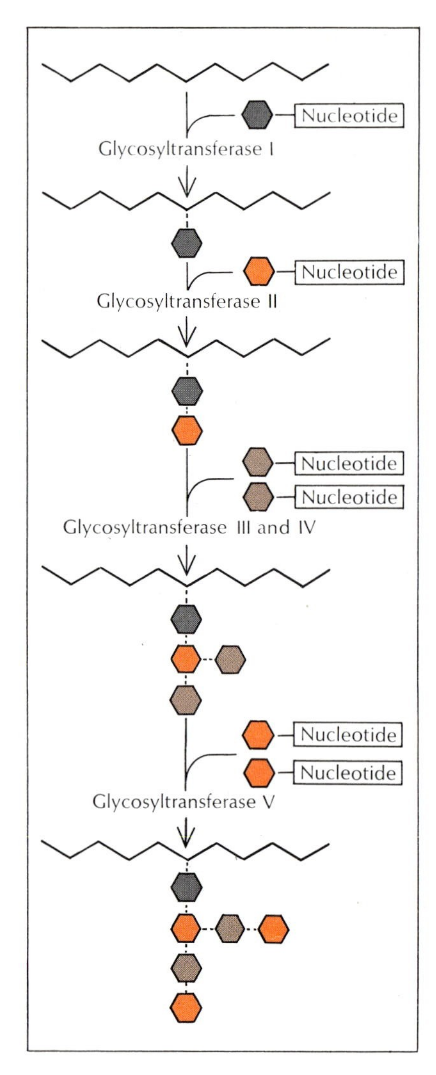

Oligosaccharides are synthesized by the sequential action of many different enzymes (glycosyltransferases), each of which is highly specific for both the donor sugar and the acceptor, which is an incomplete oligosaccharide chain in a glycoprotein or a glycolipid.

terned structures, obviously implies that the cells will not adhere at certain stages, being free to move from one location to another (singly or in sheets of cells), yet will begin adhering to their fellows when they reach the proper site in the developing structure. A simple example of this is seen in the now classic experiments carried out by Townes and Holtfreter some 40 years ago. Starting with animal embryos at an early stage of development, when only three cell types – ectoderm, mesoderm, and endoderm – were present, they dissociated the embryos into suspensions of isolated cells and then separated the cells by type.

Cells of a single type – say, mesoderm – clump together to form a ball. Mixtures of ectoderm and mesoderm would in the beginning form a single, mixed ball, but a day or so later would have sorted themselves out into a ball of two uniform, concentric layers, the ectoderm on the outside and the mesoderm on the inside. The exception was a mixture of ectodermal and endodermal cells, which would not normally be in contact; here the sorting process would give rise to two separate balls, in only peripheral contact with one another. Finally, mixtures of all three types would eventually reconstitute the original three-layered embryonic ball, which was sufficiently viable to continue normal development for a period of time.

In light of our present knowledge, it is clear that selective cell interactions of this general type must involve at least three processes. First, a cell must be able to recognize which of its fellows it should – or should not – bind to; second, the cells must adhere mechanically, and, third, they will, in many or most cases, form more elaborate connecting structures such as the gap junctions discussed elsewhere in this book (see Chapter 9, "Junctions Between Cells," by Pappas and Chapter 11, "Cellular Communication by Permeable Membrane Junctions," by Loewenstein). Concerning the first stage, recognition, one can thus far do little more than speculate on the role of membrane sugars; I shall do so later in this chapter. Concerning the second, adhesion, we have more experimental evidence.

A central problem in studying intercellular adhesion is that of assaying it. All investigators agree on a definition of adhesion – the formation of more or less mechanically stable bonds between cells – but they disagree on how the formation of these bonds can best be measured accurately. Moreover, it is clear that in experiments of the Townes-Holtfreter type, in which the "sorting out" processes took place over 24 hours or longer, there was time for all sorts of interactions to take place, some of them almost certainly involving the formation of the more elaborate cell junctions already mentioned. My associates and I have therefore concentrated on measuring what might be called "immediate" adhesion – the cell-cell interactions that take place among dissociated cells during the first hour or so.

An early assay method involved shaking suspensions of single cells under standard conditions and then, at various intervals, examining portions of the suspension with a Coulter counter, which can determine the number of single cells remaining in the sample and therefore, by subtraction, the number of clusters (ranging from two cells to hundreds) that have formed.

Though useful, this method was not wholly satisfactory, among other reasons because mixed populations could not be used. That is, this method could not distinguish between homologous and heterologous interactions in the mixed population. Lately we have been employing a more precise method, in which a layer of cells is grown in a Petri dish, after which a suspension of single cells grown in a radioactive medium is added to the dish. When the suspension is removed after varying intervals, the radioactivity of the cell layer gives a quite accurate measure of the number of suspended cells that have stuck to it.

Even before developing our improved assay technique, we set about investigating whether there was any relationship between the cells' "sugar-coating" and what might be called their adhesiveness. An important point here is that even the gentlest method of dissociating single cells from tissues or cell clumps, treatment with trypsin, is in fact not particularly gentle: it strips away considerable material from the cell surfaces. Presumably for this reason the resulting cells do not adhere to one another in any automatic, mechanical fashion; specifically, they must remain alive or healthy in their culture medium. This, in turn, suggests that they must be capable of regenerating the destroyed surface material before adhesion can take place. The question we set out to answer was whether that material was in fact oligosaccharide.

Our first approach was through varying the culture medium, which was a complex artificial mixture of more than 50 components – glucose, salts, amino acids, vitamins, steroids, and so on. By systematically varying the concentrations of different components, we found that only one of them had any discernible relationship to intercellular adhesion: the amino acid L-glutamine. Without this substance, the cells remained healthy so far as one could determine, but would not clump together. Moreover, quantitative experiments showed a clear correlation between glutamine concentration (up to an optimal level, of course) and rate of adhesion. The next question, then, was whether glutamine was necessary because of its role in oligosaccharide synthesis.

In earlier experiments, we had explored the biosynthesis of oligosaccharides – in particular, that of the amino sugars or hexosamines, N-acetylglucosamine, N-acetylgalactosamine, and sialic acid – and had found that L-glutamine does indeed play a key role in these processes. Specifically, the amino acid, which is one of those with an "extra" amide ($-NH_2$) group in its structure, serves as a nitrogen donor in the formation of hexosamines from the simple, neutral hexose sugars.

While this was suggestive, it was not conclusive so far as the adhesion findings were concerned, since L-glutamine is known to act as a nitrogen donor in many other biochemical reactions. We therefore tried to find donor compounds that would act more specifically in hexosamine synthesis. We discovered that either glucosamine or mannosamine could substitute for L-glutamine. Without L-glutamine (or glucosamine or mannosamine) the membrane oligosaccharides could not be synthesized. Incorporation of the hexosamines occurs at the terminal ends where the oligosaccharides attach to the protein, as noted earlier, and without this first step, synthesis of the oligosaccharide chains cannot get started at all.

Clearly, then, without oligosaccharide synthesis the cells could not

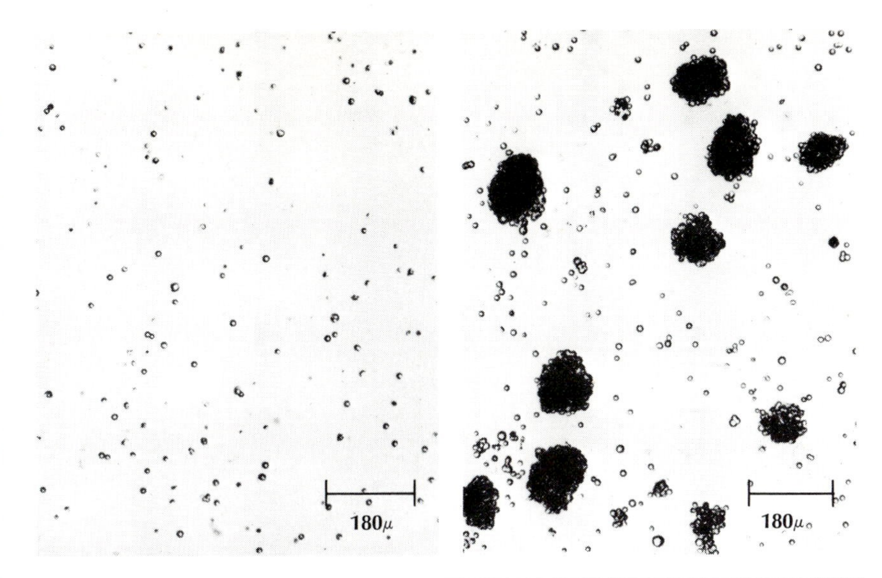

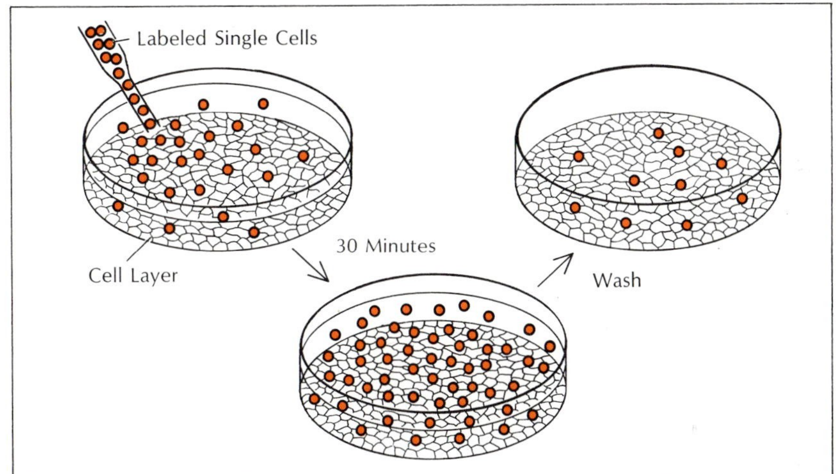

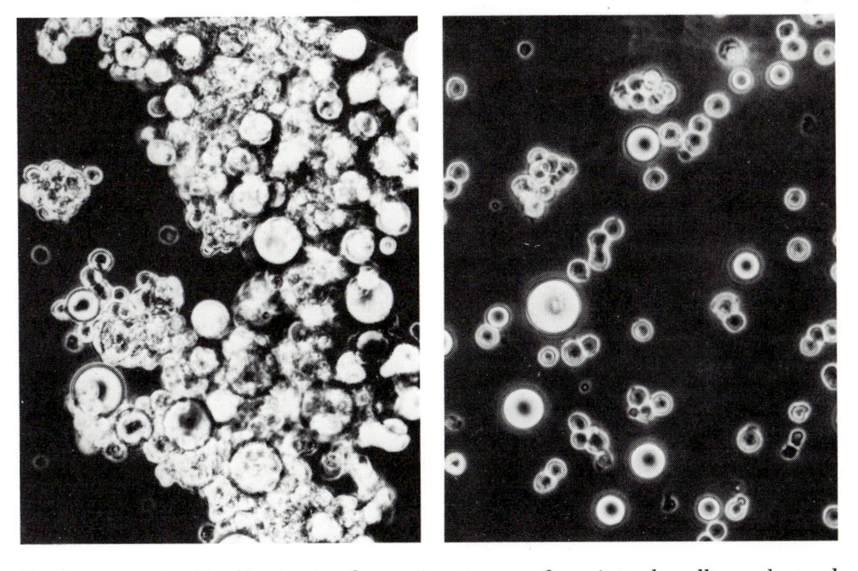

Crude assay of cell adhesion involves estimating number of single cells vs clumped cells after given time interval (top). In more sophisticated assay (center), labeled cells are added to layer of unlabeled cells; residual radioactivity after washing shows number of cells adhering. Role of sugars in adhesion was clarified by experiments with "sugar-coated" beads; cells adhere to beads with galactose on surface (bottom left) but not to those with other simple sugars such as glucose (bottom right).

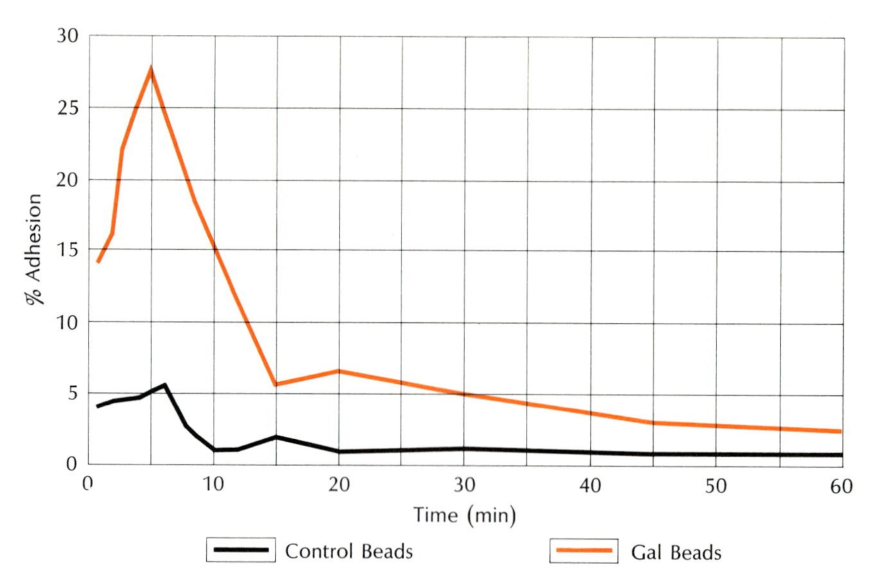

Galactose-coated beads readily adhered to fibroblast layer, but only temporarily; within an hour nearly all beads had detached from cells. Control beads without galactose showed similar pattern, but the initial degree of adhesion was far smaller.

adhere, but this still did not precisely define the role of sugars in adhesion. We decided to see whether cells would react with artificial surfaces containing sugars. Sephadex beads were used to whose surfaces we had attached various simple sugars and hexosamines (monosaccharides). With glucose- and N-acetylglucosamine-derivatized beads, the cells failed almost totally to stick to the beads or to clump with one another (at the dilutions used). With galactose beads, however, one could see under the microscope great clusters of thousands of cells and beads mixed together.

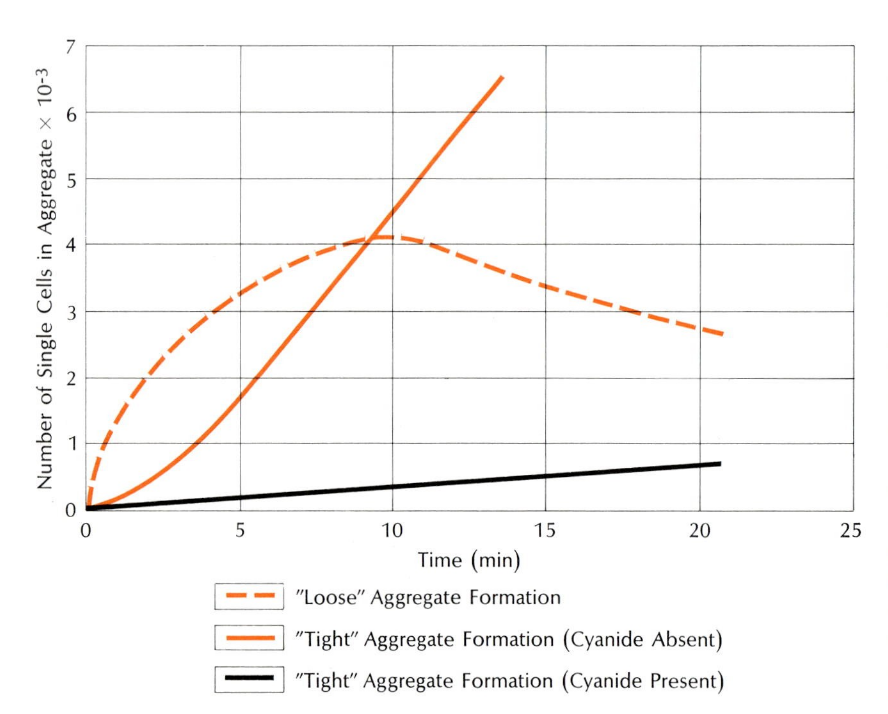

In experiments using embryonic chick retinal cells, loose aggregates formed rapidly but gradually disappeared and were replaced by tight aggregates. Addition of cyanide blocked formation of tight aggregates, indicating process is energy-dependent.

Though this technique was essentially qualitative rather than quantitative (our "assay" of clumping was made by visual examination), it clearly demonstrated two important facts. First, galactose – and, so far as is now known, only galactose – can trigger cellular adhesion of these particular cells. The second point was more subtle, resting on the fact, apparent even on visual examination, that most of the cells were not adhering directly to the beads but to one another. The galactose on the beads, in other words, not only made them sticky with respect to the cells, but also induced some change in the cells that made them sticky with respect to one another.

In an effort to explain this rather puzzling event, we performed the converse experiment. If loose cells will stick to galactose-coated beads, we reasoned, then such coated beads should stick to a fixed layer of cells, such as we had already employed for our quantitative adhesion assays. The experiment showed that they would indeed do so – but only temporarily. Almost instantaneously a large fraction (some 50%) of the beads would stick to the cells, but shortly thereafter they would begin to detach, and within five to 10 minutes virtually all of them would have ceased to adhere to the cell layer. Evidently, not only could the beads change the cells by making them more adhesive to each other (as in the first experiment) but the cells could also change the beads by making them lose adhesiveness. The beads may well have become unsticky in the first experiment, but, since they were mostly trapped within large clusters of cells, there would have been no way of detecting their loss of adhesion.

The obvious explanation – that the beads had lost their galactose coating – could be ruled out for a number of reasons, among them the fact that no known enzyme is capable of detaching the coating. When we repeated the experiments, this time using a layer of highly radioactive cells, it turned out that rather than losing something the beads had themselves picked up part of the cells' surface material. Moreover, further experiments have shown that the transferred substance is not loosely and fortuitously adsorbed on the beads but bound covalently to them. Thus far, however, we have not yet managed to determine just what this substance is; if and when we manage to do so, we should know a good deal more about the specific chemical interactions involved in adhesion.

And there, for the moment, the experimental phase of the adhesion question rests. What follows is largely hypothesis, which, though based on these (and some other) experimental findings, may provide a plausible explanation of adhesion.

One of the earliest adhesion hypotheses, put forward more than 25 years ago, explained the phenomenon in terms of a "lock and key" mechanism similar to the well-known antigen-antibody reaction, with an antigen on one cell binding to the equivalent of an antibody on another. In its favor is the fact that cells are indeed covered with antigenic compounds, so that one need only postulate the appropriate antibody-like molecules. Against it is the fact that in certain situations, at least, adhesive bonds between cells are reversible. During differentiation, for example, various groups of cells adhere (as in ectoderm, endoderm, and mesoderm of the early embryo), then dissociate and migrate, then readhere as they form distinct tissues and organs. This would imply, first, that either the antigens or the antibodies were being destroyed or transformed, to permit dissociation; subsequently, after migration, the same antigens or antibodies were being re-formed in other cells. And this, while certainly possible, seems inordinately complex.

A second possibility is that the bonds underlying adhesion are not of the lock-and-key, or antigen-antibody, type but rather involve interactions between oligosaccharides of the two cells, which would bind together by establishing hydrogen bonds between sugars of two chains. Bonds of this sort form quite easily and, moreover, can produce extremely stable structures; for example, the fibers of cellulose are held together by bonds between the long, parallel polysaccharide chains that compose them.

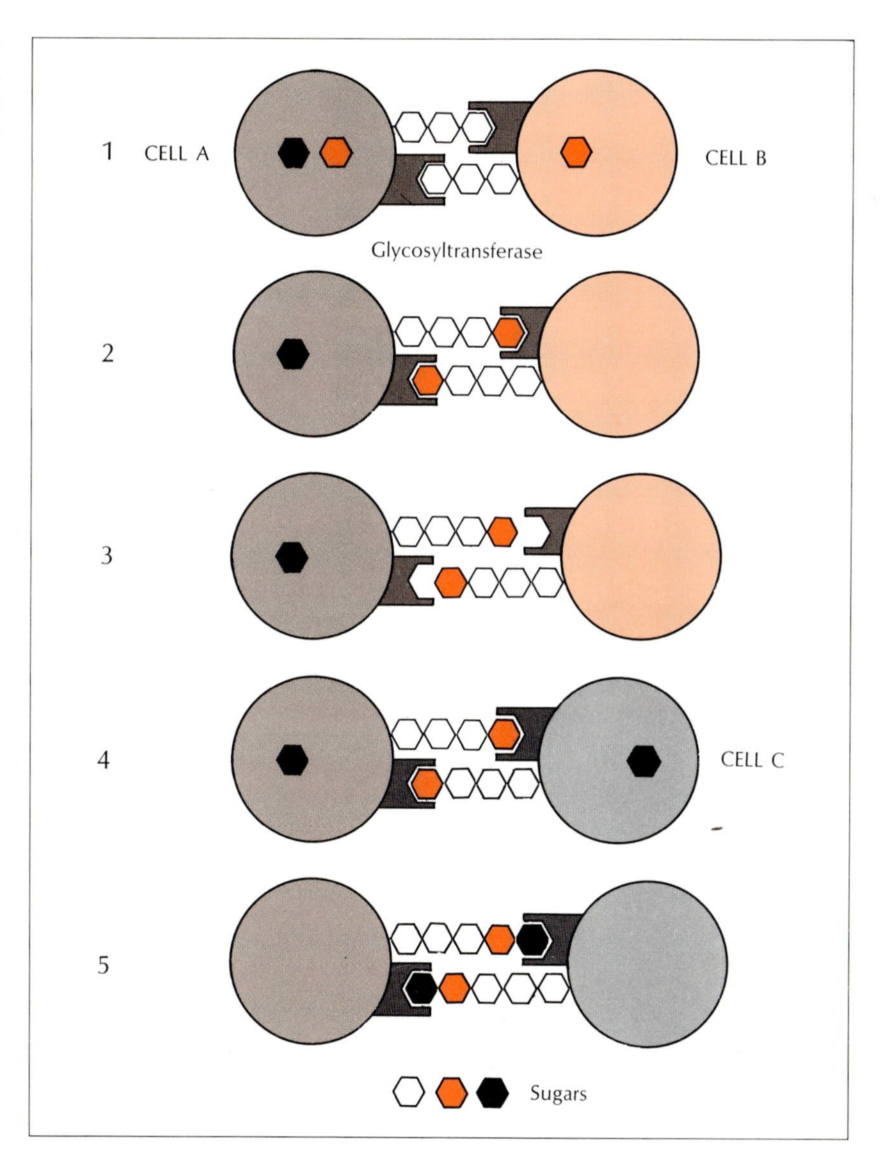

Hypothetical process of reversible adhesion begins with sugar on one cell binding to glycosyltransferase on another (1). Enzyme adds another sugar to the chain (2) lessening enzyme-sugar affinity and permitting cells to detach (3). Process can then be repeated with another type of cell (4), perhaps several times. Eventually, addition of new terminal sugar increases *sugar-enzyme affinity, producing permanent adhesion (5).*

The sugar-sugar hypothesis, however, does not explain *reversible* adhesion much more adequately than the antigen-antibody hypothesis. Moreover, it is at present a rather unproductive hypothesis, in that there seems no way of confirming (or disproving) it on the basis of present knowledge. The fact that like cells adhere and unlike ones do not would imply a high degree of structural specificity in their respective oligosaccharides, resembling the structural specificity of enzymes and other proteins. And here we are in terra incognita. Given the known variables in oligosaccharide structures cited earlier, involving the identity and sequence of the component sugars, the presence or absence of branching in the chains, and the many different ways in which one sugar can attach to another, it is clear that oligosaccharides can, in principle, assume an enormous variety of structures, quite sufficient to account for the known specificity of cellular adhesion. The trouble is that we do not know what those structures are; while the *sequence* of sugars is known for several oligosaccharides, the tertiary, three-dimensional structures of their chains are not. Nor do any of the methods developed for ascertaining the tertiary structures of proteins (x-ray crystallography, for example) show any promise of yielding this information in the near future. If, then, one adopts the sugar-sugar hypothesis as a basis for further experiment – which is, after all, the main practical point of any scientific hypothesis – one rather quickly reaches a dead end.

Yet another hypothesis serves to explain some of the more troublesome facts about cellular adhesion – notably, its reversible character. This postulates not an antigen-antibody reaction, nor yet a sugar-sugar reaction, but a protein-sugar reaction between an enzyme (protein) on one cell and its substrate (oligosaccharide) on another. (There is, of course, no reason why each cell surface might not embody *both* enzyme and substrate.) As we know, sugars are present in profusion on cell exteriors and can assume a great variety of configurations that could fit an equally diverse variety of enzymes.

The evidence is still very tenuous as to whether suitable enzymes are present on the cell surface. Perhaps the most likely enzymatic candidates are the glycosyltransferases, which, because of their known high specificity in reacting with sugars in the synthesis of oligosaccharides, could presumably act no less specifically as receptors for surface oligosaccharides. And there is a certain amount of evidence that suggests that glycosyltransferases may be present on the surfaces of at least some cells.

For example, when brain tissue is homogenized to form the bodies known as synaptosomes (see Chapter 17, "Membranes in Synaptic Function," by Whittaker), glycosyltransferases have been detected in the membranes of these bodies, though there is no way that one can tell whether they are located on the inside or the outside of the membrane. Again, the enzymes have been detected in the Golgi bodies of rat liver cells. And since the Golgi membrane is known to fuse with the cell's plasma membrane during secretion, it seems perfectly possible that the fusing membrane would carry with it various glycosyltransferases.

Studies in our laboratory, subsequently followed by similar work in a number of laboratories, have presented evidence that indeed does support the conclusion that cell surfaces contain glycosyltransferases. The technical problems associated with these kinds of studies are such that we cannot draw a definitive conclusion at this time. In view of the considerable effort being expended by many workers in this area of research, the unambiguous evidence should soon be obtained one way or another.

If one is willing to postulate the presence of the enzymes for the sake of argument, one can then visualize reversible adhesion as occurring somewhat along the following lines:

First, an oligosaccharide on one cell would bind with the appropriate glycosyltransferase on another.

Second, the enzyme would perform its normal enzymatic function by adding another sugar to the end of the oligosaccharide chain. This would lessen the affinity of the altered chain for the enzyme, so that chain and enzyme would become detached from one another.

Third, the altered chain would bind with a different glycosyltransferase – on the same cell or one of another type – and go through the same process; this could be repeated as many times as necessary.

The hypothetical dissociation could equally well be used to explain increased adhesiveness. The new oligosaccharide chains would bind more tightly to the next transferases in the sequence.

We do not, I should emphasize, have any persuasive evidence that processes of this type do in fact underlie intercellular adhesion. The most we can say is that such processes do not contradict any known facts, and do explain, fairly simply, how adhesive bonds between cells could form, dissolve, and re-form.

We might also note that the process of adhesion, whatever its mechanisms, does seem capable both of decreasing the affinity of sugars for a particular receptor and of increasing it. The loss of adhesion of the beads to the cell layer, previously described, showed a decrease in affinity, and one, moreover, coupled with the addition of something to the beads, as the enzyme-substrate hypothesis postulates (though we do not yet know whether that "something" was a sugar). Likewise, some careful quantitative experiments on adhesion between cells have shown that it takes place in two stages, a loose and more or less instantaneous bond, which is succeeded by a tighter bond, arguing an increase in affinity. Formation of the tight bond, moreover, is energy-dependent (i.e., it can be blocked by ATP inhibitors), which would also be the case with the enzymatic addition of another sugar to the chain.

Let me repeat, all we have at present is a hypothesis that fits the very modest number of facts we now possess; we will need far more facts before the hypothesis can be considered proved or disproved. As is well known from studies of proteins, biochemical function depends ultimately on chemical structure, and when it comes to the structures of complex sugars, both in the membrane and elsewhere in the cell, we know about as much as we did about protein structures a generation ago. Nonetheless, I am convinced that as our knowledge of these still mysterious compounds expands, they will be shown to be involved not merely in adhesion but in such processes as cell recognition, immunity, and malignant transformation – playing biochemical roles hardly less vital than those of the proteins themselves.

7

Isolation and Properties Of Natural Membranes

LEONARD WARREN

University of Pennsylvania

Readers who have followed the earlier chapters in this book will have noted that by and large the authors were concerned either with "model" membranes — both real and imaginary – or with particular chemical components of natural (and model) membranes. It seems appropriate at this point that we begin focusing our attention on the natural membrane, not as a collection of molecules but as a functioning entity. There is, of course, an immense amount to be said on this subject, and most of the succeeding chapters will be devoted to one or another aspect of it. The following discussion, therefore, will be of a predominantly prefatory nature, and will be necessarily couched in rather more general terms than might otherwise be the case.

Before considering the ways in which membrane functions are investigated, and some of the findings concerning them, let us briefly review the nature of these functions. The most obvious is what might be called compartmentalization. Biologic membranes form a closed compartment within which various biochemical reactions can proceed with minimal interference from other (and quite possibly competitive) reactions that may be going on in the vicinity. The compartment may optimize conditions, such as pH, salt concentration, etc., for certain reactions to occur. That the plasma membrane performs this function for the cell vis-à-vis its environment is well known and has, indeed, already been emphasized by earlier contributors. Less widely appreciated, perhaps, is the fact that most, if not all, of the cell's interior membranes perform similar functions for particular organelles. Thus, the nuclear membrane separates nucleus from cytoplasm, the mitochondrial membrane separates the mitochondrion from the cytoplasmic fluid in which it resides, and so on.

Implicit in the concept of compartmentalization is the principle of a degree of biologic organization, in the sense that particular activities will proceed in one compartment and not in another – e.g., oxygen binding in the erythrocyte, epinephrine synthesis in cells of the adrenal medulla, and so forth. On the subcellular level, DNA synthesis is compartmentalized in the nucleus, protein synthesis in the endoplasmic reticulum, and so on. In addition it appears that membranes may organize biologic activities by many more detailed mechanisms. There is considerable evidence,

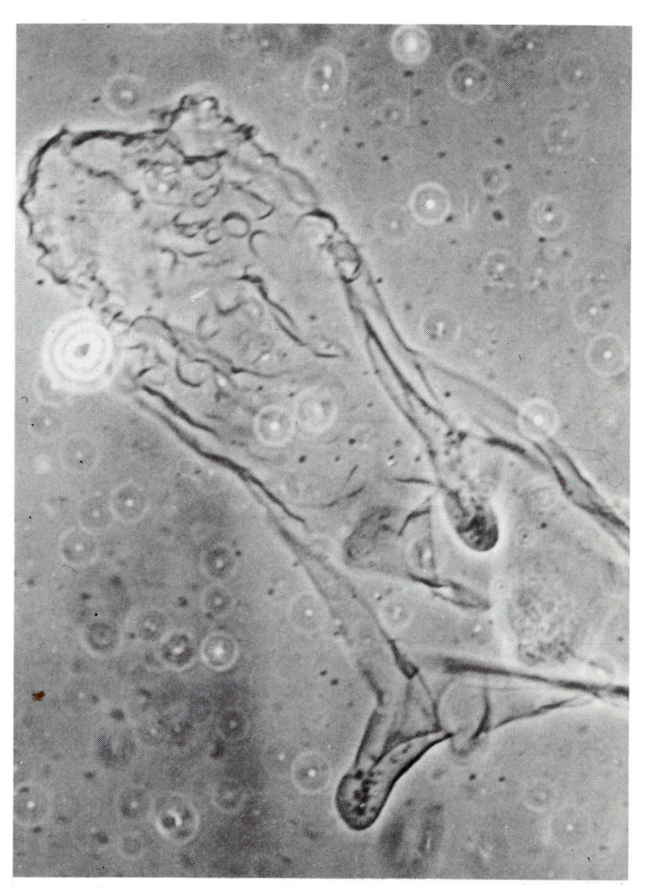

Isolated cell membranes (here of an amoeba) are produced by treating cells with "toughening" reagent, then homogenizing them.

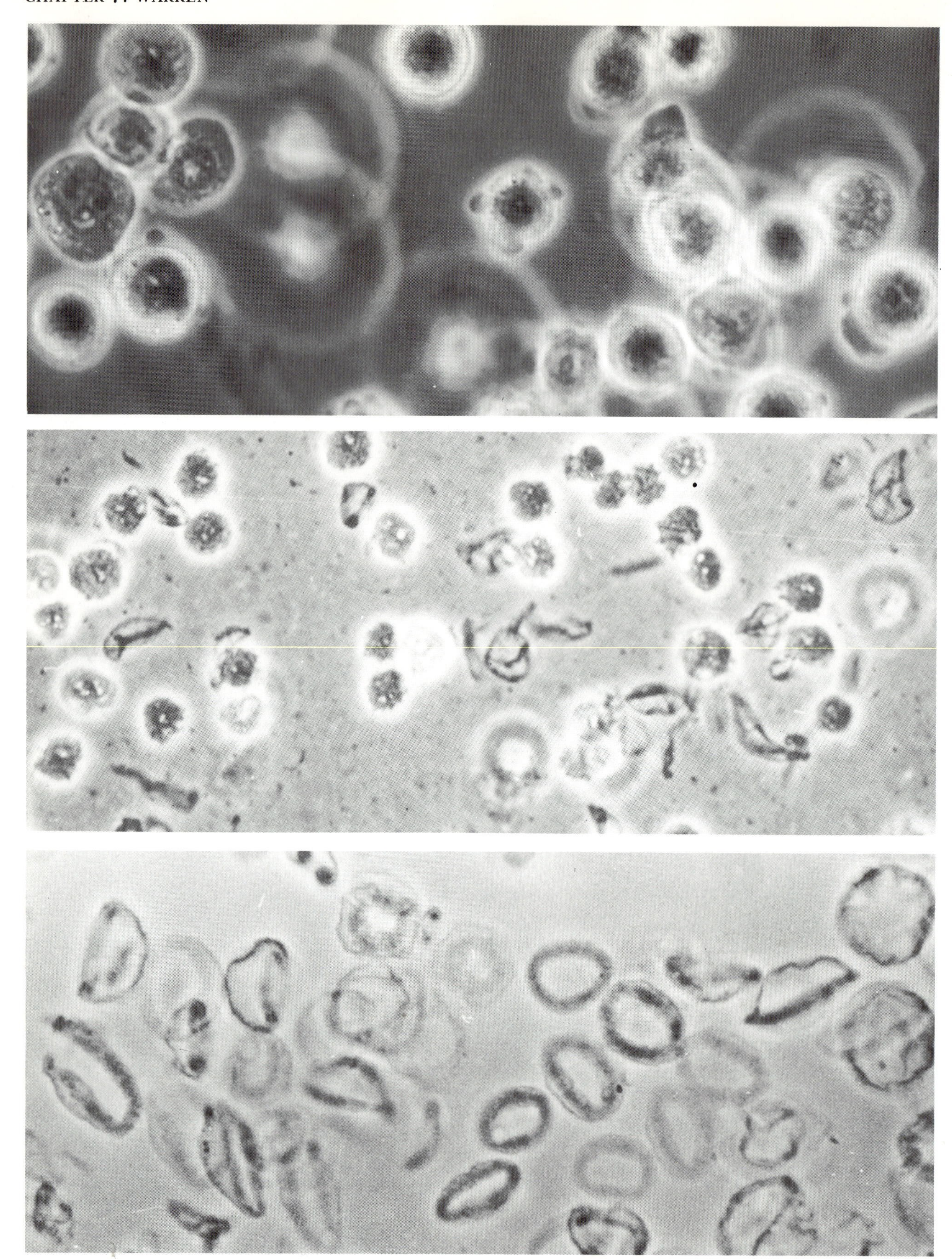

Membrane isolation process begins with treatment of cells with toughening reagent, in hypotonic solution that swells the cells, detaching membrane from cytoplasm (top). Homogenizer then breaks the membranes away from the "naked" cytoplasm (center). After centrifugation has separated membranes from cytoplasm, almost pure preparation of isolated membrane is obtained.

for instance, that the intricate energy-yielding reactions that take place in the mitochondrion are governed not by the mere presence of the necessary enzymes within that organelle but also by the specific arrangement of enzyme molecules on and in the mitochondrial membrane. Similar spatial organization of enzymes and other biologically active substances may also occur within the endoplasmic reticulum and within other cell membranes. This ordering permits sequential, geographically advantageous, assembly of certain macromolecules.

Interaction

The third major category of membrane activities can be summed up under the heading of interaction. In other words, not only does the membrane divide and organize the organism and its individual cells into compartments, it also presides over interactions among these compartments. A large group of these interactions involves the transport of various chemicals. The membrane, insofar as external conditions permit, must "arrange" for the transport into a particular compartment of whatever compounds are needed for the reactions proceeding there and for the exclusion of substances that would interfere with these reactions. Conversely, the membrane must also transport, in the reverse direction, metabolic products, some of which are not needed in the compartment wherein they are formed but may be of use in other compartments.

Other types of membrane-mediated interaction, however, involve the transport, not of materials but of information – though in some cases the information is transmitted by the passage from one membrane to another of particular molecules (e.g., acetylcholine at the neuromuscular junction). Another example of transport of information is the possibility that when the cell surface and its exterior glycoprotein structures come into contact with the surface of another cell, mutual alterations or rearrangements result that may have many consequences. Thus, for example, when a normal cell comes into contact with another cell as a result of increased cell density, it is subject to "contact inhibition," which leads it, first, to cease moving and, later, to metabolize more slowly, so that it ceases to divide; one of the outstanding features of malignant cells is their loss of contact inhibition, so that they multiply unchecked.

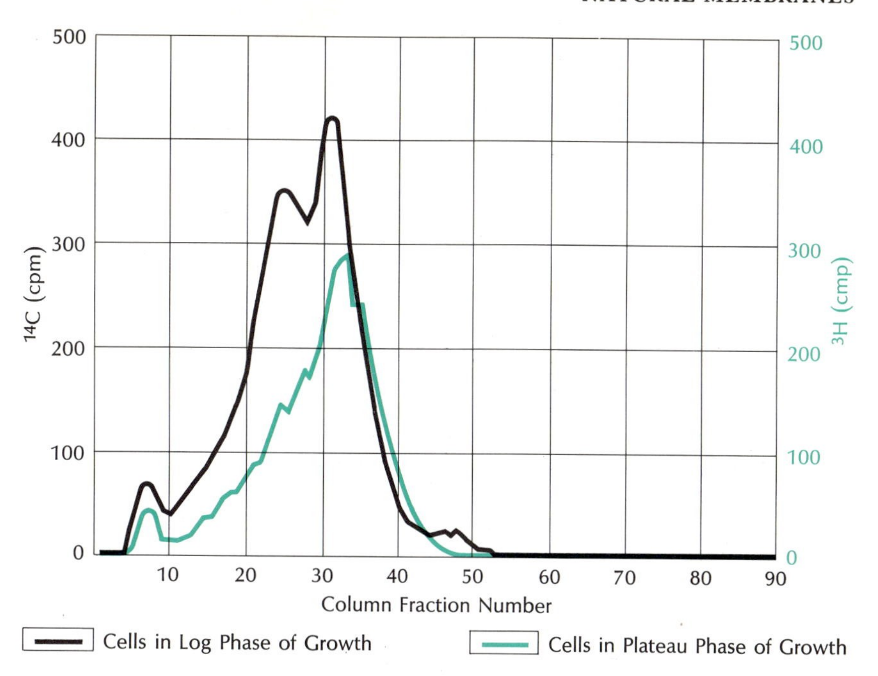

"Growth substance" in plasma membrane appears as extra peak in column chromatogphy graph of digested membrane from rapidly growing labeled cells; similar graph from a preparation of nongrowing cells shows only a single high peak.

Similar processes of information transfer appear to be involved in the phenomena of "cell recognition." For example, if labile lymphocytes are injected into an experimental animal, they will rapidly concentrate in the lymph nodes, which evidently recognize them as cells that belong there. If, however, the lymphocyte surfaces are chemically stripped of the sugars normally attached to them, the cells are no longer recognized by the nodes, but continue to float about in the general circulation. Similar recognition processes, often involving surface sugars, play an obvious role in immunologic reactions. A simple example is that of the blood group determinants. Complex carbohydrate chains are attached to protein and lipid molecules that protrude from the outer surfaces of erythrocyte membranes. In type A cells the carbohydrate chains, or some of them, end in the sugar N-acetylgalactosamine linked to an underlying sugar; in type B cells the chains end in D-galactose. Each of these sugars is recognized by the appropriate antibodies, if they are present, which thereupon combine with the sugar as a first step in the agglutination of the cells. (In type O cells, neither of these two sugars is present on the ends of the chains).

More obscure are the processes by which a cell will recognize other cells of the same type. If, for example, cultures of embryonic liver and kidney cells are thoroughly mixed, the two types of cell will gradually segregate themselves, liver with liver and kidney with kidney, and in some cases will even organize themselves into characteristic arrays – e.g., the kidney cells into an approximation of a tubule. Recognition processes of this sort probably play a role in embryogenesis.

Functional Changes

A key point for the understanding of membrane functions is that these can change over time. Often such changes occur during maturation. For example, the blood-group substances are present only in small amounts at birth (and earlier, during fetal life, are not detectable at all), but within a few days they appear on the membrane in amounts present in adult life. Similarly, canine kidney cells suspended in tissue culture will remain separated for a day or so; they then agglomerate into little balls of cells, which, in another day or so, develop into tiny, hollow cysts resembling kidney tubules. It seems likely that these changes in some sense parallel those that occur during embryonic differentiation. In any case, they obviously involve changes in the cells' capacity to respond to one another – i.e., changes in the plasma membrane and surface structure.

Other membrane changes reflect the organism's adaptation to changing conditions and, as such, are often reversible. A good example is the phenomenon called enzyme induction, which has been studied in the endoplasmic reticulum of rat liver cells (among other places). These specific membranes can be isolated by disrupting the cell mechanically by homogenization, or sudden decompression, and then subjecting the resulting fragments to density-gradient centrifugation, whereby different types of membranes with different densities layer out on solutions of sucrose of different densities. If the rat has been exposed to drugs such as barbiturates, certain enzymes (hydroxylating enzymes containing cytochrome P_{450}) whose physiologic function is to detoxify the drug will appear in the reticular membranes, though in a normal animal they are present only in relatively small amounts or not at all. Subsequently, if the drug is withdrawn, enzyme production ceases and the membranes will lose most or all of the enzymes in question.

For fairly obvious reasons, the type of functional change in membranes that has attracted much recent attention from researchers is that which occurs in the plasma membrane when normal cells are transformed into malignant cells – in particular, the loss of contact inhibition previously mentioned. Before recounting some of the findings in this area, however, I must say something about how cell membranes are isolated for microscopic and chemical examination.

As already noted, fragments of cell membranes can be obtained quite easily by disrupting the cells and centrifuging out the desired type of fragments, according to differences in density. Such fragments usually occur in the form of small, closed vesicles or as ragged pieces. Obtaining more or less intact plasma membranes is a more difficult business, since in most animal cells the membranes are quite fragile. Moreover, the presence of the nucleus within this delicate sac creates a problem rather like that of removing a rock from inside a sealed paper bag without completely destroying the bag. Until a few years ago, the only plasma membranes that had been obtained intact were those of erythrocytes, which of course have no nucleus, so that the "ghost" membrane could be produced simply by swelling the cell, thereby opening up pores in the surface through which hemoglobin and other internal material escaped. The membrane was then permitted to shrink back to normal

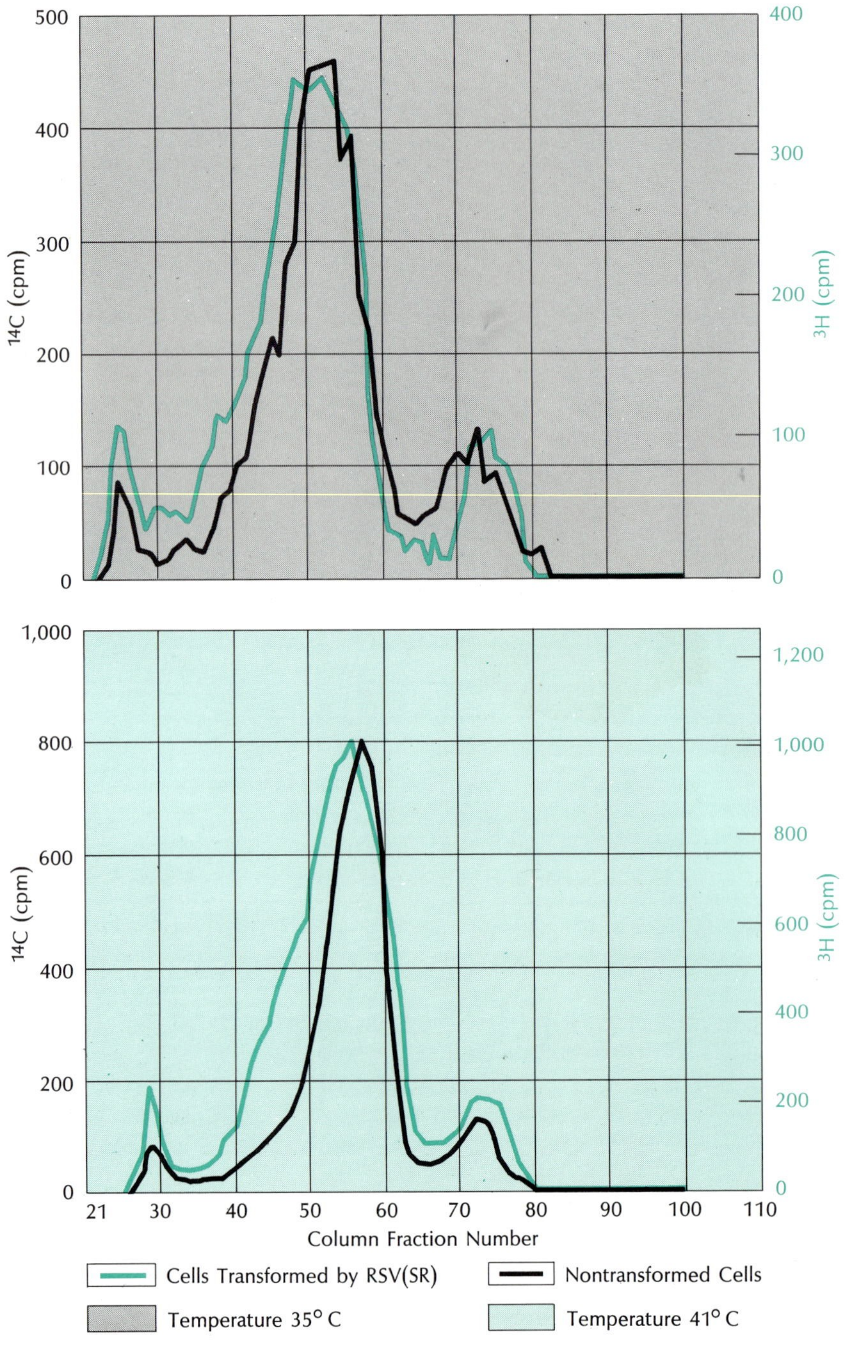

Transformation brought about in chick embryo by "wild" type virus is independent of temperature at which cells are grown. Chromatography of membranes from such cells reveals clear differences from normal cell membranes, whether cultures are maintained at 35°C (top) or at 41° C (bottom).

size with closing of the transient pores.

More recently, my associate Dr. M. C. Glick and I have worked out several techniques for isolating plasma membranes, which, with modifications, seem applicable to many types of animal cells. These involve immersing the cells in solutions of substances – often heavy metal compounds such as zinc chloride or fluorescein mercuric acetate – that in some little-understood way toughen the membrane, probably by attaching themselves to sulfhydryl (-SH) groups in the membrane proteins. The toughening process, moreover, takes place in a hypotonic solution so that the cells swell in such a way that the surface membrane rises from the underlying cytoplasm, leaving a clear area between the two. Gentle homogenation then "pops" the loosened cytoplasm (and nucleus) out of the membrane, which can be separated from them by density centrifugation in solutions of sucrose. (Salt solutions destroy the membrane.)

The materials my associates, Drs. C. A. Buck, Glick, and J. P. Fuhrer, and I have worked with are various types of animal cells that can be oncogenically transformed by certain viruses. (Such virally "transformed" cells, which can be distinguished from normal cells by their loss of contact inhibition and by various morphologic characteristics, have become a staple of tumor research.) In cultures of normal cells, the carbohydrates on the exterior of the cell membranes are labeled with carbon-14; in transformed cells, with tritium. The membranes are isolated as already described, the two populations are then mixed and the membranes digested with proteolytic enzymes such as pronase. This frees the labeled carbohydrates from the membrane proteins to which they are attached. The individual components of carbohydrate fraction from the glycoproteins are analyzed by column chromatography. Two elution patterns can be plotted, one from normal cells (followed by carbon-14) and the other from malignant cells labeled with tritium. Each is the other's control, since they were both processed together and went onto and came off the column in each other's company. The two isotopes can be easily distinguished in a liquid scintillation spectrometer.

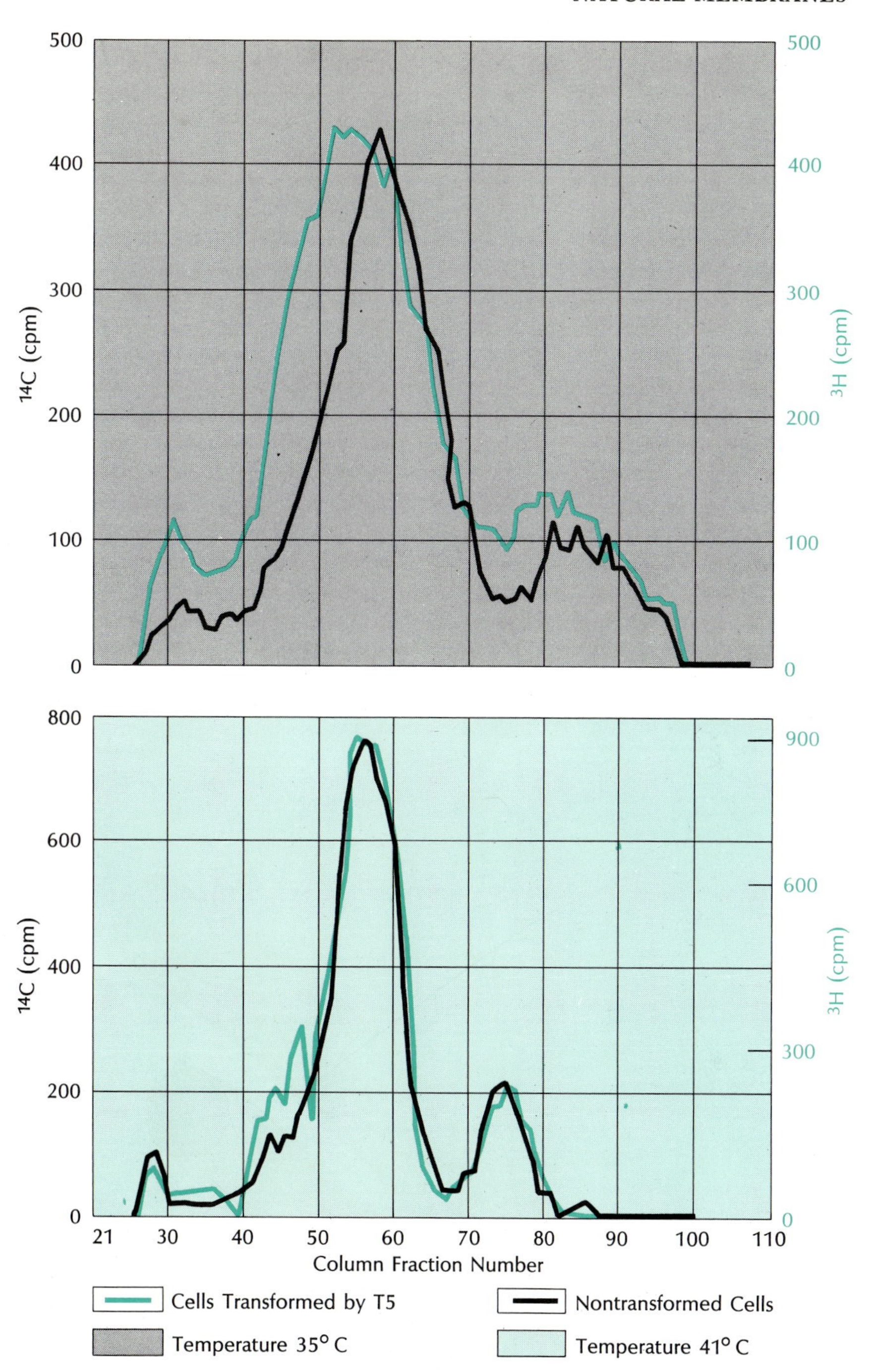

In contrast to effect of "wild" type viruses (opposite page), transformation by mutant (T5) virus is reversible. Mutant produces similar differences from normal cells at permissive temperature of 35° C (top), but nonpermissive temperature of 41° C reverses transformation in terms of morphology, behavior, and membrane constituents (bottom).

When these readings are plotted, we find that whereas the radiocarbon (normal) and tritium (transformed) curves are quite similar, the latter consistently show an extra peak, corresponding to the presence of a more-or-less distinctive polysaccharide (or more than one) that had formerly been attached to glycoproteins in the membranes of transformed cells. More extensive studies of this substance or substances have revealed some very interesting properties.

In the first place, these compartments are located on the outside of the cell membrane in life. We have

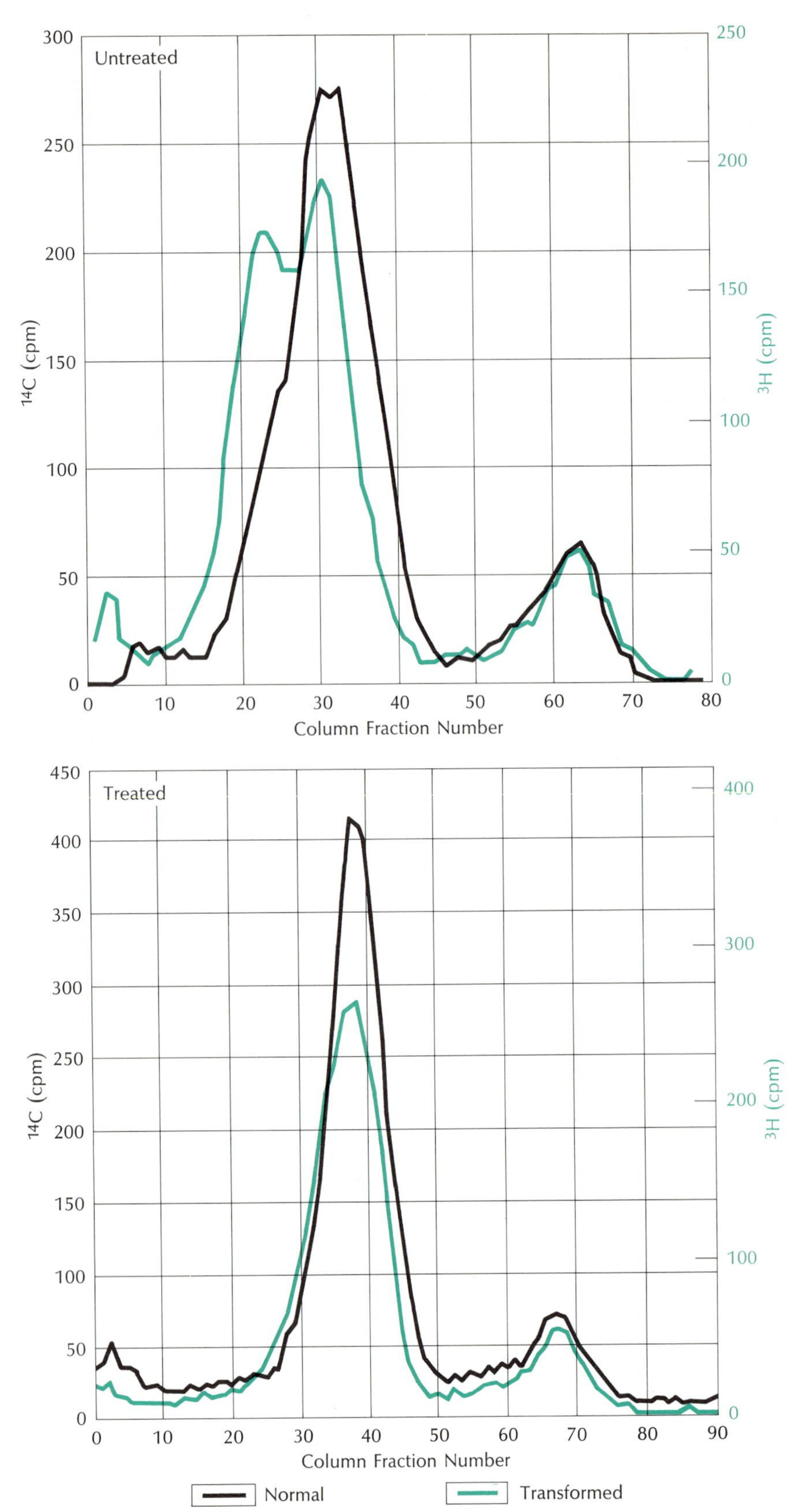

"Growth substance" present in relatively large quantity in transformed cells is shown to contain sialic acid by treatment of surface material with neuraminidase. Chromatograms of untreated surface materials (top) show extra peak in transformed cells contrasted with normal ones, but treatment removes sialic acid, largely eliminating difference between them (bottom). This change can be brought about by neuraminidase treatment before or after surface structures are removed from living cells.

been able to "shave" the labeled sugars off the exterior of intact, living cells with trypsin, the resulting preparations showing the same contrast between normal and transformed cells as seen in the whole surface membrane structure. In the second place, the substance(s) turn(s) out to be present in both normal and transformed (malignant) cells, but far more copiously in the latter.

Perhaps most interesting of all, the "transformation substance" is growth dependent. That is, if the growth of either normal or transformed cells is arrested, the substance will be virtually undetectable. When growth is allowed to proceed, it will appear in both types of membrane, but, as already noted, in far greater quantities in the transformed cells. However, though the presence of the substance is in some way related to growth, it does not seem to act as a growth controller. At least, its high concentration in the transformed cell membranes does not reflect any special rapidity of growth in these cells; like many other malignant cells, they divide no more rapidly than normal cells (their pathogenic properties stem not from their speed of growth but *from their failure to stop dividing*).

Yet the presence of the substance is evidently very closely linked to some special property of the transformed cells. This has been shown in a very subtle experiment involving a temperature-sensitive strain of the Rous sarcoma virus called T5. If cells transformed with this mutant virus are kept at 35°C, they retain their transformed morphology – and the transformation substance is present in its expected large quantities. If, however, the temperature of the cell culture is increased to 40°, the cells revert to an apparently quite normal appearance and behavior and the transformation substance drops to its usual low normal levels.

What we have identified, then, is a substance whose presence in relatively large quantities on the membrane surface is strictly correlated with cell transformation (and presumed malignancy), whether that transformation is reversible (as with the T5 strain of Rous sarcoma virus) or irreversible (as with other viruses). Moreover, its presence in significant quantities is also strictly correlated with growth, in either normal or transformed cells. Therefore, we have good reason to hope that the substance, whatever it is, will be found to play some significant role in the malignant process, since malignancy, whatever else it may be, is above all a disorder of growth.

Further attempts to characterize the transformation substance have involved some quite complex experiments whose details need not be recounted here. That the alteration is in the carbohydrate we knew already, both because our techniques were specifically designed to isolate carbohydrates from the membrane proteins to which they were attached and because the radiocarbon or tritium label was introduced into the membrane by growing the cells in a medium containing the labeled sugars, fucose or glucosamine.

It has been found that the carbohydrate material in question contains a relatively large amount of the amino sugar sialic acid and that the presence of this sialic acid-rich material in malignant cells correlates with the level of an enzyme, sialyl transferase, which attaches the amino sugar to the carbohydrate material. This enzyme, we have found, shows a pattern of occurrence similar to that of the transformation substance itself – very little in nongrowing cells, greater amounts in growing normal cells, but much greater quantities (2.5- to 11-fold or more) in growing, transformed cells, and dropping to normal levels if the transformation is reversed by temperature in T5-transformed cells. And there, for the moment, the matter rests. It should be noted, incidentally, that we are not discussing sialic acid generally, which several investigators have found to *decrease* in transformed cells, but rather that small portion of total membrane sialic acid that is attached to a particular type of receptor, through the action of a relatively specific sialyl transferase.

Further light on the growth process has been shed by studies of metabolic turnover in cell membranes. Since the 1940's, it has been a matter of more or less common conviction among physiologists that the structural and functional molecules of any organism, or cell, are in a state of constant flux; though the organism itself retains a continuous identity over time, this is in fact merely a resultant of the active and competitive processes of anabolism and catabolism.

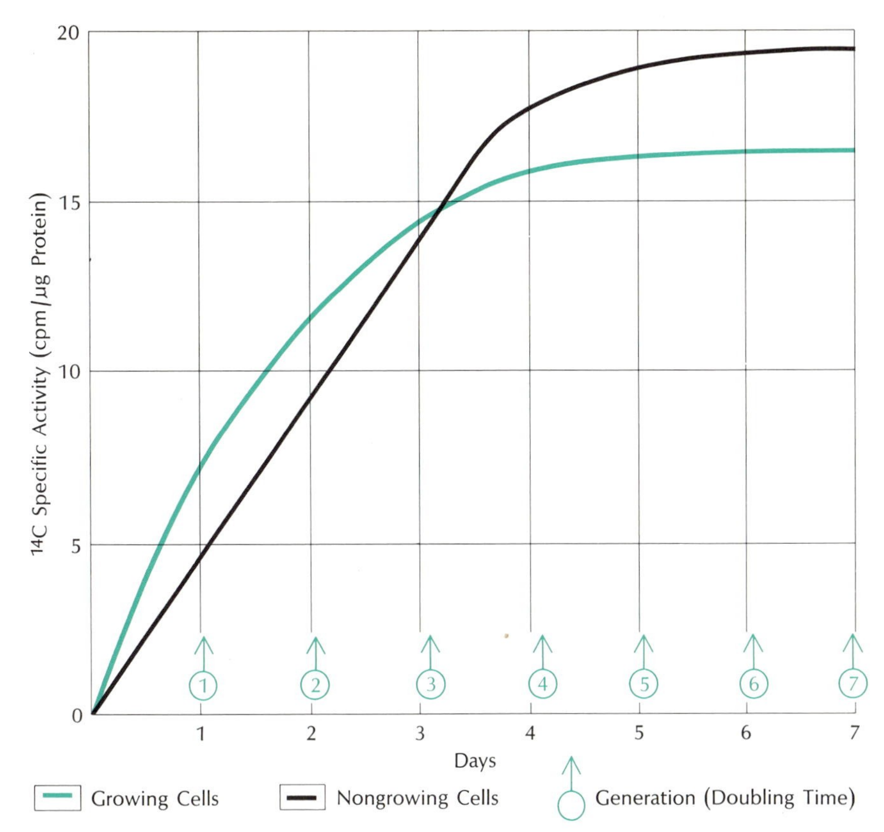

Incorporation of labeled glucose into membrane occurs at essentially the same rate in nongrowing and growing cells, showing that both categories of cell actively synthesize new membrane constituents. The reason why uptake is more rapid in the nongrowing cells toward the end of this experiment is not wholly understood.

As applied to cell membranes, this metabolic turnover can be measured in two complementary ways. Cells may be cultured in a medium containing some radioactively labeled membrane precursor; by withdrawing portions of the cell population at successive intervals and isolating the membranes, one can determine how rapidly the label is incorporated into them. From the opposite standpoint, one can start with isotopically labeled cells, place them in an unlabeled medium and determine how rapidly the label is eliminated from the membranes or (which may amount to much the same thing) is excreted into the medium. The labeled material eliminated from the membrane may, however, be reutilized. This introduces a complication in the assessment of the rate of turnover.

My associates and I have performed experiments of both types on cultures of mouse fibroblast cells, under two different conditions. In the first, the cells were actively dividing, at a rate such as approximately to double their numbers every 21 hours. Here, there was a rapid uptake of the labeled precursors, which was only to be expected, since the cells were busily manufacturing new membrane for new cells. We also studied uptake in cultures where cell division had been inhibited by allowing cell concentrations to reach a critical value or by adding inhibitors of mitosis such as vinblastine sulfate. Much to our surprise, uptake and incorporation in these nondividing cells occurred at almost the same rate as in the rapidly dividing cells, the figures ranging from 80% to 100% of the "dividing" values.

Since both dividing and nondividing cells were incorporating membrane precursors at much the same rate, it seemed evident that the metabolic difference between the two (so far as membrane manufacture was

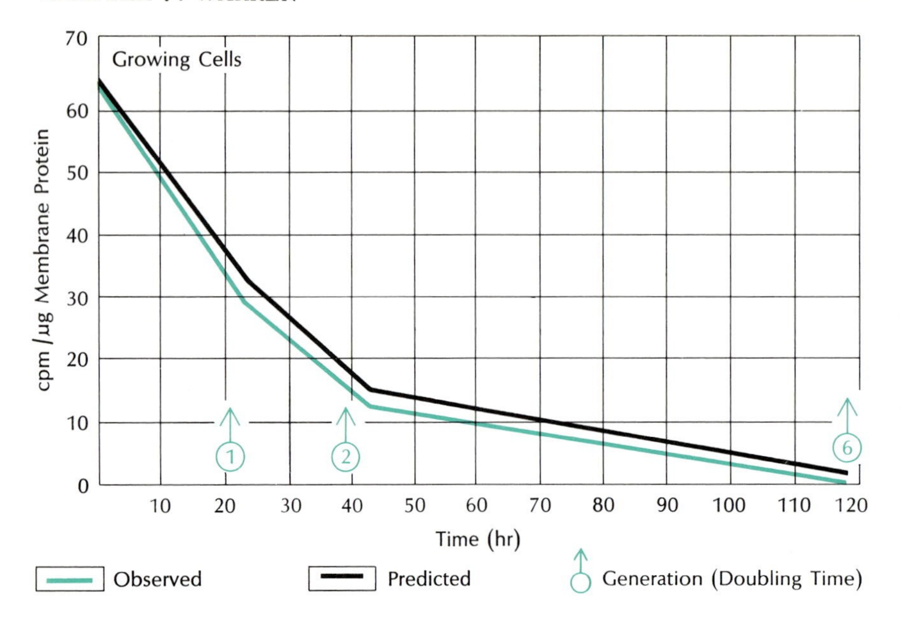

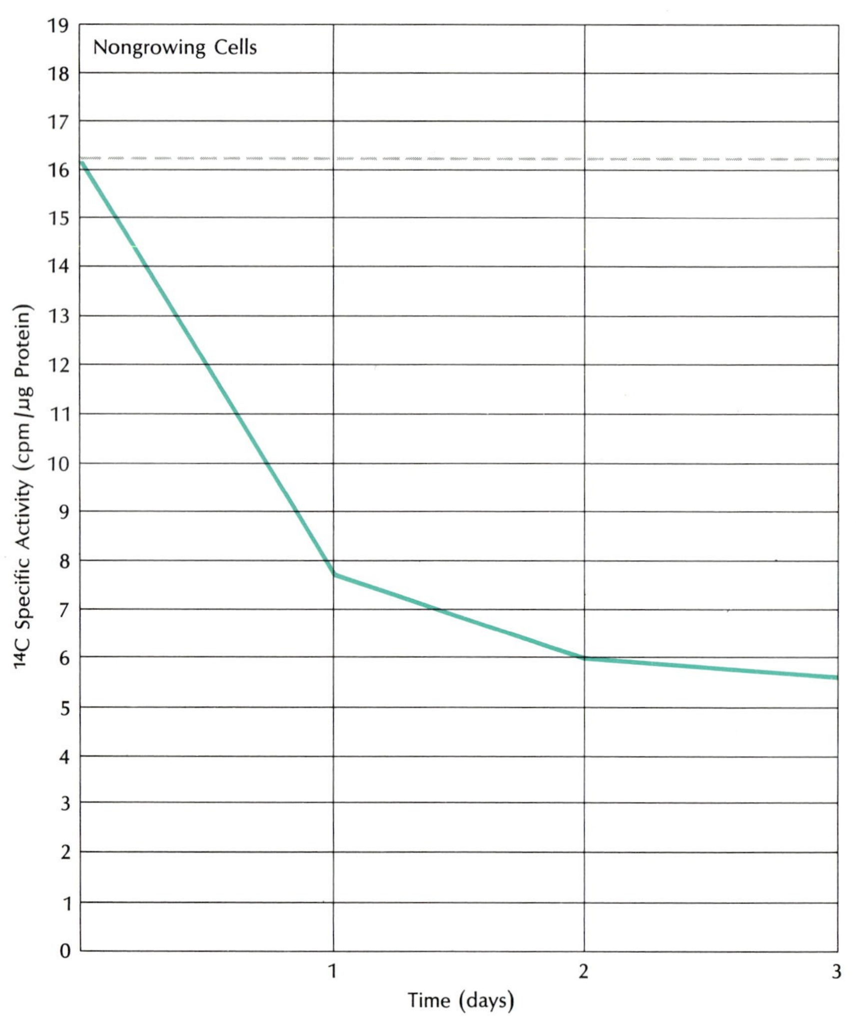

In radioactive cells growing in nonradioactive medium (top), predicted radioactivity per unit membrane mass declines rapidly owing to dilution of label by increasing membrane mass, which is derived from nonradioactive precursors; observed values are very close to predicted, showing that little membrane breakdown is occurring. In nongrowing cells (bottom) breakdown at similar slow rate would produce essentially static level of radioactivity (dashed black line); in fact, radioactivity drops sharply, showing that in these cells, membrane breakdown and turnover are very rapid.

concerned) most probably was controlled by catabolism rather than anabolism. That is, most or all of the precursor substances in dividing cells must remain "fixed" in new membrane. In contrast, nondividing cultures, which manufactured membrane at almost the same rate, did not grow and did not increase their mass but must have broken the substance down just as rapidly. And, in fact, experiments with labeled cells supported this assumption. In cells previously labeled and now growing in nonradioactive medium, the amount of label per cell, or per gram of cell protein, naturally underwent dilution with time, since a given amount of label was being "spread" over more and more cells. The radioactivity per cell fell by 50% when the cells divided in a nonradioactive medium in the absence of turnover. The observed rate of dilution (57%) was very close to the figure calculated simply on the basis of increasing cell numbers. Putting it another way, the dividing cells were replacing only about 7% (57-50) of their membrane per 21-hour generation. In the nondividing cells, as expected, the turnover rate was much more rapid; approximately 50% of the total mass of membrane was replaced over the same period of time.

It did not seem to make much difference whether the precursors were those of protein (amino acids), of lipids (choline), of sugars (glucosamine), or even the "all-purpose" metabolite, glucose. Evidently, at least as a first crude approximation, the different membrane constituents renew themselves at about the same rate. I should note, however, that other researchers have found that lipids of the endoplasmic reticulum turn over more rapidly than other constituents.

I should also note that the rates in question must be considered minimums. That is, the actual turnover rates, for both growing and nongrowing cells, may be considerably higher, but are masked by "recycling" of labeled components. Our calculations were based on a simple model in which broken-down membrane components are not reutilized to an appreciable extent and are simply ejected into the medium; conceivably, however, portions of the membrane might be only partially degraded, taken into the cell, and reutilized. But we have not yet been able to determine to what extent, or even whether, such recycling takes place.

It is evident then, at least in these experiments, that the essential difference between dividing and nondividing cells (and perhaps, by extension, between malignant and nonmalignant cells) is that in the former the catabolic processes are in some fashion depressed, allowing growth and division to proceed, while in the latter catabolism and anabolism – biosynthesis and biodegradation – are in balance. Interestingly, we have confirmed an observation made several years before that the balance remains true in the nondividing cell even when biosynthesis is artificially depressed, e.g., by adding inhibitors of protein synthesis to the medium or by withholding essential nutrients; the reduced rate of membrane synthesis somehow engenders a correspondingly reduced rate of membrane breakdown. On the other hand, if the membrane is artificially damaged (as by treatment with trypsin, which removes surface protein) the rate of biosynthesis does not appear to be accelerated. The cell surface is repaired, but apparently by the normal processes of replacement (turnover); there is no special burst of activity in response to the trypsin.

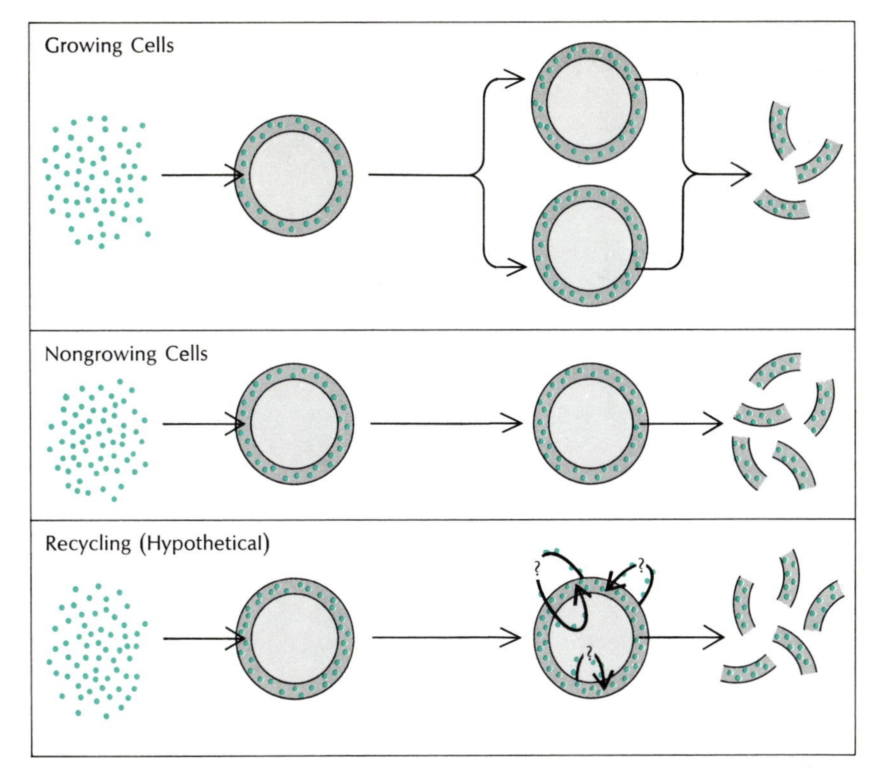

Schematic summary of graphs on preceding pages suggests how, in growing cells (top), rapid uptake of label mostly remains fixed in newly synthesized membrane, with little escaping to medium. In nongrowing cells (center), uptake is equally rapid, but is balanced by rapid breakdown and escape of label. In both cases, actual turnover rate may be considerably higher than those calculated from simple models, since constituents may be reused for synthesis of membrane instead of being ejected into medium(bottom).

Concerning the mechanisms that maintain the balance between biosynthesis and biodegradation (or, under other circumstances, tip the balance toward biosynthesis by slowing biodegradation), we are altogether in the dark. Presumably any normal cell will contain two sets of enzymes, one involved in membrane synthesis, the other in membrane breakdown; but when it comes to the control systems that link and govern the activities of these enzyme systems, we can only speculate.

Just as cellular growth and division appear to result from a falling off of biodegradation (rather than, as one might expect, an acceleration of biosynthesis), so cellular death could result from an acceleration of biodegradation beyond the cell's biosynthetic capacities. Alternatively, it could result from a decline in biosynthesis coupled with continued high-level biodegradation. In fact, neither process appears to occur. As cell cultures age, they show a falling off in *both* biosynthesis and biodegradation, such as might be described in "clinical" terms as a general loss of vigor.

I should stress that all these observations have thus far been limited to cells in culture; the extent to which similar membrane processes occur and are relevant to such phenomena as growth, malignancy, and aging in the intact organism – in cells in a tissue situation – is a matter for future research. Nonetheless, I, for one, find it somewhat mind-boggling that in these "simplified" systems the cell membrane can undergo such constant and rapid reconstruction, while at the same time retaining its physiologic identity: its characteristic functions of compartmentalization, organization, and interaction.

8

The Fusion of Cell Membranes

J. A. LUCY
University of London

By now we know that the cell membrane plays a far more important role than that of an inert envelope enclosing the cytoplasm. Not only does it continuously preside over the transport of chemicals and information into and out of the cell but, in addition, its own constituents are in a constant state of dynamic turnover, with the membrane's own substance being continuously broken down and renewed. But even this picture does not do full justice to the dynamics of cellular membranes. Almost routinely, portions of them – sometimes from different cells, sometimes from the same cell – fuse together and then separate into new conformations that enable the cell (or cells) to perform certain physiologic functions.

These processes, which for simplicity can be subsumed under the term "membrane fusion," are interesting in their own right, since they are basic to cellular physiology. In addition, the artificially induced fusion of cells (and membranes) has for some years served as a valuable research tool in such fields as oncology and cytogenetics. Before recounting some of the experiments in which my associates and I have attempted to elucidate the mechanisms of membrane fusion, therefore, I should like to describe some of the situations in which it occurs, both in and out of the laboratory.

It is well known that multinucleated cells occur in a number of physiologic contexts, both normal and pathologic. Some of these are formed by the fusion of several, ordinary, mononucleated cells. Such fusion is particularly frequent in embryonic development, as when, during the growth of muscle, one finds myoblasts fusing to form myotubes. Moreover, this is by no means a random process but under close physiologic control; DNA synthesis is active in myoblasts before cell fusion but ceases when fusion occurs. We find other embryonic multinucleated cells in the trophoblast, the so-called syncytial cells. Osteoclasts, which play such an important role in reshaping and absorbing bone both pre- and postnatally, are also multinucleated. No less well known is the occurrence of multinucleated cells in a variety of pathologies – most conspicuously, perhaps, in malignant diseases but also in tuberculosis and a variety of inflammatory diseases such as rheumatoid arthritis. While some of these multinucleated cells may be formed by cell fusion, they are admittedly rather unusual examples of the fusion process.

But we find membrane fusion in many everyday situations, quite apart from those that engender multinucleated cells. To begin with, the reproduction of most multicellular animals begins with the fusion of two cells, an egg and a sperm, and continues through cell division, a process in which the separation of the two daughter cells necessarily involves the fusion of portions of the parental cell's membrane to pinch off the daughters from one another. We find a similar pinching-off process in pinocytosis and phagocytosis, in which a portion of the plasma membrane invaginates, carrying with it various substances to be drawn into the cell. The surrounding membrane then tightens like a drawstring purse and fuses together, leaving the invaginated portion – whose membrane is also closed off by fusion – as a free vacuole in the cytoplasm. Pinocytosis and phagocytosis are the standard methods by which cells absorb large molecules or other substances which, for reasons of size or from other causes, cannot pass, by various mechanisms, *through* the membrane as can many ions and such compounds as glucose and water. Exocytosis, by which the cell ejects certain of its wastes or characteristic secretion products (e.g., some hormones), involves the reverse process; a membrane-enclosed vacuole or storage granule fuses with the plasma membrane, which thereupon opens at the point of fusion to allow release of the enclosed material without the loss of other cell contents.

Thus membrane fusion of one sort or another, if not precisely a universal in cell physiology, is certainly a commonplace. Attempts to study the process have mainly centered on two approaches, one chemical, the other viral.

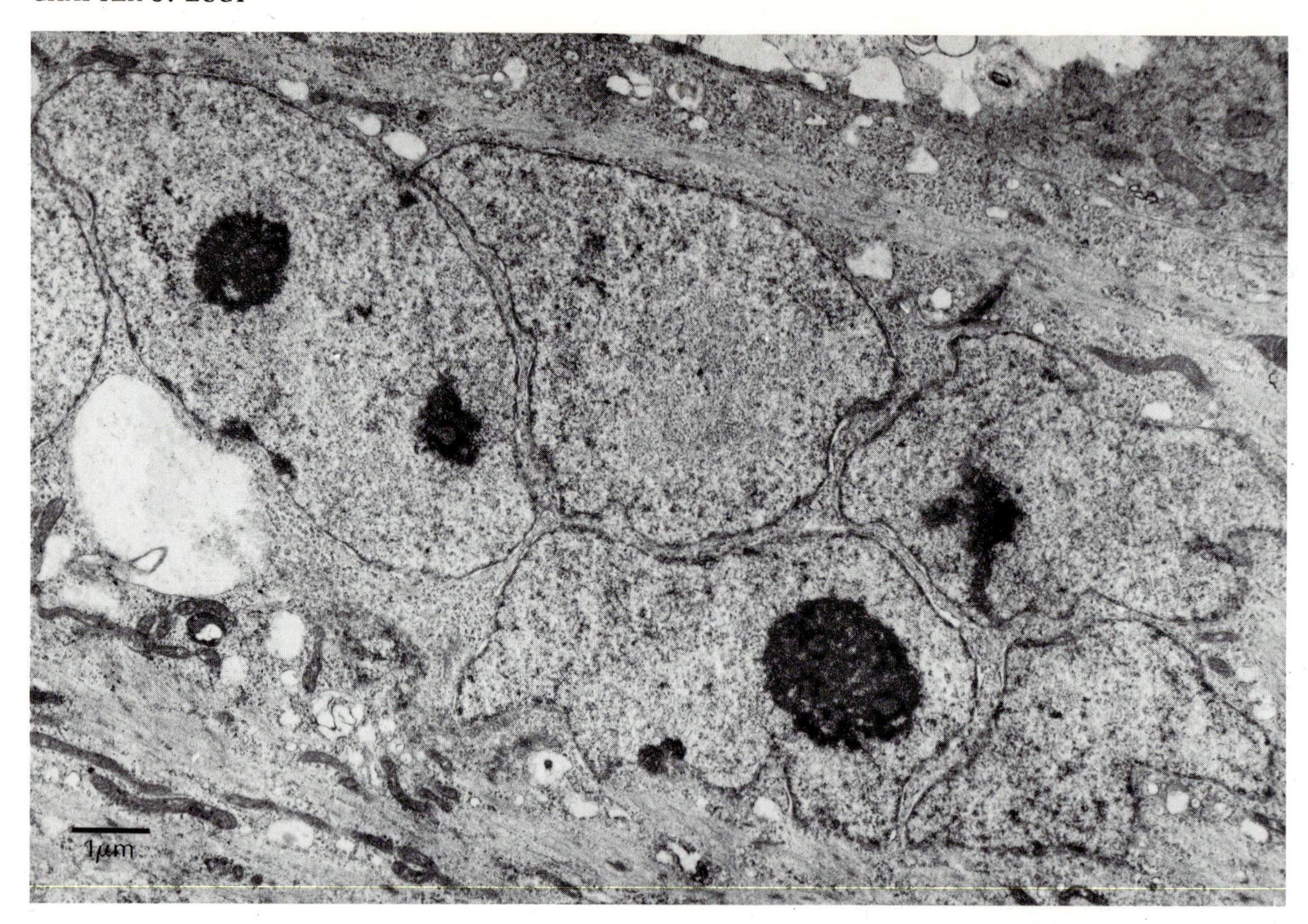

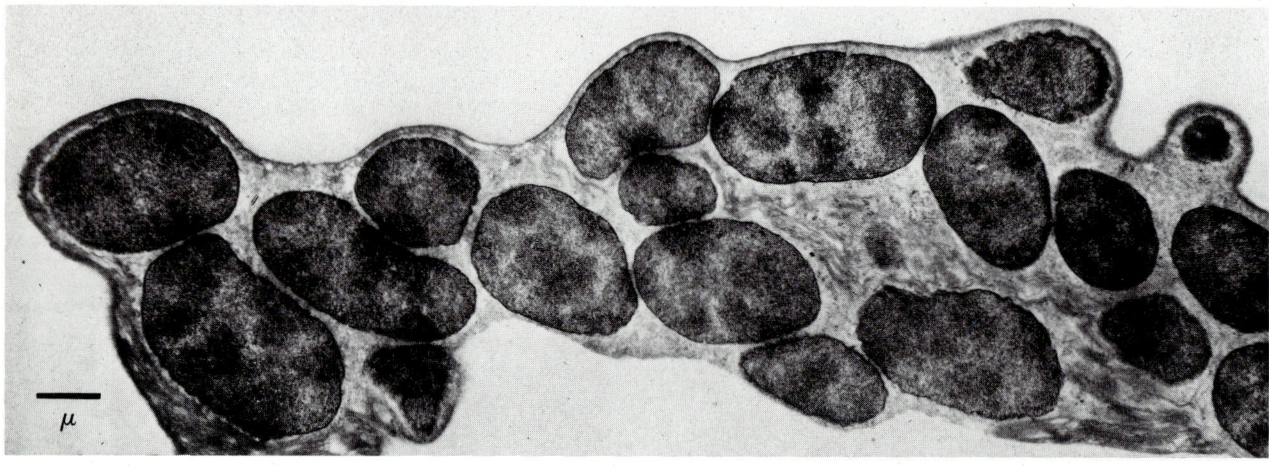

Fusion of cells into natural syncytia is important in embryonic development; electron micrograph at top, by de la Haba and Amundsen of University of Pennsylvania, shows five nuclei in a syncytial area in breast-muscle myoblasts from 11-day chick embryo after three days in culture (original x4,015). Below it is an electron micrograph from the author's laboratory of a section through an artificial syncytium formed from hen erythrocytes by treatment with lysolecithin; 19 nuclei can be seen.

That cells can fuse in certain infections has been known for many years; in measles, for example, the appearance of large, syncytial cells has long been noted as a common characteristic of the disease. Of course, when multinucleated cells were observed in inflammatory states a century ago, nothing was known of viruses; not until fairly recently has it become apparent that the fusion is induced in some fashion by the virus. Perhaps the best known of the cell-fusing viruses is the so-called Sendai virus, also known as HVJ (hemagglutinating virus of Japan). Cell fusion by the Sendai virus was discovered by the Japanese worker Okada, who showed that fusion occurs only at physiologic temperatures; at 4°C, the cells merely agglutinate without fusing. Subsequently it was reported that viral fusion requires the presence of calcium, and also the expenditure of energy, so that if oxidative phosphorylation is blocked chemically, the cells lyse instead of fusing.

Sendai virus can fuse cells of different types and even of different species – the latter phenomenon having been studied extensively by such workers as Henry Harris at Oxford. Harris and others produced heterokaryons – single cells with nuclei from different species – and also showed that if both nuclei of the heterokaryon divide synchronously, hybrid daughter cells are produced, with only one nucleus but a complete set of chromosomes from both parent species. Working with heterokaryons, Harris also found that if a hen erythrocyte is fused with a mouse fibroblast certain of the physiologic processes of the erythrocyte nucleus are reactivated. That is, a mature avian erythrocyte does not normally synthesize its characteristic proteins (e.g., surface antigens), but fusion "rejuvenates" the nucleus so that protein synthesis begins again. Such findings are of obvious relevance to a number of basic research problems – for example, the still mysterious processes whereby the genes that control protein synthesis are turned on and turned off at different times and in different types of cells. Similar questions are involved in experiments where a differentiated cell is fused with an undifferentiated one, to determine whether differentiation, i.e., the capacity to synthesize a particular protein, survives the fusion.

Cell fusion can be used to detect viruses in transformed cells. For example, if a hamster cell is transformed by SV40, the virus may be transmuted into an occult form, nonreproducing and undetectable. If, however, the transformed cell is fused with normal, noninfected monkey cells, the virus begins reproducing and virus particles can be recovered from the heterokaryons. Virus-induced cell fusion has also proved to be useful in chromosome studies. To cite an example: if a human cell and a mouse cell are fused and the hybrid cells obtained are cultured, human chromosomes will gradually be lost. As this occurs, one can test the culture for the continued presence of some characteristically human enzyme and, by a number of such experiments, attempt to localize the genes coding for synthesis of that enzyme to a particular chromosome.

Cell fusion has also been explored as a possible therapeutic technique against cancer. The Ehrlich ascites cell, for example, is notoriously highly malignant in mice of any strain, none of them possessing any significant immunologic defense against it. Watkins and Chen at Oxford wondered whether it might not be possible to develop immunity by fusing Ehrlich cells with other cells foreign to mice, and in fact found that Ehrlich-hamster hybrid cells could induce a degree of immunity against subsequent challenge by ordinary Ehrlich cells.

Thus viral cell fusion has led to interesting and valuable results in a number of research areas, but its significance has been rather as a tool than as a thing in itself; when it comes to the mechanisms by which it works, we know very little. The viral nucleic acid does not seem to be involved; at least the Sendai virus still causes fusion even when inactivated by ultraviolet light or a chemical agent. Rather, fusion seems to be induced by some factor in the viral envelope or capsule – and possibly a factor, moreover, that is in some manner derived from the last cell the virus infected. Thus if Sendai virus is grown in porcine kidney cells instead of chick embryo cells, it will not induce fusion; indeed, its activity may even vary from one chicken egg to another. This fact has complicated its use as a research tool, since two batches of virus prepared by apparently identical techniques may possess quite different specific activities in cell fusion.

For this reason, among many others, my associates and I at the University of London have focused on chemical rather than viral fusion techniques, since a consistently reproducible fusion method would possess wide utility even apart from its intrinsic interest. Our initial work centered on the compound lysolecithin, which was known to be capable of perturbing membrane structures to the point of lysis. It occurred to us that in smaller quantities it might instead merely alter the membrane structure just enough to permit cell fusion without cell destruction.

We were more interested in lysolecithin because it is not an artificial, laboratory curiosity but a natural sub-

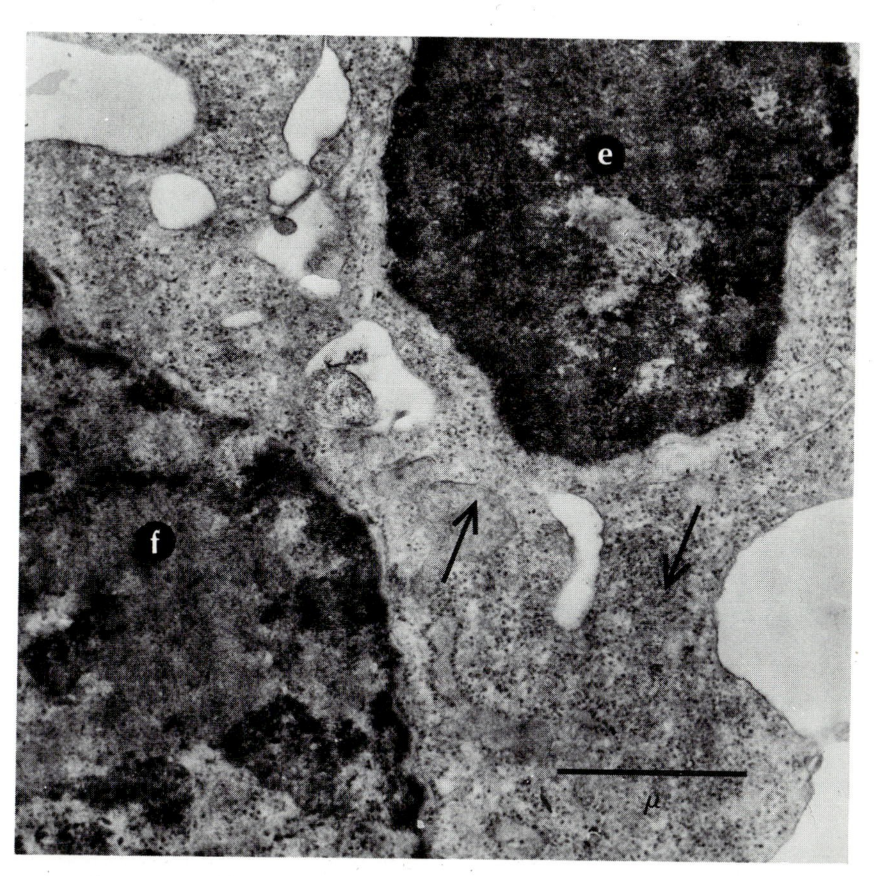

Heterokaryon of hen erythrocyte and mouse fibroblast was formed by treating parent cells with glyceryl mono-oleate; nuclei are designated respectively e and f. Bridges linking the cytoplasms of the two parent cells are present (arrows). Electron micrograph (original x21,400) was prepared by Dr. J. I. Howell of the author's laboratory.

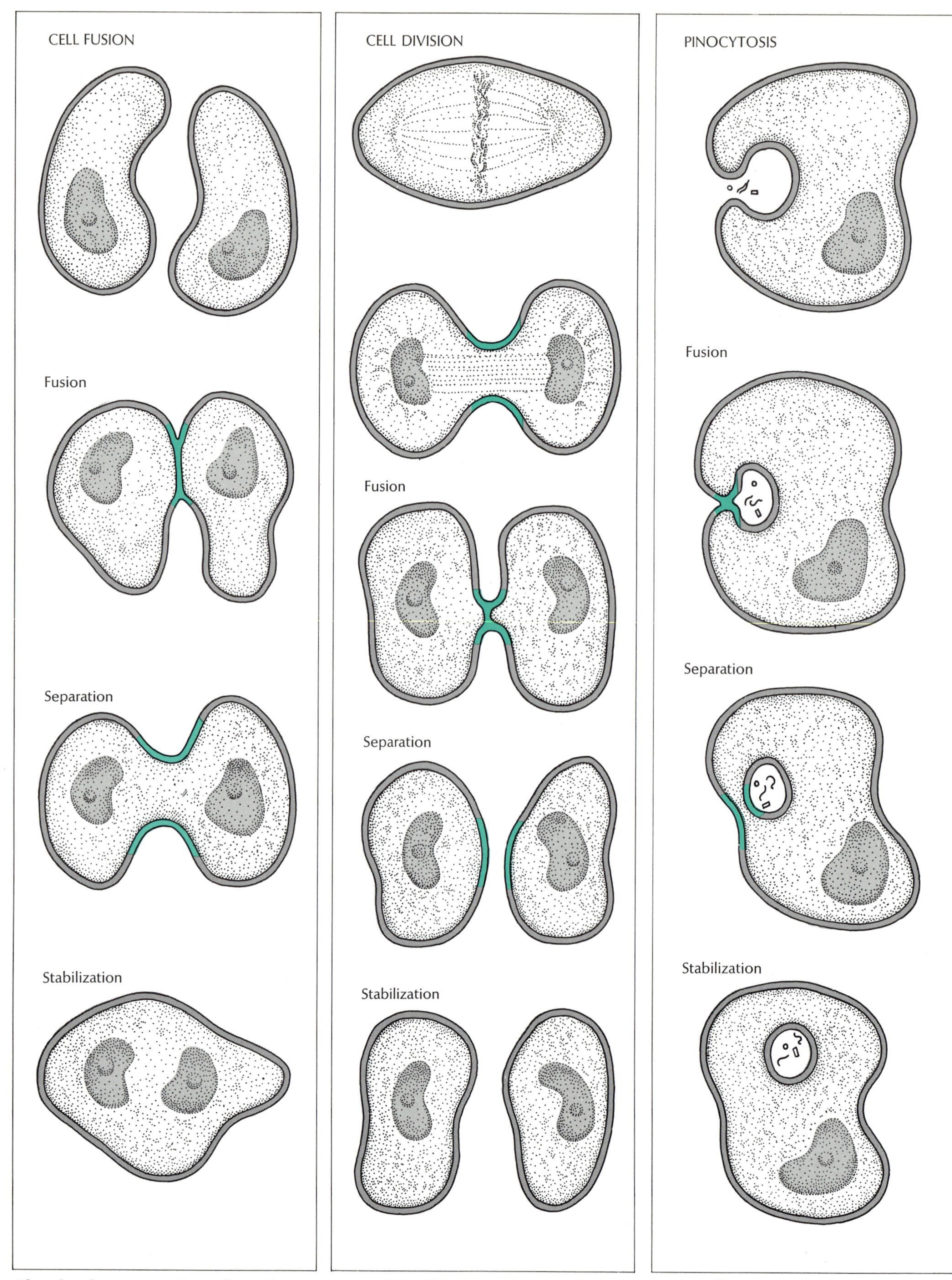

That the phenomenon of membrane fusion is essential to cellular physiology is suggested in the schematics above showing its participation in a variety of cell processes. In each situation color indicates the site of membrane fusion. Though the "aim"

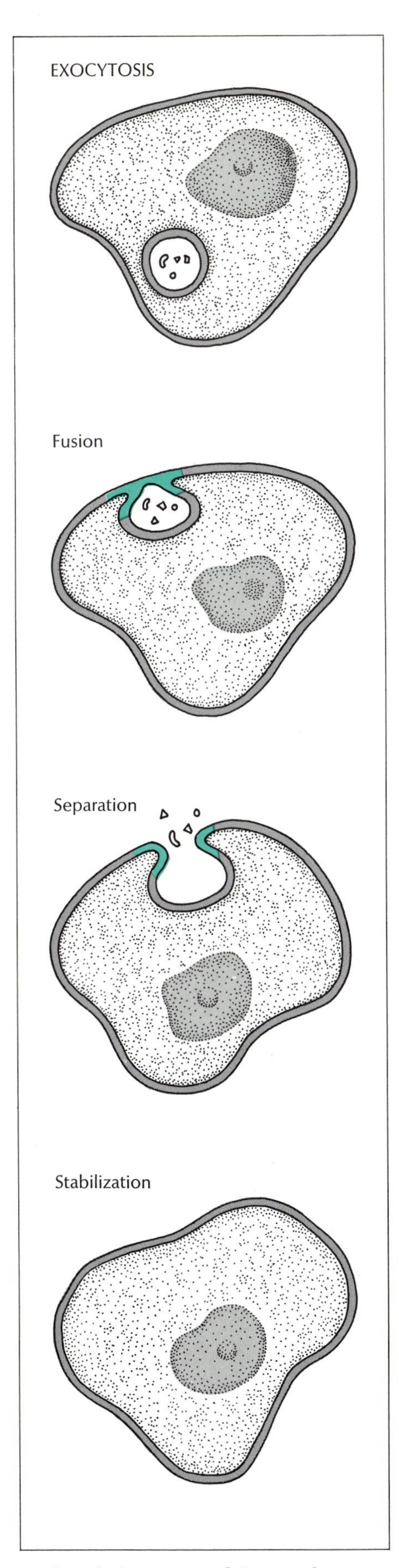

of each function is different, the processes are basically similar.

stance, metabolically important, produced by many living cells and quite widely distributed in the tissues, albeit in small amounts. As its name indicates, it is chemically related to lecithin, which is of course one of the main phospholipid constituents of membranes and present in virtually all of them. Lecithin follows the standard phospholipid structure, with a phosphate-containing polar group (in this case, phosphatidylcholine) attached to the carbon atom at one end of the three-carbon glyceryl framework, and two fatty-acid chains linked with the other two carbons (see Chapter 2, "Lipid Dynamics in Cell Membranes," by Chapman). However, in lysolecithin, one of these fatty acids is missing – either from position 1 (at the end) or from position 2 (in the middle).

Lysolecithin is formed in tissues by the action of an enzyme originally called phospholipase A; we now realize that there are actually two similar but distinct enzymes, A1 and A2, which detach the fatty acids at positions 1 and 2 respectively. Even more interesting is the fact that cells, along with the means for making lysolecithin, also possess mechanisms for unmaking it; it can be converted back to lecithin or, alternatively, further degraded into simpler compounds. In lysolecithin, therefore, we have – in theory, anyway – a compound with the basic properties needed for any system of membrane fusion: it can perturb membrane structures, it can be manufactured within the cell when perturbation (i.e., fusion) is required, and it can then be removed to restore membrane stability when fusion is completed.

A number of researchers have conducted experiments designed to ascertain whether such a lysolecithin system is involved in viral cell fusion, by determining whether the compound is present at such times, and some of them claim that it is apparently not. My own feeling is that negative findings of this sort should be interpreted with great caution. If lysolecithin is involved, it will still be present only in very small quantities, sufficient to perturb a small area of membrane, and also very transiently, since fusion is complete in a few minutes or even less.

Our own experiments, at any rate,

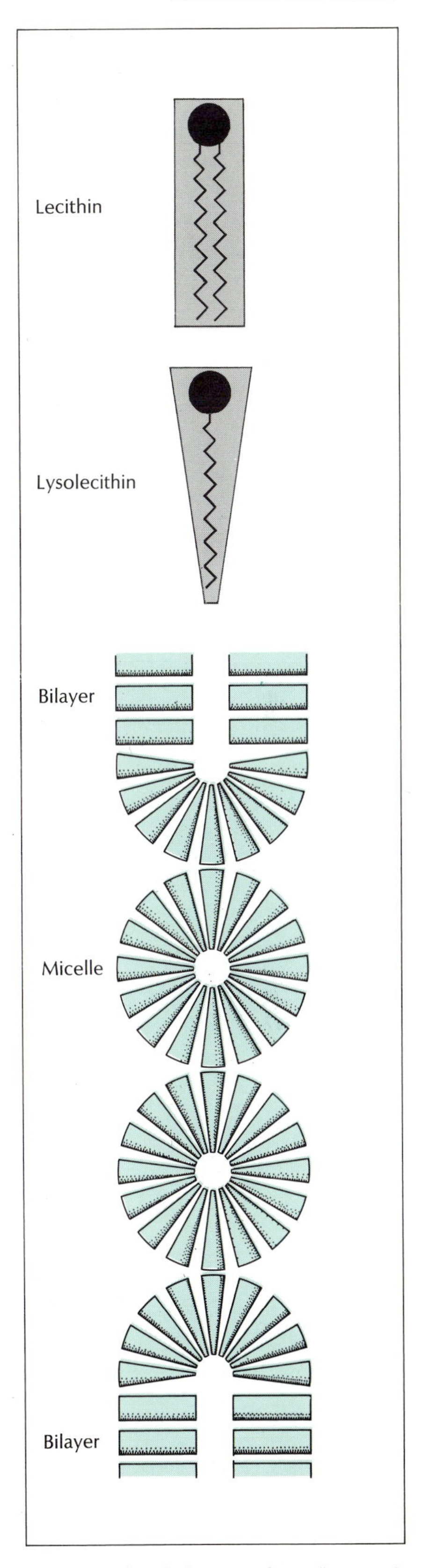

When a phospholipase splits off one of the hydrophobic tails of lecithin, lysolecithin is formed; its property of perturbing cell membranes may arise from this change, which facilitates a (reversible) shift of membrane structure from a lipid bilayer to a micellar aggregate.

have clearly demonstrated that whether or not lysolecithin participates in membrane fusion in vivo, it can certainly induce fusion in vitro. As our main experimental subjects we have used hen erythrocytes, which are cheaply and easily obtained. They also possess several other properties that make them suitable for fusion experiments. For one thing, unlike mammalian erythrocytes, they are nucleated, so that the progress of fusion can be easily followed simply by counting the number of nuclei within each cell; mammalian fused cells are merely larger and cannot so easily be distinguished from normal ones. For another thing, the mature avian erythrocyte, as compared with the fibroblast (for example), possesses very few organelles apart from the nucleus; specifically, it contains virtually no lysosomes or mitochondria. Thus, if one observes fusion after adding a chemical to an erythrocyte preparation, one can be fairly certain that it is the *direct* result of the agent's action on the plasma membranes, rather than (for example) an indirect result produced by the agent's action on the lysosomes to release lysosomal enzymes.

We prepared our lysolecithin by treating egg lecithin with snake venom, which contains the required phospholipase. The purified compound reliably fused the erythrocytes, forming large syncytia containing many nuclei. Subsequently, we were able to fuse mouse fibroblasts with one another, as well as with hen erythrocytes to form heterokaryons.

A problem in all these experiments was that even where fusion was successful the resulting cells were exceedingly fragile, their membranes tending to lyse within a few minutes (the fusion process itself occurred in only about 30 seconds). It is possible that the cells in question lacked enzymatic mechanisms for removing unwanted lysolecithin, such as we have discussed above. This would be particularly likely in the case of the hen erythrocytes, which would have no "use" for such a mechanism, since when mature they neither divide nor engage in such processes as pinocytosis that involve membrane fusion. More certainly, we can say that even if the mechanisms existed, they would very probably be overwhelmed by the

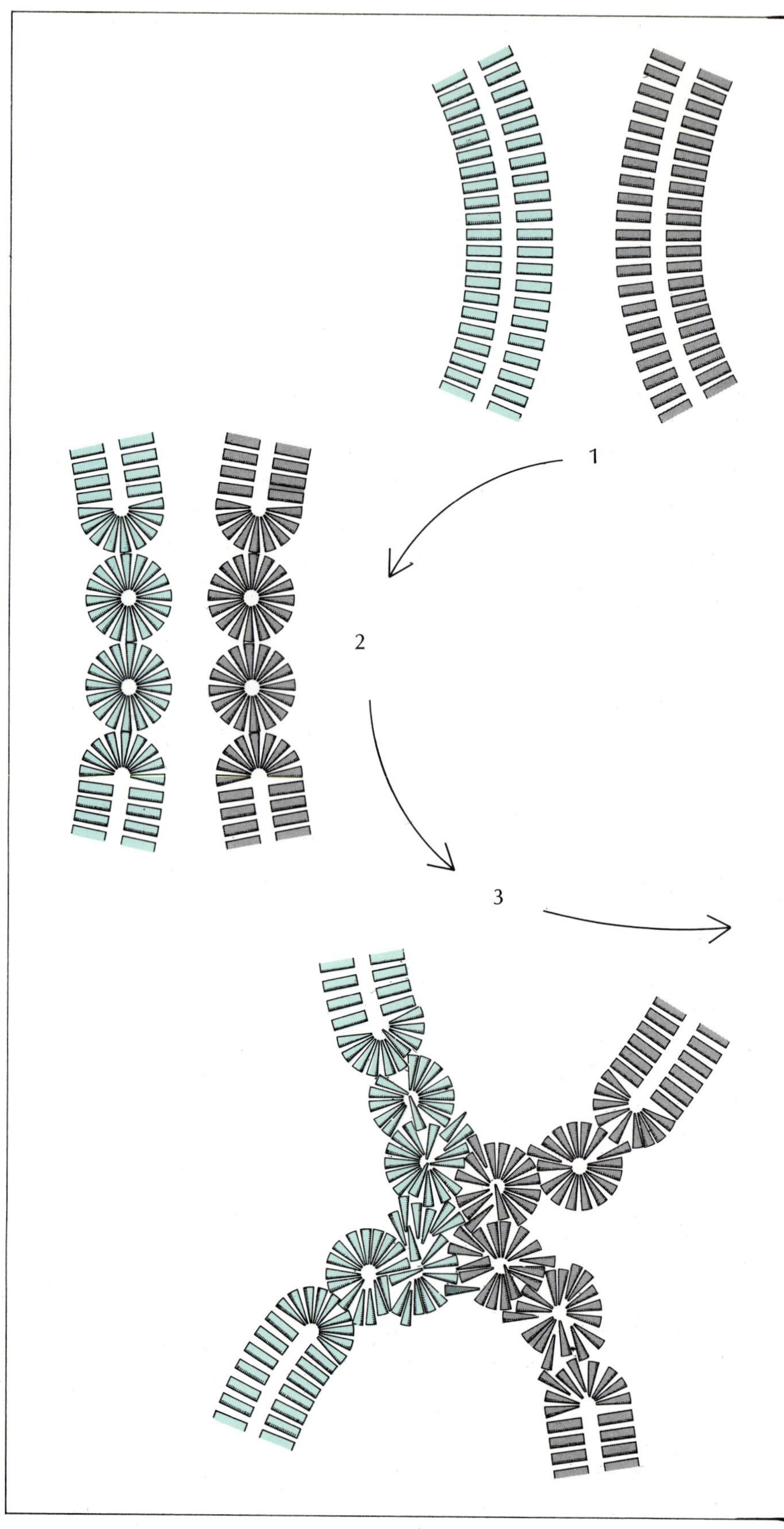

The possible involvement of micellar configurations of lipid molecules in fusion of membranes is illustrated diagrammatically. In stage 1, two adjacent membranes are shown wholly in the form of lipid bilayers, or bimolecular leaflets. In stage 2, as the result of the presence of an exogenous perturbing molecule, a part of each bimolecular leaflet becomes organized into globular micelles. (It is held that fusion will probably

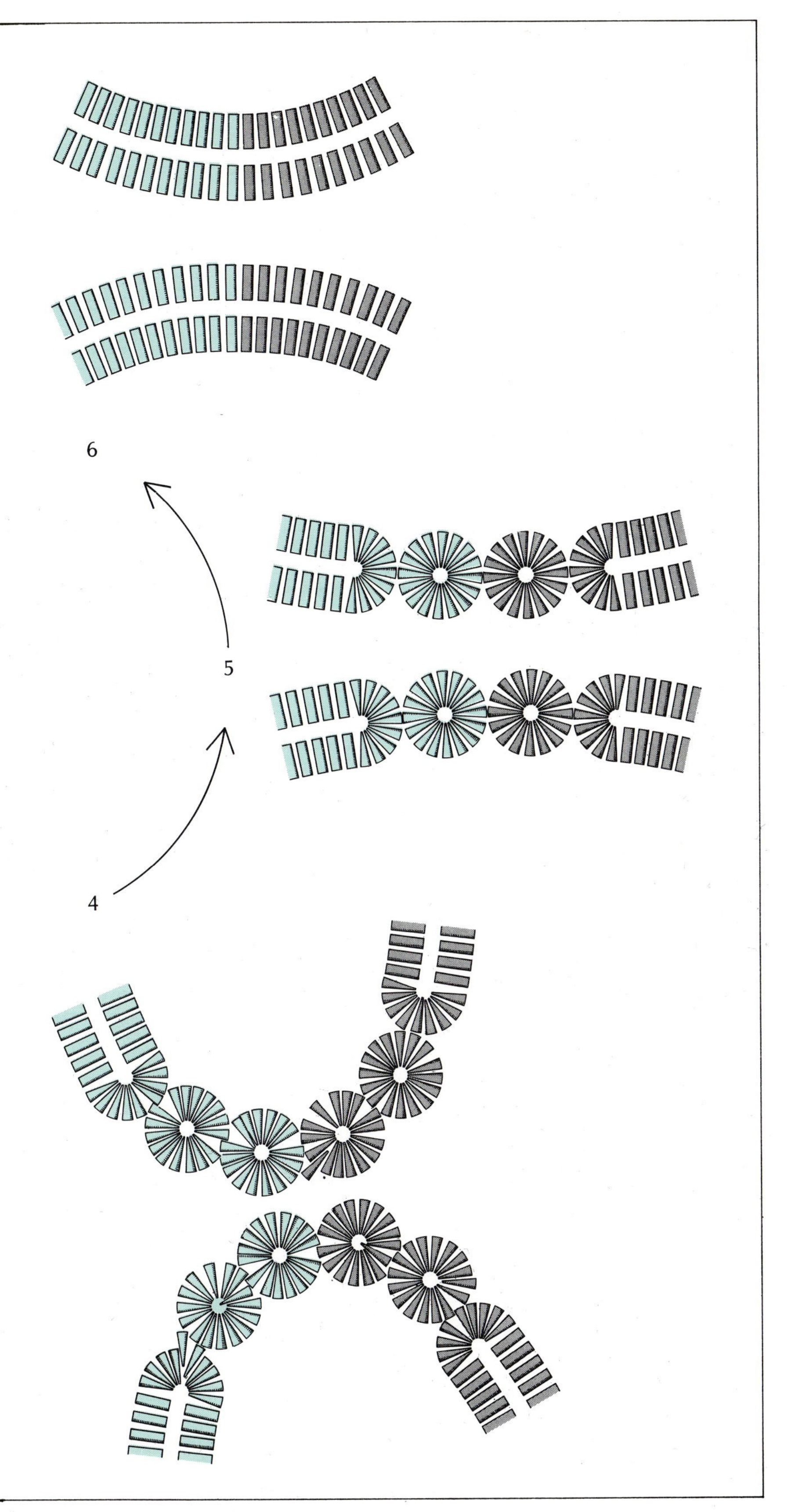

not occur if one or both of the membranes has its lipid molecules predominantly arranged in the bilayer mode.) Interdigitation of micelles can now occur, leading to local fusion of the two membranes into a single entity (3), as in the processes shown on pages 78 and 79. The unstable junction then breaks down (4, 5). Finally, stabilization occurs (6) with reversion to the bilayer state.

amount of lysolecithin (and membrane damage) in our experiments, since most natural fusion processes involve only small portions of the membrane and presumably equally small quantities of lysolecithin.

Subsequently, we have found that lysis can be largely prevented by suspending the lysolecithin in microdroplets of lipid rather than dispersing it directly in the aqueous medium. In structure, these microdroplets are not unlike the fat droplets in milk (which, interestingly, contain small quantities of lysolecithin), and in fact it was the latter that suggested the microdroplet approach to us. These tiny "containers" of lysolecithin, which are considerably smaller than the cells, agglutinate them by adhering to two cells at once, and one gets fusion at the point of adhesion. However, even in this system cell damage is still considerable, so that it leaves much to be desired as a technique for inducing cell fusion to order.

It was gratifying to find that lysolecithin could induce membrane fusion, as I had previously suggested that this substance might cause membranes to fuse in view of its physicochemical properties. The basic problem in explaining fusion is that it necessarily implies interpenetration between the two membranes, or membrane segments, involved. And if the membranes are visualized as normal lipid bilayers, hydrophilic on the outside and hydrophobic on the inside, it is hard to see how the hydrophilic layers could be forced into the hydrophobic interior, for reasons made clear in early chapters of this volume. It has been known for some time, however, that in some systems at least the presence of lysolecithin tends to favor an alternative type of lipid organization, the so-called micellar conformation, in which the lipid molecules form spheres rather than bilayers, with their hydrophobic tails inside, the polar head-groups outside; this was demonstrated over 10 years ago by Dr. Alec Bangham of Cambridge in his studies of artificial membrane models. The shift has been explained on the ground that whereas the molecules of normal membrane lipids (e.g., lecithin) are roughly cylindrical in shape, and hence would tend to pack into relatively flat layers, lysolecithin, having lost one of its

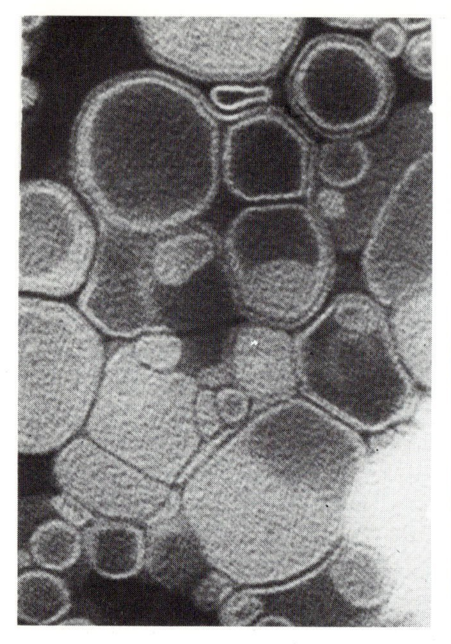

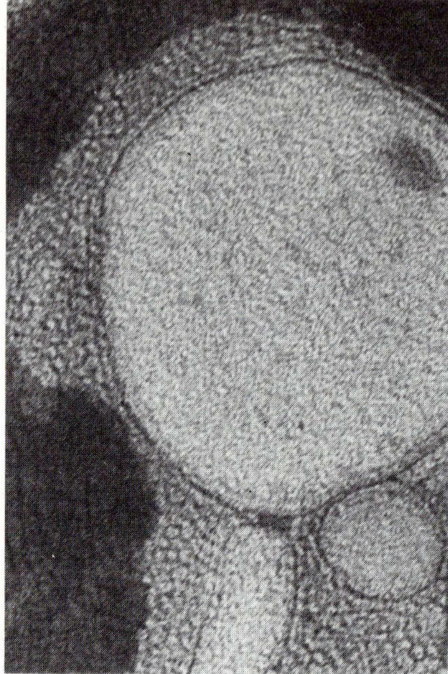

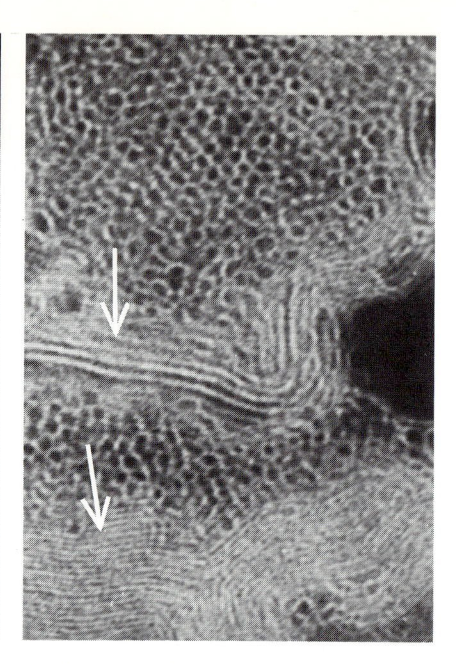

Introducing lysolecithin into lecithin liposome membrane "models" (left, pretreatment) causes these structures to disintegrate into micelles (center). Unsaturated fatty acids and their derivatives that cause cell fusion do not produce this effect; instead (right), following glyceryl mono-oleate treatment, the result is that widely and narrowly spaced lamellae (arrows) appear.

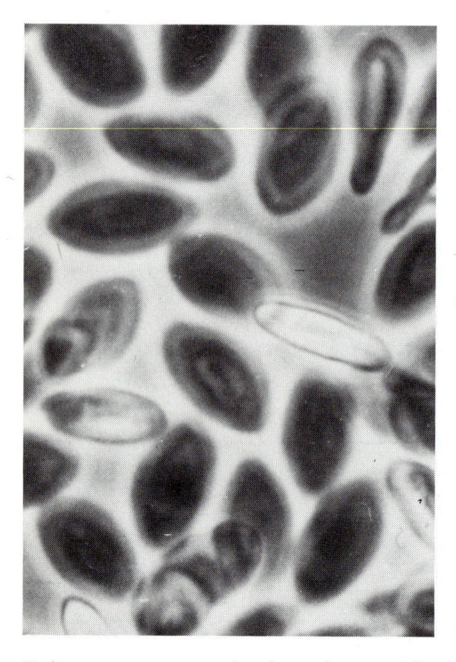

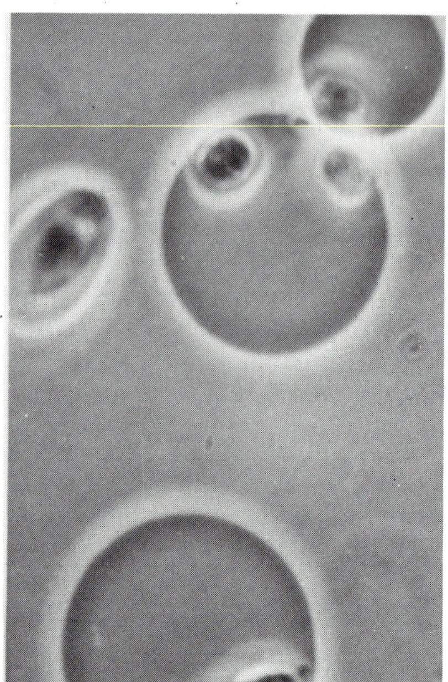

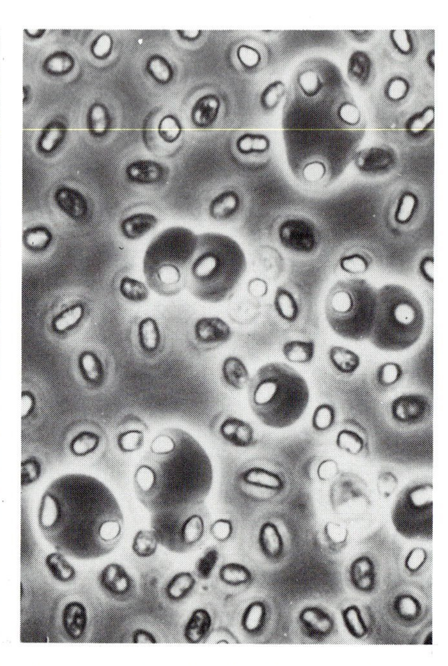

Before treatment with glyceryl mono-oleate, hen erythrocytes have the usual oval shape (left); incubation with the ester for 10 minutes at 37° C results in fusion (center); note binucleate cell. Fusion can also be obtained (right) by heating alone, to about 48°. Multinucleate cells can be seen as well as cells in various stages of fusion; paler cells are lysed.

fatty-acid tails, is rather wedge- or cone-shaped, and hence would fit more easily into a sphere, at the same time, as it were, encouraging other lipids to enter into the same configuration.

One can then visualize membrane fusion as beginning with the entry of a number of lysolecithin molecules into localized regions of the membrane (or perhaps being formed in situ from lecithin). These regions then shift from the bilayer to the spherical micellar conformation; when two such regions come into contact, the spheres, being entirely hydrophilic on their surfaces, can easily interdigitate with one another. An opening across the interdigitating region then permits formation of two new micellar regions, more or less at right angles to the original fused region, each of which contains micelles from *both* the original membranes. The final step would be removal or reprocessing of the lysolecithin and reversion of the micelles to normal, stable bilayers.

This model seems plausible enough in the case of lysolecithin, the more so in that in the artificial system employing Bangham's liposomes one can actually see the spherical micelles in electron micrographs. But it is almost certainly not the whole story. Subsequent experiments have shown that other substances of rather different molecular conformation can also induce membrane fusion, probably by quite different mechanisms.

The substances in question are, like lysolecithin, lipids, and one group of them embodies certain chemical similarities to that substance. These are monoesters of certain fatty acids, meaning that like lysolecithin they include a single fatty-acid chain attached to a hydrophilic head – either the glyceryl framework (but without the complex phosphate-plus-choline group of lysolecithin and lecithin) or some other suitable structure, such as mannose or sucrose. The other group consists simply of the fatty acids themselves, which in suitable concentrations will also induce cell fusion. An important finding, however, is that not just any fatty acid will do, whether in the pure form or as an ester; what is required is either a short-chain fatty acid, such as capric and myristic, or longer-chain but unsaturated acids, such as oleic and linoleic. This contrasts sharply with our findings on lysolecithin; the fatty acid moieties in our preparations of this substance were palmitic and stearic (predominantly the former), both of them long-chain and saturated. Neither of these fatty acids can induce fusion by itself.

Some of the monoesters are more or less wedge-shaped, and it is perhaps significant that the diesters, which are not, also will not induce fusion. As against the "wedge" theory, however, we have the fact that none of the unsaturated acids or their esters, when introduced into preparations of lecithin liposome "models," will induce formation of micelles as lysolecithin clearly does. Instead we get collections of lamellae – thin, roughly parallel layers that, in electron micrographs, somewhat resemble part of a contour-plowed field viewed from the air. And though the identity and significance of these structures is not yet clear, they are unquestionably nothing like globular micelles.

The findings on short-chain and unsaturated fatty acids nevertheless tie in with a basic property of membranes that has been established by a great variety of evidence: these substances, at least when incorporated into membrane phospholipids, loosen the structure of the bilayer, thereby lowering the transition temperature above which the membrane exists in the liquid crystal rather than the more rigid gel state (see Chapter 3, "Models of Cell Membranes," by Bangham). These substances in effect make the membrane – or that portion of it in which they happen to predominate – more fluid. The same effect, of course, can be produced simply by raising membrane temperature, and it is therefore interesting that we have been able to fuse hen erythrocytes merely by incubating them at 48° to 50°C – a temperature considerably above the physiologic range.

It seems possible, therefore, that lysolecithin, if it really operates by our postulated wedge mechanism, may represent something of a special case. Indeed, there is no a priori reason why the various types of membrane fusion in different kinds of cells should all proceed according to a single chemical mechanism, whether or not it involves lysolecithin – though it would be esthetically satisfying if this turned out to be the case. We must note also, however, that while both lysolecithin and the enzyme systems for making it and unmaking it are known to be present in many living cells, as mentioned earlier, there is as yet no evidence that this is the case with many of the other fusion-inducing substances we have tested.

Thus, when it comes to specifying the actual mechanisms of membrane fusion in cells we still have a large gap to fill between in vitro and in vivo.

We do not know what chemicals actually preside over fusion in the living cell; we do not know whether calcium is necessary for all such processes, as it is for most test-tube fusion, both chemical and viral, and also for exocytosis – and even there we do not know *why* it is necessary. We cannot be certain why energy is required, at least in some instances, for the process of fusion, though indications are that its main or only role is not in inducing fusion but rather in restoring the membrane to its normal, stable condition when fusion is completed. Nor, for that matter, have we achieved our goal of devising a simple, reliable technique for fusing cells to order, though the past few years' work has brought us closer to it. We know, as we knew at the beginning, that cell membranes fuse, that fusion is an integral part of many routine cellular processes, and we now know of several ways in which they *might* fuse. But to ascertain how they *do* fuse in vivo requires many more experiments.

Section Two

Cell Biology

9

Junctions Between Cells

GEORGE D. PAPPAS
Albert Einstein College of Medicine

As authors of preceding chapters have made abundantly clear, the external membranes of cells are involved in interactions of the cell with its environment, including interactions with other cells. Less widely known, perhaps, is the fact that some cell membranes embody actual physical connections between cells. These connections serve a variety of purposes, such as adherence, electrical communication, or tissue organization. These intercellular junctions, viewed primarily from the anatomic standpoint, will be the subject of this chapter.

The simplest and probably best known of these intercellular "bridges" is the desmosome, which is simply an area in which two cells are strongly adhesive. The adhesion was demonstrated in a classic series of experiments by Robert Chambers during the 1920's, in which he employed the techniques of microdissection to pull cells apart from one another. He found that in epithelial tissue the cells could not be separated without destroying them, and – along with other investigators of the period – concluded that these cells were in fact a syncytium, that is a single, multinucleated giant cell with continuous cytoplasm. This view was consistent with what could be seen with the light microscope.

More recently, electron microscopic studies have established that this view is incorrect. The desmosome, or area of adhesion, does not include cellular cytoplasm; rather, one can observe on either side of it the intact membranes of the connected cells as normal bilayers 75 to 90 Å thick. The intercellular space, too, is of normal thickness – about 200 Å – for well-packed epithelial cells. It is within this space that one observes regions of electron-dense material extending from one cell to another, which are the desmosomes.

The desmosomes are evidently protein of some sort; they can be denatured, and the cells caused to separate, by treatment with proteases such as trypsin, or by removing calcium from the medium. As proteins they can be grouped with the so-called peripheral proteins (such as spectrin) mentioned in earlier chapters of this book. Interestingly, one finds complementary areas of electron-dense, fibrous material on the opposite sides of the membranes, i.e., within the cytoplasm of the joined cells, but no one knows what this is. Conceivably, this material may be keratin-like or it may represent the metabolic apparatus that synthesizes desmosome protein or perhaps another sort of structure that "anchors" the ends of the desmosome within the membranes of the cells it joins.

The electron-dense areas on both sides of the membrane are sometimes interpreted as suggesting that the desmosome is a thicker-than-normal section of membrane. If one considers the term "membrane" as including the peripheral proteins, this is correct, but it is important to remember that the basic membrane bilayer is no thicker than normal; whatever the desmosome is (and we still don't know very much about its chemistry), it seems to be essentially external to the bilayer.

As already noted, desmosomes are found primarily between epithelial cells, which is to say those of the body's external surface and of the internal surfaces of most of its organs. (Topologically speaking, the entire respiratory, gastrointestinal, and genitourinary epithelium is continuous with the skin and can be viewed as an extension of it – this in contrast with the cardiovascular endothelium, which cannot.) The function of the desmosomes seems to be purely mechanical: that of holding together surface cells, which, because of their position, are subjected to a certain amount of mechanical stress. Repeated, even modest, stresses could, in the absence of specialized connecting structures, destroy the continuity of the surface. Desmosomes are sometimes referred to as "intercellular cement," but this metaphor is somewhat misleading in that it suggests a sort of bricks-in-mortar arrangement. The junctions can be more accurately compared to rivets or spot welds – discrete points holding together two otherwise unconnected structures. A less common type of desmosome,

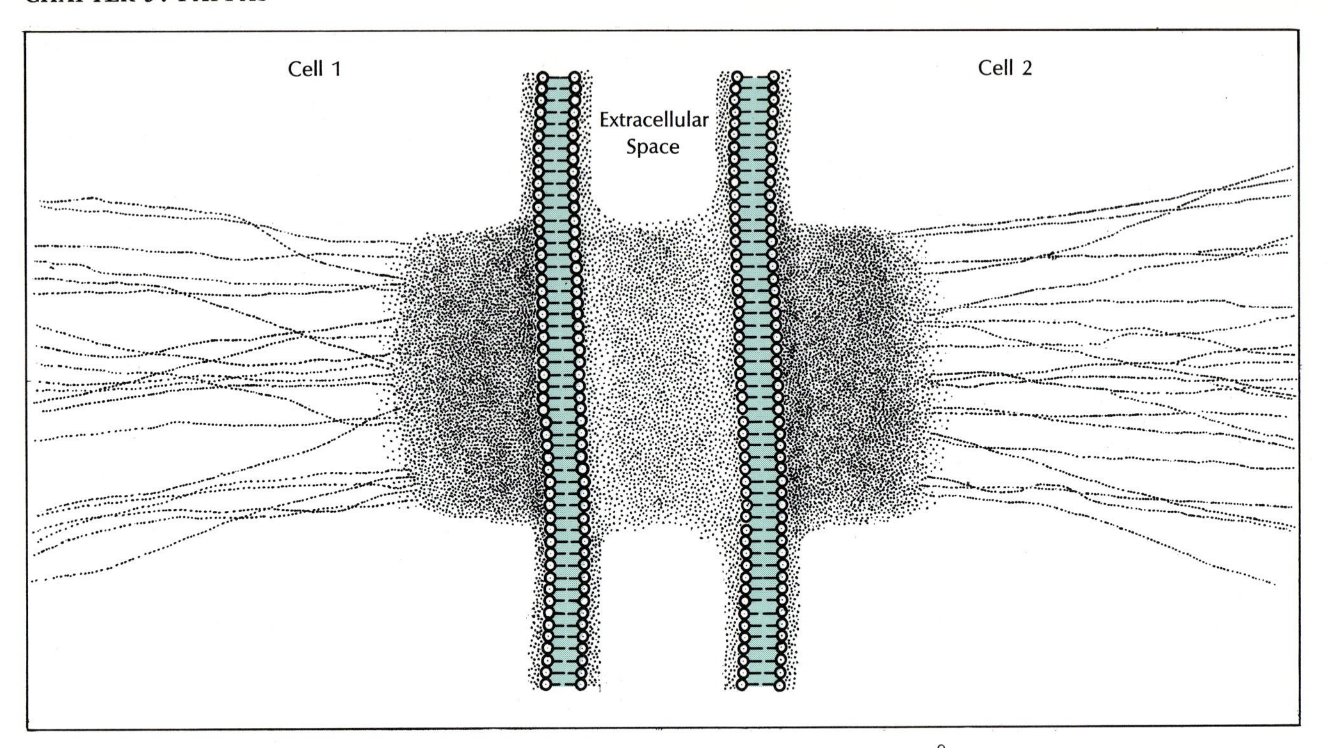

The desmosome is a specialized area of cellular adhesion; its "spot weld" character is suggested in the drawing above and demonstrated in the related electron micrograph at right (x118,000). Area of electron-dense fibrous material in the intercellular space (about 200 Å) separates the two cell membranes, which have a normal bilayer structure. Denser material appears on the inner side of cell membranes, with fine keratin-like fibrils projecting into the cytoplasm of each cell. Electron micrograph

found especially in intestinal epithelium, is linear rather than punctate in form, as reflected in its name, the zonula adherens; it can be compared to a seam weld rather than a spot weld.

A quite different type of intercellular connection, with physiologic rather than merely mechanical functions, is the so-called tight junction. This is exactly what its name implies: an area of tight connection between two adjacent cell membranes – so tight, in fact, that the junction is essentially impermeable. In contrast to the desmosome, which bridges an intercellular space of normal width, the tight junction involves the complete disappearance of that space; each cell is literally up against the adjacent cell wall. The occlusion of the extracellular space gave these structures their original name, the zonula occludens. At that time, perhaps 10 years ago, other supposedly tight junctions, of punctate rather than linear form, were called maculae occludens, but it has subsequently been determined that the latter are actually structures of another sort – so-called gap junctions – which we shall discuss later.

Electron micrograph and freeze-etch studies have established that in the true tight junction the apposition of the cell walls does not merely occlude the intercellular space but actually involves fusion of the two membranes. Thus an electron micrograph of the region of apposition shows not four dark lines, indicating the presence of two normal bilayers (such as one observes in both the desmosome and the gap junction), but only three. That is, the outer surfaces of the two bilayers have in some fashion "melted" together into a single layer. The zone, or "seam" of fusion, completely encircles each cell, meaning that a layer of such cells is in effect embedded in one essentially continuous plasma membrane, with no gaps or intercellular spaces.

Since cells connected by tight junctions would constitute a continuous barrier, it will probably suggest to the reader just where such junctions should be and are found: in the apical portions of lining cells that separate certain lumens from serosae or from connective tissue. *Tight junctions abound in anatomic situations where a sharp physical separation between the two compartments is essential.* A good example is the well-known blood-brain barrier, which consists of the lining of the cerebral blood vessels, the cells of which, unlike those of vascular endothelium elsewhere in the body, are joined by tight junctions. Here the action of the tight junction is essentially protective, since it blocks the passage of many molecules from the circulation into the brain tissue whose complex and delicate operations they might disrupt.

Of course the blood-brain barrier is not, and cannot be, completely impermeable; molecules such as glucose and some neurohumors routinely pass through it as does carbon dioxide in the opposite direction. It does not even exclude all noxious molecules; ethyl alcohol and diacetylmorphine (to take only two examples) are notoriously capable of escaping from the circulation into the brain and there altering its functions in various ways. The point is not that the tight junctions of the blood-brain barrier make it impossible for molecules to enter (or leave) the brain through intercellular spaces; rather, these molecules must pass through the membranes of the lining cells into the cytoplasm and across the membranes on the other side. The implication is that molecules entering the brain are under a degree of physio-

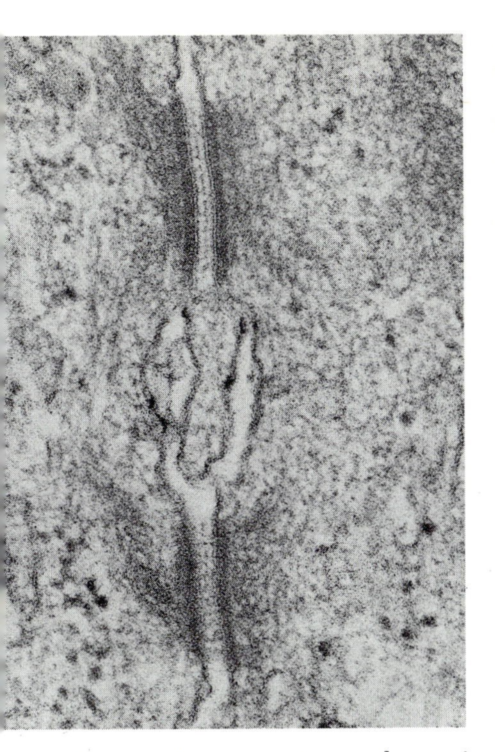

shows two desmosomes from epidermis of an axolotl larva; note the apposed densities formed by fibrillar meshwork, which is confined within the cytoplasm.

logic control by the metabolic actions of the lining cells. Some, including large-molecular-weight proteins, are excluded absolutely, while others, such as precursors to myelin formation in the newborn, are admitted or excluded by the barrier cells in accordance with the organ's physiologic needs at a given time.

In certain infections or intoxications (e.g., meningitis), the tight junctions of the blood-brain barrier have turned out to be rather too tight for the physician's or patient's convenience. Thus it is impossible to use antibiotics to control CNS infections, since they will not pass the barrier. And when George Cotzias sought to control parkinsonism by increasing brain levels of dopamine, he found it impossible to do so directly, since this substance will not pass the barrier; his success came only when he discovered the value of the dopamine precursor, L-dopa, which is not blocked in this manner.

In other organs, the function of the tight-junction barrier is often not to keep substances out of a given physiologic compartment but rather to keep them in. Thus, in the anterior chamber of the eye, for example, the aqueous humor that fills it contains abnormally high concentrations of both sodium ion and ascorbic acid – in the latter case, up to 15 times the plasma level. These concentrations are produced by the physiologic "pumping" action of epithelial lining cells of the ciliary body, which of course requires energy. It is therefore not surprising that these cells are joined by tight junctions that minimize leakage of sodium and ascorbic acid back into the circulation and thereby reduce the energy expenditure required to keep the cornea in functioning condition. We also find tight junctions between cells that serve to segregate particular metabolic products, as in the epithelium of the gallbladder and in the endocrine glands.

The third type of intercellular connection, the gap junction, is in one sense intermediate between the desmosome on the one hand and the tight junction on the other. Like the former, it bridges the space between two cells, but that space is considerably narrowed, from its normal 200 Å or so to a width of 20 Å to 40 Å. In another sense, however, the gap junction differs markedly from both the other types in that it does not merely connect the exterior membranes of cells but also their cytoplasms.

This has been demonstrated by experiments in which certain marker substances – their molecular weight must be less than 200 – are injected intracellularly into one of several cells connected by gap junctions. Immediately afterwards, the marker is seen to pass rapidly into adjacent cells but *not* into the intercellular spaces. If, on the other hand, one adds to the fixative of a cell preparation a heavy metal such as lanthanum, which is electron-opaque and hence visible with the electron microscope, and which we know cannot cross the plasma membrane, it will still penetrate the gap junction, insinuating itself between the 20 Å to 40 Å extracellular space or gap. However, if one takes a section *across* the gap junction – i.e., more or less perpendicular to the cell walls that it connects – the electron microscope reveals a hexagonally arranged mosaic of more-or-less circular areas into which the lanthanum has not penetrated.

Evidently, then, the gap junction consists of an array of channels, or pores, passing through the cell mem-

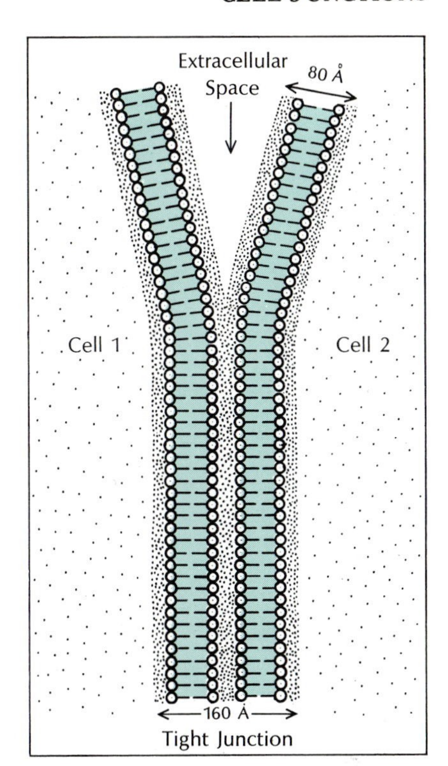

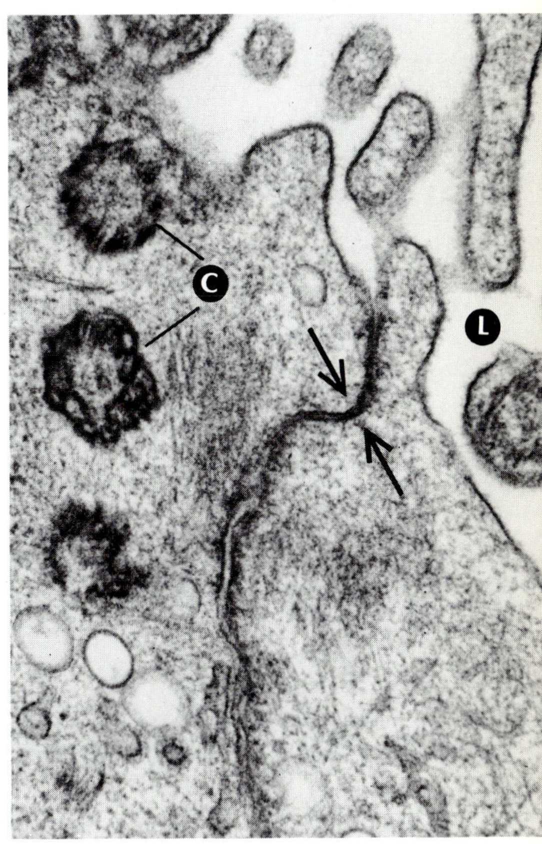

In the tight junction, the intercellular space disappears, as drawing indicates. In the EM of a portion of the junction between two ciliated tracheal epithelial cells of monkey (x20,000), two apposing membranes have come together (arrows) to form a tight barrier that seals off the lumen (L). Sections of cilia can be seen in the lumen as well as ciliary roots (C).

brane, across the narrow intercellular space and through the membrane of the adjacent cell. Passing among these channels where they cross the intercellular gap is another, complementary, set of channels through which extracellular substance can move from one part of the extracellular space to another. What function, if any, is served by these extracellular channels is not known. It seems likely that they are merely incidental formations resulting from the geometry of the cell-to-cell channels. If by analogy one visualizes the system as a pair of metal tanks (cells) connected by a close-set array of short pipes (channels) it is obvious that substances external to the tanks will be able to pass through the spaces between the pipes, though they will of course be unable to get into either the pipes or the tanks.

The function of the "pipes" themselves – the cell-to-cell channels – is clear: they permit the passage of certain substances from one cell to the next. One can even say that a group of cells connected by gap junctions actually shares the same cytoplasm, albeit in a limited way – limited in that the narrowness of the connecting channels (no more than about 15 Å) prevents passage from cell to cell of large molecules and, even more obviously, of intracellular structures such as mitochondria. But the tubes are still large enough to permit passage of molecules such as sucrose (which, as a disaccharide, is considerably larger than glucose), fluorescein, Procion Yellow, and various other dyes, as well as various sorts of hydrated ions, whose passage from cell to cell can be followed by employing tracers such as radioactive potassium and radioactive chlorine.

Comparing the three types of junctions in terms of membrane structure, we see that whereas the desmosome is composed of peripheral protein bridging between normal lipid bilayers, and the tight junction is composed of two partially fused bilayers with little protein between them, the gap junction must involve neither lipid nor peripheral protein but integral protein (see Chapter 4, "Architecture and Topography of Biologic Membranes," by Singer). In order to permit the sort of intercytoplasmic communication we have been describing, the tubes – which are undoubtedly composed of protein – must be embedded in the lipid bilayer, forming a channel entirely through it. Singer has indeed postulated the existence of such channels through which substances can be moved into and out of the cell. It is significant that in freeze-etched preparations the gap junction appears as precisely the sort of structure one would expect under these circumstances: a large agglomeration of protein shared by two cells.

Gap junctions are always of punc-

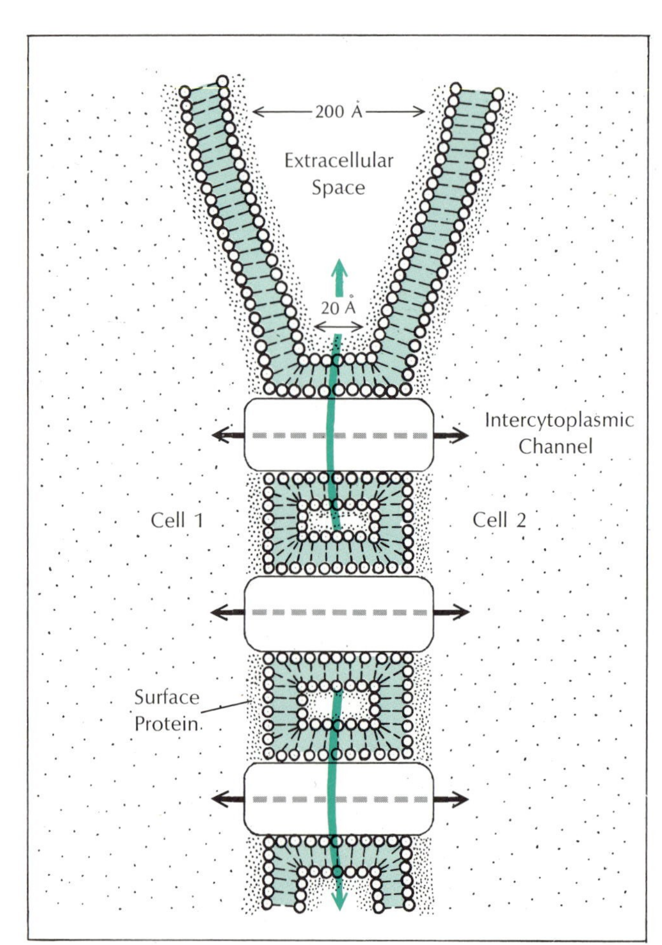

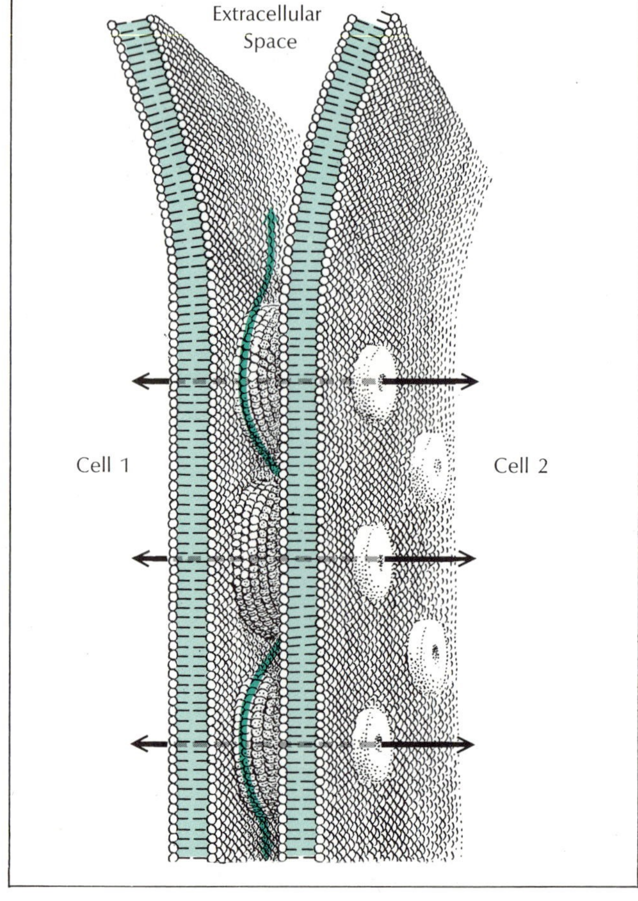

The gap junction not only connects exterior membranes of cells but permits passage of certain molecules into their cytoplasms through channels that form a mosaic of circular units when seen three-dimensionally (right). There is also an extracellular channel (color) through which substances can move through the extracellular space without entering cells. Protein layers on the inner and outer surfaces of the membrane bilayers, shown as fine stippling in the diagram at left, are omitted in the diagram at right.

tate form, never linear or circumferential as are some desmosomes (zonulae adherens) and tight junctions (zonulae occludens), though their diameter ranges from 0.1 to 10 μm; this is of course wholly consistent with their function, which is to permit direct communication between the cytoplasm of adjacent cells, enabling them to act as a sort of syncytium. Perhaps the most interesting example of this "group action" is that of certain specialized nerve cells, which we will discuss later. But gap junctions occur in other types of tissue as well – most notably in smooth (involuntary) muscles and in cardiac muscle.

As has been known for a long time, in voluntary (striated) muscle each cell has its own nerve input at the neuromuscular junction or motor end plate, whereas in smooth muscle this is not the case. Instead, the cells are electrically coupled to one another through gap junctions, through which a flow of potassium ions successively sets off contraction in each cell. The result is not the rapid contraction of striated muscle cells, but a slow, sustained wave of contraction passing through the smooth muscle tissue; intestinal peristalsis is just such a wave. So is the contraction of the myocardium, though here the picture is complicated by the existence of a second "communication network," the specialized muscle cells of the heart's own conductive tissues. It is worth noting, incidentally, that heart muscle cells are also connected by specially large and dense *desmosomes* – large enough to show up under the light microscope as the so-called intercalated discs. This peculiarly strong system of cell-to-cell "bracing" is no more than one would expect, considering the powerful and continuous mechanical stresses to which the myocardium is subjected.

Especially interesting is the finding that gap junctions appear to serve some special function during embryonic life, at least in certain species. In the fundulus, or killifish, for example – a small, shallow-water marine fish much used in biologic work – such junctions are present in the blastula or "hollow-sphere" stage of embryogenesis, but disappear as the embryo enlarges and differentiates. It is thought that this primitive intercellular communication apparatus controls or synchronizes differentiation in some manner. Further evidence comes from some work that my student Joe Keeter has been doing on the axolotl, a large salamander found in Mexico. In their early larval stages, before hatching, and at just about the time that myofibrils begin to develop in the somatic muscle tissue, gap junctions form between adjacent muscle fibers, and even between the adjacent embryonic muscle segments or somites. These junctions persist until neuromuscular synapses have become well differentiated. During this early period, then, striated muscle cells are electrically coupled, and in fact one can observe some movement in these embryos long before there is any neuronal control over the cells (i.e., in the absence of neuromuscular synapses); significantly, the motion is thought to play some role in guiding the further differentiation of the muscle cells. Similar findings have been reported in studies of the development of the retina and optic nerve tract in the frog.

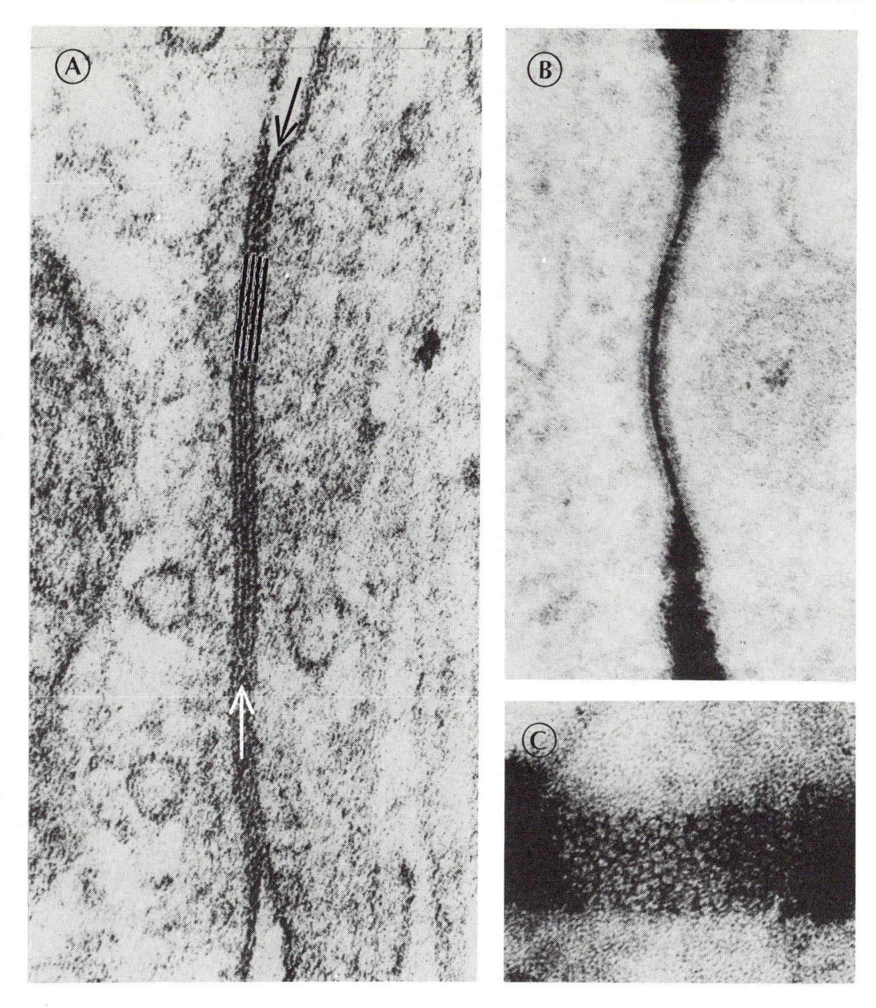

Gap junctions between neurons in the axolotl salamander are seen in these EMs (x220,000). In A, two unit membranes (two bilayers each or four dense lines) can be seen separated by a gap of about 20Å (arrows). This space represents the extracellular channel; the intercytoplasmic channels are not visualized. B shows a similar gap junction prepared with electron-opaque lanthanum, which has penetrated into the narrow extracellular space but has not crossed the apposing, unstained, cell membranes. When the section plane is rotated 90° (C), one sees a mosaic of roughly circular areas into which the extracellular lanthanum has not penetrated. These facets outline an array of interlacing channels that connect the cytoplasms of the two cells.

The putative role of electrical coupling through gap junctions in the differentiation of axolotl muscle may not exhaust the role of these structures in embryogenesis. They could well be channels through which various small molecules or ions could pass to members of a particular group of cells, thereby ensuring that all of them differentiate in the same sense. The molecules might be the mysterious "inducer substances" that have been postulated by

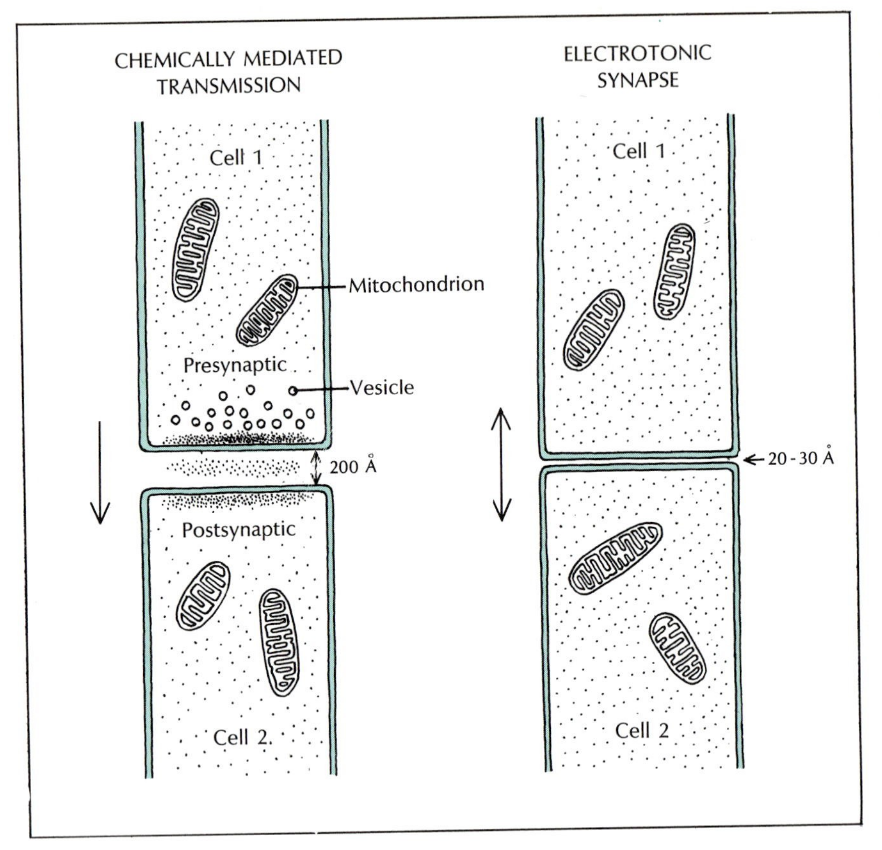

Difference between the gap junctions of the electrotonic synapse and that of the more typical, chemically mediated synapse is schematized in the drawing; electrotonic gap is much narrower and signal may move in either direction. Electron micrograph below (x13,500) shows two axons, A_1 and A_2, forming a synaptic contact with a dendrite taken from the oculomotor nucleus of the frog. Increased cytoplasmic density in the axons labels the site of release of transmitter substance from associated synaptic vesicles.

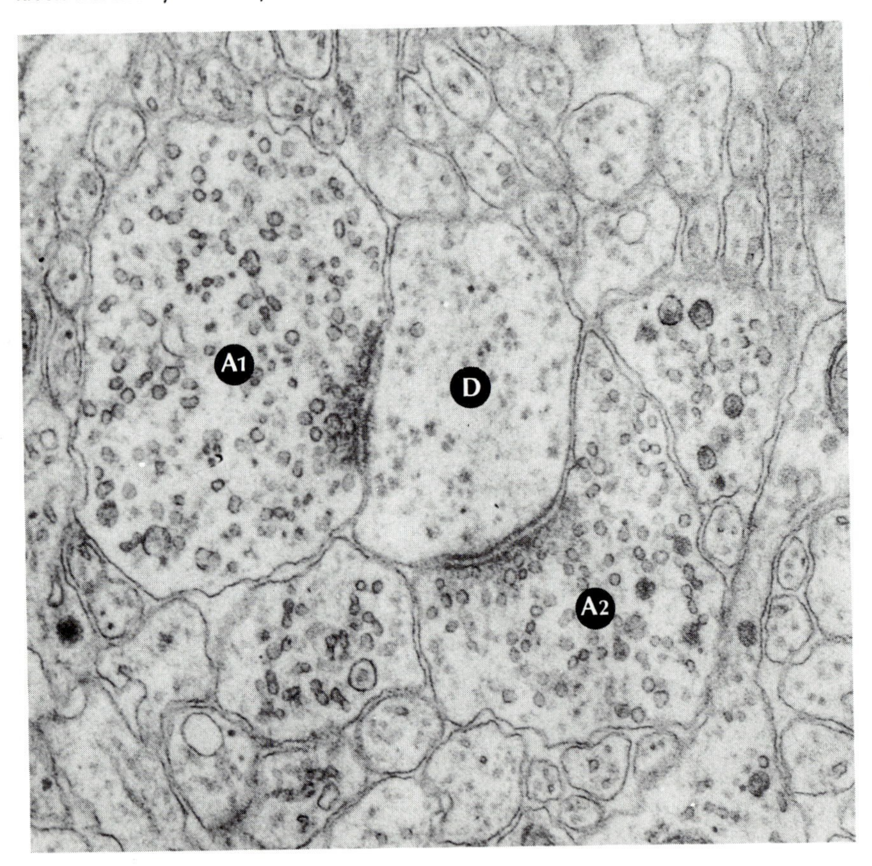

some embryologists and/or substances that repress or de-repress particular genes. Thus far there is no direct evidence supporting this possibility, but that is not surprising; there is hardly more direct evidence on the nature of the "inducer" or other differentiation-controlling substances, though their existence has been postulated on what seem to be very persuasive grounds.

Many experiments have established that the gap junction in some species acts as a special sort of synapse that plays a part in the neurologic control of certain organs. As is well known, the typical synapse that functionally connects neurons, or a neuron and its receptor cell, is chemically activated. That is, electrical activity in the membrane of the initiating cell releases transmitter molecules such as acetylcholine; these cross the synaptic gap to the membrane of the receiving cell, where they launch events that set off another wave of electrical activity, thereby transmitting a signal from one cell to the next. In what my colleagues and I have called the electrotonic synapse, by contrast, the cells are electrically coupled through a gap junction, through which the signal can be transmitted (as a surge of potassium ions) without the need of a chemical intermediary.

Much of the evidence for the existence of electrotonic synapses has come from studies of various electric fish. These animals – they include the electric catfish of Africa, the electric eel, the torpedo, and several other species – are well known for their ability to deliver electric shocks, often powerful enough to stun or kill a small or, in some cases, even a large animal. (In addition, some species can generate weak but continuous electric fields whose interaction with nearby objects can apparently be perceived by the fish, thereby presumably helping it to orient itself in its environment – often turbid waters – and perhaps to find prey.) And electrotonic synapses play a role in the control of the fishes' electric organs.

The electric organs are composed of modified somatic muscle cells, specialized by evolution to produce an electric discharge. These individual electrocytes of the electric organ are innervated typically by chemically transmitting synaptic inputs from spinal motor neurons. The electrocytes are not coupled to each other, yet they all fire synchronously. When my colleague, Michael V. L. Bennett, examined spinal motor neurons that control electric organ discharge in the South American mormyrid fish, he found that these neurons are electrically coupled; the stimulation of one neuron caused all the others to fire. As far back as 1887, Fritsch suggested that because of the fact that these neurons are closely bunched together, they may be single, large multinucleated cells forming syncytial masses. We showed with the electron microscope that this was not the case, but that the cells, thanks to the gap junctions connecting them, *acted* as syncytia.

We find a somewhat different arrangement in the electric eel. These large fish – they grow up to four feet – can generate a discharge of as much as 500 volts. The cells of the electric organs are controlled by motor neurons that are not immediately adjacent to each other and have no apparent direct connections. In other words, the neurons are not coupled directly to one another but, rather, indirectly, through a "pre-fiber" or third element that forms gap junctions with both.

The presence of electrical coupling by means of gap junctions in the control systems of these and of other electric fish suggests that one of the functions of this system is the precise synchronization of the electric cells' discharge. The chemical synapse is intrinsically slower than the electrotonic synapse, since the molecules of transmitter must be released and diffuse across the synaptic gap – about 200 Å and as much as 1,200 Å in the neuromuscular synapse – before they can trigger the receptor cell. The resulting delay is about one millisecond, depending on many factors, one of which is temperature. In the electrotonic synapse, in which the intercellular gap junction produces what is almost literally a short circuit, the much shorter latency of action means that there is less "margin for error," and hence, potentially, a much finer synchrony. On the face of it, a difference one way or another of only a fraction of a millisecond would not seem of much importance, but when we note that the duration of the discharge itself is of the same order of magnitude, it becomes obvious that

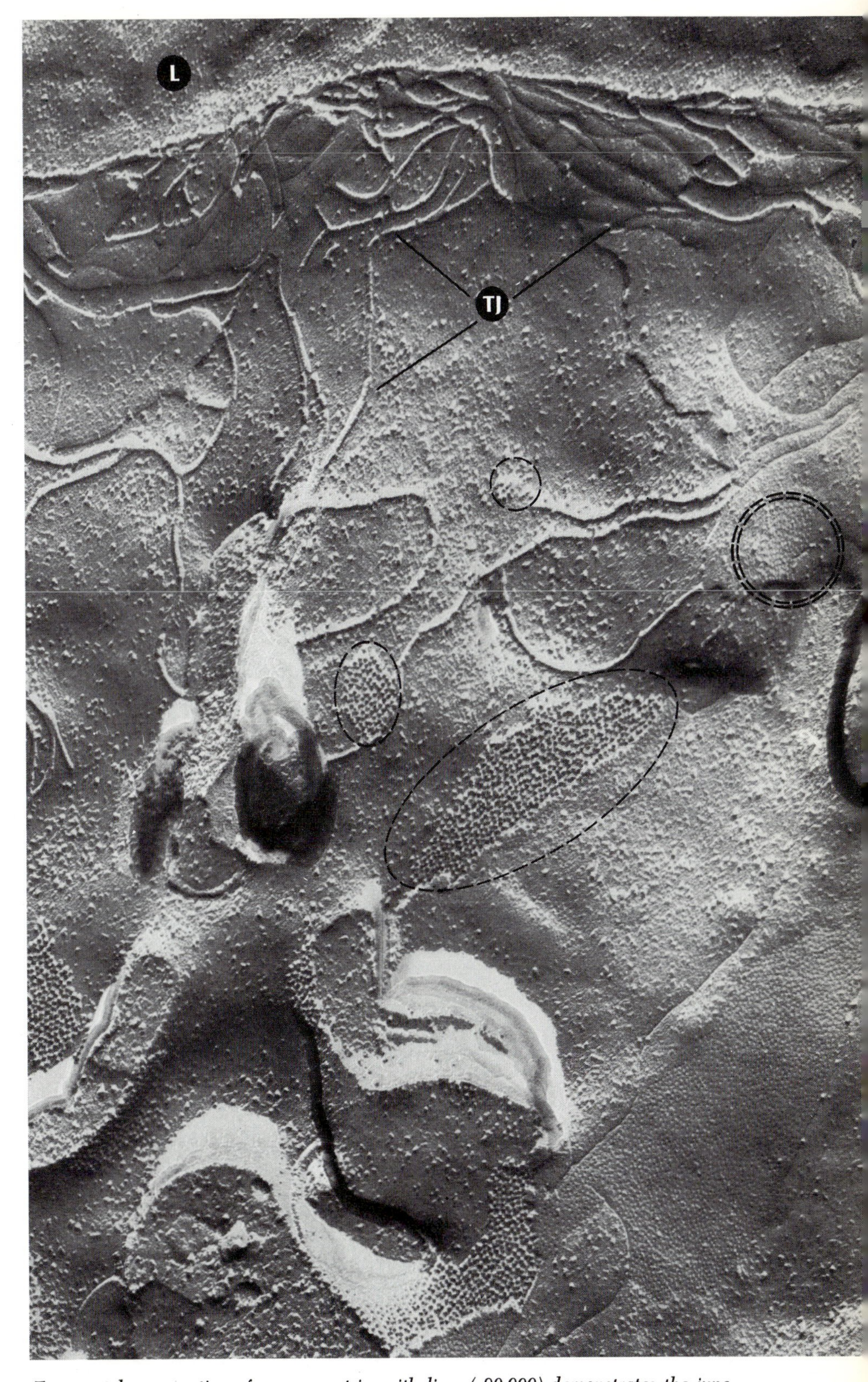

Freeze-etch preparation of mouse gastric epithelium (x90,000) demonstrates the junctions between two adjacent cells lining the lumen (L). Tight junctions (TJ) sealing lumen off from the adjacent extracellular space can be seen as a tight woven network of either ridges or depressions, depending on the exposed membrane faces. Below the tight junctions, small and large areas of gap junctions can be seen as either bumps (circles) or pits (double circle). These globular arrays, which are distinct modifications in the structure of the plasma membranes, represent the area of hydrophilic channels connecting the adjacent cells at these sites (micrograph courtesy of Dr. Jean-Paul Revel, California Institute of Technology).

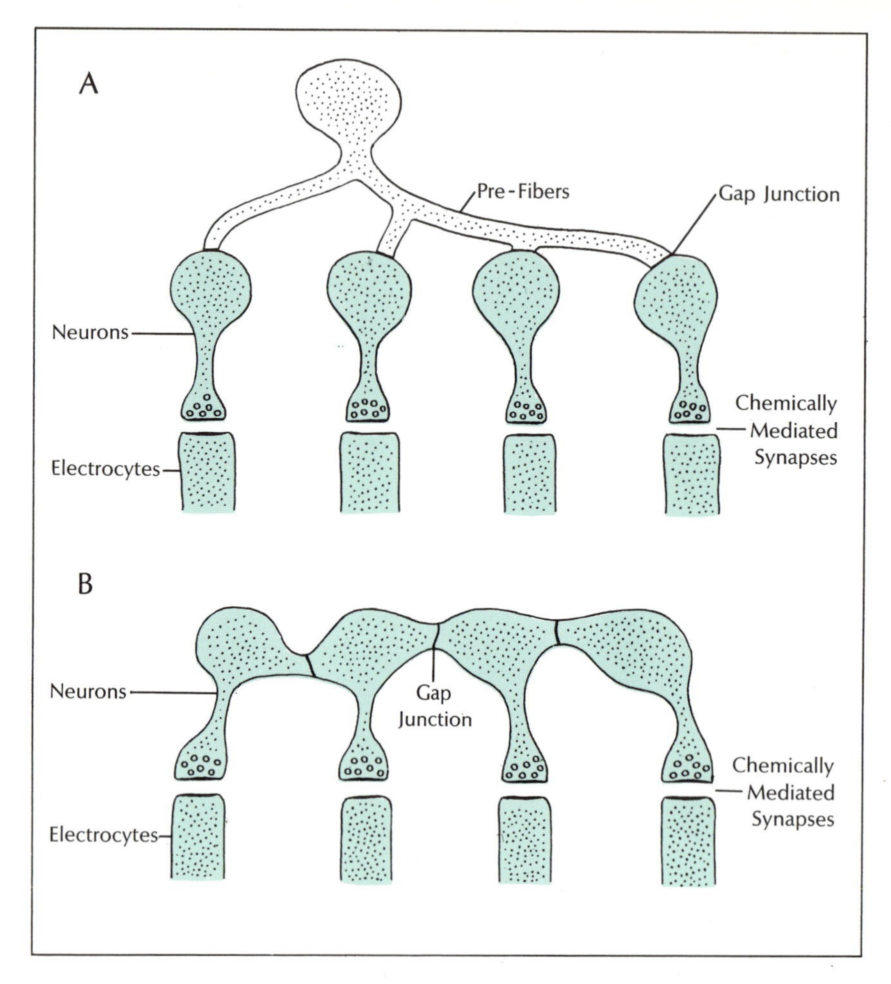

The synchronization of the electrical discharges produced by certain fish is accomplished by electrotonic synapses. Neurons may be coupled indirectly through a "pre-fiber" (A, as in the electric eel) or directly (B, as in the mormyrid fish of South America).

imperfect synchronization could drastically cut the peak voltage and current, with one group of cells having already shot its bolt by the time another group was ready to fire.

Another functional aspect of the electrotonic synapse may be related directly to its speed of action, which could make for a more rapid defensive or escape reaction by the animal possessing it. It is perhaps significant, therefore, that we find such synapses not only in the electric organs of fish (whose discharges in some cases probably serve a defensive function) but also in the crayfish, where the control of the powerful tail muscles is mediated electrotonically rather than chemically. It is through a sharp contraction of these muscles that the animal characteristically flips itself backward in case of danger (the same phenomenon has been seen by most people who have cooked a live lobster).

As already noted, the electrotonic synapse is *not* characteristic of the mammalian nervous system. Gap junctions have recently been observed in a few localized areas of the mammalian CNS, including retina, cerebellum, and vestibular nucleus, but we do not yet know what functions they serve there. In general, however, mammals rely overwhelmingly on the chemical synapse. This, though slower than the electrotonic synapse, can inhibit as well as facilitate cell activity; moreover, it seems far more capable of modification as a result of prior synaptic activity. And the modification of nerve activity as the result of prior activity is one crude, functional definition of learning – which is, of course, a much more complicated process than indicated by this definition. One might say, therefore, that in largely abandoning the all-or-nothing electrotonic synapse for the chemical synapse, we mammals have sacrificed a degree of speed for selectivity and plasticity – which is, of course, consistent with what we know of our actions.

For all that, I believe it would be distinctly premature to categorize the electrotonic synapse as merely an interesting but relatively unimportant evolutionary survival in mammals and man. It is only about 10 years since its existence was discovered, and another 10 years may well expand its known role in the mammalian CNS. One can be certain that as our knowledge of the physiologic significance of other, nonsynaptic gap junctions expands, their role in both embryonic and adult life will help our understanding of cellular communication.

10

Ionic Transport Across The Plasma Membrane

JOSEPH F. HOFFMAN

Yale University

As is well known, the ionic composition of cells differs markedly from that of the extracellular fluids in which they are bathed, especially in regard to sodium (Na^+) and potassium (K^+) ions. Virtually all cells, whether they are free-living or form part of a larger organism, are low in Na^+ and high in K^+, with concentrations of 10 to 20 μmoles/ml for the former as compared with 90 to 100 for the latter. Extracellular fluids, by contrast, contain these ions in quite different proportions – for normal plasma, some 140 μmoles/ml of Na^+ and 4 to 5 of K^+. It is probably significant that this proportion approximates that in seawater, in which all life is believed to have originated, though the actual concentrations in today's oceans are considerably greater than those in plasma.

Assuming that these ions are in any sense free to cross the cell's exterior (plasma) membrane, it is apparent that maintaining the existing state of ionic imbalance between cell and medium should require considerable energy. Otherwise the forces of ordinary diffusion would rapidly equalize concentrations on both sides of the membrane. For many years, indeed, it was thought that the membrane was impermeable to ions having a positive charge (cations) and that the intracellular concentrations of these ions were fixed at the time of maturation, with the cell thereafter functioning as a closed compartment so far as they were concerned. In the late 1930's, however, experimenters demonstrated that, for example, if an animal were fed a low-potassium diet, the concentrations of potassium ions in the tissues would eventually drop. Thus it could and did "leak" out of the cells. Soon after, some of the first experiments with radioisotopes revealed an active ionic exchange between cell and medium, with radiopotassium and radiosodium being used to trace the rapid movement of the ions across the cell membrane.

We now know, in fact, that the mere maintenance of the ionic imbalance with its medium accounts for some 25% of the cell's basal energy expenditure. One would expect that a process so expensive in metabolic terms would subserve important physiologic functions, and this has in fact been demonstrated. For example, ionic transport is the primary way in which the cell can regulate its water content, in other words, its volume. Water is known to diffuse quite rapidly across the plasma membrane, and it is not, so far as anyone knows, subject to any form of active transport. The movement of water into or out of the cell appears to be wholly governed by osmotic forces, which in turn are dependent on ionic concentrations on either side of the plasma membrane. As is well known, cells placed in hypotonic media will swell, gaining water to the point of disruption, if the process continues long enough. More recently it has been shown that similar results can be produced by drugs that impair the cell's capacity to regulate the transport of ions across the plasma membrane.

The cellular requirement for ionic transport is relatively unselective with respect to water movement since, from this standpoint, it is the *total* ionic concentration that counts – one ion is as good as another. Much more selective is the process of protein synthesis, which is carried on at least intermittently by nearly all cells. Here a relatively high K^+ concentration appears to be an essential requirement. In fact, it has been shown that several types of cells fluctuate between a high-potassium phase, corresponding to cell growth and division (which of course involves protein synthesis), and a low-potassium phase when the division cycle ends. Even more selective in ionic terms is cellular excitability, which is of central importance in the two types of cell – muscle and nerve – that make up much of the body's cellular mass. Here the re-

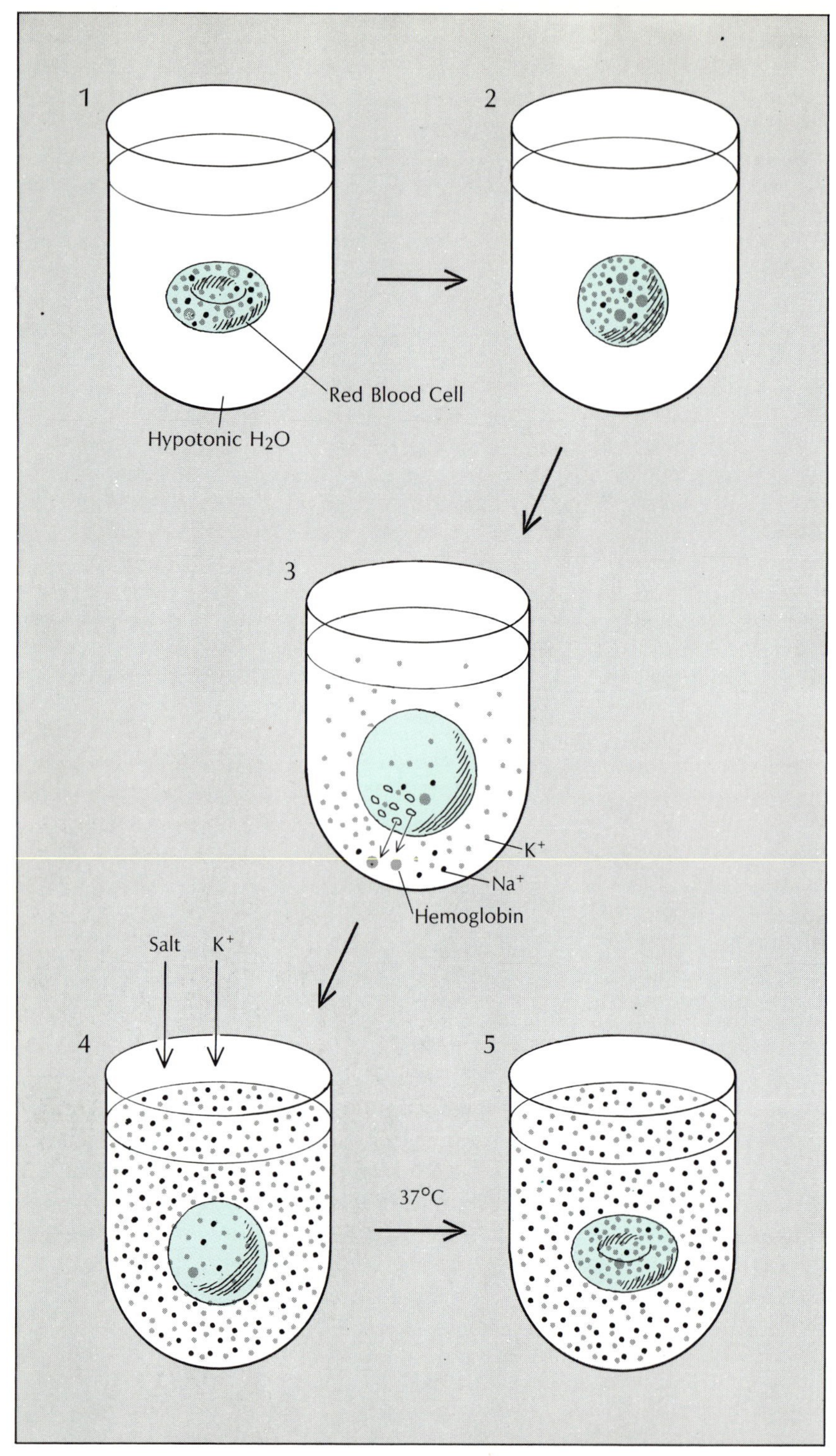

Red cell "ghosts," which greatly simplify the study of cation transport across cell membranes, are prepared as shown schematically above. When erythrocytes are placed in a hypotonic medium, water passes rapidly into the cell, causing it to swell (2) and resulting in leakage of cations and hemoglobin through stretched pores of the membrane (3). Substances (such as MgATP) present in the hemolysis solution can diffuse inside the ghost during the time the membrane is permeable to hemoglobin. Resealing of the membrane is possible after the addition of salt to the hypotonic medium, allowing the internal ionic composition to be varied by the choice of the particular salts added (4). By incubation of the suspension at 37° C for a short period of time, the ghosts regain their original low permeability to cations and also their original shape (5), but they are now hemoglobin-poor. The activity of the pump can now be studied as a function of these altered environments.

quirement is that the cell interior be high in K^+ and low in Na^+. This sets the stage for the rapid, sequential flow of ions across the membrane, which underlies, for instance, depolarization when the cell is appropriately stimulated.

These and many comparable findings have forced us to infer the existence of an ionic pump within the plasma membrane, which, through the expenditure of energy, transports K^+ into the cell and Na^+ out of it, in both cases, of course, against its electrochemical gradient. The nature and operation of this pump is the chief concern of the balance of this chapter.

To set the stage for discussion of the pump, however, we must first consider what the flow of cations would be under purely passive conditions, i.e., if no energy were being expended on ionic transport. In this situation, ion movement would be governed by two different forces. The first is simple diffusion, which, of course, tends to equalize concentrations of a given ion on both sides of the membrane. The second is electrical potential – the difference in electrical charge across the membrane. Cells typically contain not only cations but also anions (notably, Cl^- and HCO_3^-) whose negative charges may or may not balance the positive charges of whatever cations are present. A slight excess of negative charges within the cell (which exists in most animal cells) creates a potential that tends to drive cations into the cell and/or anions out of it.

It is possible to calculate the flux of a given ion under passive conditions, knowing its concentration and the potential difference across the membrane, and taking into account the membrane's relative permeability to that ion. The plasma membrane is, in fact, somewhat less permeable to Na^+ than to K^+ (presumably because of the former's greater size) and, for some cells at least, far more permeable to anions than to either cation. This difference is thought to reflect in part the existence of fixed positive charges on the membrane so that it acts to repel cations even as it attracts anions. In the erythrocyte, for example – the cell with which much of the work on ion transport has been done – the half-time for anion permeability is on the order of a few hundred milliseconds, while that of cations is measured in hours.

In the erythrocyte as well as in some other types of cells, the concentration and movement of anions appear to be controlled only by passive forces. This is the case, for example, in the physiologically crucial process whereby the erythrocyte, in passing through the alveolar capillaries, gives up HCO_3^- in exchange for Cl^-.

For the situation with cations, this is far from the case. Neither their concentration in the cell nor their movement into and out of it – which can be measured by using radioisotopes – is what would be predicted from the passive forces alone; the only way to balance the equation is to add some energy-dependent process, which, operationally speaking, is what is meant by active transport.

One question concerning active transport of cations is whether it involves both sodium and potassium or only one of them. Conceivably, the pump might actively transport only Na^+, with K^+ flowing in passively to maintain electroneutrality within the cell. In some types of cells this may in fact occur, but not in the erythrocyte. Because of the high anion permeability of its membrane, the loss of positive charge produced by the efflux of Na^+ would be balanced by an equal efflux of anions, rather than by the much more sluggish influx of K^+ that has been observed. It is necessary, therefore, to infer a pump (or pumps) that transports both cations, though, of course, in opposite directions.

In the erythrocyte membrane at least, we are evidently dealing with a single pump, in which the outward transport of Na^+ is coupled to inward transport of K^+. This has been shown by a number of experiments involving either intact cells or what are called reconstituted erythrocyte ghosts – cells that have been hemolyzed by exposure to a hypotonic medium, with a momentary increase in the permeability of their membranes before being resealed, with full recovery of pump activity. By suitably altering the composition of the hemolysis medium under these conditions, the ionic content of these ghosts can be manipulated, and the subsequent operations of the pump studied. It turns out that if the cell interior is depleted of sodium the pump will not work – that is, Na^+ *must* be pumped out if K^+ is to be pumped in. Equally, if a cell with normal sodium content is placed in a potassium-free medium, the pump still does not work; unless K^+ can be pumped in, Na^+ cannot be pumped out. The reverse is not true, however: potassium depletion *inside* the cell does not inhibit the efflux of Na^+, nor does sodium depletion in the medium inhibit the influx of K^+. Clearly, the Na^+-K^+ pump is one of the many asymmetric features of the plasma membrane, whose outside – as several earlier chapters have made clear – possess quite different properties from its inside.

Interestingly, the same asymmetry obtains even in species whose erythrocytes have a quite different ionic composition from those of the human cell. For instance, the erythrocytes (but not other body cells) of some sheep and goats are high in Na^+ and low in K^+. Thus, different sheep or goats contain erythrocytes either rich in Na^+ or in K^+. The two forms appear to be controlled by a single gene that segregates in simple mendelian fashion. One might expect that in the reversed cell, the ionic pump would also be operating in reverse, but this is not the case. In experiments performed with Daniel C. Tosteson, we have determined that the pump still functions in the normal sense, but with a much reduced rate of ionic transport. The major change is in the leaking of cations across the plasma membrane, which is much higher in the high-Na^+ erythrocyte. The result is that whereas in the normal cell the pump is able to maintain an ionic imbalance against the passive forces that cause cations to leak across the plasma membrane, in the reversed cell the increased leakage (and reduced pumping) produces an intracellular ionic content much closer to that obtaining in the medium.

What evolutionary function (if any) is served by the reversed erythrocyte in the species (or individual animals) that possess it is unclear; it

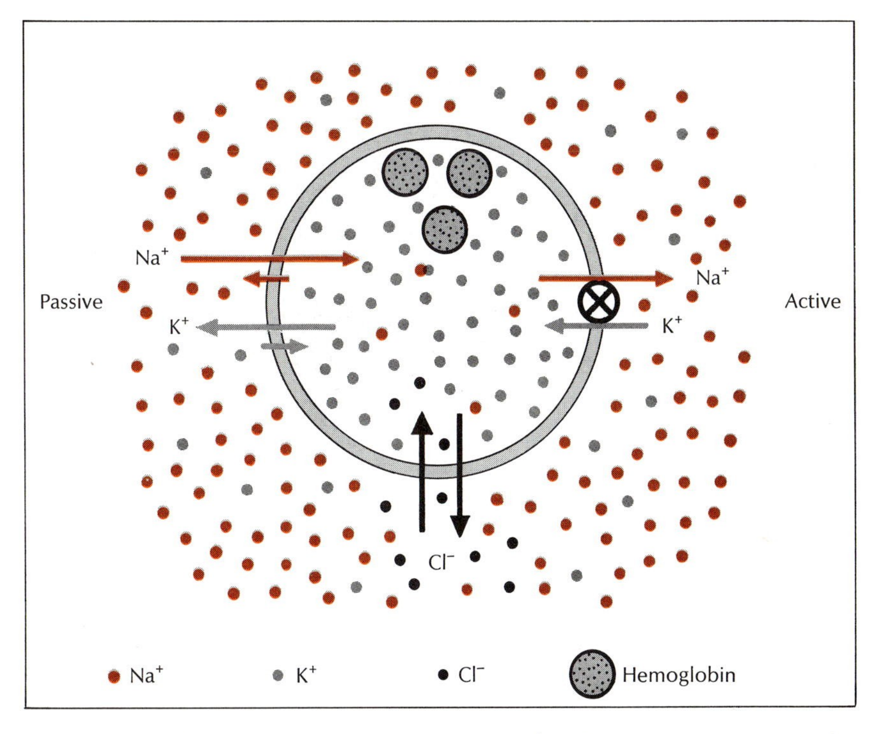

Diagram of erythrocyte summarizes the primary modes of cation transport across the cell membrane. Under conditions of passive transport there is an excess of sodium entry into cell over exit and an excess of potassium exit over entry (left); action of Na^+-K^+ pump (right) simultaneously overcomes both imbalances, producing a K^+-Na^+ ratio inside the cell of 10:1 (outside the cell, this ratio is 1:20). Maintenance of chloride or bicarbonate at equilibrium (less inside than out) does not require energy expenditure, the distribution ratio reflecting the cell's membrane potential: negative 6 to 8 millivolts inside relative to the outside.

may well be that we are dealing with one of the many dimorphic and polymorphic traits that do not confer any known functional advantage on the genotype over others (a common example in man is straight vs wavy hair). The mature mammalian erythrocyte, it should be remembered, is neither an excitable cell nor does it carry on protein synthesis—both processes in which a high K^+ content is critical. In this connection, it is probably significant that the immature erythrocyte – which does carry out synthesis – is invariably high in potassium regardless of the state of the adult cell in the species or individual animal; the low-potassium state of the reversed cell appears only as the cell matures.

Another significant aspect of this ionic dimorphism in the erythrocyte is its apparent control by a single gene. This implies that it is due to changes in a single macromolecule or macromolecular complex in the membrane. As already noted, the dimorphism affects *both* pumping and leakage, suggesting that these processes, which we tend to think of as separate, may in fact be interrelated.

Turning now to the actual operation of the pump, we note first of all that though, as stated earlier, the transfer of Na^+ and K^+ are obligatorily coupled, in the sense that neither can proceed without the other, the coupling is not (or not necessarily) on a one-to-one basis. Specifically, in the erythrocyte – and perhaps some other cells—the pump appears to move three sodium ions for every two potassium. Along with Na^+ inside and K^+ outside, another requirement for the pump's operation is a source of energy—specifically ATP, the cell's universal medium of exchange for energy transfer and utilization. Moreover, ATP is required *inside* the cell; if the cell interior is depleted of ATP, the addition of high-energy compound to the medium will not induce the pump to operate, indicating another asymmetric feature of the membrane. Quantitative experiments on erythrocytes and nerve cells have shown that one ATP molecule is expended per 3 Na^+ and 2 K^+ pumped.

A further requirement for the pump's operation is the presence of magnesium ion (Mg^{++}) within the cell. This seems reasonable, since magnesium catalyzes the hydrolysis of ATP (to ADP plus inorganic phosphate)—the process whereby it gives up its energy. The presence of Mg^{++} would appear to be a necessary concomitant of the pump's use of ATP as an energy source. However, it should be remembered that in the in vitro hydrolysis of ATP other ATPases of the membrane can be catalyzed by replacement with divalent cations such as calcium (Ca^{++}). The pump, however, will *not* function if intracellular Mg^{++} is replaced by Ca^{++}. We are forced to conclude either that Mg^{++}, along with catalyzing ATP hydrolysis, also participates in the pump's operation in some other manner or that Ca^{++}, though it can hydrolyze ATP quite normally, inhibits the pump's operation at some other point in the process. Leaving aside the reasons for the moment, it appears that the latter is the case.

The apparent ability of Ca^{++} to block the pump's action recalls an earlier theory, which held that the Na^+-K^+ pump can also pump Ca^{++}, and in fact does so exclusively when intracellular levels of the latter ion are high, shifting back to Na^+-K^+ transport when Ca^{++} has been reduced to its normal low concentration. Though this theory would obviously explain how Ca^{++} "blocks" the Na^+-K^+ pump – by taking over its mechanism for another purpose – recent studies indicate that we are in fact dealing not with a single pump but with two. To understand the evidence for this, we must consider the pump's operations in somewhat more detail.

By labeling the third, or terminal, phosphate group in ATP, one can demonstrate that the pumping process proceeds in two steps. In the first, stimulated by intracellular Na^+, the terminal phosphate is detached from ATP (converting it into ADP) and binds to a membrane protein, forming a phosphorylated intermediate. If K^+ is not present in the medium, the process stops there, and the labeled intermediate can be separated from a solubilized preparation of the membrane by acrylamide-gel electrophoresis as a phosphoprotein weighing some 103,000 daltons. If, however, K^+ is present in the medium (or added to it), the phosphate is detached from the intermediate (becoming inorganic phosphate), which can still be identified by electrophoresis as a protein of the same weight, and in the same quantity per mg of membrane, lacking only the phosphate label. At this point the pump has returned to its initial state, ready for another cycle.

If one does the same experiment with Ca^{++} present, the first stage of the process proceeds unchecked; one then can recover the same phosphorylated intermediate of 103,000 daltons. But the second stage is blocked; even with K^+ in the medium, dephosphorylation cannot take place. At the same time, however, one finds a second phosphoprotein present. This one has a weight of about 150,000 daltons and appears to be identical with or akin to the enzyme calcium ATPase – the Ca^{++} pump, in fact. Thus along with setting its "own" pump into action, Ca^{++} also blocks the dephosphorylation step of the Na^+-K^+ pump, though we do not yet know how it does this.

The spatial and temporal aspects of the pump, whereby phosphorylation and dephosphorylation are linked to the reciprocal transfer of Na^+ and K^+, have not yet been worked out, apart from the fact that the pump does *not* transport either of the breakdown products of ATP (ADP and inorganic phosphate). Both of these remain within the cell, presumably to be recycled into ATP. The "simultaneous" model of pump action visualizes the Na^+ and K^+ as binding simultaneously to opposite sites on the pump molecule (or molecular complex). ATP breakdown, by phosphorylating the pump protein, changes its conformation in such a way that the pump protein shifts the ions across the membrane in opposite directions, the K^+ dephosphorylating the protein as it passes through. A "sequential" model has also been proposed, in which only a single type of site is involved. This site first accepts Na^+, then (presumably by becoming phosphorylated) "flips" to the outside of the cell and releases its Na^+, where, because phosphorylation has altered its configuration, it accepts a K^+ in exchange. The two models are very difficult to distinguish experimentally, and so far there is no very persuasive evidence in favor of either. Nor does either model explain how

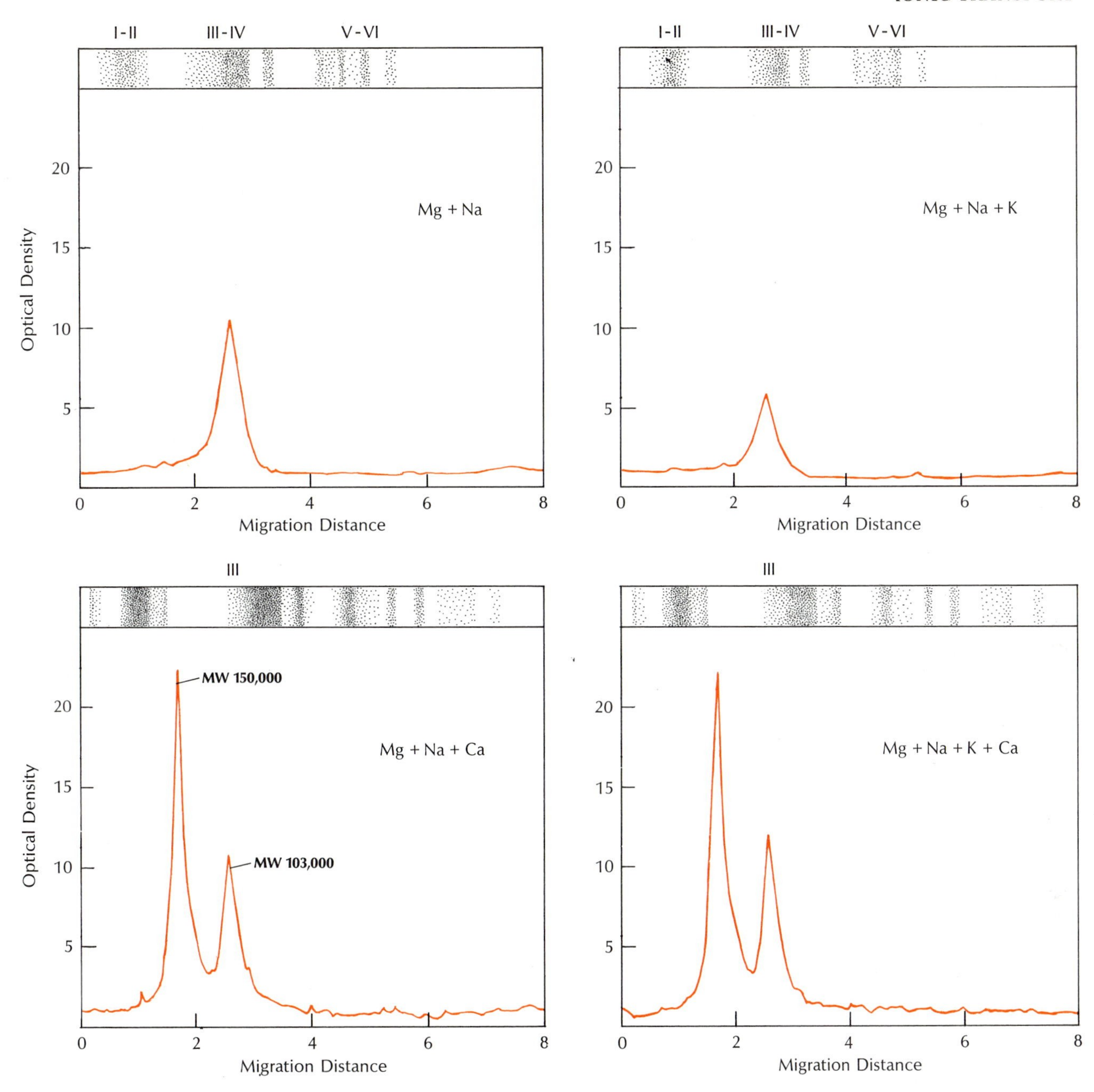

Densitometer traces of autoradiograms of gels on which phosphorylated red cell ghosts have been electrophoresed help demonstrate that the Ca++ pump is not the same as the Na+-K+ pump and that the action of the latter proceeds in two steps, activated by Mg++. As seen at top left, when labeled terminal phosphate is detached from ATP under stimulus of intracellular Na+, it binds to a membrane protein, forming a phosphorylated intermediate weighing about 103,000 daltons. With K+ added (top right), the phosphate is detached from the intermediate (becoming inorganic phosphate), the radioactivity remaining in the peak representing the baseline level of phosphorylation. The pump is now ready for another cycle. When experiment is repeated with Ca++ present (lower left and right), dephosphorylation does not occur even when K+ is added. A second phosphoprotein is found, weighing about 150,000 daltons, which appears identical with calcium ATPase, i.e., the Ca++ pump.

the pump transfers, not (as one might expect) a single ion of each species, but rather three of one and two of the other.

Further clues to the pump's action – not all of which, however, can yet be clearly interpreted – have been obtained by studying the effect on it of other inhibitory substances. Fluoride (F^-), for example, blocks the pump in the living erythrocyte, which seems easily explicable, since this ion is known to inhibit glycolysis, the process by which these cells metabolize glucose to synthesize ATP. Cut off the energy supply and the pump is obviously going to stop. However, in erythrocyte ghosts containing incorporated ATP, F^- still inhibits the pump, indicating that it somehow blocks the pump mechanism directly, quite apart from cutting off its energy supply. This contrasts with the action of other glycolysis inhibitors such as iodoacetate, which can work only indirectly by controlling ATP levels.

Among the pump inhibitors that have been studied, some of the most interesting–from both the clinical and general physiologic standpoints – are the cardiotonic glycosides, including ouabain and digitalis. The effects of these substances in strengthening cardiac contraction and stepping up cardiac output are well known. It is debatable whether they do so solely by inhibiting the Na^+-K^+ pump (which they unquestionably do) or with another mechanism involved. One possibility is that they step up the transport of Ca^{++} into the cytoplasm from outside or from the sarcoplasmic reticulum (which is believed to act as a storage depot for that ion) by alteration of the Na^+ concentration gradient across the membrane, but more work is necessary to define specifically the inotropic effects of the cardiotonic steroids. Our particular interest in these compounds, however, is to use them as tools to dissect the pump apparatus rather than to study their pharmacologic action.

It is of interest, then, to consider various aspects of the way the cardiac glycosides block the Na^+-K^+ pump. In the first place, they can only do so from outside; that is, when they are present in the medium. This has been demonstrated with both erythrocyte

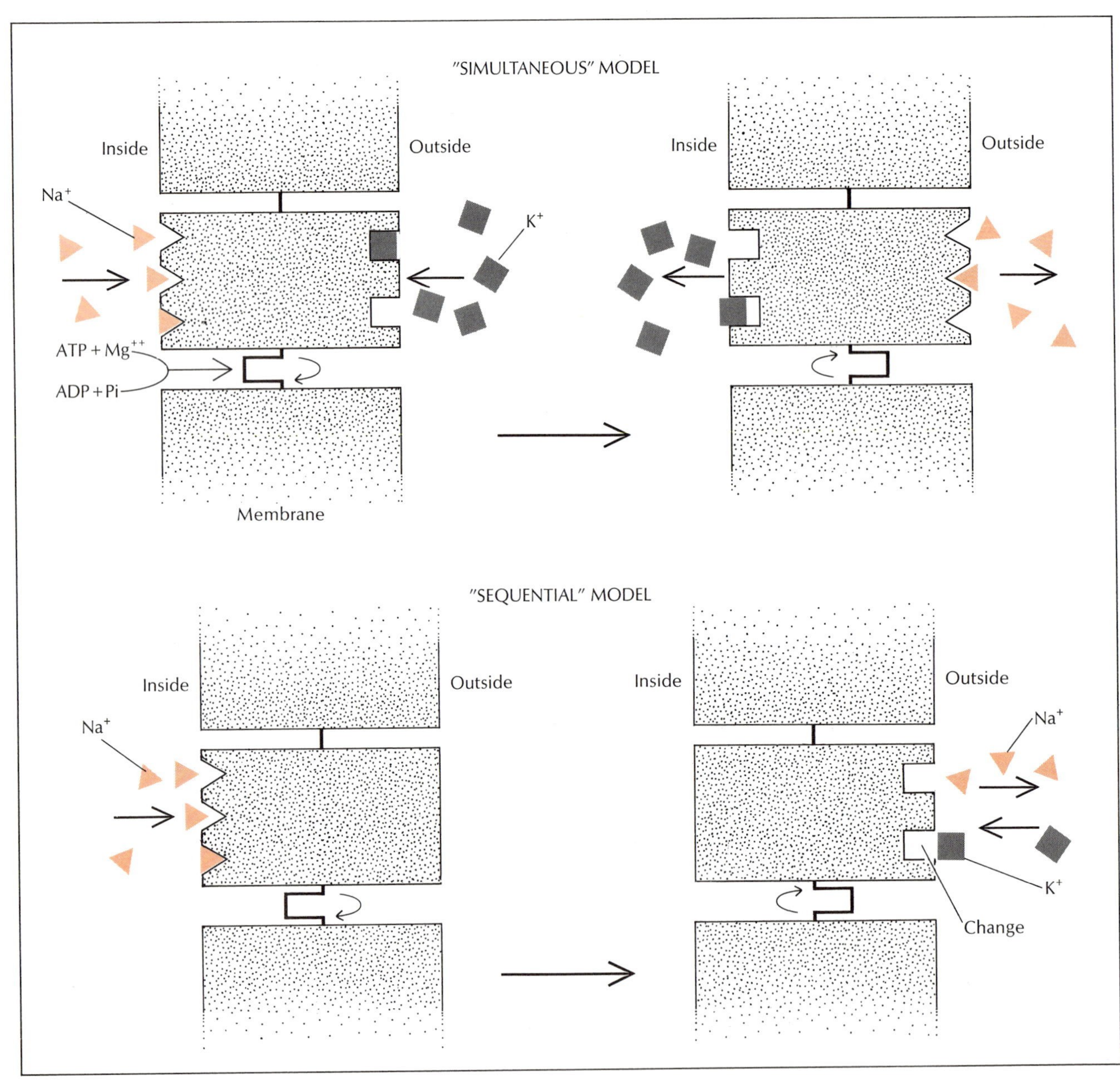

Both "simultaneous" and "sequential" models of action have been proposed for the Na+-K+ pump, as schematics suggest. In simultaneous model, Na+ binds to inside of cell membrane while K+ binds to exterior. ATP breakdown changes configuration of pump protein in such a way that the protein shifts the ions across the membrane (somewhat like a revolving door), with the K+ presumably dephosphorylating the protein as the ions pass through. In the sequential model, only one type of site is involved. This first accepts Na+ from interior of cell, then, perhaps by becoming phosphorylated, moves Na+ to the outside. Since its shape has been changed by phosphorylation, the site now accepts K+. Additional clues to the pump's action have been obtained by studying the effect of inhibitory substances, such as calcium, that block pump action by preventing dephosphorylation of the phosphoproteins. Cardiotonic glycosides inhibit the pump only from the cell's outside; it is thought that by binding to K+-sensitive sites they impede the "cooperative" action of Na+ and K+ needed for the pump to operate.

ghosts and giant axons into which cardiotonic steroids have been placed. The pump operates normally with the steroids inside but is inhibited when they are present outside.

Additional work has made clear that glycosides in the medium block the pump by interfering with the interaction of K+ on the outside surface; as noted earlier, the pump cannot operate normally unless both K+ and Na+ are being transported. The mechanism by which it does so, however, is quite complicated. A good deal of evidence indicates that activation of the pump involves the "cooperative" action of at least two K+ ions outside. Since glycoside binding is antagonized by external K+ it may be that the two K+ ions bind to separate sites on the membrane, only one of which can interact with glycosides. However, the two types of sites, though evidently distinct, are spatially very close. When glycoside-labeled membrane is solubilized and subjected to sucrose gradient centrifugation, the glycoside binding site remains linked to the pump proper, the entire complex having a molecular weight of 300,000 to 400,000 daltons.

From the standpoint of the glycosides' cardiotonic mechanism, an interesting point is that by stepping up K+ in the medium, one can reduce the quantity of bound glycoside. And, significantly, stepping up serum K+ is also known to reduce the glycosides' clinical effect – it is, in fact, a standard treatment for over-digitalization. Thus, there is evidence that glycoside binding to the membrane exterior, and the resultant interference with the Na+-K+ pump, may be at the root of the compounds' cardiotonic activity. Another significant aspect is that glycosides are bound to the cell exterior only under certain conditions – specifically, if ATP or some similar substance is present inside the cell. The important point here is that, as regards glycoside binding, other nucleotide triphosphates (inosine TP, uridine TP, and the like) seem quite as effective as ATP itself, yet these substances, unlike ATP, cannot activate or even be hydrolyzed by the pump to any significant extent. This means that they do not release their phosphate group for phosphorylation of the pump apparatus. What this

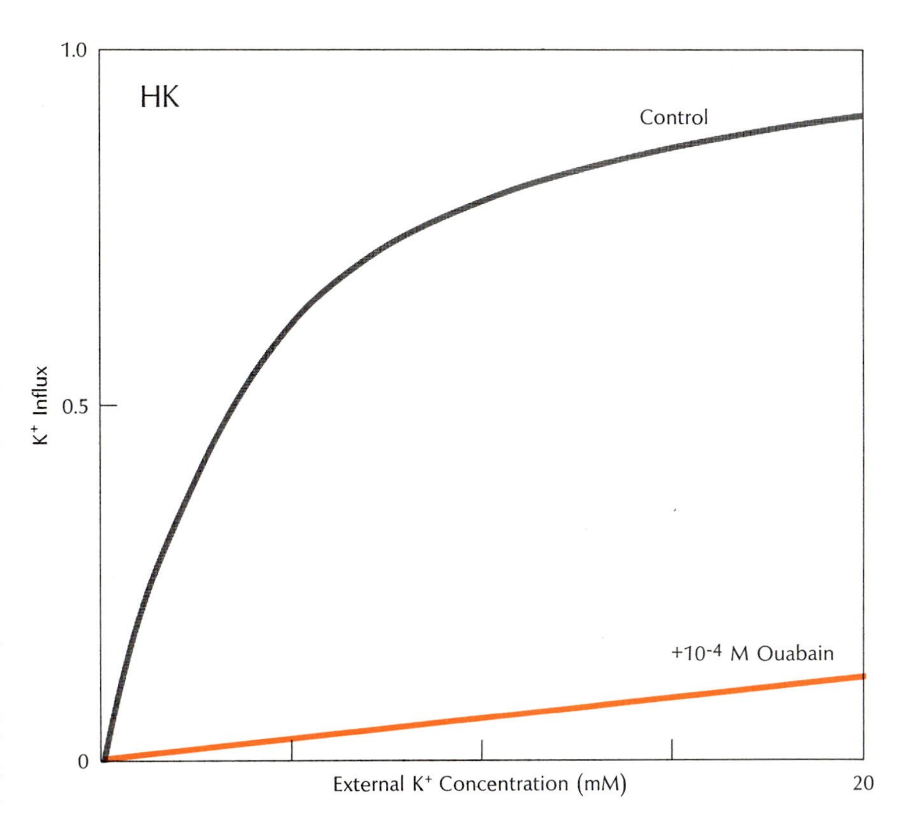

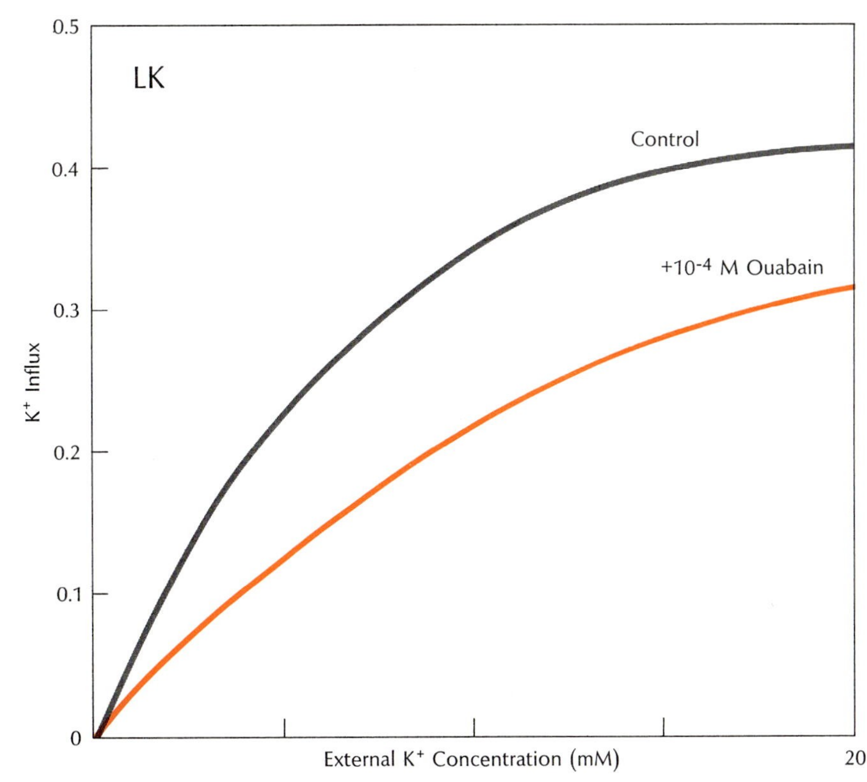

Studies of K+ influx in high-potassium (HK) and low-potassium (LK) red blood cells of sheep, utilizing ouabain as a pump inhibitor, show that the magnitude of the K+ pump flux (indicated as the difference between the control and ouabain curves) is much greater in HK cells than in LK cells. Measurement of the pump flux of K+ as a function of the external K+ concentration indicates both the activation and the saturation characteristics of the pumps in the two cell types. The differences in the rate of K+ transport of the two types of cell have been shown in other experiments to reflect primarily the number of pumps per cell rather than differences in the way the pumps operate. Relationship between number of ouabain molecules bound and percent concomitant inhibition of the pump was approximately linear for both HK and LK cells.

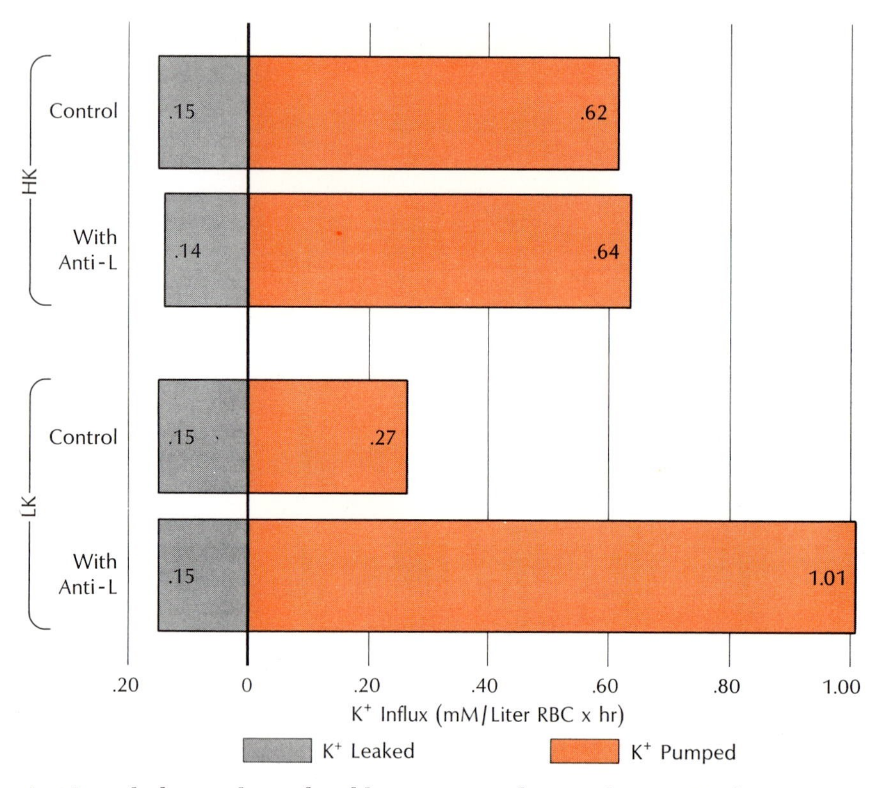

Anti-L antibodies can be produced by injecting erythrocytes from an LK sheep into an HK animal; when LK erythrocytes are treated with anti-L, their capacity to pump K+ rises substantially, as shown in graph above. Rather than uncovering or increasing the number of K+ pumping sites, the action of anti-L antibodies on LK cells is thought to be that of stepping up K+ pumping at existing sites. The basis for this effect of anti-L can be seen in the graph below. In untreated (control) LK goat red cells, inside K+ inhibits the pump influx of K+. Addition of anti-L relieves this inhibition by lowering the pumps' affinity for inside K+, thus allowing the pumps to operate at faster rate.

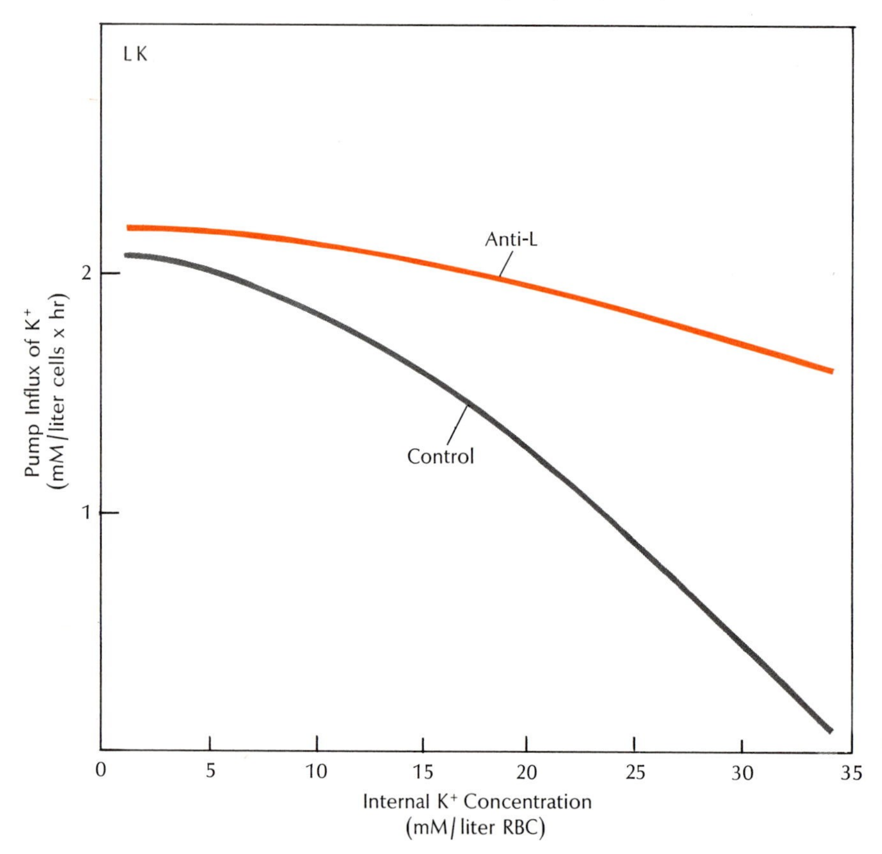

seems to imply is that ATP plays at least two distinct roles in the pump's operation. If we assume that the glycoside binding site is under normal conditions a K+ binding site, which certainly seems reasonable, it would seem that ATP, simply by binding to the membrane interior, induces a conformational change in the exterior that enables K+ (or glycoside) to bind there, which presumably sets the stage for the pump's further operations. The cycling of the pump requires a reversible phosphorylation of a membrane protein in association with the translocation of Na+ and K+.

From a different point of view, one can determine the number of glycoside molecules bound per cell – in the human erythrocyte, 200 to 300. If we then assume that each binding site corresponds to a single pump complex (which seems very plausible), we can calculate the turnover number of each pump. From the known rate of active ion transport in these cells, it thus appears that each pump must be handling some 6,000 K+ per minute or about 100 per second. These figures may appear high, but are in fact rather low as such things go; some enzymes are known to process their substrates at a rate several orders of magnitude faster.

This rate of turnover seems to be essentially the same regardless of cell type. In other words, differences in the rate of ion transport between cells reflect primarily the number of pumps per cell, rather than any major difference in the operation of the pumps. We find, for example, that in the two varieties of sheep erythrocytes mentioned earlier, the normal (high K+ or HK) cell has some 42 pumps per cell (the sheep erythrocyte is considerably smaller than its human counterpart), while the "reversed" (low K+ or LK) cell has only six to eight.

Further experiments have shown, however, that the situation can be more complicated than a mere difference in the number of pumps, since the operation of each pump is subject to modulation. For example, if erythrocytes from an LK sheep are injected into a normal, HK, animal they produce what we call anti-L antibodies. When LK erythrocytes are treated with these antibodies, the cells' pumping capacity rises to levels approximating those of HK cells. One can interpret this in two ways. Perhaps the LK cells contain occult pumping sites, comparable to the occult insulin binding sites described later in this volume by Cuatrecasas (see Chapter 18, "Hormone-Receptor Interactions and the Plasma Membrane") that are somehow uncovered by antibody action. Perhaps the more likely explanation is that the sites are not occult but altered, and the antibody acts to put the sites in proper working order. In support of this interpretation a number of fairly intricate experiments indicate that anti-L serum does not in either sheep or goat erythrocytes increase the number of active pumping sites–as measured by the number of glycoside binding sites – but rather steps up pumping at the existing sites. It does so by reducing the affinity of the pump for intracellular K^+.

In the normal (HK) cell, pumping activity appears to be regulated by the intracellular concentration of K^+ – the higher the concentration, the less active the pump. Functionally, this operates as a quite straightforward feedback system that holds intracellular K^+ within normal limits. The immediate cause of the pump's inhibition by K^+, however, is not (as one might expect) a reduction of its affinity for extracellular K^+ but rather of its affinity for intracellular Na^+. In practical terms, of course, this comes to the same thing. Since transport of the two ions is coupled, a reduction of Na^+ efflux will necessarily reduce K^+ influx, which is presumably the physiologic "point" of the arrangement.

In the goat LK cell, however, the pump's affinity for intracellular K^+ is greater than normal, meaning that it is inhibited at lower levels of K^+ ion. That is, it can pump just as vigorously as the HK pump but is turned off by intracellular K^+ more easily. Anti-L serum, it appears, reduces the pump's K^+ affinity on the inside to normal, relieves the inhibition, and thereby permits the pump to operate at faster rates.

Two other groups of experiments are not without interest in elucidating the pump's operations. In the first place, by changing the ionic concentrations outside the cell–placing it in a medium free of K^+ but containing a high concentration of Na^+ – one can force the pump to operate in reverse, moving Na^+ inward and K^+ outward. As shown by Ian Glynn and his colleagues, the effect of this is to *generate* energy, by forming ATP from ADP and inorganic phosphate; just as the normal "uphill" (against the gradient) movement of ions consumes energy, so their "downhill" movement produces energy. This type of experiment also indicates the complete reversibility of the pump and signifies in a different way that the pump's substrate is ATP.

In addition, by placing erythrocytes in a potassium-free medium, meaning that there is no K^+ to be pumped into the cell, the pump apparatus can be induced to exchange Na^+ for Na^+. Here, as one might expect, energy is neither expended nor consumed. Specifically, the outward movement of Na^+ involves the hydrolysis of ATP to ADP and inorganic phosphate, while the inward movement of the ion reconverts the ADP and inorganic phosphate to ATP. This can be demonstrated, for example, by depleting the cell of ADP; if this compound is not present to act as an acceptor for inorganic phosphate, sodium-sodium exchange will not occur.

The possibility of sodium-sodium exchange may seem to contradict what has been said earlier about the coupling of Na^+ to K^+ pumping. Note, however, that in the sodium-sodium case, actual pumping is not taking place; the mechanism is turning over, but no energy is being expended.

At this point, the reader may well feel somewhat bewildered by all these facts; if it is any comfort, the bewilderment is shared by workers in the field. On the one hand, we know a great deal about how the pump operates, both normally and abnormally (e.g., in sodium-sodium exchange) in terms of kinetics and ionic dependencies. On the other hand, many of the details of its operation are still anything but clear. Perhaps the molecular mechanism will be elucidated when the relationship between transphosphorylation and translocation becomes established. Certainly, taking the pump apart and identifying the roles of its components, using specific labels, will help. Obviously, reconstitution experiments of purified pumps embedded in lipid bilayer model systems could provide the circumstances for evaluating the critical molecular steps that are responsible for active transport.

Much as one would like to present a neat, clear, and detailed picture of how this key physiologic mechanism operates, present data do not permit it. In another few years, this situation may have changed.

11

Cellular Communication by Permeable Membrane Junctions

WERNER R. LOEWENSTEIN

University of Miami

For well over a century, biology and medicine have been greatly influenced by the Cell Theory. In one of the theory's earliest formulations, the botanist Schleiden writes in 1838: "Every higher organism is an aggregate of fully individual independent units, the cells." He uses the precise German wording "*in sich selbst abgeschlossene Einheiten*," circumscribed, self-contained units. This has been an extraordinarily productive theory. We are now aware, to be sure, that the self-containment is not absolute. In particular (as earlier chapters in this volume have discussed at length), cells communicate with and regulate other cells by releasing various signal molecules (e.g., hormones, neurohumors) that trigger receptors on the membrane of the target cell. For all that, however, we continue to think of cells as essentially discrete bodies, whose many interactions with one another do not negate their basically separate character. Indeed, the past 20 years of electron microscopy have given us the picture that all cells are completely surrounded by a membrane, and this membrane was assumed to be a continuous diffusion barrier.

It now appears that this aspect of cell theory must be revised. Cells, it is becoming evident, are typically *not* wholly separate from their neighbors; rather, large masses of cells are connected by fine channels built into the plasma membranes where the cells are joined. These "permeable junctions" enable the cells in a tissue to rapidly exchange a variety of substances, with the result that they can interact and function as a unit rather than as discrete bodies. This finding has obvious and major implications for biology and medicine, many of which are yet to be explored. In particular, it may be relevant to our understanding of growth processes, both normal (i.e., embryonic growth and differentiation, wound healing) and abnormal (e.g., cancer).

My involvement in this field began with a chance observation in 1963, when Yoshinobu Kanno and I were studying the permeability of the nuclear membrane. This entailed inserting micropipettes into the nucleus and cytoplasm of large salivary cells, passing a current through them, and measuring the resulting potential. Much to our surprise, we found that it made little difference whether we measured the potential in the same cell or in an adjacent one, despite the interposition of two plasma membranes in addition to the nuclear membrane itself. Since electrical currents within cells are carried by ions, our finding implied that plasma membranes at the cell junction interfered very little with ionic flow, which contradicted flatly the prevailing ideas of how cell membranes operated. This finding was all the more surprising, since salivary cells, like all epithelial cells, do not normally carry electrical signals.

To be sure, Silvio Weidmann at Cambridge had found in 1952 that certain heart cells are electrically coupled; and Edwin Furshpan and David Potter, then at University College, London, had elegantly demonstrated this in 1957 for certain nerve synapses. These important findings showed a mode of transmission of electrical signals from one heart or nerve cell to another. But, since the heart and nerve cells were long known to specialize in communication by electrical impulses, these investigators properly considered their findings membrane specializations adapted to electrical signal transmission. (Actually, heart muscle, since the turn of the century, had been thought of as a syncytium, and until 1951, when chemical transmission was established, synaptic transmission in the CNS was held to be electrical.) Thus

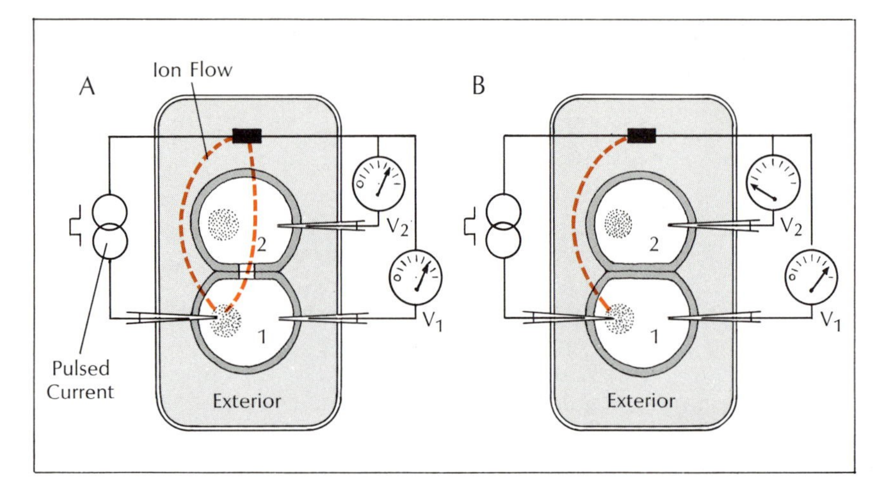

Initial clue to existence of intercellular communication came from experiment in which electrical resistance between two contiguous cells proved to be relatively low. This implied a rather free flow of current-carrying ions across the junctional membranes between the cells. The basic demonstration consists of passing a pulse of current into one cell and measuring resulting voltage in both. When cells are fully coupled by a junctional path of low resistance (A), pulse produces a voltage in cell 2 nearly as high as in cell 1 (see also illustration below). When junction is blocked and cells are uncoupled (B), pulse produces a large voltage in cell 1 but a barely detectable one in cell 2.

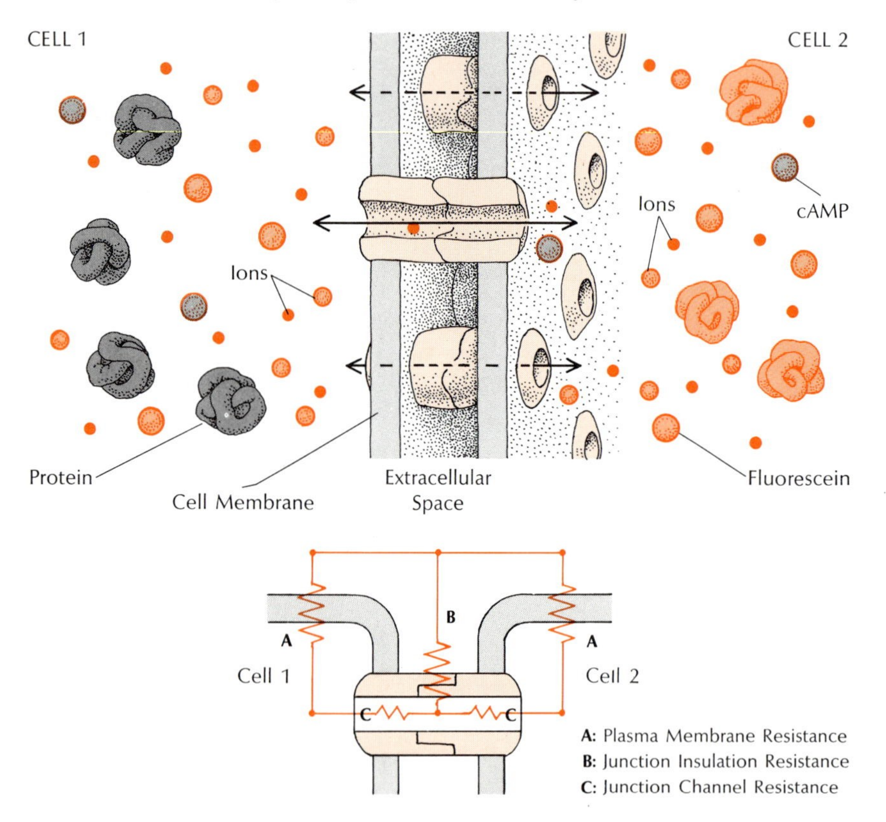

In this schematic view of a permeable junction, pairs of abutting membrane channels form cell-to-cell passageways insulated from the exterior. Channels are pictured as built into protein particles, according to the prevailing concept (see Chapter 4, "Architecture and Topography of Biologic Membranes," by S. J. Singer). This, as well as the spatial interrelationship between the particles, is merely one of several possible representations of the cell-to-cell passageways, which are defined precisely in their resistive properties by electrical measurement (lower drawing). The size of the channels permits rather free passage of small molecules, such as inorganic ions, fluorescein, or cyclic AMP, but excludes larger ones, such as DNA and most proteins.

from the viewpoint of general cell physiology, no one had found these phenomena especially surprising.

Within a few months, Kanno and I were able to show that the coupling in the epithelial cells was not limited to the small inorganic ions. Studies with the organic molecule fluorescein demonstrated that this molecule, when injected into a salivary cell, could pass from the interior of one cell to that of an adjacent cell, its progress showing clearly in sequential photographs taken under ultraviolet light. The cell-to-cell passage of fluorescein was soon confirmed in other laboratories for several kinds of cells, including the electrically transmitting nerve synapse. As organic molecules go, fluorescein is not especially large, but its molecular weight – about 300 – is still considerably greater than that of the largest inorganic ion. It appeared, therefore, that the cytoplasms of the adjacent cells were in direct communication with one another through channels big enough to permit the passage of the fluorescein molecule.

In subsequent studies, we showed that such channels are present in a wide variety of epithelial tissues including those of liver, kidney, thyroid, skin, urinary bladder, and pancreas. Indeed, the only tissues in which they are not found are those of skeletal muscle and the nervous system, in which the proper transmission of information by electrical impulses demands insulation between parallel information lines. It seems fair to conclude that direct communication via junctional channels represents a very primitive cellular mechanism (it is present even in sponges); humoral transmission in the nerve and muscle systems is probably of more recent vintage.

In a more detailed exploration of the properties of the channels by means of electrical techniques, we have found that these channels are at least 10,000 times more permeable to small inorganic ions than is the bulk of the cell membrane facing the exterior. Moreover, the channels are effectively insulated from the exterior so that the permeating molecules pass from cell to cell with little leakage to the outside. Finally, by the use of

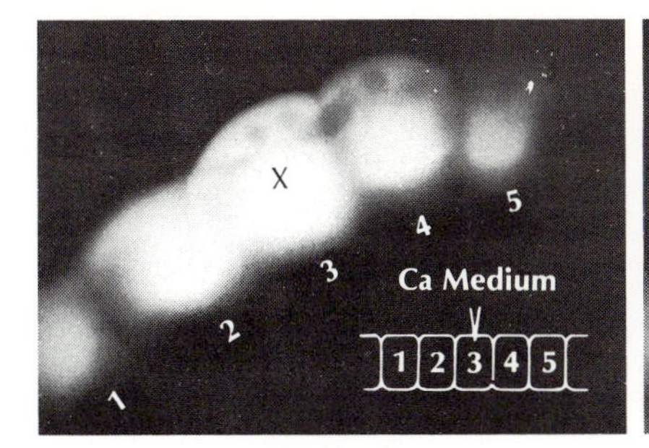

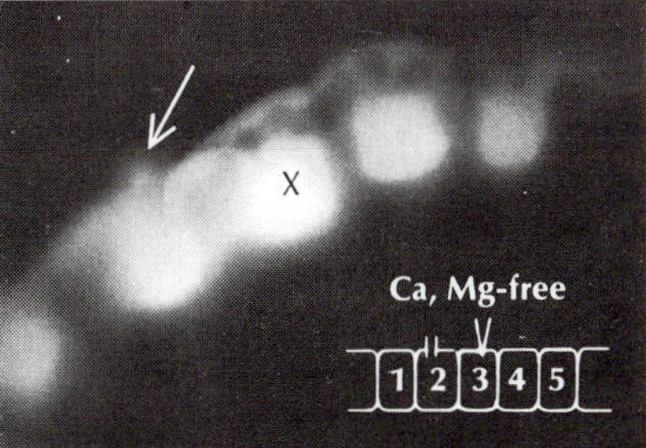

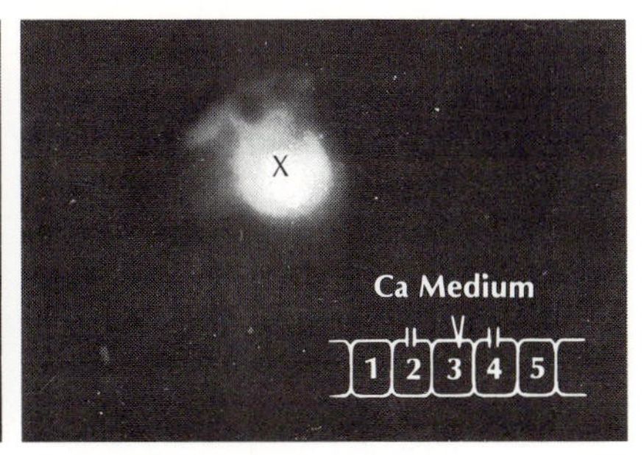

Fluorescent dye injected into one of five adjacent salivary cells (left) diffuses rapidly to the other four. A hole in cell 2 produces no abnormalities of dye flow in a medium free of Ca^{++} (and Mg^{++}), apart from visible leakage (arrow) of dye through the hole (middle). In a Ca-containing medium, with holes punched in both cells 2 and 4, Ca^{++} enters through the holes and the cell-to-cell flow of dye is blocked (from Oliveira-Castro GM and Loewenstein WR, Jn Membr Biol *5:51, 1971).*

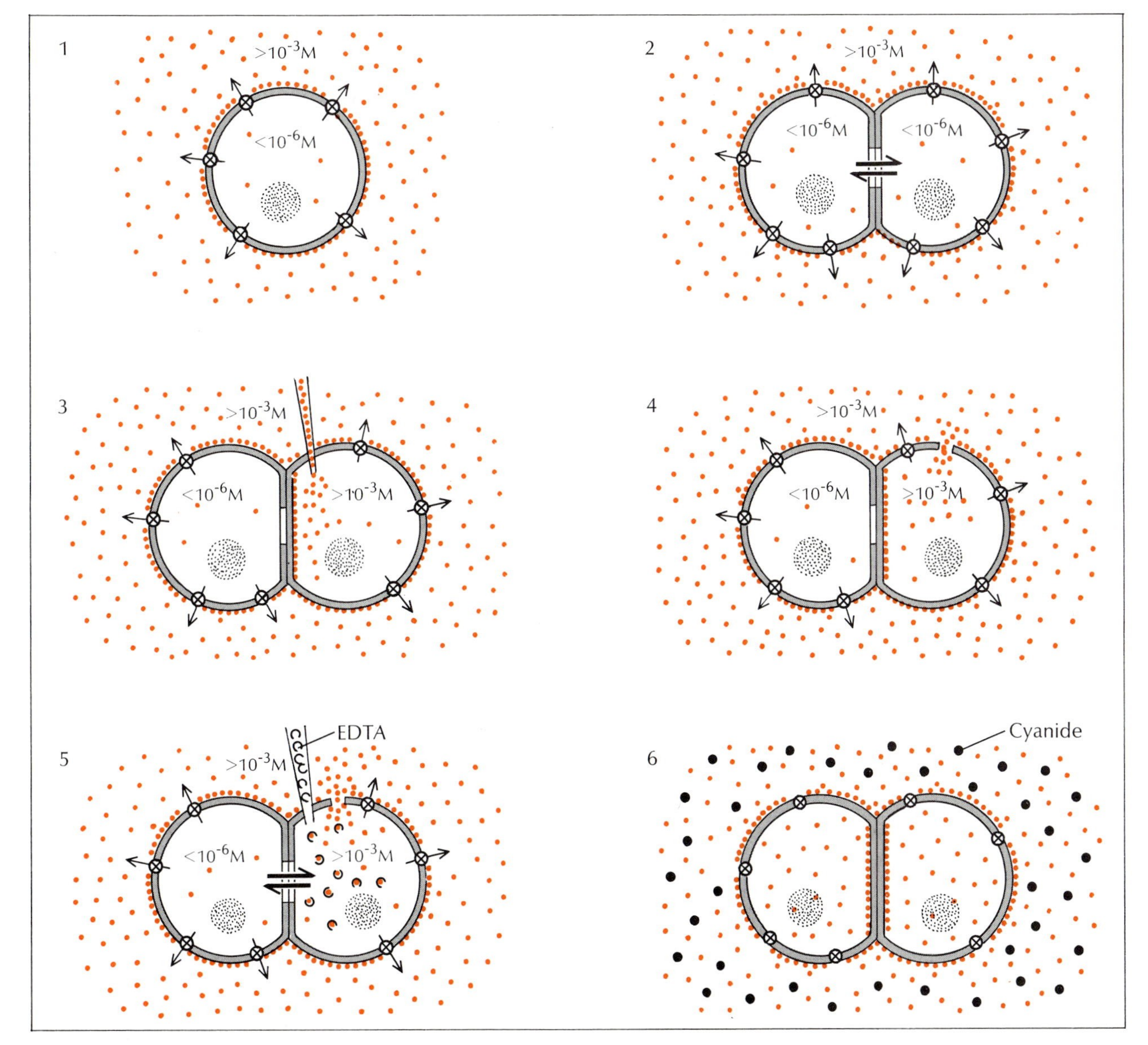

Key role of Ca^{++} ion in intercellular communication is shown in series of experiments. In normal cell (1), Ca^{++} is held at very low levels by calcium "pump"; such cells rapidly set up communication channels when brought into contact (2). Injection of Ca^{++} (3) or leakage of the ion through a hole in the membrane (4) blocks the channels; this can be reversed by injecting EDTA (5), which chelates the Ca. Similar blockage is produced by cyanide treatment (6), which stops the pump.

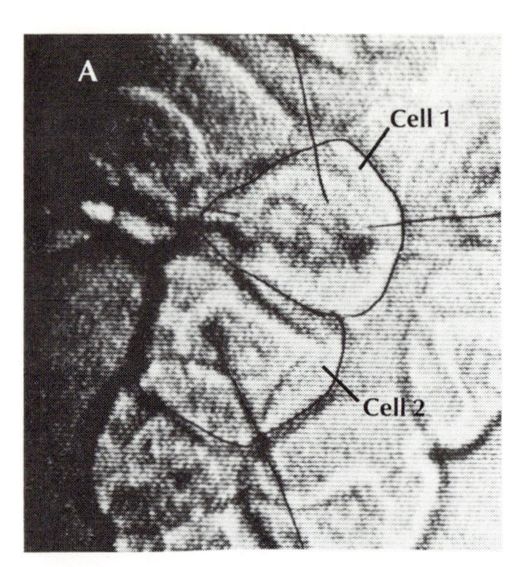

Calcium ions inside a cell are visualized by the glow of the calcium-sensitive protein aequorin. In photo A the cells are shown in bright field with electrodes in position for electrical measurement. A small local injection of Ca^{++} into center of one cell (B, dark field) does not affect electrical coupling, as the oscilloscope record makes clear. A larger injection (C) reduces the channel permeability somewhat, while an even greater amount (D), flooding the junction, blocks the channels altogether (Rose B, Loewenstein WR, XVIth International Congress of Physiological Sciences 1974, Lectures).

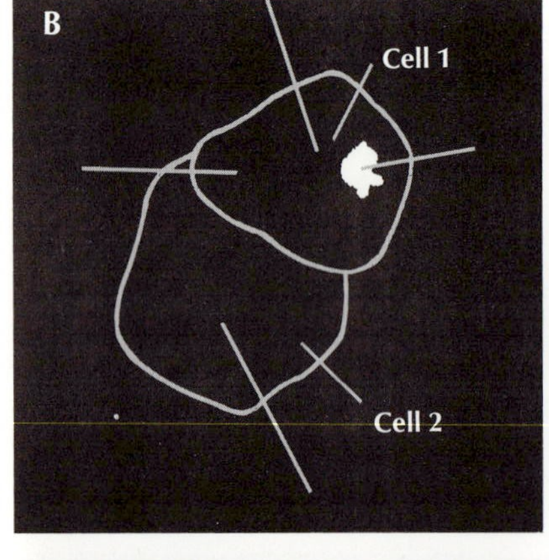

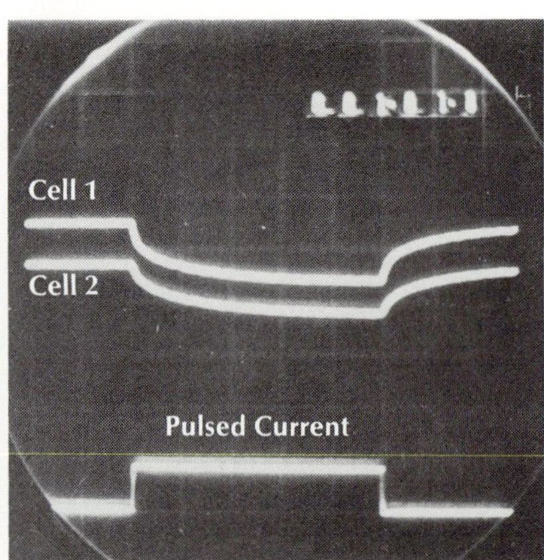

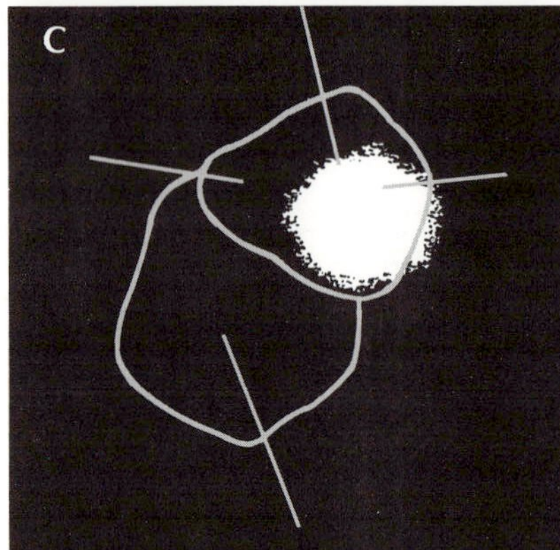

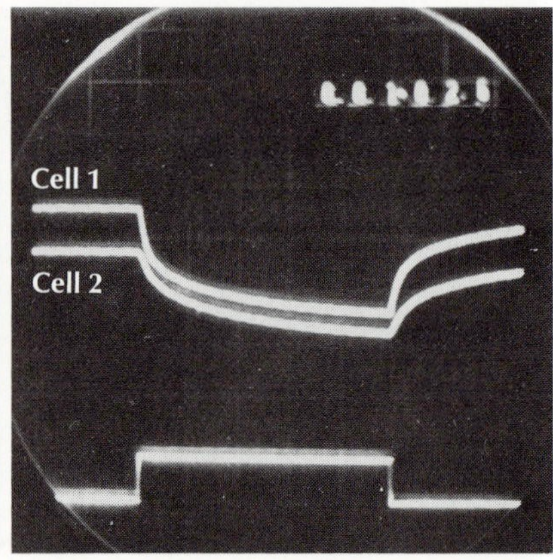

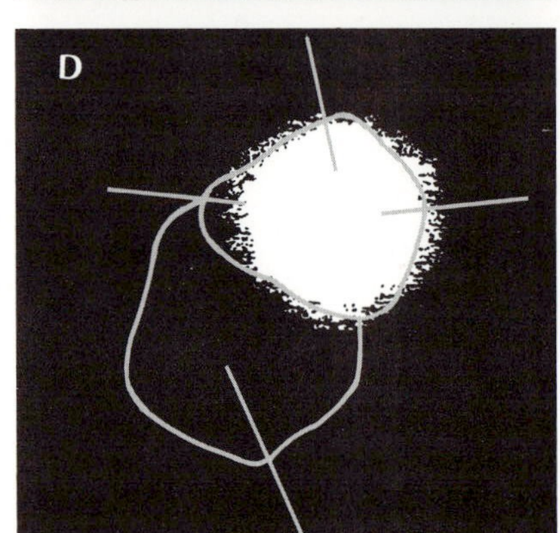

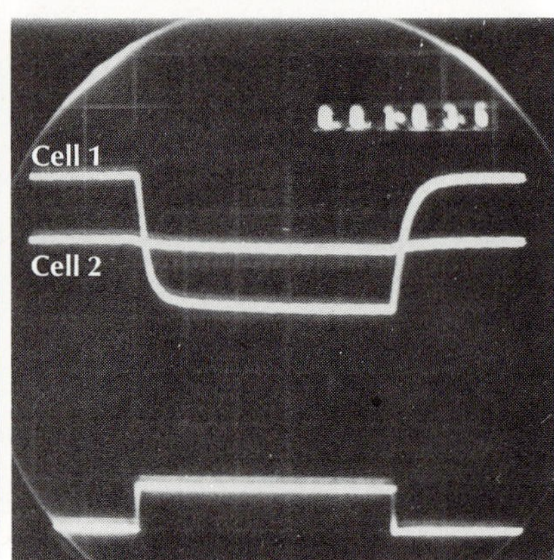

several tracer molecules, we learned that the channels will pass molecules with weights of at least up to 1,000. Into this range fall a wide variety of cellular molecules: the steroid hormones, cyclic adenosine monophosphate (cAMP), all of the common metabolites, such as sugar, amino acids, nucleotides, vitamins, and so forth. On the other hand, the channels will not pass most of the larger molecules such as proteins, RNA, or DNA. With respect to these macromolecules, the carriers of the basic genetic information, the principle of cell individuality is valid.

Ever since this form of intercellular communication was described, electron microscopists have been searching for the related differentiations in membrane structure. Functionally, in terms of their resistive properties to ion diffusion, the cell-to-cell channels are reasonably well defined. They must contain at least two elements: a transmembrane channel in the two joining membranes through which molecules can move and some sort of insulation where the membranes meet, which seals off the passageway from the extracellular medium. Indeed, this is how I defined the channels in 1966 on the basis of electrical measurement made on living cells. Such measurements can be done rather easily in suitably large cell systems. It is, however, a much more difficult task for the electron microscopist, working with dead cells, to demonstrate the channels morphologically. The fixation, staining, and other procedures of present-day electron microscopy may alter the membrane structure and even induce artifactual ones. Thus, unfortunately, there is as yet little structural knowledge at the channel level. Progress has been made, however, at a coarser level, in identifying the membrane regions where such channels may be located. Here the focus has been on the differentiated structures in the membrane junctions between coupled cells that may offer the necessary continuity.

There are various kinds of differentiated structures in cell junctions, but to sort out, on morphologic grounds, which of these structures are mediators of coupling, one needs to have a coupled cell system in which, ideally, only one type of junctional structure is present at a time (see Chapter 9, "Junctions Between Cells," by Pappas). One such structure, discovered by David Robertson in an electrically transmitting nerve synapse and by Jean-Paul Revel in epithelial cells, is the "gap junction" (which also goes under the name of "nexus"). This structure appears as an aggregation of neatly arrayed membrane particles, which x-ray diffraction studies by Daniel Goodenough and Don Caspar show to be aligned on either side of the two joining membranes. It is conceivable that these particles, when properly aligned and joined, form the cell-to-cell passageways; that is, each particle contains a water channel (the membrane channel), and the insulation is given by the hydrophobic particle walls and particle junction.

The evidence implicating the gap junction in coupling is now strong: the structure is widely present among coupling cells (in some it is the only visible differentiated structure) and, as Norton Gilula and his colleagues at Berkeley have shown, it seems to be lacking in certain noncoupling cells. Moreover, when such noncoupling cell strains are rendered coupling by genetic manipulations that I shall describe further on, the gap junction clearly appears again.

I am citing here only the structures for which the evidence as a mediator of coupling is most compelling. But this by no means concludes the list of structural candidates. There are no reasons to assume that coupling particles could cluster only in one or two kinds of arrays. One possible candidate, for instance, is the septate junction, which contains highly ordered and particularly extensive arrays of membrane particles; and there are other possible coupling structures.

A specially interesting aspect of permeable junction is that the channels are not permanent functional entities in the plasma membrane. Rather, the channels form or become functional when the cells that can make them are brought into contact and disappear or are blocked even more rapidly than they form if the cells are separated, with all of the cell membrane returning to the impermeable state. Nor is their development limited to any special region of the membrane. The first clue in this respect came from experiments with sponge cells, in which I manipulated single pairs of cells into contact at random spots; permeable junctions formed wherever the contact happened to be. Later experiments, conducted with Shizuo Ito on vertebrate embryo cells, were even clearer: coupling channels formed when we brought the cells into contact; they self-sealed and then re-formed when in contact at other points.

Electron micrograph by William Larsen of three freeze-fractured gap junctions shows the typical clusters of particles in a split membrane junction of coupling human-mouse hybrid cells. The particles here all belong to one membrane; the overlying contiguous membrane has split off. These hybrid cells are also shown on pages 110 and 111.

Thus it appears that junction formation is a capacity of most and perhaps all parts of the plasma membrane in these cells; its only immediately necessary condition seems to be

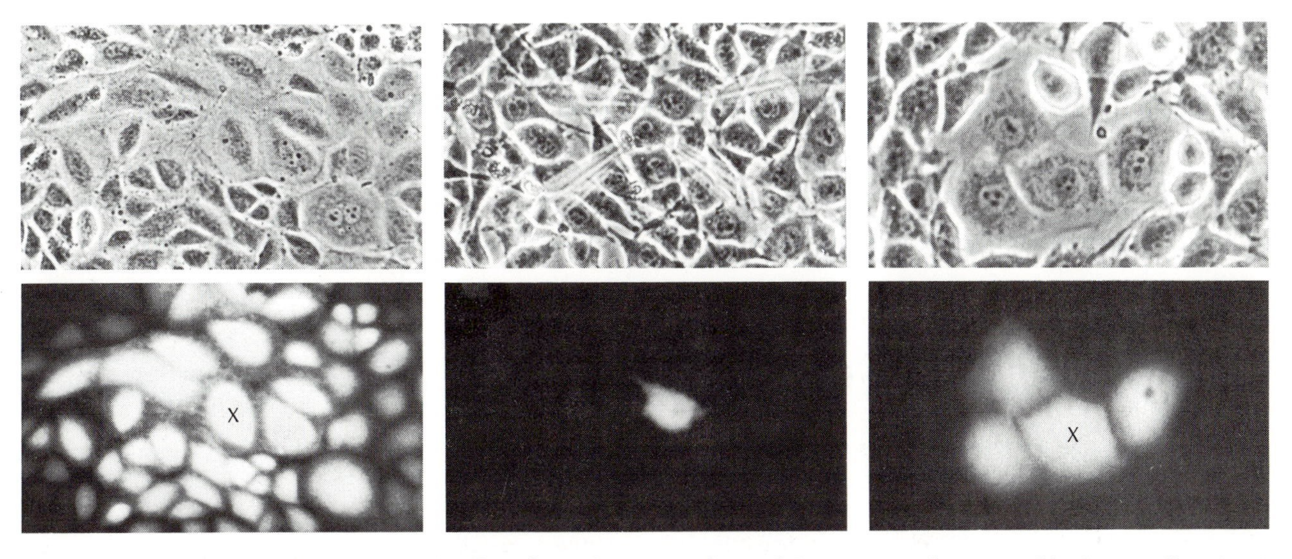

Contrast in communication between normal cells and a malignant strain of hepatoma cells is demonstrated by fluorescein injection. Same cells are shown in both bright and dark fields. In normal cells (left), dye injected into the one marked "X" rapidly diffuses into the rest, even though they are of different types and from different species (large are rabbit lens; small, rat liver). When injected into malignant cell (middle), dye does not spread beyond it. In mixed culture of normal liver and malignant cells (right) dye diffuses only in normal ones (from Azarnia R and Loewenstein WR, Jn Membr Biol *6:368, 1971).*

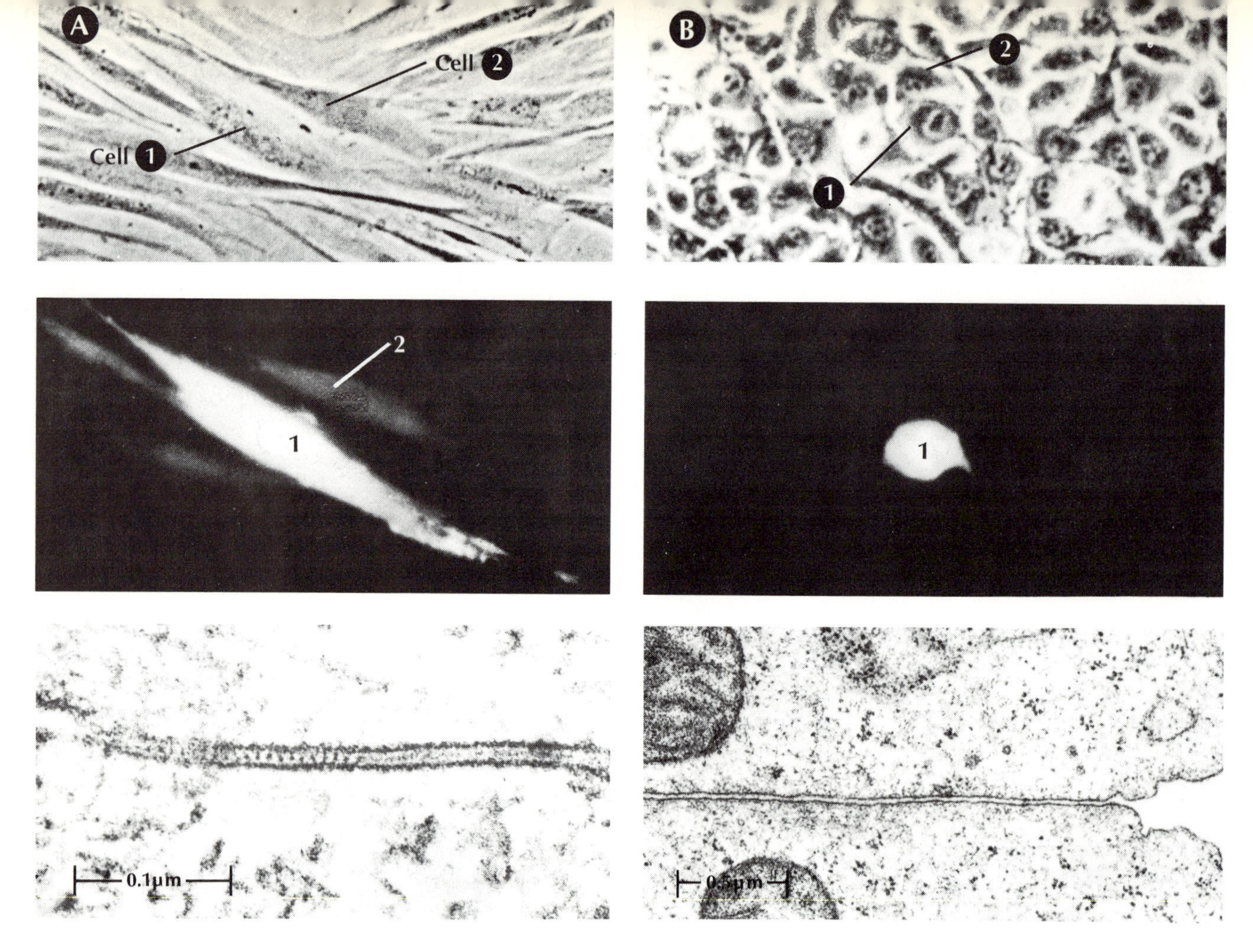

Crossing malignant, junction-defective mouse cell with normal human fibroblast can correct junctional defect. In the normal fibroblasts (A, top), cell-to-cell fluorescein flow proceeds normally (middle) and electron micrograph of contiguous plasma membranes shows typical gap junction (bottom). In the malignant cells (B), dye flow is blocked; membrane has no gap junc-

intimate contact (adhesion) between the two membranes. Surprisingly, the membranes need not be equal for junction formation. In experiments in which Wolfgang Michalke and I paired cells from different organs and from different species in culture, the cells turned out to establish good coupling. For example, a lens cell from a rabbit made a perfectly viable junction with a liver cell from a rat or a human skin fibroblast.

Thus far, these interorganic and interspecific junctions have only been demonstrated on cultured cells that may have undergone some de-differentiation. It would be dangerous therefore to conclude that cells in well-differentiated tissues necessarily behave this way too. But the fact that such heterologous junctions can form under any conditions is surely remarkable. The cultured cell pairs tested were clearly genetically different; they had different morphologic features, made different enzymes, and had different immunologic properties. Yet, they were capable of a coupling process that must require at least some membrane symmetry, such as alignment of channels. This points up once again how basic and general the mechanism of junctional formation is.

The formation of permeable junctions takes on the order of 10 minutes in various cell types. In newt embryo cells, for example, where we have recently been able to monitor the coupling process continuously and precisely by measurement of electrical resistance, the first signs of intercellular communication appear between four and 25 minutes after the cells are placed in contact, with communication attaining its full extent over the next 10 to 30 minutes; the process is equally rapid whether the cells are forced into contact by micromanipulation or are allowed to make contact by their own spontaneous movements. Two important events could be detected during the process of junction formation: the resistance to ion movement across junctional membrane fell gradually until it reached its normal low level, and the resistance to ion movement across junctional insulation increased gradually to a peak, to settle finally at a level somewhat below peak. A plausible and simple explanation for these progressive events is that more and more individual channels develop between cells during junction formation and that junctional insulation improves progressively during the early phase. Unfortunately, we have as yet no electron microscopic information to go with these electrical measurements. However, there are the suggestive observations by Gilula on septate junction in sea urchin embryo cells and by Sheridan and Johnson and their colleagues on gap

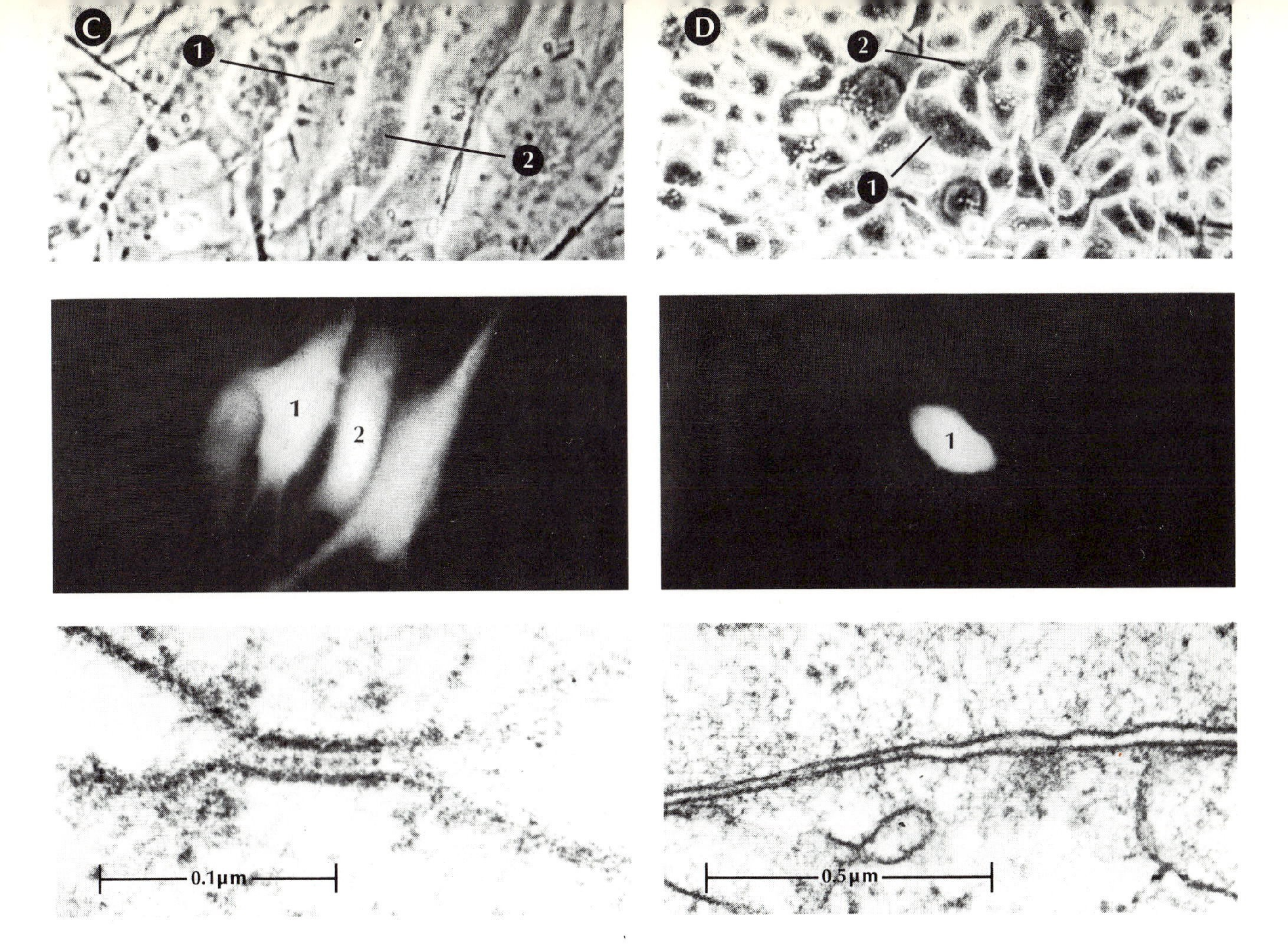

tion. Crosses between the two strains, produced by cell fusion (C), manifest both the normal dye flow and the junction-forming capacity of the human parent. As the culture reverts to malignancy through chromosome loss (D), fluorescein flow is blocked and the gap junctions disappear (Azarnia R, Larsen W, Loewenstein WR, Proc Nat Acad Sci *71:880, 1974).*

junction in cultured rat fibroblasts that the number of membrane particles increases progressively with development. To establish the correlation, one needs a cell system in which both the electron microscopy and the measurement of electrical resistance can be done (measurements of electrical coupling are not sufficient). This is technically difficult – but, no doubt, the right system will be found before long.

In the regulation of the permeability of the channels, the calcium ion seems to play a key role. Here it must be remembered that at their internal face the junctional membranes are exposed only to whatever calcium is in the cytoplasm, where its concentration (in free, ionized form) is normally below the order of 10^{-6} molar. If, however, the cytoplasmic free calcium level is raised, the permeability of the channels falls very steeply. In fact, at concentrations approximating the normal level in the extracellular fluids (10^{-3} molar), the channels virtually disappear as functional entities.

This experimental "dechannelization" can be accomplished in a number of ways – most simply, as in experiments with Gilberto Oliveira-Castro, by punching a hole into the plasma membrane with a microneedle and allowing the cytoplasm to equilibrate with the high exterior levels of calcium (swamping, so to speak, the cytoplasm with calcium ions) or, more elegantly, by injecting calcium into the cells with a micropipette. Similar results are obtained, as shown by my colleagues Alberto Politoff and Sidney Socolar, when the membrane's calcium "pump" or that of the mitochondria – which maintains the intracellular level of the ion at the normal low level – is blocked by poisoning with cyanide or dinitrophenol.

The most revealing results on the action of calcium were obtained in recent studies performed in collaboration with Birgit Rose. The experiments involve the use of aequorin, a protein that luminesces in the presence of calcium, and an electronic image intensifier to make the luminescence visible. The procedure is as follows: The aequorin is injected into a pair of coupled salivary gland cells and serves as an indicator of the cytoplasmic free calcium, and the image intensifier scans the cytoplasm and tells us where inside the cells the cytoplasmic calcium concentration is changing and by how much. With this method, a small injection of calcium is seen as a luminescent puff in the cytoplasm that does not spread much beyond the micropipette. This is because the excess calcium is rapidly swept up and sequestered by mitochondria. If such an injection is made into the cell center, the junc-

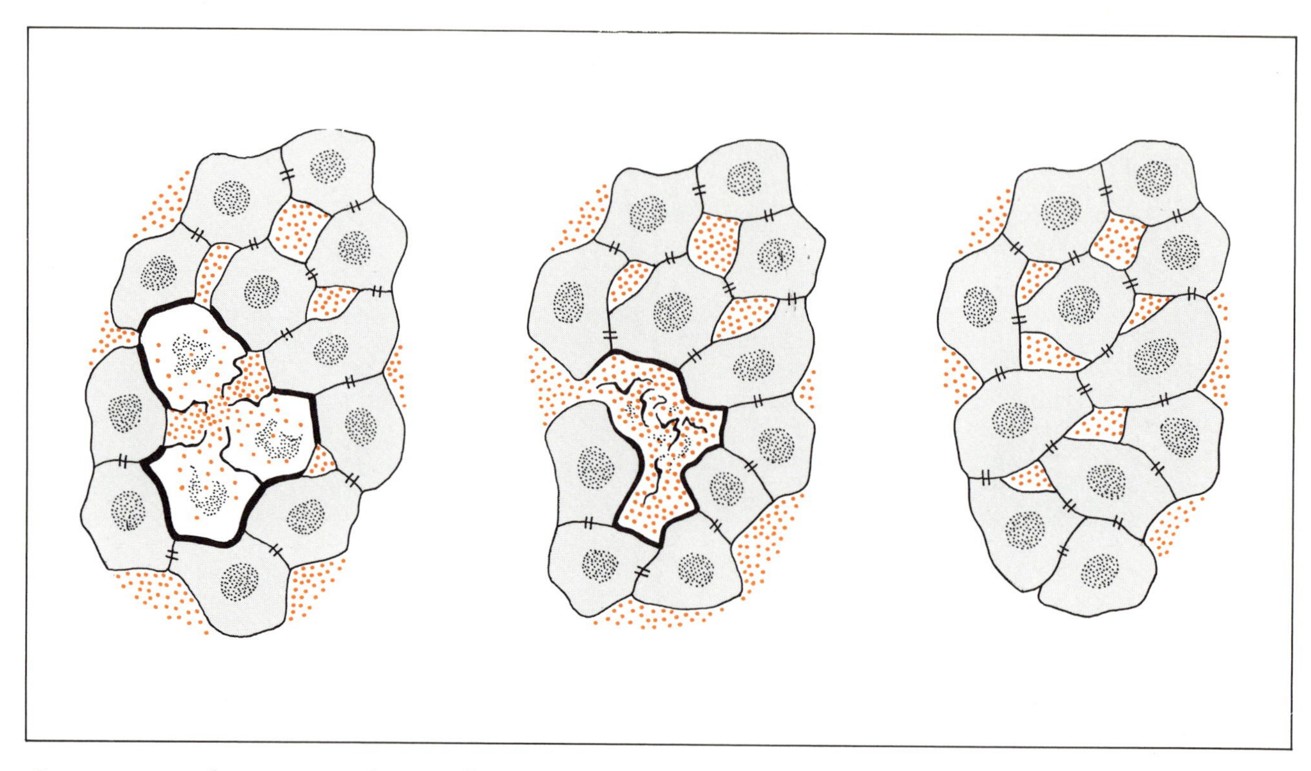

First reaction to skin injury is sealing of cell junction at the wound border by the calcium mechanism (left); this protects intact cell population against leakage. As the wound gap closes (middle and right), coupling is reestablished among intact cells.

tional channels are not affected. The channels, however, are promptly blocked if such an injection is made close to a junction, or if a large calcium injection into the cell center swamps the mitochondria, raising visibly the calcium concentration in the cytoplasm around the junction. The channels open up again spontaneously as the excess calcium is pumped out of the cell. Other experiments, in which the calcium influx is stepped up by incorporating a calcium-transporting ionophore into the plasma membrane, show that the junctional channels are blocked whenever the cytoplasmic calcium concentration rises above a certain level. Finally, the drop in permeability can be reversed by injection of EDTA, a compound that avidly binds the calcium.

Having described (to the extent that present data permit) the genesis and the physiologic properties of permeable junction, let us now consider some of their implications for cell biology and medicine. To begin with, there is every reason to think that the reaction of the junctional membrane with calcium plays an important part in the organism's response to injury. The above experiment of blocking a junction by the influx of calcium through a hole punched into the plasma membrane is a perfect micromodel of what happens during ordinary tissue injury: the intact tissue is sealed off from the injured cells. If it were not for this fast mechanism of junctional blockage, a tissue like skin with widely interconnected cells could not survive an injury, or a liver (in which most cells, perhaps all, are interconnected by junctions) could not survive the death of even a single cell; the intact tissue would be leaky at the wound borders. This mechanism has actually been demonstrated in skin wounds, where the intact cells at the wound border remain sealed off from the exterior until they make contact with one another as the wound closes, at which point they form permeable junctions within 30 minutes.

Perhaps the most interesting possibility to consider is that of the junction playing a role in the regulation of cellular growth and differentiation. Before pondering this possibility, let us briefly sum up what junctional communication implies: a common intracellular milieu, including many sizable molecules, for a large community of cells – meaning large portions of an organ and perhaps (in some cases) the entire organ. There is thus ample opportunity for concerted interaction by regulatory molecules within such a community.

The exciting possibility then to be considered is that the size range of molecules passing through the junctions includes substances involved in the regulation of gene activity – i.e., of cell growth and differentiation. Such gene regulators have not yet been identified in higher organisms, but the past 15 years of bacterial genetics have left the lesson that in these organisms, at least, genes can be regulated by a number of quite simple metabolites such as galactose or tryptophane – molecules of only a few hundred molecular weight. Moreover, the results of the past 30 years of experimental embryology clearly imply that embryonic differentiation involves interaction between cells at close range, mediated by diffusible substances. Here we have in junctional communication an obvious candidate for such close range interaction, namely a system in which molecules can flow directly from cell to cell with little loss to the exterior.

On purely *a priori* grounds, a cell system connected by junctions is in many ways ideally suited for disseminating information through the community on the number and position of its constituent cells. The key point here is the presence of a sharp boundary, namely the continuous diffusion barrier made of the plasma membrane plus the junctional insulations. Because of this boundary, the connected cell system has a finite volume and hence could respond to simple cues of chemical concentration. For instance, if the processes controlling cell growth were cued to a concentration level of a signal molecule, then the size of a growing cell population could be self-regulated as the (asynchronously) proliferating population enlarges its effective volume, diluting the molecules. Self-regulating growth models of this kind can be readily envisaged. But equally important and, perhaps, even more unusual is the potential of a junction system for conveying information on cell position. Because of the presence of the sharp diffusion boundary, cues about the location of a given cell or a group of cells within the community could be provided by simple time-dependent parameters of signal concentration relative to the boundary. The two kinds of controls, regulation of cell number and cell position, are essential to any normal growth process.

On the experimental side, there is the fact of extensive junctional connection in embryonic development. My colleague Ito and, at Harvard, Furshpan and Potter, were the first to show this. Ito showed that at the morula stage in newt embryo most and probably all the cells are interconnected. Even more interestingly, the Harvard group found at a later developmental stage in squid embryo that there is extensive interconnection between cell groups when these are acquiring visible differentiations. By now the embryos of many different species have been investigated, and it is clear that widespread junctional communication is a general feature of embryonic organisms.

All this, of course, is far from proving that permeable junctions are involved in growth control and differentiation. It is merely the condition *sine qua non* for such an involvement. To demonstrate an involvement would be simple enough if we knew the signal molecules in question. But the lack of knowledge of these signals is precisely what hampers the field of developmental research and the related field of cancer. Thus handicapped, the workers in these fields are forced to take indirect approaches; and how hard it is, under these circumstances, to sort out relevant etiologic phenomena from epiphenomena is all too obvious when one reads the literature of these fields.

My colleagues and I are in no way better off. We too had to take an indirect approach, and only recently have we been able to cut the risk that we may be chasing an epiphenomenon by the use of genetic analysis. Our approach is to search for defects in junctional connection among cancerous tissues – in which growth control is defective by definition – and to determine by genetic analysis whether the growth defect and the junction defect are correlated.

The rationale is as follows: the control of growth, like any controlled system, involves four elements – 1) the control signals, 2) the signal transmission system, 3) the signal receptor, and 4) the effector process triggered by the signal. In principle, uncontrolled growth – cancer – can arise from a defect in any of these elements. Thus, if the junction is indeed an element of transmission for growth-controlling signal molecules, transmission block (*uncoupling*) by genetic defect should produce uncontrolled growth. This simple idea has guided our work into the cancer field, where our focus is on the etiologic category of signal transmission.

Before describing our work in this area, I should like to say what we do *not* expect to find. The term cancer is an umbrella for many different forms of uncontrolled growth. Each of the four etiologic categories above (and their possible subcategories) constitutes *a* sufficient cause and, for the particular cancer form, *the* necessary cause of uncontrol. We do not expect, therefore, to find junction defects in categories 1, 3, or 4. In fact, we know now of at least seven types of cancer cells that have permeable junctions. The question that concerns us rather is whether there are junctional defects in cancer cells belonging to category 2; or stated in terms of experimental approach, we are searching for *some* types of cancer cells that are junction-defective.

The search was encouraged by our earlier findings that junctional coupling is labile; uncoupling can be readily produced by experimental physicochemical manipulations. In fact, how vulnerable a junction is to uncoupling is evident from considering only the dependence of the junctional permeability on the cytoplasmic calcium level and the many ways this level can be altered. It seemed therefore not unlikely that uncoupling could be induced by genetic defects, and we hoped that such defects might be frequent enough to give us a reasonable chance of finding some kinds of uncoupled cancer cells. In fact, we have thus far found six such cell strains – four derived from rat hepatomas, one produced by x-irradiation of hamster embryo cells, and one a malignant derivative of a mouse L cell.

All six strains are clearly abnormal in their growth patterns. They do not show the density-dependent growth of normal cells in culture, and in animals they produce fatal tumors. Their junctional coupling is no less abnormal; in contrast to normal cells, which (as already noted) will couple not only with their own kind but with cells from different organs and different species, cells from any of these strains will not couple even with one another. This has been demonstrated in three different ways. First, electrical measurements showed that transfer of small ions is blocked; second, studies with fluorescein showed that passage of this molecule is blocked; and, finally, radioautographic studies with labeled nucleotides, such as hypoxanthine, and their derivatives show these molecules too are blocked.

But these results, though they show that the cancer cells in question are undoubtedly junction-defective, still do not tell us whether this property is any way related to the growth abnormality. To find this out, Roobik Azarnia and I analyzed the defects genetically.

We hybridized through fusion the abnormal cells with normal cells and examined the ability of the hybrids to make permeable junctions and tumors. Mary Weiss, J. Todaro, and Howard Green (then at New York University), Henry Harris and George Klein and their colleagues at Oxford, and the Karolinska Institute had already established that density-dependent growth (contact inhibition) is resumed and tumorigenicity reduced when cancer cells of various types are hybridized with normal cells. The question for us then was whether the normalization of growth properties would go hand in hand with normalization of junctional properties.

In the three types of hybrids we have thus far examined, the results are simple: correction of the growth defect is paralleled by correction of the junction defect. In one system, for example, the partners for fusion (the parent cells) were a normal liver epithelial cell and an epithelial hepatoma cell, both from rat. The initial contrast between them could not have been more striking. The liver cell was "coupling" – i.e., it formed normal junctions, it grew to densities of 10^4 cells/cm^2 in cultures, and it was not tumorigenic. The hepatoma cell was noncoupling, achieved densities in excess of 10^6 cells/cm^2, and was highly tumorigenic: inocula of only 100 cells produced fatal tumors. Hybrids between these cells took after the normal parent in all these respects.

Even more persuasive evidence has come from fusion of mouse and human cells, where we could carry the analysis one step further. Here one parent cell was a skin fibroblast from a patient with Lesch-Nyhan syndrome; apart from its characteristic enzyme defect (which was useful in the selection of the hybrid) it was altogether normal in growth and junction: it was density-dependent, coupling, and, indeed, as our colleague William Larsen found under the electron microscope, it had the characteristic gap junction. The abnormal mouse parent cell – a malignant derivative of an L cell – was not density-dependent, noncoupling, and lacked gap junctions. Again, the hybrids took after the normal human parent.

We then continued to grow the hybrids for a number of generations and found that (as is characteristic for the human-mouse cross) they tended to lose the human chromosomes with continued replication. Eventually clones appeared that had reverted to the growth-defective state. These clones had also reverted to the junction-defective state, showing neither coupling nor gap junctions. In every case, the reversions to the growth defect went hand in hand with reversion to the junction defect, although these properties had segregated from a number of other morphologic and biochemical ones that originally occurred together in the mouse parent. Evidently, the normal human cell contributed a genetic factor (probably linked to one of its chromosomes) that simultaneously corrected the growth defect and the junction defect. The junction defect and the growth defect in vitro are thus clearly correlated in this cell system. It remains to be shown whether the correlation applies also to tumorigenicity, as we know it does at least in the unsegregated hybrids of the liver/hepatoma cell system. Tumorigenicity tests with heterospecific cells with unstable chromosome complements are complex and difficult, but, if all goes well, we should soon have the answer.

These findings on the relationship between cell coupling and cell growth clearly do not begin to exhaust the possible implications of communication by permeable junction, which as an old and general cellular phenomenon is likely to have adapted to a host of other cell functions. I have here selected only those functions on which we are getting the first experimental glimpses. The interested reader will find a discussion of other possibilities including control models in an article in *Perspectives in Biology and Medicine* (see Selected References section for this chapter).

However, I should not like to end this chapter without mentioning at least that junctional communication may well be at the root of the familiar phenomenon in medicine that diseases generally affect whole organs or large parts of them, rather than the individual cells. Ever since the days of the Greek philosophers, it has been customary to liken a living organism to human societies, and the comparison is apt: both consist of interacting units with organized activ ity. The very word "organism," rathe new in our language, implies this; and, indeed, it was precisely the concept of organization that led the French naturalist Buffon to introduce the term in the early 19th century.

Clearly the existence of organization requires the exchange of information between the constituent units, and one can confidently say that the health of an organism, like that of a society, depends on how well the units communicate with each other. It seems no less clear that permeable junctions must take their place as one of the primary communication systems of living organisms. Precisely what they communicate, and how, and when, will no doubt be revealed by further research, which may well develop into a whole new field of physiology.

12

Dynamics of Intracellular Membranes

PHILIP SIEKEVITZ
Rockefeller University

In considering the topic of how intracellular membranes change or turn over, the first thing that must be said is that these processes are not well understood – which no doubt is one reason so many theories have been formulated on the subject. Given the relative sparseness of pertinent data, our present understanding of this biologic area is beset with paradox. On the one hand, membranes, like all other structures, exist in time as well as in space and change both structurally and functionally over time – in the case of membranes, sometimes quite rapidly. On the other hand, and despite this temporal lability, membranes also retain a clear individual specificity and identity of structure and function over time. How membranes manage this paradoxical combination of permanence and change is a question to which there is as yet no satisfactory answer; all one can do for the moment is to note some manifestations of the paradox and suggest how it might be resolved.

For the purposes of this discussion, I will focus chiefly on the dynamic problems of intracellular membranes, in particular those of the mitochondria and endoplasmic reticulum (ER). It should be noted, however, that many of the observations – and problems – to be cited also apply to the exterior, plasma membrane.

Endoplasmic Reticulum and Mitochondrial Membranes

Both specificity and dynamic change in intracellular membranes can perhaps be observed most clearly in the endoplasmic reticulum. This structure, an interconnecting system of membrane-enclosed cisternae occupying a substantial part of the cytoplasm, performs a variety of chemical functions. As is well known, portions of the ER are the sites of protein synthesis, serving as a sort of foundation for the polysomes (aggregations of ribosomes) on which the synthesis actually proceeds according to the "directions" provided by particular sequences of messenger RNA. The proteins in question include the structural and enzymatic proteins that are incorporated into the cell's own membranes (including the ER itself), and also various secretory proteins such as blood albumin (synthesized by the liver) and gut proteolytic enzymes (synthesized by the pancreas) whose ultimate destination is outside the cell. The appearance under the electron microscope of ER with polysomes attached has given it the name of "rough" ER (see Chapter 15, "Membranes and Secretion," by Jamieson, and Chapter 13, "The Network of Intracellular Membranes," by Trump).

Both rough ER and smooth ER (ER without polysomes attached) also perform other chemical tasks. By means of enzymes embedded in their membranes they synthesize various nonprotein substances, including such compounds as the steroid hormones (for the cell's "export trade") and the phospholipids, which, along with proteins, make up the structures of all the cell's membranes. In a few types of cell – notably those of the liver – the ER performs another function that can be summed up as detoxification. That is, instead of synthesizing new molecules for transport either to other parts of the cell or to the extracellular space, it deals with a variety of potentially harmful exogenous molecules by breaking them down or otherwise modifying them to remove their cytotoxic qualities. The compounds thus processed include carcinogens, substances used as herbicides and pesticides, and drugs such as the barbiturates. A frequent feature of the detoxification process is the conversion of lipid-soluble substances (e.g., many carcinogens) into water-soluble compounds, which can then, of course, be eliminated through the kidneys.

The mitochondria, which also provide us with examples of specificity and dynamic change in their membranes, appear to be rather less versatile than the ER. Their primary function is to metabolize Krebs cycle intermediates, such as succinate, to produce the energy-yielding compound adenosine triphosphate (ATP) plus, of course, carbon dioxide and water, although mitochondria are also thought to have other functions, most of which are still not well understood (see Chapter 14,"Inner Mitochondrial Membranes," by Racker).

The membranes of the ER and mitochondria, like all biomembranes, are composed of proteins and phospholipids, the former including both the loosely bound "peripheral" proteins, which can be detached from the membranes without radically altering their structure, and the "integral" proteins, which can be isolated only by destroying the membrane and generally denaturing the protein as well (see Chapter 4, "Architecture and Topography of Biologic Membranes," by Singer). The proportion of peripheral to integral protein can at present only be guessed at. With regard to the ER, for example, an examination of the "microsomal fraction" of homogenized cells, which consists almost entirely of fragments of the ER, reveals some 30 to 40 different enzyme activities, each one presumably representing a different protein. Other proteins that serve other, nonenzymatic functions may also be present. Of this number I would guess that no more than half are essential to the structure of the membrane.

The structural relationship between proteins and phospholipids is also obscure. The "protein iceberg floating in a lipid sea" model that has been postulated by some workers seems to me less than satisfactory, at least as regards the intracellular membranes. For one thing, the proportion of protein to phospholipid in these structures is invariably greater than 1:1 – in most cases, 2:1 or even more; i.e., there appear to be more "icebergs" than "sea." For another, it is possible to remove the phospholipids from these membranes by extraction without destroying the membranes' integrity. Thus it seems to me very probable that the structural integrity of intracellular membranes must depend to a large extent on protein-protein interactions, rather than protein-phospholipid and phospholipid-phospholipid interactions – and the former, for reasons that earlier chapters in this book have made clear, seem inevitably much more complex than the latter, as the three-dimensional structures of proteins are more complex than those of phospholipids.

Time-Related Properties of Membranes

Thus far we have concerned ourselves with "idealized" membranes that are essentially static structures. This rather lengthy prologue has been necessary in order to make clear some of the problems that arise in trying to comprehend and explain the dynamic activities of real membranes. These activities are of three general types. First, membranes "multiply," in the sense that the new cells created by mitosis (especially during embryogenesis but also in adult life) contain a normal complement of membrane; in certain circumstances the amount of membrane within a single cell may also increase. Second, the biologic properties of membranes change with time; this is notably the case during differentiation but is also seen in the response of adult cells to certain kinds of environmental challenge. Finally, even when membranes are neither multiplying nor undergoing functional change, their substance is constantly being renewed. Existing protein (as well as phospholipid molecules) are extracted from the membrane and broken down into their constituent amino acids, while new proteins are simultaneously synthesized (probably mostly in the rough ER), transported, if necessary, to the intracellular sites where they are needed, and incorporated into the appropriate portions of the membrane structure in question.

The creation of additional membrane can be visualized in two ways. First, newly synthesized protein and phospholipid molecules might come together and more or less spontaneously "crystallize" into a new membrane. Alternatively, these molecules might be incorporated randomly but specifically into existing membrane, whose expanding mass would (in the case of dividing cells) eventually partition itself into the membranes of the two daughter cells. All the evidence that has been accumulated indicates that the latter process is what actually happens. For example, if one adds labeled amino acids to cultures of cells that are manufacturing membrane, one finds no tendency for the label to become incorporated preferentially into "new" membrane; in fact there is no way in which "new" and "old" membrane can be distinguished. From this and other evidence we can say that just as every cell in the adult animal is the direct descendant of the single fertilized ovum, so the membranes of every cell seem to be descendants of the ovum's original membranes. In a few cases, cells may lose their membranes, notably in the case of the mammalian erythrocyte, in which all intracellular membranes disappear as it matures, but the reverse never happens; unless membrane is already present, it never will be; as cell derives from cell, so, too, "membrane derives from membrane."

If we accept this principle, the functional change of membranes through time is clearly implicit in the process of differentiation. Though the cells of the mature organism are all descended from a single cell, their biologic activities differ enormously. These differences must to some degree reflect differences in the functions of their membranes – for example, the ER. And at least some functional differences in the ER clearly must reflect differences in its protein content.

Thus, for example, only liver cell ER has "detoxification" proteins such as the chain of at least four enzymes that oxidize nicotinamide adenine dinucleotide phosphate (NADPH) – an essential step in detoxifying barbiturates. Likewise, only the liver cell ER contains the enzyme glucose-6-phosphatase, which helps liberate glucose for use elsewhere in the body. Similar intercellular enzymatic differences in the ER must be inferred from the fact that adrenocortical cells secrete corticosteroids, gonadal cells secrete their appropriate hormones, etc., etc. Indeed, such differences can be demonstrated directly, albeit in a rather crude way, simply by electrophoresis of the microsomal (ER) fraction of various cell types; the bands showing different proteins clearly indicate not merely quantitative differences in the proportions of different proteins but qualitative ones as well.

In the case of the NADPH oxidation enzyme chain, the process of differentiation has actually been observed in embryonic rat liver cells. Morphologically, these cells are mature by the end of the second week of intrauterine life, but their biochemical activity

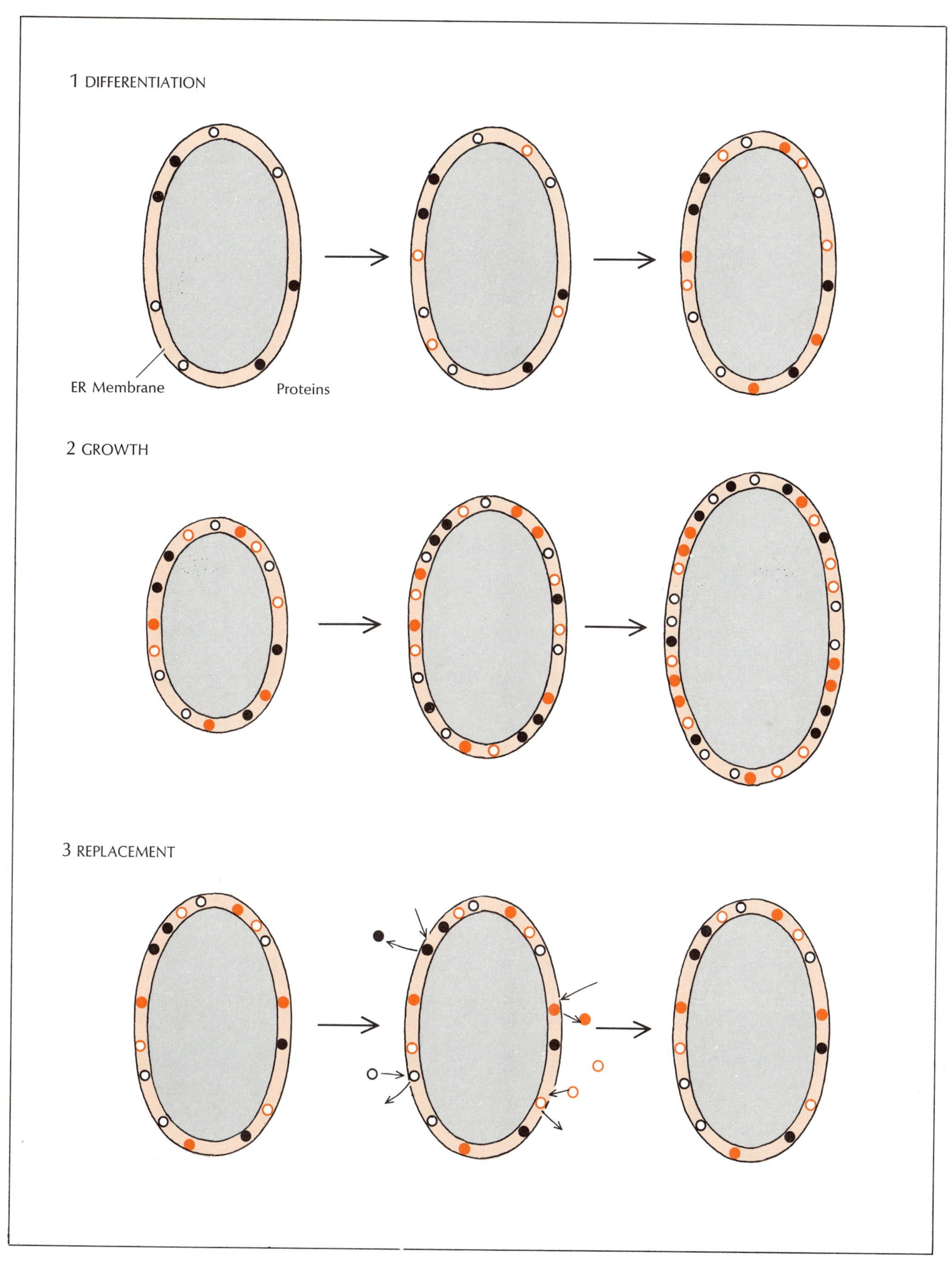

Three types of dynamic processes can be distinguished in membranes of ER. In differentiation (1), new proteins (all of them may be enzymes) are incorporated into membrane structure ("proto-enzymes" already present may also be activated). Membrane growth (2) involves mere quantitative increase in existing enzymes. In addition, enzyme molecules are constantly being removed and replaced by others of same species (3). More than one process may be going on at a given time.

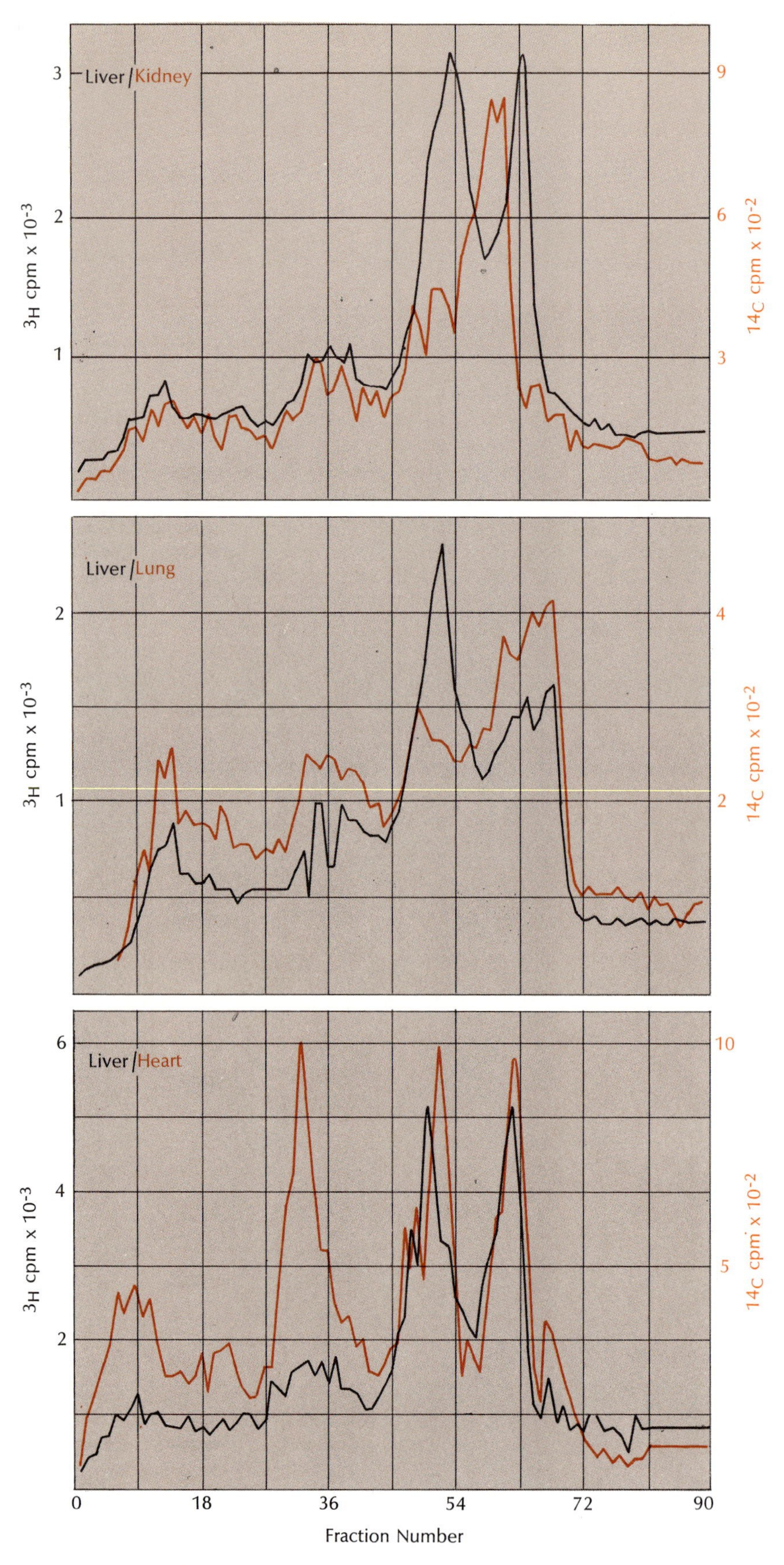

Sharp organ-specific differences in protein composition of endoplasmic membranes are shown by chromatography involving double labeling. Curves for kidney, lung, and heart membranes all differ markedly from those for liver, and also from one another. Minor differences among the three liver curves are due to variations in the experimental conditions (redrawn from E. O. Kiehn and J. J. Holland, Biochemistry *9:1729, 1970).*

develops only gradually. Specifically, the enzyme activities encompassed by the NADPH chain appear one by one, and in strict order in the differentiating ER membranes. Interestingly, the order appears to correspond to the sequence in which the enzymes will actually function in their oxidative task. A similar stepwise appearance of enzymatic activities has been observed in the electron-transfer chain in fetal mitochondria. Yet during these relatively protracted processes, neither the ER nor the mitochondrial membranes show any morphologic changes under the electron microscope, though such modifications must presumably have taken place on a submicroscopic level.

No less interesting is the finding that this stepwise appearance of ER enzymes is essentially recapitulated in the adult hepatocyte during enzyme induction by barbiturates. Here, of course, it is not a question of new enzyme activities appearing but rather of marked increases in the activity of certain ER enzymes—however, the increases of enzymes within the oxidation chain occur in the same order that the activities originally appeared during cell maturation.

A third aspect of dynamism in intracellular membranes is that of turnover. The membranes, as we have seen, are present at the very beginning of the organism's development and attain their mature appearance (though not their mature functions) shortly thereafter; they persist (apart from such special cases as the erythrocyte) for the life of the cell. But this is by no means true of the individual molecules that make up the membranes, which are constantly being replaced. Thus, for example, the lifetime of a human hepatocyte – and of its membranes – is estimated at something like 6 to 12 months; by contrast, the protein and phospholipid molecules of its ER membranes have lifetimes ranging from 1 to about 20 days, so that the structural continuity of the membrane is maintained despite the constant replacement of its constituent molecules. The membrane can be compared to a deformable, semifluid wall consisting of varicolored "bricks." Periodically, a brick of a given color is removed, to be replaced by a similar brick; nevertheless, the wall itself retains its essential structural integrity.

Thus the membrane we observe in a given cell after a lapse of a month or two is a quite different membrane from the one we originally observed, since few or none of its molecules were present (or even in existence) at that time. Yet it is also the same membrane, since its functional and (so far as we can determine) its structural properties have remained unaltered. There is an obvious parallel here with the well-known relationship between multicelled organisms and their individual cells. Most of the latter, too, are constantly being replaced – yet the organism is obviously able to retain its physical (and even its psychological) identity.

In regard to the organism, we can at least crudely rationalize this continuous replacement process: cells have a finite lifetime, and must therefore be replaced as they die if the organism is to survive. By analogy, it has been suggested that individual membrane molecules also have a finite lifetime. Thus enzyme molecules, in the course of performing their metabolic activities, are possibly being constantly deformed, a process that – by analogy with metal fatigue – might eventually produce permanent damage to their structures. But we have no notion of how such damage might occur.

However, in contrast with cellular replacement, molecular replacement within the cell appears to have no relationship to the age of the molecule in question. When membrane proteins are labeled with isotopically tagged amino acids and then removed to a normal medium, one might expect to see a continued high level of radioactivity for a period of some days, followed by a precipitous decline as the population of labeled molecules reaches its term of life. This sort of decay can be observed in the case of human erythrocytes; effete or older cells are selectively removed by the spleen. What one actually sees with membranes of the ER, however, is a standard, half-life decay curve, in which the rate of decline is highest at the beginning and decreases in direct proportion to the decreasing number of labeled molecules remaining. All the evidence we have indicates that molecular replacement proceeds

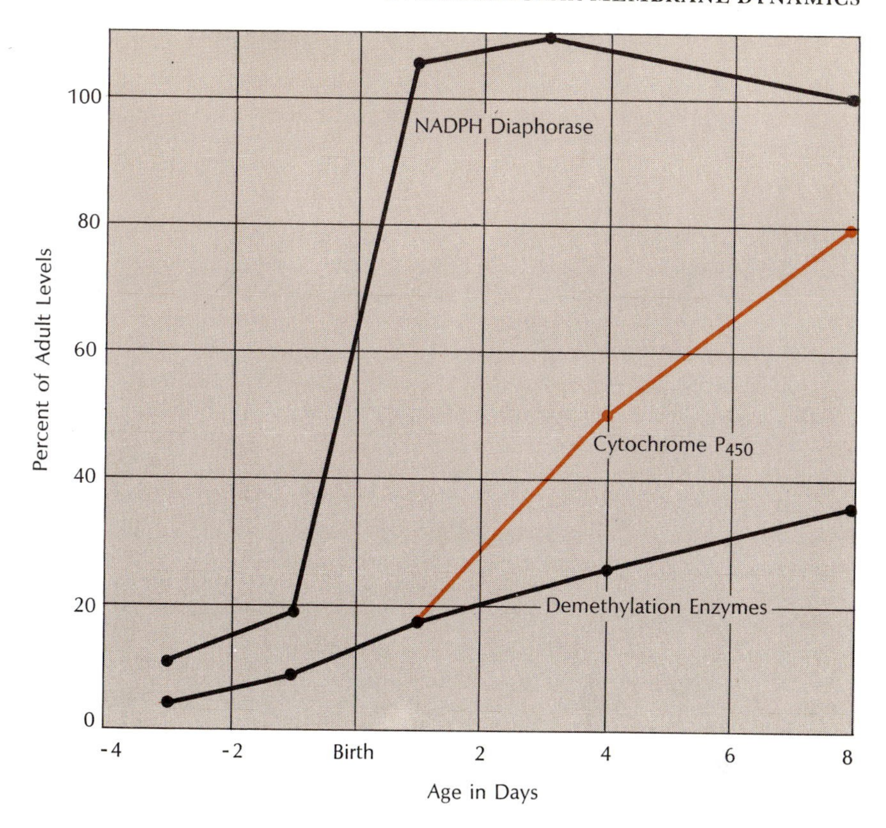

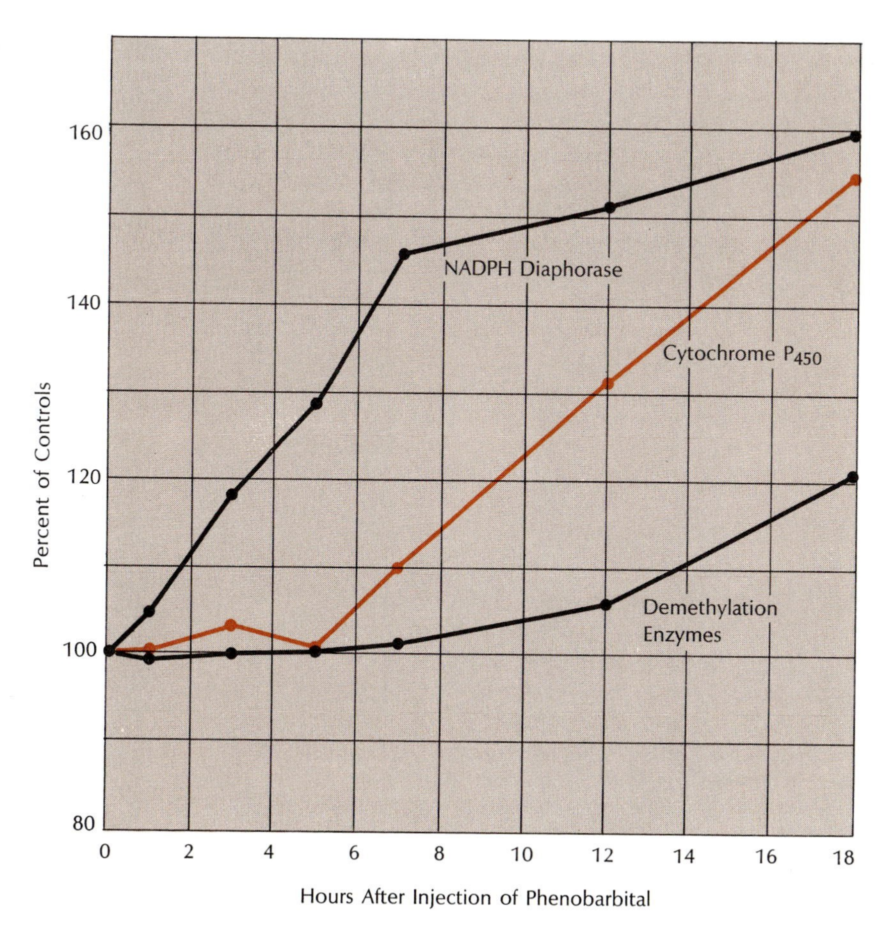

During perinatal differentiation of rat liver cells, the enzymes of the smooth endoplasmic reticulum that serve to detoxify barbiturates and other exogenous substances appear in a specific order (top). The same order of appearance is observed when these enzymes are induced by IV injection of phenobarbital into the rat (bottom).

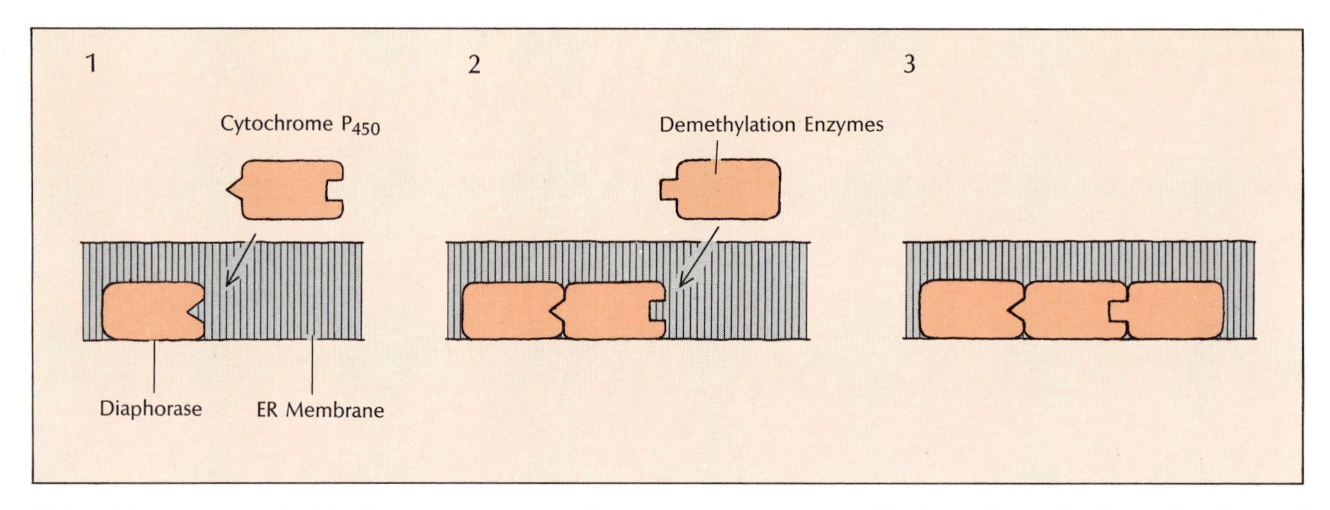

Sequential appearance of detoxification enzymes can be explained by postulating attachment sites on each member of the series keyed to complementary sites on the next to appear. Thus, a given enzyme could take its place only when its "predecessor" was already in position. This arrangement could also ensure a closed, fixed, spatial relationship among the enzymes.

on a purely random basis, with no relation to molecular age or (presumed) damage.

Thus, if we assume that molecular replacement is physiologically necessary – and it is hard to believe that such an energy-consuming process would be so universally present in living cells without that necessity – it would seem to proceed in a singularly inefficient manner, having no discernible relationship to the functional capacity of the molecules being replaced at any given moment. Of course, no evolutionary law has been passed that directs organisms to operate at 100% efficiency; they need only be efficient enough to survive. It is possible that the membrane can tolerate a certain proportion of nonfunctional molecules (assuming that loss of function is the physiologic rationale behind replacement), provided only that the replacement process proceeds with sufficient vigor to maintain an adequate number of functional ones.

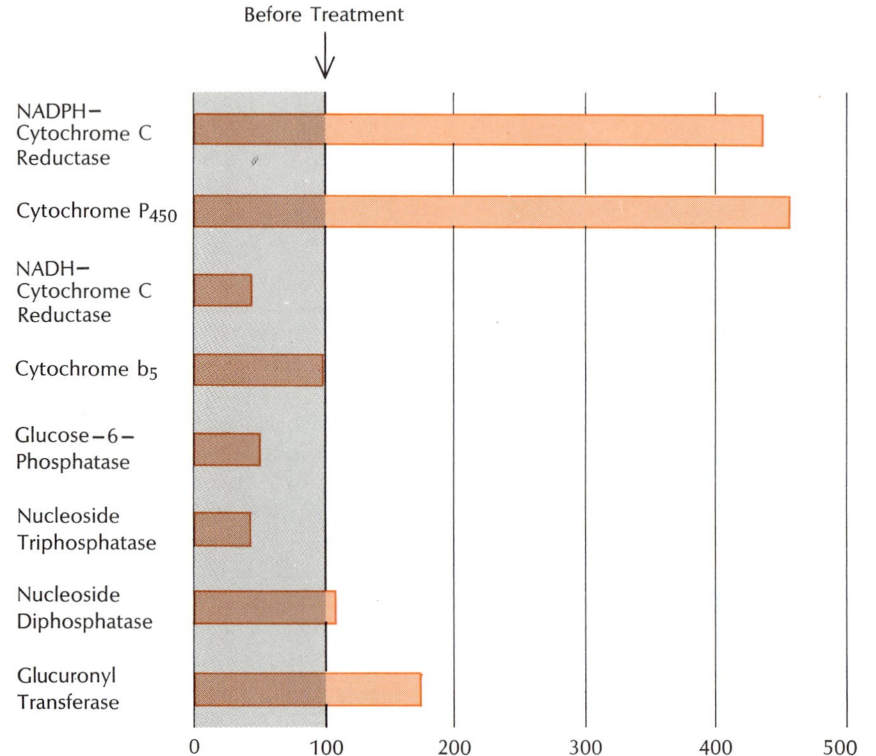

Proportions of enzymes in liver cell ER change radically after induction with barbiturates. With enzymes involved in detoxification rising, and others falling, the overall protein content–and structure–of the membrane shows no changes.

Questions Still To Be Answered

All these dynamic processes affecting the intracellular membranes raise a number of difficult and, in our present state of knowledge, mostly still unanswerable questions.

The first of these concerns the mechanics of differentiation – specifically, how "new" enzymes get into the membrane at a particular point in development. There are at least two possibilities here. One is that the enzymes in question do not "get into" the membrane because they are there all along, but in an inactive form. This is what seems to occur with some of the cytochromes, at least five of which are found in the inner mitochondrial membranes, with others in the outer mitochondrial membrane and in the ER membranes. All the cytochromes are enzymes consisting of a protein attached to an iron-containing heme group. And immunologic and other studies have shown that the protein moieties – the so-called apoproteins – of some of these cytochromes (cytochrome oxidase in mitochondria and cytochrome P_{450} in ER membranes) are present in the appropriate membranes long before any enzyme activity is detectable. They become activated when (for unknown reasons) they combine with a heme group. The same sort of activation of nonfunctional enzymes has been observed in nonheme proteins such as insulin and trypsinogen; here activation is presumed – though it is not known how – to result from the deletion of a portion of the amino acid chain.

Other enzymes, however, such as mitochondrial cytochrome C and ER-glucose-6-phosphatase, do not seem to be present in the membranes from the beginning. They cannot be detected immunologically, and if the cell is treated with substances that block protein synthesis, they never appear at all. Evidently they begin to be manufactured only at a certain stage in differentiation.

This, then, raises a second question: how does a newly manufactured enzyme "know" into which particular intracellular membrane(s) it must incorporate itself? As already noted, most enzymes, and other cellular proteins, seem to be manufactured in the rough ER, but most of them migrate elsewhere in the cell, or, in the case of secretory proteins, outside it. Cytochrome b_5, for example, is found only in the rough and smooth ER and in the outer mitochondrial membrane, never in the inner mitochondrial membrane or plasma membrane. Cytochrome C, by contrast, is found in the inner mitochondrial membrane and nowhere else, cytochrome P_{450} only in the ER, and so on. Given the constant turnover of membrane constituents described above, these migrations of newly synthesized proteins to specific sites must be proceeding continuously throughout the life of the cell. The same problem exists with respect to membrane phospholipids. These very varied molecules are by no means as specific as the membrane proteins, in the sense that most of them are found in all types of membrane, albeit in different proportions. But some are not, and for them, as for the proteins, one must enquire how they get from the smooth ER, where they are manufactured, to the particular membrane where they reside and function.

As regards the enzymes and other proteins, it has seemed necessary to postulate the existence of binding sites in the various membranes to which particular proteins, and no others, can attach themselves. (These sites should not be confused with the active sites of enzymes, which interact with the substrates of these enzymes.) By this theory, particular proteins end up in particular membranes because only in those loci can they be incorporated into the membrane structure. Though there is no direct evidence for this assumption, it seems to explain a number of facts with a minimum of hypothesis. Thus, the sequential appearance of enzymes in one of the metabolic chains cited above would suggest that each enzyme molecule possesses a binding site for the next in line, so that number two could appear only after number one, number three after number two, and so on. We know from independent evidence, moreover, that these enzyme chains can only function efficiently (if at all) when each "link" is in close juxtaposition with the next; physical junctions between the members of the chain, in the shape of the attachment sites, would obviously help maintain the necessary spatial contiguity. Concerning the phospholipid molecules, some workers have postulated "carrier" protein molecules that transport phospholipids from the smooth ER to other loci in the cell; in certain cases, at least, these carrier molecules would need to possess some sort of affinity for particular membranes in order to ensure that certain types of phospholipids became incorporated only where they belong.

But these hypotheses, plausible though they may be, only push the mystery back one step. For even if we assume that particular proteins become localized in particular membranes by their affinity for attachment sites of other proteins, there remains the question of how those proteins got there in the first place – and continue to get there as they continue to be replaced in the normal process of turnover. Likewise, if the phospholipid molecules are transported by carrier proteins, how do these molecules know where to go? It is rather like the problem posed by the little boy who asked "If God made the world, who made God?"

There is also the question of how membranes retain their specificity –

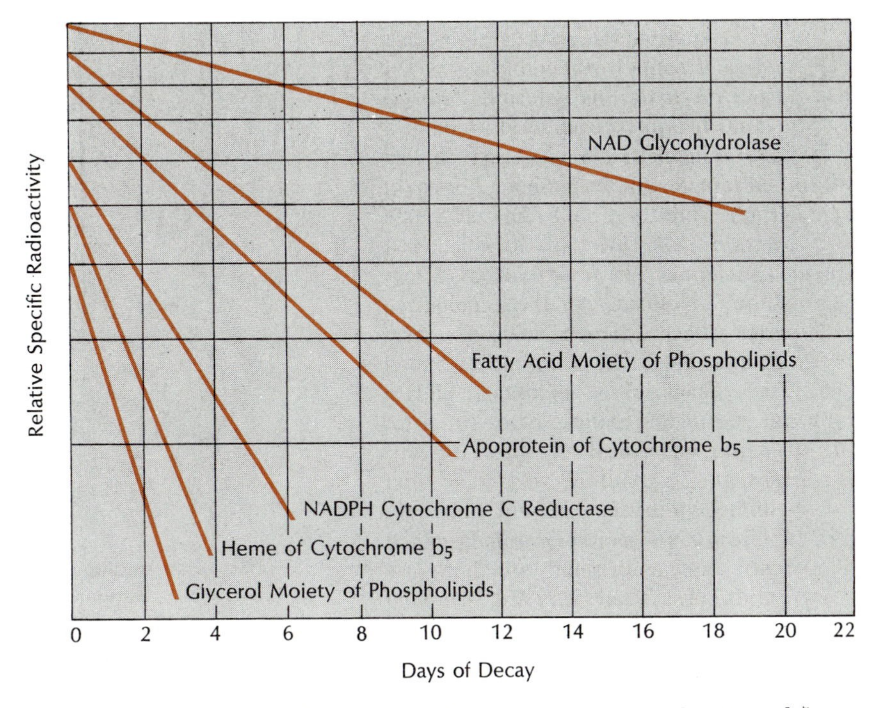

Rate of replacement varies from one membrane component to another. Note differing slopes of curves. Rate in no case depends on age of individual molecules but is essentially random, as shown by linear plots on semilogarithmic scale.

their characteristic proportions of the appropriate proteins and phospholipids – in the face of the constant breakdown and replacement of these molecules; of what keeps the synthesis and transport of a particular species of molecule in step with the loss of the same species. To some extent this problem is perhaps less intractable than those just cited, in that there seems to be a rough feedback relationship between molecular synthesis and degradation. The rate of breakdown of a particular molecule, being apparently a wholly random process, is in proportion to the concentration of that molecule; that is, should its rate of synthesis increase for some reason, its concentration – and therefore its rate of breakdown – will also rise, while the reverse will occur should synthesis slacken; thus a steady-state concentration is assured.

It is obvious, however, that this control process can only operate within certain limits; should synthesis increase or slacken unduly, the result will be an increase or decrease in concentration of the molecule in question, which can be only partially compensated by changes in its breakdown rate. One would expect, therefore, that in addition to the just-cited feedback linkage between rate of synthesis and rate of breakdown, some other control process would link rate of breakdown or (more likely) concentration of the molecule to rate of synthesis, somewhat in the same way that increased serum concentrations of hormones reduce their synthesis by the endocrines. But we cannot even guess through what mechanisms such a control process might operate.

Even more puzzling is the matter of how membranes can change their constituency to a considerable extent while undergoing no discernible change in structure. For example, cytochrome P_{450} – one of the "detoxification" enzymes – is the major protein constituent of smooth ER, amounting to about 10% of the membrane protein by weight. When the detoxification enzymes are induced, by barbiturate injection, for example, the concentration of cytochrome P_{450} rises to as much as 40%. Were this cytochrome a loosely attached, peripheral protein, its increase would pose few problems. Since the aggregate proportion of proteins in the ER, relative to lipid, does not increase during induction – the gain in cytochrome P_{450} being balanced by a lower concentration of other proteins – one would need only assume that binding sites on the membrane surface normally occupied by other enzymes were now preempted by the cytochrome. In fact, however, by every known criterion cytochrome P_{450} is *not* a peripheral but an integral protein; it is tightly bound to the membrane and for this reason has never been purified to any significant extent. What this means is that during enzyme induction a large proportion of the other ER proteins has been replaced by cytochrome P_{450} (a molecule, one must assume, of considerably different shape and perhaps electrostatic properties) *without* deranging the overall membrane structure. One ends up with a membrane having the same enzymic specificity, but variations in the amount of these specific proteins, yet still exhibiting the same structure as the membrane prior to the induction by barbiturate injection.

Finally, we have the problem of how all this molecular coming and going can take place in conformity with thermodynamic law. In the classic membrane bilayer theory, the basic structure of these membranes is hydrophilic on the exterior surfaces, hydrophobic inside. Thus, any protein molecule that is embedded in the membrane (as opposed to being loosely attached to its surface) must presumably contain a hydrophobic portion that will be compatible with the membrane interior. This has actually been shown to be the case with a few enzymes, including cytochrome b_5, whose hydrophobic section (consisting of about 40 amino acid residues) can be detached without altering its metabolic activity. But we are still in the dark as to how the hydrophobic chain of a newly synthesized protein can force its way through the hydrophilic exterior of a membrane into the hydrophobic interior where it will "feel at home."

The same problem applies to newly synthesized phospholipid molecules, many of which travel from their birthplace – the ER – to do replacement duty elsewhere. Labeling experiments have demonstrated that this transport actually occurs between, say, the ER and the mitochondria (and also, for unknown reasons, in the reverse direction). And double-labeling experiments, in which the phosphate and lipid portions of the molecule are labeled with different isotopes, indicate that the transport involves whole phospholipid molecules, not their components. If one postulates (as noted above) that at least some phospholipids are carried by transport proteins, the function of the latter might include the shielding of their lipid, hydrophobic moieties from the aqueous medium through which the phospholipids are carried. But however they get to where they are going, the phospholipids must then be inserted into an existing and presumably intact membrane. Which seems thermodynamically unlikely, if not impossible – yet it not only demonstrably occurs, but does so routinely.

Enough has been said, I think, to indicate, first that the dynamic processes of intracellular membranes are exceedingly complex, and second, that they have engendered far more questions than can at present be answered. But that, of course, is what makes this field of biologic research so very fascinating – if occasionally frustrating as well.

13

The Network of Intracellular Membranes

BENJAMIN F. TRUMP
University of Maryland

While authors of previous chapters have made it clear that the term "cell membranes" refers to the inner as well as the outer (plasma) membranes of the cells, their discussions have been concerned mainly with the outer membranes. Although, as we shall see in a moment, there is really no hard and fast line between the intracellular membrane and the plasma membrane, those within cells constitute, in fact, a subject in themselves. This chapter will attempt to provide an overview of this area, while subsequent chapters will concentrate on particular types of these interior formations.

Another goal of this chapter is to begin to illustrate how the detailed consideration of cell membranes will lead to increased understanding of human disease. In studies based on immediate autopsies performed in collaboration with the Maryland Institute for Emergency Medicine, which is directed by Dr. R Adams Cowley, we have developed a technique that permits a subcellular approach to the study of human disease. When Dr. Cowley and I began this work it was surprising to us to discover that, because of limitations of technique, many of the obvious subcellular properties quite clear in vertebrate and invertebrate model systems had not been carefully studied in the human.

When we speak of intracellular membranes, we mean the membranes of virtually all the organelles that occur in the "cell sap" or cytoplasm. The organelles, mitochondria, endoplasmic reticulum, lysosomes, microbodies or peroxisomes, Golgi apparatus, and so on are defined by membranes of various types, which, in addition to serving particular functions, serve to separate them from the undifferentiated cell sap in which they reside.

Organelles are characteristic of all eukaryotic cells, which is to say virtually all cells of metazoan organisms (the mature erythrocyte being a notable exception) and a number of unicellular organisms as well. They are less well developed, however, in bacteria, whose small size in comparison with other cells is probably related to this fact. While most biochemical activities in bacteria are thought to take place in and around their plasma membranes, in larger cells this limitation does not apply; if it did, they would be unable, because of their smaller proportion of surface to volume, to carry on their business. By extending the membrane into the cell interior and thus markedly expanding the cells' "working surface," evolution has, as it were, provided them with an alternative. In addition, of course, the structural and functional differentiation of the internal membranes into various types of organelles – somewhat paralleling differentiation among the cells of many-celled organisms – makes possible intracellular activities of much greater variety and complexity.

It is convenient to think of the entire system of interior membranes as a network – sometimes called the "cytocavitary network," in recognition of the fact that the organelles form cavities within the cytoplasm. All the parts of the network, i.e., all the organelles of a given cell, can be regarded as continuous in time, in the sense that all of them are in communication with one another at some point, directly or indirectly, though no single organelle ever links up directly with more than a few of the others. To what extent the cytocavitary network is also continuous in space, in the sense of a number of organelles being linked together simultaneously, is uncertain; recent studies indicate that such spatial continuity is considerably more extensive than was once thought.

Older concepts of how organelles were disposed led us to believe that they were spatially discrete bodies having only occasional connections with one another. Such concepts now seem to be in considerable measure the consequence of the technique from which they were derived – thin-section electron microscopy, in which the portions of the cell visualized were only about a thousand angstroms thick. The inevitable result of this limitation was to overemphasize the discreteness of individual organelles, or even parts of organelles, in much the same way as a "thin section" of the human body would show the intestines as a series of discrete loops rather than a continuous tube. Nowadays, using high-voltage electrons, we can visualize cell sections up to 2 μm thick or even more; also, by tilting the specimen slightly we can obtain stereo pairs of micrographs that show three-dimensional structure when viewed through appropriate equipment. In addition, scanning microscopy is currently giving much more information about intracellular membranes.

A clearer understanding of the "network" concept of intracellular membranes can be obtained by considering some of the network's characteristic activities. In phagocytosis (or pinocytosis), the organelle called a phagocyte vacuole, or phagosome, is formed when a portion of the plasma membrane invaginates, producing a sort of pocket around some body in the extracellular medium. Fusion of the plasma membranes around the mouth of the pocket permits the vacuole to pinch off and float free within the cell. The phagosome then merges with a lysosome – a similar vacuole originating inside the cell, containing acid hydrolases and other materials. The enzymes then break the extraneous substance into much smaller aggregates – assuming, of course, that it is digestible – which can pass through the wall of the consolidated vacuole into the cell sap. Phagocytosis of this sort is a very common cellular process. We find it in unicellular organisms such as amebae, which derive virtually their entire nourishment in this manner, and also in the leukocytes of higher animals. Many other cells employ the same process to ingest and digest macromolecules such as proteins, as well as to deal with nondigestible materials like carbon particles (e.g., in pulmonary anthracosis), finely divided silica, urate, and so on.

When digestion, if any, is complete, a certain amount of debris is often left within the phagosome-lysosome vacuole. Under certain circumstances (i.e., presumably when the lysosomal enzymes within the vacuole are still functional) the vacuole may merge with another phagosome and repeat the digestive process. In other cases the debris-laden vacuole will move back to the plasma membrane and, in a process reversing that by which the phagosomes are formed, merge with it and eject its contents, much of them

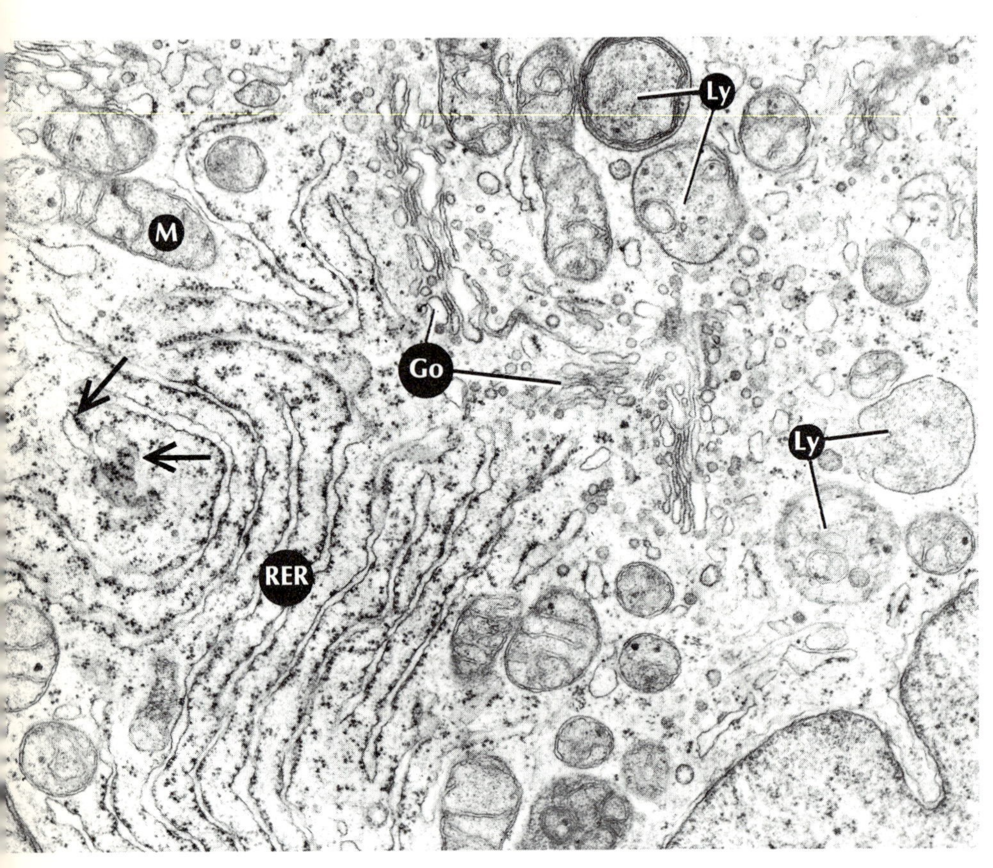

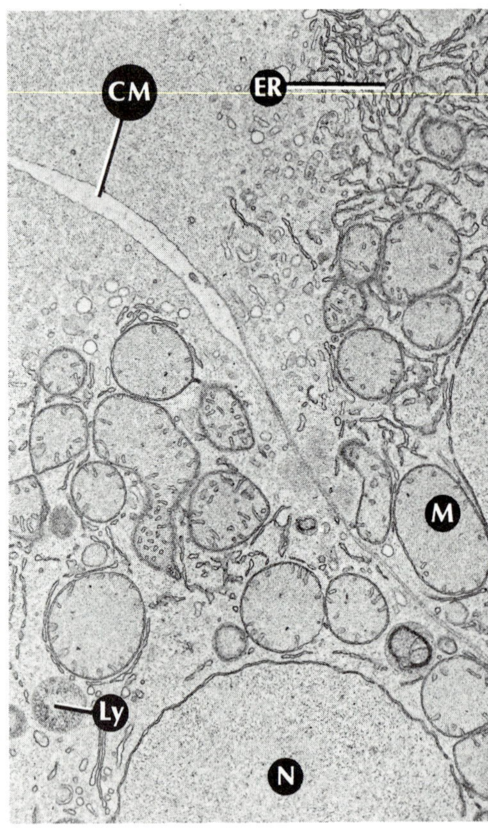

Several elements of the cytocavitary network are seen in a portion of a human kidney tubule cell (left): the Golgi apparatus (Go), which, in thin sections like this one, appears as stacks of membrane-bounded saccules; lysosomes (Ly) containing various types of debris on the exoplasmic side of the Golgi; rough endoplasmic reticulum (RER) on its other side, and mitochondria (M). The RER is seen in profile; ribosomes arranged into polysomes are visible on its surfaces. It is thought that the proteins secreted by the polysomes pass into the lumens of the ER and thence into the Golgi. Free arrows indicate tangential or face-on (rather than profile) cuts of the RER; here the polysomes appear to be arranged in parallel rows or spirals. At right: a fish kidney tubule fixed in potassium permanganate, which extracts most of the cytoplasmic proteins without affecting the membranes. This makes it possible to see that the membranes form a series of channels through the cytoplasm.

indigestible debris, into the extracellular space. It is not certain whether all cells do this. Those of the kidney tubules and liver certainly do, ejecting the contents of their lysosomes into the lumens of the tubules and bile duct respectively. Other cells, such as neurons, may lack this capacity; it is suspected that the "aging pigment" (lipofuscin) that accumulates with time in these cells may derive from just such debris.

Thus far I have described phagocytosis in the "old" style, as a series of more or less discrete steps. In fact, there is now considerable evidence that in some cases at least the fusion of phagosome and lysosome occurs *before* the former has pinched off from the plasma membrane, while the mouth of the pocket is still open to the extracellular space. This process has been termed "regurgitation during feeding" as the digestive ferments can escape to the outside of the cell. Time-lapse films by light microscopy of bacteria being ingested by a polymorphonuclear leukocyte show just such an overlap of vacuole formation and vacuole fusion. In other cases, involving what is called "frustrated phagocytosis" of antigen-antibody complexes on a surface, the polymorphs seem incapable of internalizing the complexes at all; instead, the lysosomal granules discharge their enzymes directly into the extracellular space.

Biochemical studies have confirmed that even in the former case the fusion of lysosome and "incomplete" phagosome leads to a certain leakage of the acid hydrolases into the extracellular space – as occurs, a fortiori, when the phagosomes do not form at all, as with the antigen-antibody complexes. Processes of these sorts may play an important role in the pathogenesis of injury and inflammation. For example, Weissmann has shown that injection of the hydrolases into the joints of an animal can produce something very much like rheumatoid arthritis; it is perhaps significant that antigen-antibody complexes, whose presence would lead to maximum hydrolase leakages, are almost certainly present in human rheumatoid arthritis. The less marked leakage produced by lysosome fusion with an incomplete phagosome could produce much of the destruction at any inflammatory focus.

Consideration of where the lysosomes come from takes us deeper into the cytocavitary network. They principally originate in the Golgi apparatus, a series of roughly lenticular, membranous saccules or cavities, connected by narrow anastamoses, though the latter do not show up in most micrographs. Originally it was thought that the lysosomes budded off from the Golgi membranes much as the phagosomes bud off from the plasma membrane, but here, too, recent studies have cast doubt on this simplistic picture. In some cells at least (such as kidney tubules), thick-section EMG's by Dr. Elizabeth McDowell in our lab have revealed long, enzyme-filled tubules extending out from the Golgi apparatus to the point where they may even merge with a phagosome. In other cases, however, the budding process seems to be an accurate description. This is notably true of various secretory cells (e.g., those of the pancreas); the secretory granules, budded off by the Golgi apparatus, move directly to the plasma membrane where they merge with it and eject their contents into the extracellular space.

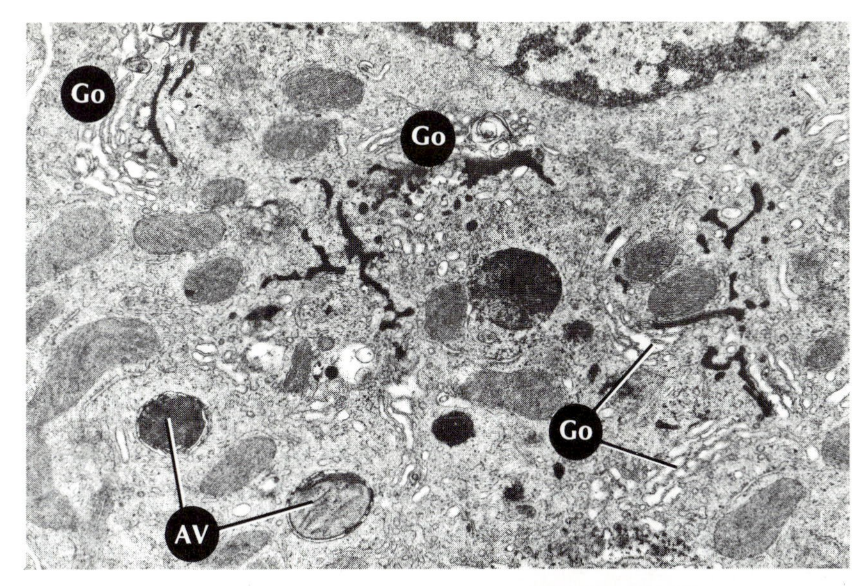

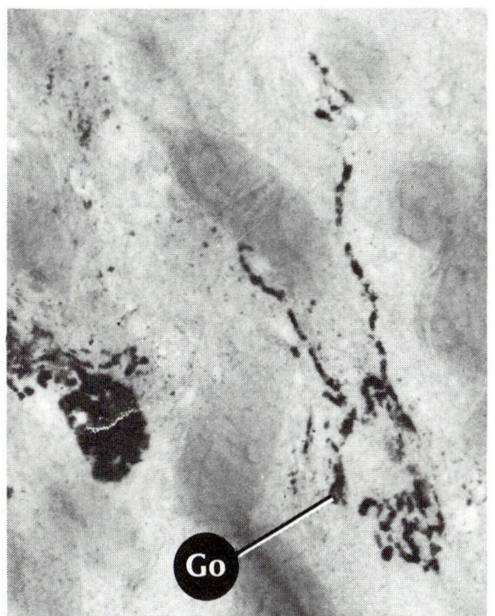

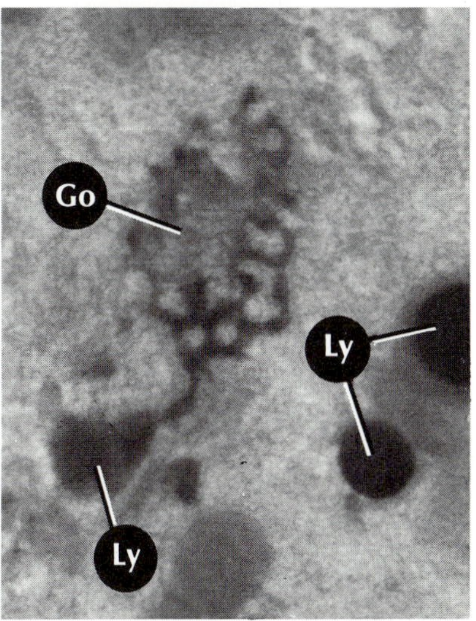

When lead is used as a label in studies to demonstrate acid phosphatase activity, the connections leading from the Golgi apparatus to other organelles are visualized as electron dense reaction products. These EM's, all of proximal rat kidney tubule, are by Dr. Elizabeth McDowell of the author's laboratory. Top: acid phosphatase activity is localized along one side of Golgi and in a variety of vesicles representing extensions from it, in secondary lysosomes, and in autophagic vacuoles (AV). Below, left, thick section (0.5 μ) EM shows network of tubules comprising the Golgi, some extending far from the apparatus itself. Right: continuity of the Golgi and lysosome is evident.

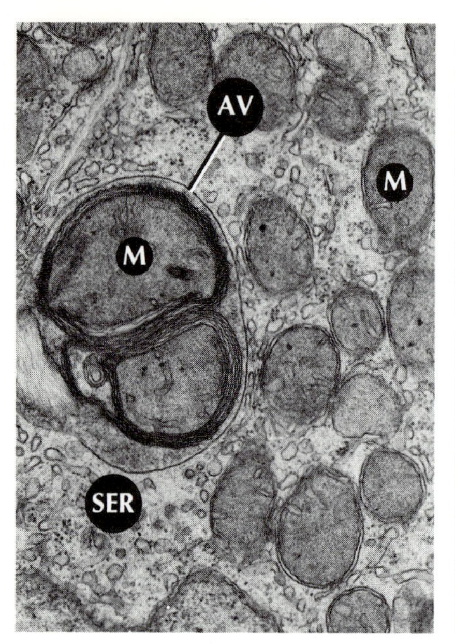

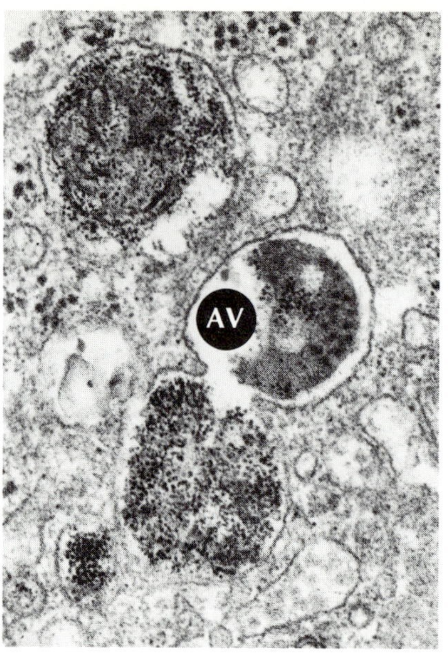

Metabolic functions of ER include autophagocytosis; at left, an autophagic vacuole adjacent to smooth ER contains mitochondria, lamellar membranous material and other debris. At right, in rat hepatocyte 72 hours after administration of iron dextran, dense particles inside lysosomes are ferritin and hemosiderin resulting from iron stimulation; autophagic vacuole is fusing with a lysosome. In human hemosiderosis and hemochromatosis, the lysosomes are the site of storage of ferritin and hemosiderin.

Probing still deeper into the network, we find that the enzymes and other substances that are, as it were, packaged by the Golgi apparatus and shipped either to the phagosomes or outside the cell do not originate in the apparatus. Rather, they are manufactured by the endoplasmic reticulum (ER), which, like the Golgi apparatus, consists of a series of anastomosing cisternae, permeating much of the cell substance. Transport of the ER's products to the Golgi apparatus occurs either through the formation of vacuoles, which then merge with the Golgi cisternae at one pole, or by means of actual tubules; again, we do not yet know which (if either) of these processes predominates under different conditions and in different cells.

In addition to these transport vacuoles or tubules the ER seems to form other types of organelles, often by budding portions of its substance away from the endoplasmic reticulum as microbodies. These microbodies or peroxisomes are still of unknown significance in both the normal cell economy and in disease, though Svoboda and his coworkers have recently been studying the striking induction of peroxisomes which occurs after administration of the lipolytic drug CPIB. The endoplasmic reticulum also has potentially important, though at present rather poorly understood, relationships to the mitochondria, since there are indications that the outer mitochondrial membrane is continuous with the membranes of the endoplasmic reticulum and that the endoplasmic cisternae are continuous with the space between the outer and inner mitochondrial membranes. These relationships, if established, might indicate that mitochondria are well-adapted intracellular parasites, which, according to the endosymbiotic theory, have set up permanent "residence" within the lumen of the endoplasmic reticulum.

The fact that so many products – digestive enzymes, peroxidases, and the like – and structures such as the peroxisome appear to originate from the ER suggests that a great deal of biosynthesis must take place in and around these structures, and in fact this is the case. The ER serves as a physical support for the membrane-bound polysomes: aggregations of ribosomes that can "read off" the coded information on a strand of messenger RNA and use it to synthesize specific sequences of amino acids to form proteins; in fact, this membrane relationship may influence protein synthesis. Those portions of the ER with adherent polysomes are called, from their appearance, rough ER. This area synthesizes many proteins that are "exported" from the cell. But some cells also contain much smooth ER,

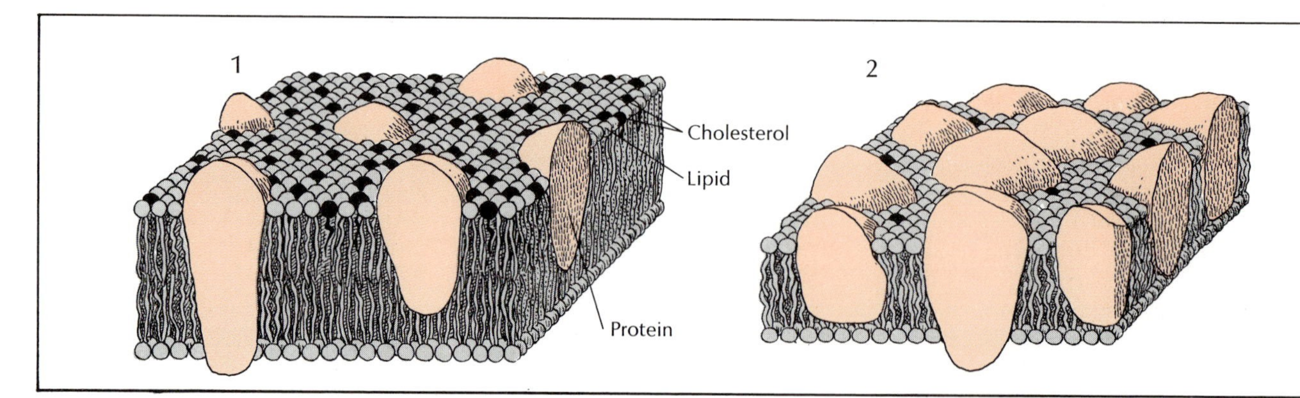

Postulated molecular structure of exoplasmic membrane (1) is thicker than that of endoplasmic (2); latter also has higher ratio of protein to lipid and contains less cholesterol. Endoplasmic membranes have greater structural stability, which they retain, though in a different conformation, (3) even after treatment with a bacterial exotoxin like the phospholipase C produced by Clostridia perfringens, *important in gas gangrene. Phospholipase C modifies lipid components of the membrane, releasing phosphate*

which is known to be involved in the synthesis of nonprotein cell components such as cholesterol and steroid hormones, as well as in certain steps of other lipid syntheses. The synthesis of cell membranes themselves appears to involve a collaboration between rough ER, smooth ER, and Golgi, which is precisely what one would expect given the composite, protein-lipid character of membrane structure.

Biosynthesis does not exhaust the role of the ER; sets of enzymes embedded in the (smooth) ER membranes are known to metabolize and detoxify such extraneous compounds as phenobarbital. When cells are stimulated by the presence of such substances, we first observe a proliferation of rough ER, evidently representing the stage of enzyme synthesis (induction), and later a similar proliferation of smooth ER, in which the induced enzymes perform their biochemical function. The ER's metabolism of extraneous compounds does not always successfully detoxify them; indeed, in the case of certain carcinogens as well as chemical toxins like carbon tetrachloride it is suspected that the damage is done not by the original substance but by its metabolically modified descendents.

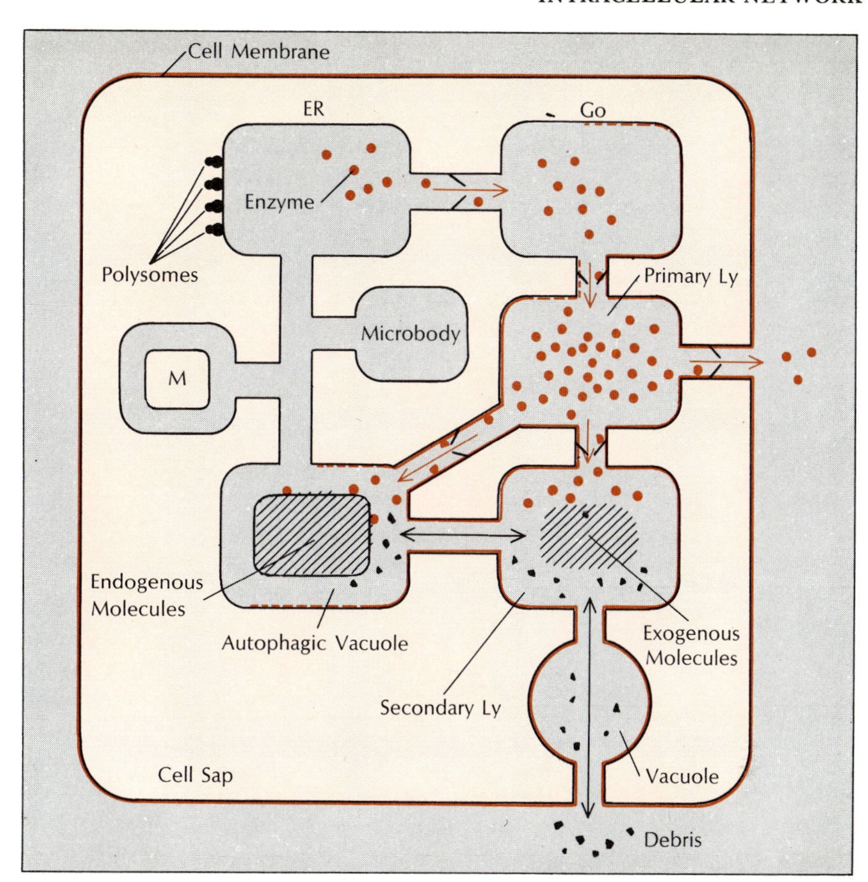

In this conceptualization of the cytocavitary network, the endoplasmic membranes are shown in black, the exoplasmic in color, and the emphasis is placed on the continuity among the various structures; note that the postulated valves are all one-way.

The ER's metabolic functions are not limited to the breakdown and detoxification of exogenous compounds; they can also involve the substance of the cell itself. Often a portion of the ER membrane will bud into its cavity, carrying with it a portion of cell sap; on occasion, this may include another organelle such as a mitochondrion. The resulting autophagosomes (which may also be formed by similar budding of the cell sap into an ordinary phagosome) often then merge with lysosomes, with the result that whatever structures they contain are digested in the same way as exogenous substances in ordinary phagosomes. This process of autophagocytosis is believed to be responsible for much, if not all, of the cell's regular turnover of its substance (the normal half-life of mitochondria, for example, is estimated at about 10 days) and thus acts as a sort of recycling apparatus.

Recently Dr. John Shelburne, Dr. Antti Arstila, and I have studied the process of autophagocytosis, which can be readily induced by the hormone glucagon in experimental animals as first described by Ashford and Porter. We have found that it is mediated by the chemical messenger cyclic adenosine monophosphate (CAMP) in a process that is dependent on intracellular calcium levels and probably modified by the function of intracellular microtubules. The process requires adenosine triphosphate (ATP) and may be related to certain chemical influences that can be demonstrated with simpler membrane systems, such as the erythrocyte, with compounds like primaquine, as discussed earlier.

These, then, are some of the ways in which membranes and other materials flow through the cytocavitary network, by means of membrane folding, fusion, and budding. I should emphasize that the flow is anything but random; that is, it occurs in certain directions only under certain conditions, and certain types of flow do not occur at all. For example, the mitochondria do not at any point merge with the Golgi apparatus or with the lysosomes, whereas the latter do not fuse with the ER. Likewise, enzymes synthesized in the ER flow to the Golgi apparatus (among other places), but enzymes in the Golgi apparatus do not flow back to the ER. One can visualize the network as roughly resembling a hydraulic system, in which pipes constrain the flow of water and wastes into certain pathways, while valves regulate the flow at different points in

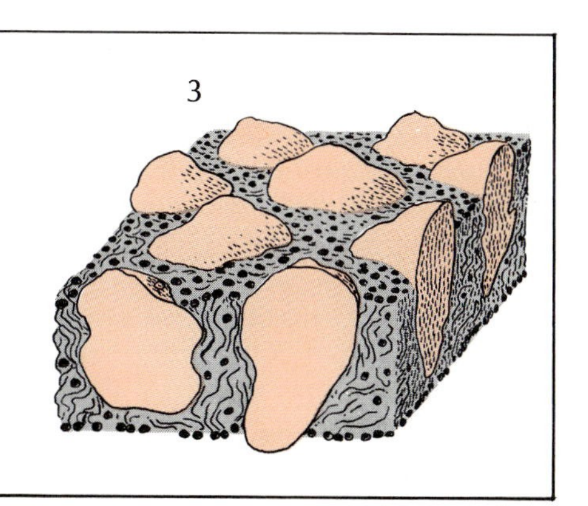

but retaining diglyceride. Similar changes are seen in patients dying of gas gangrene; in isolated membrane systems they can be reversed by adding phospholipid.

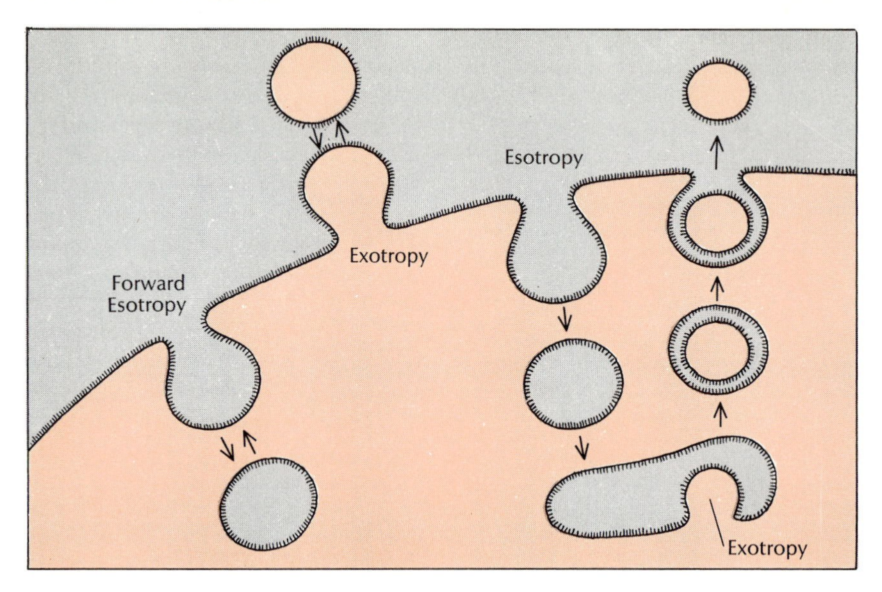

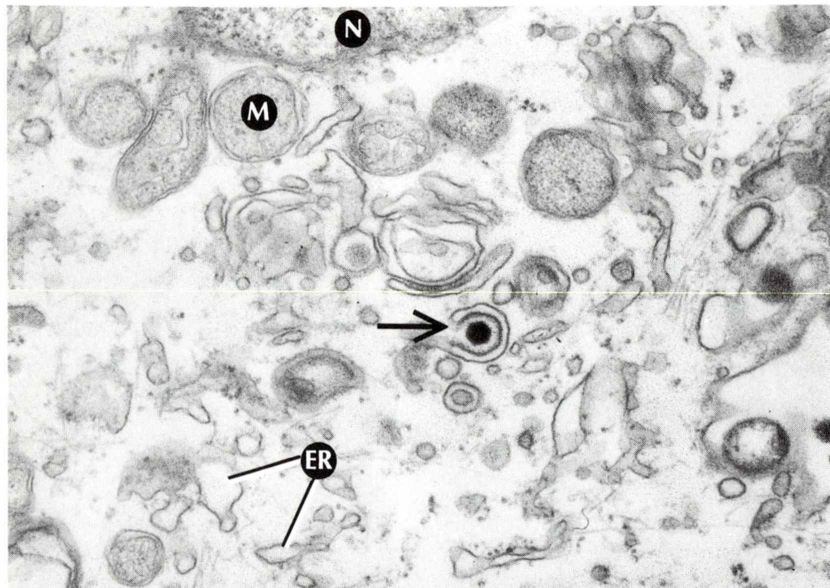

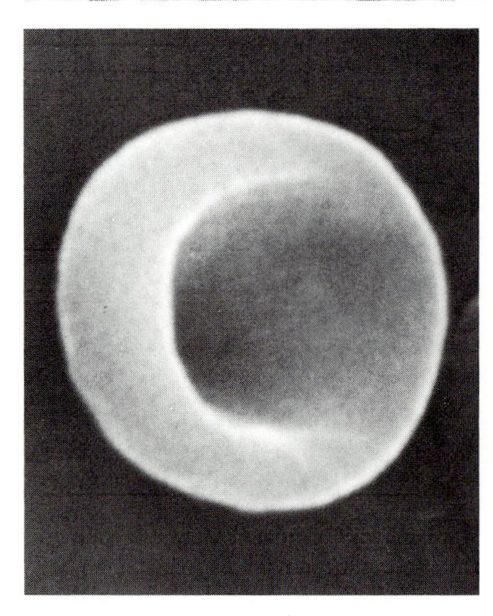

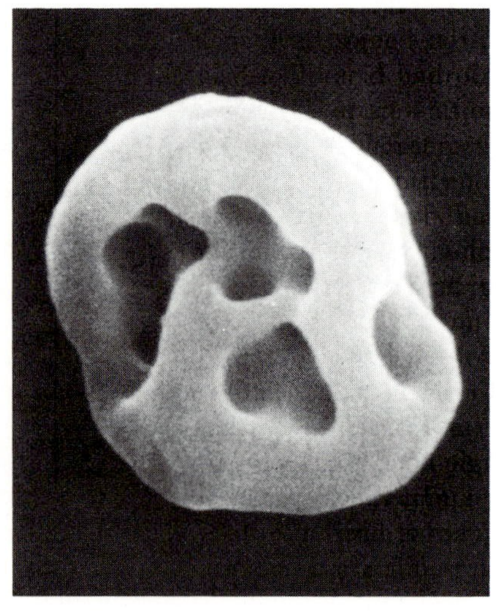

Exotropy and esotropy are the two basic processes by which substances (including membranes) enter or leave the cytocavitary network (diagram at top). Middle: Cultured HeLa cell infected with herpes shows nucleocapsid of virion (arrow) budding by exotropy into a cisterna (photo by B. B. Poeschel). Bottom: Esotropy by deep invagination occurs in normal human RBC after treatment with primaquine in presence of glucose.

time. The valves, moreover, can in this case be set to permit flow in one direction only, like the "flap valve" in the old-fashioned kitchen pump, whereby water could be lifted from one compartment (the well) into another (the pump barrel) without flowing back into the well on the return stroke.

Of course the "valves" in the cytocavitary network are nothing like hydraulic valves in any physical sense; rather, they may consist of the processes of fusion and budding already described. The question of why these processes occur at certain times and not at others, and between certain organelles and not others – in other words, *why* the valves open and close – is one of the most fascinating, poorly defined, problems in membrane research.

Before leaving this subject I should note another important if perhaps obvious point: the cytocavitary network is not the *only* way by which substances enter the cell or move about inside it. As earlier contributors have noted, many small molecules (glucose being a prime example) can pass through the plasma membrane either by simple diffusion or by some active transport mechanism without any obvious deformation of the membrane; at least at the ultrastructural level. The same molecules, as well as others synthesized within the cell, typically or mainly reside in the cell sap, from whence they pass or are transported into the membranous organelles as needed.

If we visualize the cytocavitary network as a system of tubes and valves, then there can be little doubt that the Golgi apparatus serves as a "master valve," in the sense that it is involved in the flow of most (though not all) of the substances that pass through the network. Not surprisingly, therefore, we find that the Golgi apparatus also serves as a sort of boundary between types of intracellular membrane. On the "outer" side of the apparatus we find the *exoplasmic* membranes of the phagocytic vacuoles, lysosomes, and so forth. They are about 100 Å thick and are characterized by a relatively high lipid-to-protein ratio – something like 1:1 – including a sizable proportion of cholesterol. In their general structure, they resemble the plasma membrane itself. This is not surprising, inasmuch as they are either derived from the membrane by invagination or will ultimately merge with it, as do debris-filled vacuoles and secretion vesicles. We even find that the exoplasmic membranes possess, on one side, the characteristic polysaccharide chains that are responsible for so many surface properties of the plasma membrane. In this case the polysaccharides are attached to the "wrong" side – i.e., the inner side – of the vacuolar membrane, which becomes the "right" side when this membrane merges with the plasma membrane.

This addition of the polysaccharide coat, i.e., formation of the "glycocalyx," involves the addition of strands of sugar polymers to membrane proteins such as glycophorin, which has important implications for cell-cell interactions, including their colonial behavior and cell differentiation. This places the function of the Golgi apparatus in a key position in terms of modulating cell behavior and differentiation and, indeed, in the carcinogenic process. DeLuca has recently provided indications that vitamin A, concentrated in liver Golgi, may exhibit an influence over these synthetic reactions in the Golgi apparatus, possibly acting as some sort of cofactor, at least in the intestine. This is an extremely interesting observation in view of the fact that vitamin A can modify malignant transformation and even reverse experimental premalignant lesions of the bronchus produced by benzo[*a*]pyrene in the hamster.

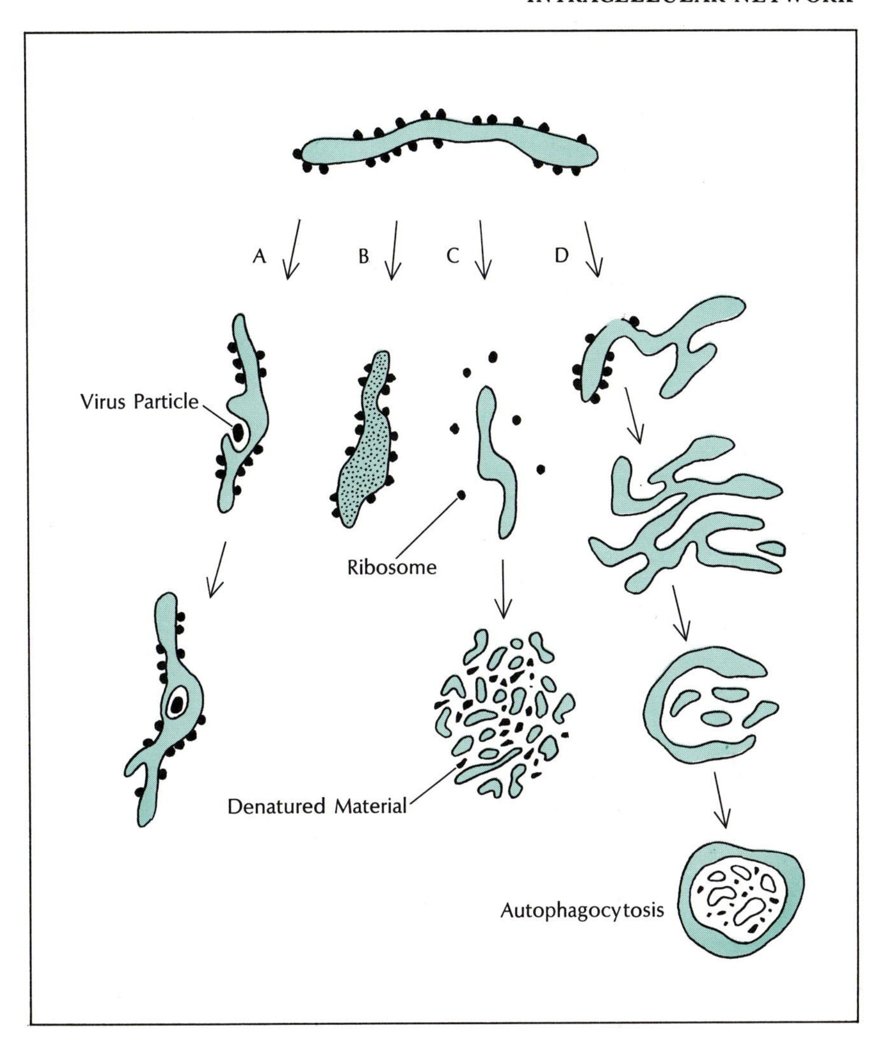

Four pathways through which the endoplasmic reticulum reacts when a cell is diseased or injured are diagrammed. In A, it first buds, then surrounds a virus particle. In B, it dilates in response to altered ion and water fluxes. In C, reacting to a chemical toxin such as carbon tetrachloride, ribosomes detach from membrane surface; protein synthesis and export are accordingly decreased. Subsequently the membrane lipid peroxidizes, yielding clusters of vesicles with dense denatured material between them. D shows induction of new endoplasmic membrane, as occurs with phenobarbital; if compound is removed, the induced membranes disappear, largely through autophagocytosis, shown as budding of bits of membrane containing cytoplasm into ER cisternae.

Upstream from the Golgi apparatus, we find the *endoplasmic* membranes of the mitochondria, ER, and so on. These are considerably thinner than exoplasmic membranes – about 70 Å – and also contain considerably more protein and less lipid (especially cholesterol), with protein-lipid ratios up to 4:1. This much higher ratio seems to be reflected in some basic structural differences. Singer and many other investigators conceptualize membranes as consisting of protein "icebergs" floating in a lipid "sea," and so far as the plasma membrane and exoplasmic membranes are concerned, the picture seems plausible (see Chapter 4, by Singer).

There is also the concept introduced by Chapman, in Chapter 2, that the sea can freeze and thaw, thus modifying the motion of the icebergs. An interesting detail added by Bretscher is that the lipid sea contains a top and a bottom in the sense that the extracellular side differs markedly from the cytoplasmic side in the type of lipids concerned. Further, as mentioned by Chapman, antidepressant drugs can apparently modify the temperature at which the lipid sea freezes and in which the protein icebergs lose their mobility and, inferentially, some of their important functions. Therefore, this freezing and thawing of the lipid sea, apart from its interesting theoretical implications, may be of great therapeutic importance.

By contrast, many if not all the plasma membrane proteins demonstrably float freely in their lipid medium, and it seems likely that they function with the same efficiency (as in material transport, for example) in any part of the membrane in which they happen to be located.

In the case of the endoplasmic membranes, the decisive structural component seems to be protein rather than lipid, so that the membrane apparently consists of "pools" of lipid scattered over an "ice field" of proteins. This can be shown directly; Arstila and I, while we were working at Duke University together with Dr. Sue Duttera and Dr. William Byrne, extracted virtually all the lipid from various types of endoplasmic membrane without destroying their structural integrity, though the membrane thickness increased. If one does this to plasma membranes, they simply fall apart.

A structure such as endoplasmic membrane, consisting of a mosaic of protein molecules interspersed with discrete areas of lipid, would almost inevitably be more rigid and less labile than the other types of membrane, and this seems quite expectable in view of the functions of at least some endoplasmic membranes. We know that the chains of enzymes responsible for electron transport in the mitochondrion can function efficiently only to the extent that they maintain a particular spatial relationship to one another. This the relatively rigid structure of the mitochondrial membrane would enable them to do. In fact, through efforts spearheaded by Mitchell, Wallach, Green, and Racker, many investigators now seriously consider integrative models of mitochondrial function related more to solid state physics than to the familiar solutions of biochemistry.

However, though the lipids of the endoplasmic membranes are structurally dispensable, they are functionally essential. For example, my associates and I have treated microsomal membranes with phospholipase C, an enzyme that liberates as much as 90% of the membrane phospholipids. Structurally, the membranes change relatively little (at least so far as we can see in the electron microscope), but the loss of lipid completely inactivates glucose-6-phosphatase. If we then restored the lipid component, by adding either purified microsomal lipid or an artificial phospholipid obtained from soybeans, the enzyme activity returned to normal. As in so many other instances, this type of test tube approach to the study of membranes gives at the same time important new clues for bedside medicine. It is well known that phospholipase C represents an important toxin from the bacillus *Clostridium perfringens*, one organism that causes gas gangrene. And it is, indeed, quite exciting to find from our patients in Baltimore who have died with gas gangrene and then been studied at autopsy that their cells show many similarities in membrane changes to those seen in the model experiments.

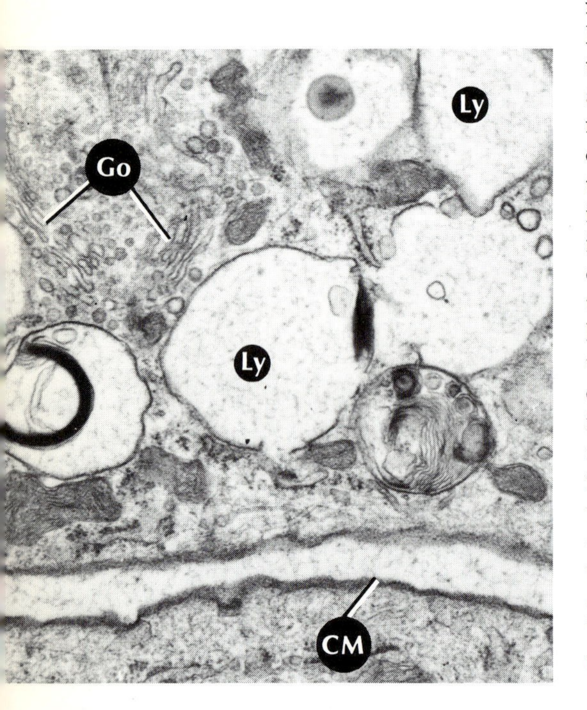

Lysosomes engorged with mucopolysaccharide adjoin Golgi apparatus in glomerular epithelial cell from patient with neurovisceral lipidosis, a storage disease.

To complete our overall picture of the cytocavitary network, we need to include the two basic processes through which substances, including membranes, enter or leave the system. It should be kept in mind in this context that membranes have an "outside" and an "inside". One type of movement is a "turning in" (*esotropy*), the other a "turning out" (*exotropy*) of the membrane upon itself. In esotropy invagination results in a vesicle the inside of which is lined by the outer or glycocalyx side of the membrane – that is, the side directed toward the lumen of the cytocavitary network, the topologic equivalent of the extracellular space. Esotropy can proceed in a forward direction, pinching off of a vacuole or vesicle; or it may proceed in the reverse direction, with the vesicle approaching the membrane and fusing with it to establish continuity of the two contents. The process of exotropy is akin to budding. When it proceeds in the forward direction, the structure formed has cell sap within and the outer or glycocalyx side of the membrane is directed outwardly. Such budding can occur either directly into the extracellular space or into the cytocavitary network. Exotropy can also proceed in the reverse direction, by fusion of the structure with the membrane surface of the cytocavitary network. This brings the cell sap compartments into continuity (see diagram, page 128).

Examples of forward esotropy include not only phagocytosis and pinocytosis but also the elaboration of transport vesicles from the ER and the detachment of lysosomes (to whatever extent detachment occurs) from the Golgi apparatus. Examples of reverse esotropy include the fusion of phagosomes and secretory granules with the plasma membrane and the fusion of Golgi vesicles with phagosomes.

When a portion of the cell sap forms a membrane-bound vesicle within the ER as the first step in autophagocytosis, this is, of course, "forward" exotropy. Other examples involve certain viruses that acquire their outer envelope by budding into the ER, the nucleus, or from the plasma membrane into the extracellular space. Mitotic division can be considered as an extreme case of forward exotropy, with the "bud" here being of equal size with the cell. Concerning reverse exotropy, in which a bud or similar structure fuses with another, we know of no good examples occurring altogether within the cell. We can observe it, however, in certain types of viral entry into the cell, in which the viral envelope fuses with the plasma membrane, allowing the "naked" virus to pass into the cell sap, and also in the fusion of sperm and ovum during fertilization. Again we have an extreme case at the cellular level in the fusion of two buds of equal size – i.e., two or more cells – to form syncytial giant cells. There are no known examples of the fusion of exotropic structures (those with cell sap inside) with esotropic ones (those with cell sap outside).

The triggering mechanisms that lead to esotropy and exotropy, either forward or reverse, are poorly understood. However, the available evidence strongly indicates that both processes are energy-dependent, involving such high-energy compounds as ATP, and mediated by the chemical messenger cAMP (see page 127). Also possibly involved are ion-dependent conformational changes in surface membrane proteins such as spectrin and interactions with transmembrane proteins as discussed by Singer.

Both esotropy and exotropy, as it happens, can be studied in the very simple system provided by the human erythrocyte. The mature erythrocyte is, as we know, perhaps the simplest of all mammalian cells, having neither nucleus nor any other type of organelle. Nonetheless, under certain circumstances, erythrocyte plasma membranes become involved in the processes we have been describing.

It has long been known that hemolysis involves a change in erythrocyte shape from the normal biconcave disc to a sphere, following which the cell lyses. It has been known as long, as shown by Ponder, that in most cases this change in shape involves a loss of part of the surface membrane. The loss, indeed, is a geometric necessity; a sphere of a given volume has a considerably smaller surface area than a disc of the same volume. Hence, unless the internal volume of the lysing erythrocyte increases – as occurs in hemolysis induced by hypotonic solutions, in which the cell becomes bloated with water – the shift from a discoidal to a spherical conformation inevitably involves a decrease in surface area, i.e., a decrease in the amount of surface membrane.

Dr. Fred Ginn, Dr. Paul Hochstein, and I found that one way in which the loss can occur is by internalization – i.e., by esotropy. When normal erythrocytes are treated with the antimalarial drug primaquine, the plasma membrane bends inward, eventually forming a series of membranous vesicles within the cell that resemble phagosomes. The vesicles are not digested; for one thing, the erythrocyte, so far as we know, lacks the necessary enzymes, but in any case the loss of membrane is quickly followed by dissolution of the cell. Exactly how the primaquine affects the membrane is unclear; it may exert a detergent action on the membrane lipids, or may affect the proteins and/or the glycocalyx. Whatever the process, it is clearly ATP-dependent and also requires the presence of either Ca^{++} or Mg^{++} ion, or perhaps both. It is perhaps significant that primaquine is among the drugs that can induce hemolysis in sensitive individuals who suffer from certain inherited hematologic abnormalities; it would

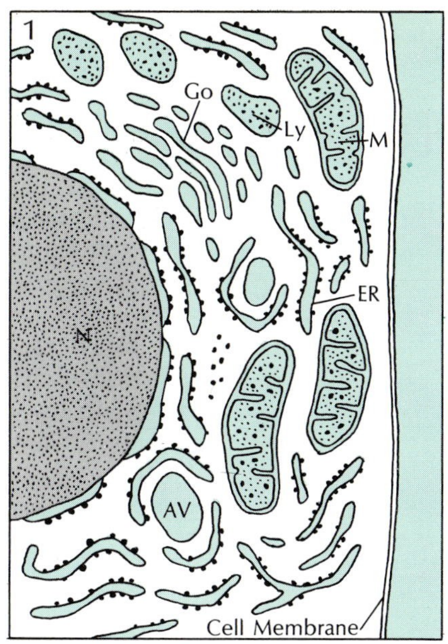

Sequence of changes following lethal cell injury is diagrammed; stage 1, immediate left, is normal. Initially (2) there is only dilatation of the endoplasmic reticulum; then (3) the cell sap expands and mitochondria show condensation. Subsequently (4) some mitochondria are condensed, others swollen, and still others show differences in the inner compartment, some being condensed, others swollen. Finally (5), the outer membranes rupture, flocculent densities and/or calcifications appear, and the nucleus undergoes karyolysis. Lysosomes cannot be seen, presumably having ruptured and participated in cell degradation. Cf *also photographs on next page.*

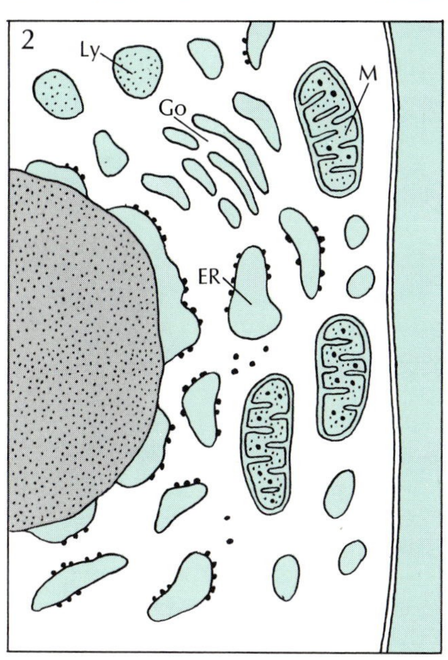

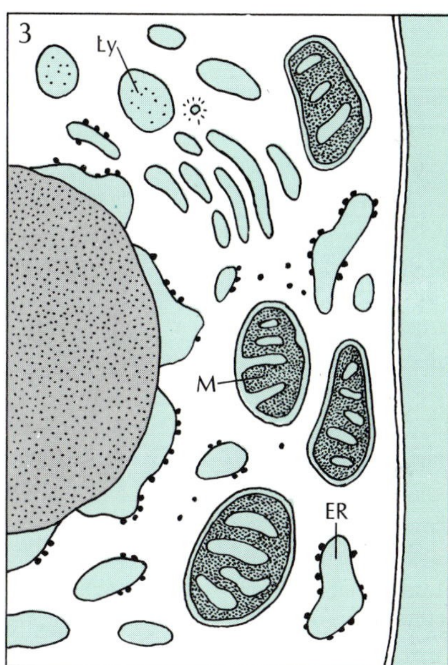

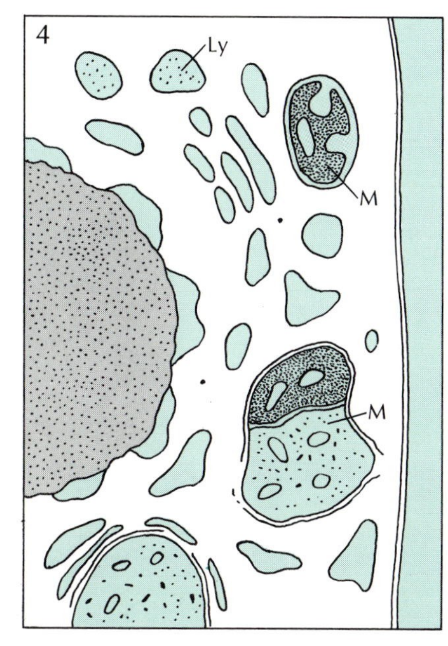

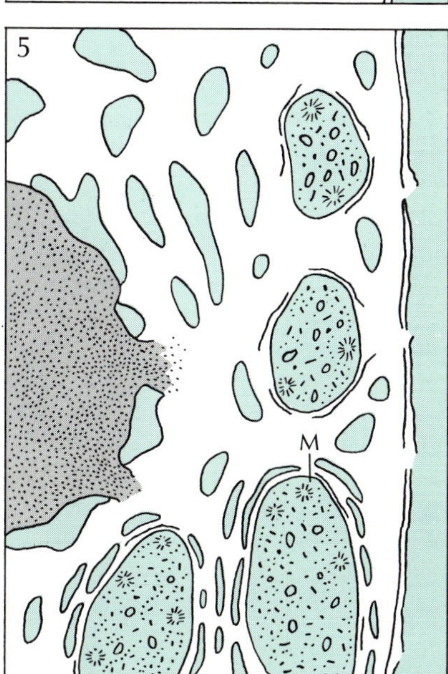

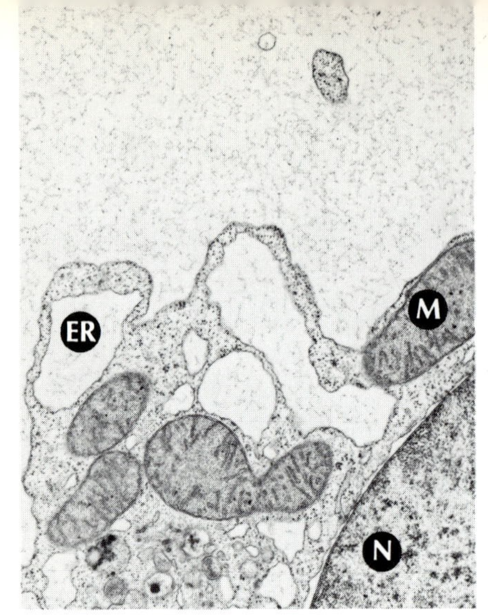

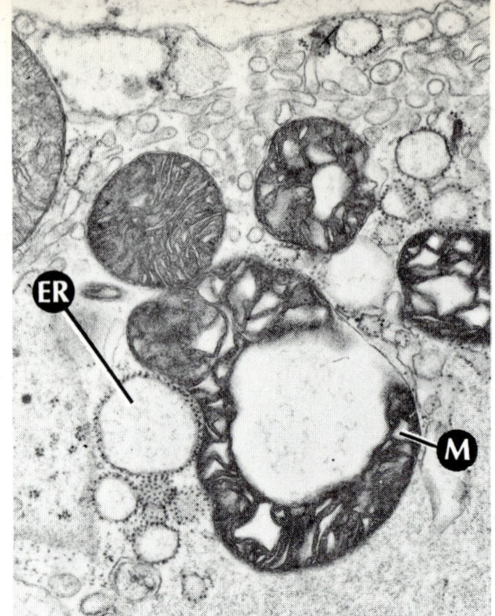

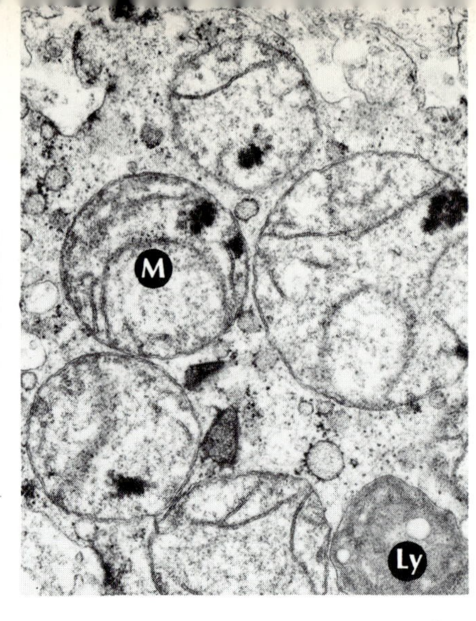

Cellular dilatation of ER is apparent in EM at left above of a portion of a hepatic parenchymal cell soon after CCl_4 administration; this alteration is reversible, as are the changes apparent in the next EM, of ischemic rat kidney slice, in which dilated ER and condensed mitochondria are seen (cf stage 3 of diagram on preceding page). Next two EM's show irreversible cell injury and correspond to stage 5 of diagram. First shows a kidney tubule cell from immediate autopsy of individual who died in hemorrhagic shock. Mitochondria are extremely swollen, with flocculent densities; such mitochondria lack ability for oxidative phosphor-

be interesting to discover whether the hemolysis in these cases occurs by the esotropic mechanism observed with the drug in vitro.

In 1973, Robert Pendergrass, Helen Liu, and Gary Cohen studied this process in our laboratory by scanning electron microscopy and freeze etching, and they produced some striking illustrations of this phenomenon, as shown at the bottom of page 128.

Erythrocytes can also lose membrane by exotropy – by budding – though this process has been less extensively studied. Here the formation and detachment of a small sphere (the bud) from the cell will involve a certain loss of internal substance but a proportionately much greater loss of membrane. This mechanism is believed to be involved in the hemolysis observed in many patients with artificial heart valves of the plastic-ball type, which are known to traumatize the erythrocytes, as shown by Weed. There is also reason to suspect that the same process may be involved in the normal destruction of senescent erythrocytes in the spleen; at any rate the cells seem to be producing buds or blebs of the sort just described.

Let me conclude this account of cytocavitary network by briefly describing some of the disturbances of its functions in pathologic states. We note first that in many cases of sublethal injury to cells there is a massive increase in the rate of autophagocytosis – the budding of mitochondria and other structures into the ER and their subsequent destruction within these vesicles. This occurs, for example, in both kidney and liver cells during shock; in fact, the many cellular inclusions observed by pathologists under these circumstances, as shown by Dr. Jan Valigorsky in liver studies in our human shock series, represent the accumulated sequelae of a number of autophagic events. Even if the cell recovers, it retains abnormally high quantities of aging pigment.

It is unclear whether these pigment accumulations, as the result of sublethal injury or of normal aging processes, can affect the functioning and viability of the cell when carried beyond a certain point. In crude terms, the more intracellular space is occupied by aging pigment (or any other nonfunctional substance or structure) the less room is left in which the cell can carry on its normal activities. Whether this in fact occurs with aging pigments is not certain; it may be the case in a whole series of lysosomal storage diseases, including the mucopolysaccharidoses, Tay-Sachs disease, and certain types of glycogen storage disease. The common factor in all these conditions seems to be a defect in some lysosomal enzyme that blocks the digestion, and therefore the elimination, of substances such as mucopolysaccharide and glycogen, whose gradual accumulation within the cells renders them unable to function.

Derangements of this type obviously suggest that if appropriate exogenous enzymes could be induced to enter the cell by esotropy, the cells could purge themselves of the deranging accumulations. In fact, this has been accomplished with some success in Pompe's disease and also in the mucopolysaccharidoses. Hypothetically, it seems possible that if cells could be induced to purge themselves of their aging pigments, they might become rejuvenated, at least to some degree.

I have already mentioned another pathologic derangement of the cytocavitary network: the regurgitation of enzyme into the extracellular space in inflammation and (probably) rheumatoid arthritis. Another type of derangement involves defects in the transport of certain substances through the network. It appears, for example, that certain types of fatty liver are caused by a breakdown in the transport of lipoproteins formed from triglycerides in the ER into the Golgi apparatus; instead they accumulate in globules within the cell sap. The mechanism here is uncertain, but we can see a similar process at work in cellular poisoning by carbon tetrachloride. Smuckler found that here protein synthesis is inhibited so that the lipids cannot form lipoproteins – the form in which many are transported – and therefore accumulate in the cell sap. Ethionine can produce similar results by yet another mechanism: trapping adenine needed for ATP and thereby blocking the energy-dependent transport from ER to Golgi. Chediak-Higashi syndrome in

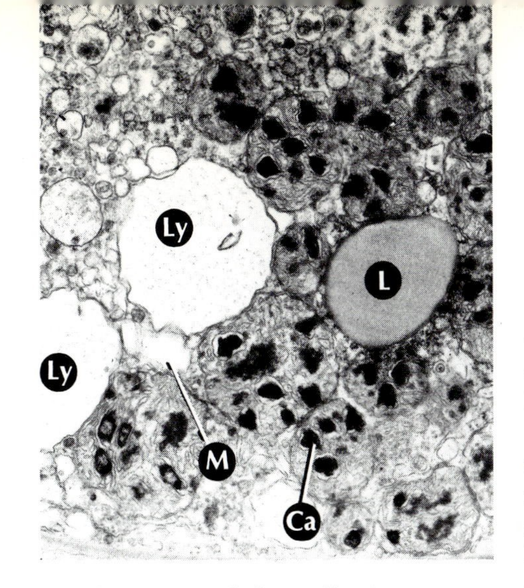

ylation. In rat kidney cells after mercuric chloride, calcifications (Ca) appear adjacent to inner membrane and lysosomes are swollen but their membranes are intact and a lipid droplet (L) can be seen.

man and related conditions in mink, cattle, and mice, all characterized by so-called partial albinism, all seem to involve some defect in the Golgi apparatus, whereby it elaborates a few large lysosomes instead of many small ones. It would not at all surprise me if, considering the Golgi's central position in the cytocavitary network, a number of other pathologies were eventually related to abnormalities in its functioning.

The phenomenon of enzyme induction, involving the proliferation of ER, is seen in such conditions as chronic barbiturate poisoning. Here the synthesis of ER in liver cells can eventually reach the point where the cells themselves become hypertrophied. The resulting enlargement of the liver may or may not be harmful, but an indirect result of the cells' enlarged component of ER is certainly dangerous. The enzymes contained in the ER, as we know, can efficiently metabolize barbiturates into a nontoxic form, but there is no guarantee that the same enzymes will not also metabolize other substances in quite a different sense. In fact, this is precisely what they do with carbon tetrachloride, converting it into free radicals that induce membrane destruction through peroxidation. Thus, an animal pretreated with barbiturate for a few days becomes susceptible to a dose of CCl_4 that would be sublethal in a normal animal – a clear example, on the subcellular level, of drug synergism.

A somewhat comparable process occurs with the carcinogen benzo[*a*]pyrene. This compound can induce a similar enzyme system (and ER proliferation) in some cells but not others, and it has been shown that it is toxic only in cells where induction occurs. In other words, cell damage and death are evidently produced not by benzo[*a*]pyrene itself but by some metabolite of it. Quite possibly the same may be true of benzo[*a*]pyrene's known carcinogenic effects: the proximal carcinogen may be a metabolite, not the compound itself. It has been shown that the enzyme system in question can be induced in some strains of mice but not in others, and the same may well be true of human beings, which might explain why benzo[*a*]pyrene is not carcinogenic in all individuals exposed to it.

Changes in the ribosome-ER relationship are prominent in many types of cell injury. Detachment of these particles from the ER surface – which would presumably prevent them from synthesizing protein – is an early change induced by many chemical toxins, though it seems to be generally reversible. Other chemicals induce various sorts of whorls and comparable abnormal configurations, often related to cell hypertrophy. Finally, a number of compounds, including aspirin, can induce the formation of microbodies in, e.g., liver cells.

Until now we have been discussing principally the sublethal or reversible changes of the network in diseased cells; however, it is striking that these intracellular membranes also go through a remarkably reproducible sequence of changes following lethal cell injury such as occurs in stroke, myocardial infarction, or acute tubular necrosis of the kidney. This sequence of changes is the subject of intensive study in our laboratory with a view toward defining the pathophysiologic principles involved in each stage, i.e., the stages that are reversible as compared with those that are not, and, furthermore, the exploration of pharmacologic methods to modify this otherwise disastrous progression toward cell death and necrosis. These stages are shown in the schematic on page 131. Note that in general these changes are characterized by differential expansions and contractions of various cell compartments as a function of time, reflecting the sequential change in cell membrane properties affecting each compartment's control of its volume and composition. Interestingly, as our knowledge of membrane structure and function progresses, it appears more and more certain that structural changes are basic to functional derangements and that, for example, in the case of the mitochondrion, knowledge of its volume control is quite indicative or even synonymous with changes in energy coupling.

Without belaboring the point further, I think it is clear that the cytocavitary network of intracellular membranes plays a central role in both health and disease. Our expanding comprehension of how and why it functions brings us to the threshold of completely new concepts of disease involving a consideration of it as involving the interactions of membranes as macromolecular assemblies rather than enzymes in solution or vague integrated physiologic effects. This promises to lead us to an incisive extension of Virchow's cellular pathology to a pathology of cell membranes.

14

Inner Mitochondrial Membranes: Basic and Applied Aspects

EFRAIM RACKER
Cornell University

The inner mitochondrial membrane is the seat of oxidative phosphorylation, a process of primary importance to the energy metabolism of cells, and a source of confusion for physicians, students, and even for professional mitochondriacs. In 1963, I remarked that anyone who is not confused about oxidative phosphorylation just does not understand the situation. Since then, a great deal has changed. There is now experimental evidence for the existence of a nonphosphorylated high-energy state in animal cells, generated by the oxidation process and utilizable directly for such energy-dependent processes as ion movements or the generation of adenosine triphosphate (ATP) from adenosine diphosphate (ADP) and inorganic phosphate (P_i). These experimental findings have eliminated a multitude of confusing hypotheses (some based on chemical models) that had invoked the formation of a phosphorylated intermediate prior to oxidation. Studies of ion movements have broadened the base of mitochondrial bioenergetics and have led Peter Mitchell of the Glyn Research Laboratories in England to formulate a hypothesis that has profoundly influenced our thinking about oxidative phosphorylation – the chemiosmotic hypothesis.

According to Mitchell, two processes take place in the mitochondrion independently. The first is a proton translocation, via the respiratory chain, that goes from the inside of the membrane to the outside and thereby establishes a proton concentration gradient as well as a membrane potential. This is the oxidative component. Secondly, there is a proton pump that reverses the proton flux – from outside to inside – and in doing so allows the generation of ATP from ADP and P_i—the phosphorylation step.

It should be stressed that a central tenet of Mitchell's formulation is that the mitochondrial membrane is intrinsically impermeable to protons and that any movement of these ions across the membrane is dependent on these two processes. Otherwise, it would not be possible to create either the membrane potential or the proton gradient. Therefore, when one speaks of coupling in the context of oxidative phosphorylation, the reference is to the sequential translocation of protons from the inside of the membrane to the outside and then from the outside to the inside in the presence of an impermeable membrane. Chemical compounds such as dinitrophenol, which will be discussed later, "uncouple" the process by carrying protons ionophorically from one side of the membrane to the other, effectively making the membrane permeable and eliminating the proton concentration gradient and the membrane potential. In this way, the formation of ATP is obviated and the energy of oxidation is dissipated as heat.

The approach followed in various laboratories in which individual membrane components are physically isolated has proved fruitful in providing support for Mitchell's formulation of the separate operation of an asymetric respiratory chain and of a proton pump that leads to ATP formation. In fact, it would appear that essentially the problem of oxidative phosphorylation has been solved.

Basic Aspects of Oxidative Phosphorylation

The machinery responsible for most of the energy production in animal cells, contained as noted above in the inner membrane of the mitochondrion, is activated when food is delivered in the form of pyruvate (from carbohydrates), amino acids (from proteins), and fatty acids (from fats). These substances are oxidized in the respiratory chain, a multienzyme system, diagrammed on page 137 (top). During the past few years it became apparent that the components of the respiratory chain are arranged vec-

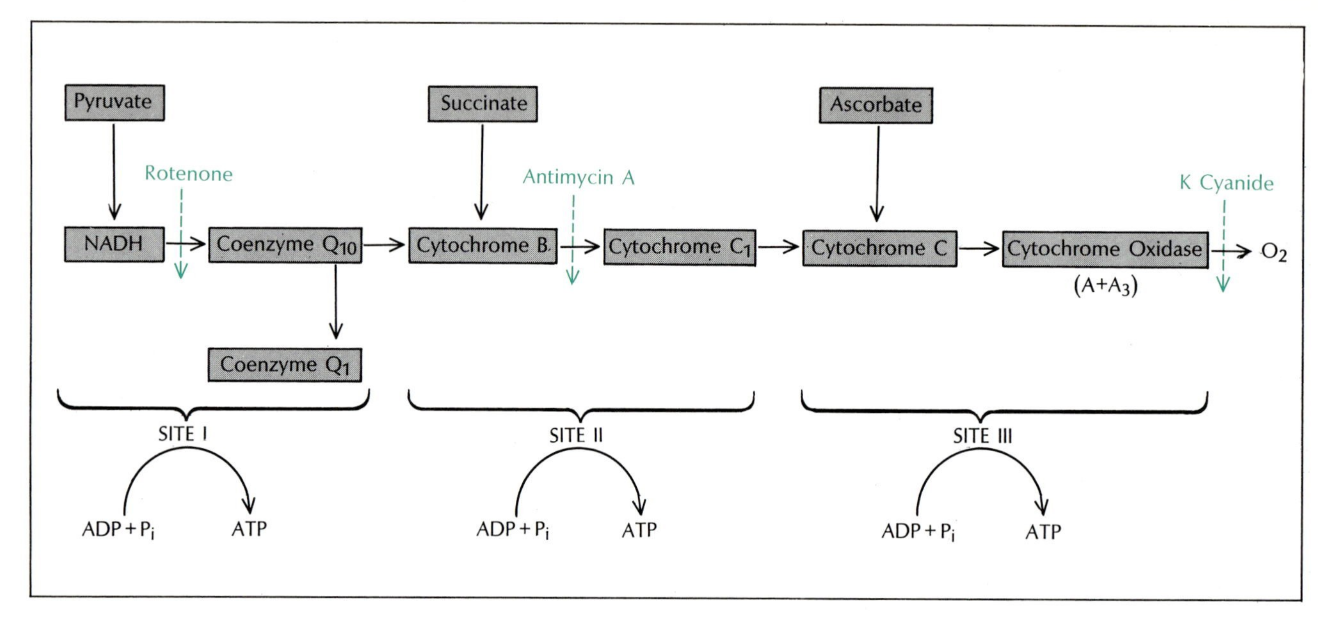

This diagram of the respiratory chain shows three sites of oxidative phosphorylation. In mitochondrial energy production, the respiratory chain functions with a P:0 ratio of 3, that is, three molecules of ATP are produced for every 0 atom used. By use of metabolic poisons or inhibitors (dashed arrows), the three sites can be functionally isolated from each other and it can be shown that each site is responsible for synthesis of one ATP per 0 atom. Thus, with succinate as substrate, the P:0 ratio is reduced to 2.

torially in the inner membrane. Some components face the inside or matrix of the mitochondrion, some face the outer membrane, and some are embedded within the inner membrane. Mitchell's chemiosmotic hypothesis postulates that this organization of the respiratory chain gives rise to a vectorial movement of electrons and protons released during the oxidation process. As a result, a proton concentration gradient and a membrane potential are formed. These are the driving forces for the generation of ATP from ADP and inorganic phosphate.

The agency through which this generation is accomplished is a proton pump, a complex structure made up of several proteins, called coupling factors, at least four of which have been identified and for convenience are called F_1, F_2, F_5 (or OSCP), and F_6. When the membrane is treated so as to remove the coupling factors, the membrane particles that remain catalyze oxidation without phosphorylation. Conversely, when these proteins are restored, coupled phosphorylation takes place. F_1 can be visualized by electron microscopy; it is an 85 Å sphere attached to the inner surface (matrix side) of the inner mitochondrial membrane. Thus, the removal and restoration of this coupling factor can be monitored with the electron microscope.

The schematic drawing of the inner mitochondrial membrane on page 137 (bottom) depicts a structure that is folded into a series of loops or cristae. Physiologically, this conformation serves vastly to increase the operative surface of the membrane. Experimentally, these foldings have a most interesting consequence. When mitochondria are disrupted by either mechanical or sonic forces, the cristae are broken; the phospholipid membrane then re-forms into smaller vesicular structures, called submitochondrial particles. These submitochondrial particles, however, are "inside-out" when compared with the intact inner membrane. This is most readily demonstrated under the electron microscope where F_1 appears on the outside surface of the particles.

Although submitochondrial particles are purely artifacts of a preparative procedure, their curious anatomic configuration serves to allow biochemists to analyze the components of the respiratory chain from both sides of the membrane under conditions of functional oxidative phosphorylation. The effects of specific reagents – e.g., antibodies against specific components of either the respiratory chain or of the phosphorylating apparatus – can be evaluated in the context of their alterations of mitochondrial function. Some examples: an antibody against cytochrome C (in the respiratory chain) inhibits only intact mitochondria; an antibody against F_1 inhibits only the submitochondrial particles; and, interestingly enough, an antibody against cytochrome oxidase inhibits both intact mitochondria and submitochondrial particles, a finding that immediately suggested that this enzyme spans the membrane from one side to the other.

These and similar studies have yielded a tentative picture of the organization of the system of oxidative phosphorylation, as schematized on page 138. In the "inside-out" submitochondrial particle, the vectorial assembly of the respiratory chain results in a movement of the protons from the outside to the inside, a reverse of the flux in the mitochondrion itself. The oligomycin-sensitive ATPase in the figure consists of a transmembranous portion, made up of so-called hydrophobic proteins, and the membrane surface coupling factors. (It is of interest that F_1 detached from the membrane catalyzes an oligomycin-insensitive hydrolysis of ATP. This observation has helped to identify the membrane components that convert the reaction to oligomycin sensitivity.) Mitchell views this ATPase as a device to translocate the protons generated by the respiratory chain back through the membrane, producing a proton current that drives the formation of ATP. Exactly how this is accomplished remains controversial; this will be discussed later.

Having outlined the overall configuration of the membrane, and some of the information now available on the factors affecting the proton flux bidirectionally across the membrane, let us now focus down on the respiratory chain and its role in oxidative phosphorylation. It has long been known that oxidation of substrates such as pyruvate or malate, which donate hydrogens to the hydrogen carrier nicotinamide dinucleotide (NAD), gives rise to three molecules of ATP for each atom of oxygen. We speak of a P:O ratio of 3. There are three sites of phosphorylation on the respiratory chain (see page 136): one between NADH (the reduced form of NAD) and Q_{10}; the second between cytochrome B and cytochrome C_1; and the third between cytochrome C and oxygen. By addition of a suitable hydrogen acceptor, such as coenzyme Q_1, site I can be measured experimentally. To do this, further electron flux down the respiratory chain must be inhibited by substances such as KCN and antimycin A. With succinate as the substrate, site II and site III operate with a P:O ratio of 2. In this experiment rotenone, a known fish poison, is added to prevent the reversal of site I and the consequent drain of some of the energy. With ascorbate as substrate, site III can be measured independently of the other sites, and the P:O ratio approaches 1. Thus the P:O ratio decreases as one moves down the respiratory chain, and Mitchell and his collaborators have shown that there is a decreased movement of protons paralleling the decrease in P:O ratio: six per NADH, four per succinate, two per ascorbate.

Additional insights into the mechanisms of oxidative phosphorylation have been gained during the past few years through newly developed methodology for the physical separation and functional reconstitution of the first and third phosphorylation sites. The isolated protein, cytochrome oxidase (containing cytochrome A and cytochrome A_3), was incorporated along with a preparation of the oligomycin-sensitive ATPase (the hydrophobic proteins plus coupling factors) into artificial phospholipid vesicles by

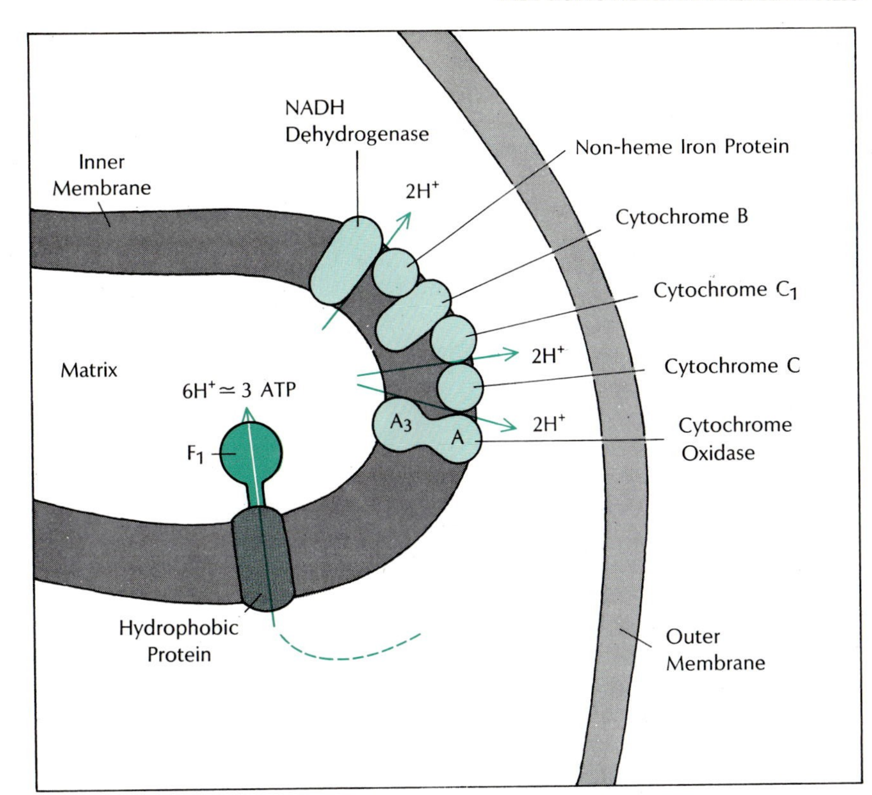

"Map" of the inner mitochondrial membrane indicates the vectorial arrangement of respiratory chain enzymes. The respiratory chain translocates protons (H^+) from matrix, thus creating a proton concentration gradient between the space separating the inner and outer membrane and the matrix, which gives rise to a membrane potential. Proton pump moves protons back to the matrix, which results in generation of additional ATP.

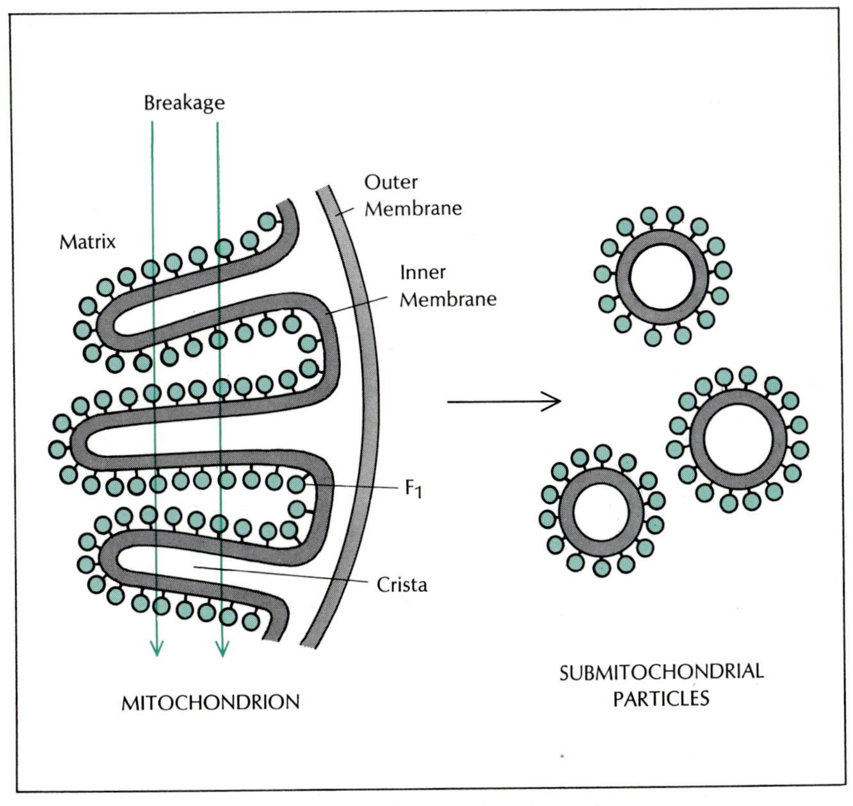

When the cristae of the natural mitochondrial membrane are disrupted by sonic or mechanical forces (left), the fragments of inner membrane that are broken off re-form as "inside-out" submitochondrial particles.

a relatively simple procedure. Appropriate mixtures of phospholipids were dissolved in cholate by sonic oscillation. To these mixtures, cytochrome C, cytochrome oxidase, and a preparation of hydrophobic proteins were added, and excessive cholate was removed by dialysis. After 18 hours particles were formed; when suitable coupling factors were added to these particles, they were able to catalyze site III oxidative phosphorylation. Similar reconstitution of site I oxidative phosphorylation was accomplished by a parallel procedure utilizing the respiratory enzymes that catalyze the reduction of coenzyme Q_1 by NADH.

There is an important and informative difference between the reconstituted particles and the native mitochondrial membrane. In contrast with the organization of the components of the latter vectorially and unidirectionally – i.e., all molecules of cytochrome C face the outer membrane, all molecules of F_1 face the matrix – the components of the reconstituted vesicles are arranged bidirectionally. The vesicles can oxidize reduced cytochrome C on either their outside or inside surfaces. And significantly, when reduced cytochrome C is present on both sides, there is no ATP formation. This was supportive of the prediction inherent in Mitchell's chemiosmotic hypothesis, since a membrane potential cannot be established by electron flow going in both directions. In order to impose the asymmetry necessary for such a potential to develop, we found that experimentally it was necessary to remove the excess cytochrome C that had been present during the reconstitution. This could be done after formation of the vesicles by sedimentation or by various other procedures.

We have also produced an artificial system employing phospholipid vesicles and a variety of rhodopsin derived from "purple" bacteria. When light energy is applied to this system, it catalyzes proton flow across the membrane, as shown by a decrease in extravesicular pH; and when mitochondrial, oligomycin-sensitive ATPase is incorporated into the vesicles during their formation, there is a resultant generation of ATP from ADP and P_i when light of an appropriate wavelength is applied.

These experiments serve to bolster the concept that the prime function of the electron transport chain is, in fact, the creation of a proton gradient. Nor can there be any doubt that a proton gradient and a membrane potential can give rise to ATP formation. Let us turn now to the question of *how*.

Among the most fruitful sources of possible answers to this question are studies of the various other biologic ion pumps, including the sodium-potassium pump of the plasma membrane and the calcium pump of the sarcoplasmic reticulum of muscle. The plasma membrane of cells contains a K^+Na^+ ATPase that is responsible for the translocation of K^+ into the cell and the excretion of Na^+. The sarcoplasmic reticulum contains a Ca^{++} ATPase that can translocate calcium ions with the expenditure of energy. It has been demonstrated that in both of these systems, the establishment of an ion concentration gradient can force the pumps to operate in reverse, i.e., to generate ATP from ADP and inorganic phosphate. We thus have precedents for ATP generation by ion gradients.

Although there are several possible explanations for this phenomenon, the most widely accepted is the conformational hypothesis. This postulates that a change in conformation of the protein (K^+Na^+ ATPase, Ca^{++} ATPase, or the Mg^{++} ATPase of mitochondria [F_1]), perhaps induced by a membrane potential, allows the uptake of a molecule of inorganic phosphate and its transference to ADP for ATP formation. Phosphorylated intermediates of the protein have actually been demonstrated for both Na^+K^+ ATPase and Ca^{++} ATPase. No one has yet demonstrated such intermediates in the case of mitochondrial ATPase. Exactly how the proton concentration gradient or the membrane potential changes the conformation of the protein to a high-energy state is still unknown.

The availability of simple reconstituted systems of oxidative phosphorylation and of some ion pumps

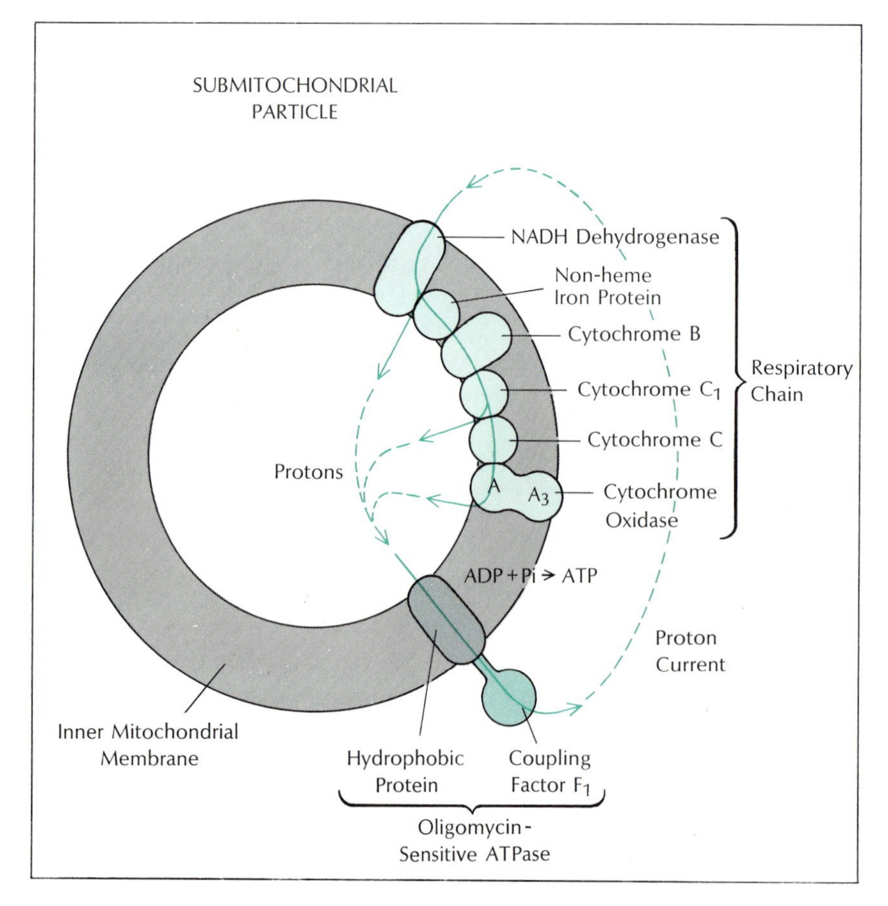

"Map" of submitochondrial particle shows reversed orientation of respiratory chain and proton pump components. The proton current that is established in this experimental preparation is also reversed, as indicated by arrows.

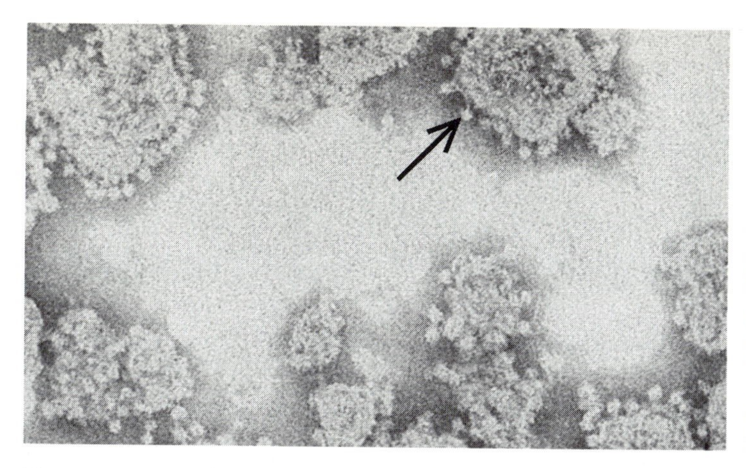

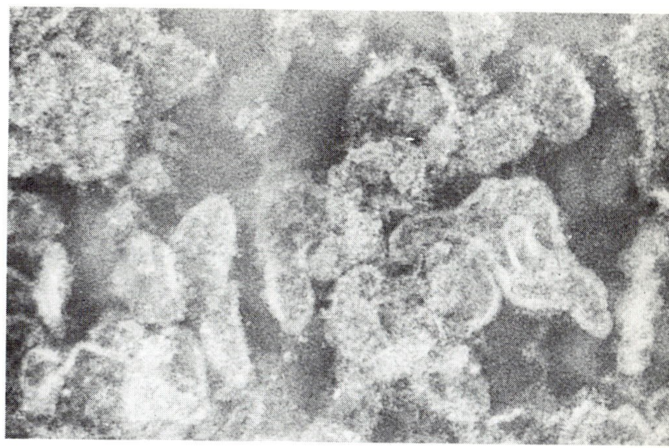

Electron micrograph of submitochondrial particles (magnification 175,000 ×) clearly shows F_1 molecules (arrow) on surfaces. *EM (right) shows particles from which F_1 has been removed by urea treatment. Such treated particles lack F_1 ATPase activity.*

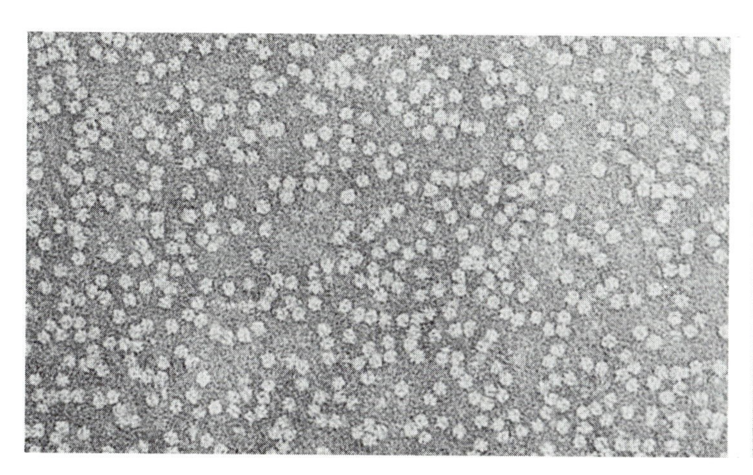

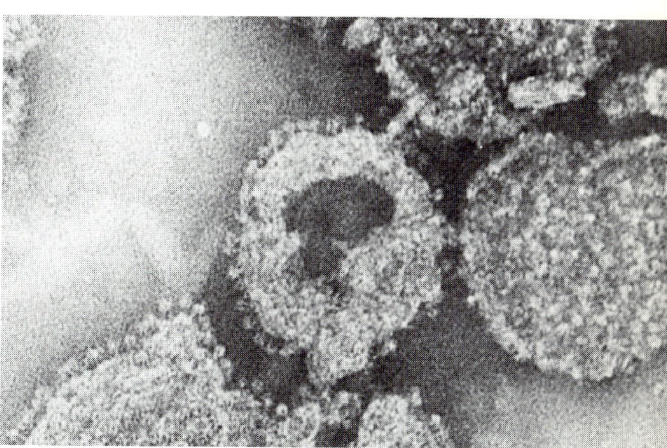

F_1 fragments can be visualized, as above, after being removed from submitochondrial particles. The particles can then be reconstituted morphologically with F_1 "reattached" to the membranes from which they had been removed by urea treatment (right).

should allow more rigorous experimentation and elucidation of this problem. But it appears that the major mystery of oxidative phosphorylation, namely how the energy of oxidation is channelled into ATP energy, has been solved and that the road to exact physical chemical experimentation with respect to definition of biologic pumps in general is now open.

Applied Aspects of Oxidative Phosphorylation

It seems appropriate at a time when doubts are expressed about the significance of basic research to recall the comment of Faraday when he was challenged about the usefulness of his experiments on electricity. "One day," he said, "you may tax it." Perhaps one day we may learn how to utilize the principles of oxidative photophosphorylation to solve the problems of the energy crisis. But there are more immediate applied problems we should attend to. Among them are the primary diseases of energy metabolism and the diseases that indirectly influence mitochondrial function. And there is the clinical problem of obesity, which involves an unbalanced energy metabolism.

It was during World War I that the pronounced effects of the nitrophenols were discovered in toxicity studies of workers who had contact with these chemicals during the production of explosives. It was only 30 years later that the mode of action of 2,4-dinitrophenol as an uncoupler of oxidative phosphorylation was discovered. (It will be recalled that uncoupling is accomplished by agents that make the inner mitochondrial membrane permeable to protons, thus obviating the proton concentration gradient and membrane potential.) Prior to this discovery, physicians tried to use dinitrophenol to combat obesity. Often these efforts were only too successful, eliminating the patient as well as his excess fat. Clearly it was not possible to control these powerful agents without endangering the life of the patient. During the following decade, another uncoupler of oxidative phosphorylation (when used in high, unphysiologic doses), thyroxin, became popular for the treatment of obesity. It was discovered here, too, that effective and toxic doses are uncomfortably close.

One of the early best sellers in the food-fad market was a book by Herman Taller, *Calories Don't Count*. One evening, we had visitors and this book came up for discussion. A young cousin of my wife, at the time an assistant professor at Princeton, vigorously defended Taller's diet, largely on the pragmatic ground of having himself used it with great success. At the time, I was not even at Cornell and in no position to argue with a

Princeton man. But the next day I wrote a review of Taller's book (which I hadn't read) and it was published as an editorial in the *American Journal of Medicine* under the title, "Calories Don't Count—If You Don't Use Them." I tried to explain in this article that unsaturated fatty acids, the key components of Taller's diet, are uncouplers of oxidative phosphorylation and therefore could be expected to act like either dinitrophenol or large doses of thyroxin. Therefore, they could be effective in some individuals if used in proper doses, ineffective in others with different rates of fatty-acid removal, and possibly toxic in those who for one reason or another might be particularly sensitive. It is remarkable how little we know about the metabolisms that control the intracellular steady-state levels of fatty acids and thereby influence the efficiency of our energy balance. Could some form of obesity be controlled by these factors? Could we influence the fine regulatory mechanisms that deliver or remove fatty acids?

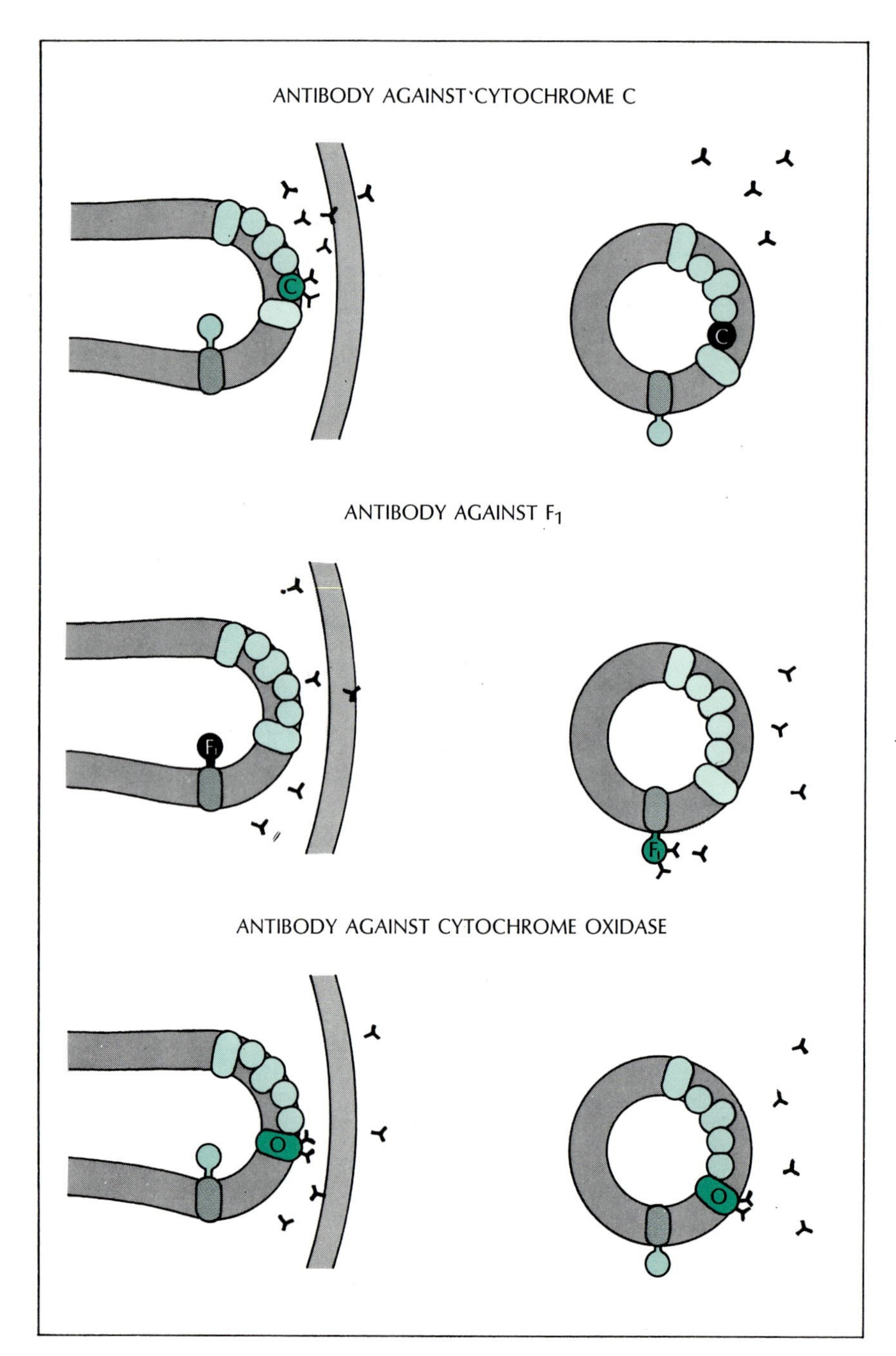

Technique for mapping components of oxidative phosphorylation system involves introduction of antibodies against specific inner membrane components in preparations of natural membrane and of submitochondrial particles and then comparing the effects. For example, anti-cytochrome C antibodies will inactivate the cytochrome in natural membranes but not in particles, establishing that the cytochrome is located on the outer surface of the natural membrane. With F_1, the results are opposite. Antibodies against cytochrome oxidase inactivate the enzyme in either preparation, indicating that cytochrome oxidase extends through the membrane.

When we isolate mitochondria from tissue and wash them free of the cytoplasmic "debris," we find that oxidation and phosphorylation are "tightly coupled." This, in essence, means that there is little or no oxidation unless both inorganic phosphate and ADP, or another energy-dissipating mechanism, are available. But are mitochondria tightly coupled in vivo, and how much partial uncoupling or loose coupling (e.g., by fatty acids) can we tolerate?

Luft et al recorded a case of a loosely coupled woman—or stated more properly—of a woman with loosely coupled mitochondria. The woman was incapable of performing any work and analysis of her muscle mitochondria revealed very poor "respiratory control." In other words, oxidative energy could not be conserved effectively and was continuously being converted to heat. During the last several years a great deal has been learned about the phenomenon of respiratory control; examination of mitochondrial membrane proton permeability and phospholipid composition may help us to understand better the nature of conditions such as that of the patient just described.

It is quite possible that some forms of cancer may be mitochondrial diseases. Changes in mitochondrial DNA have been observed in some malignant diseases. It will be recalled that, many years ago, Otto Warburg theorized that cancer is caused by a dedifferentiation of normal cells to more primitive cells that derive most of their energy by fermentation because they have lost respiratory capacity. As a generalization, this is clearly incorrect, since many tumor cells have been shown to possess normal oxidative capacity. However, the observations of a direct relationship between the rate of aerobic glycolysis and the degree of malignancy of experimental tumors have been confirmed; they deserve more attention. Even if aerobic glycolysis is only a secondary feature of these tumors resulting from genetic, chemical, or viral transformation of a cell, it may be expected to cause derangement of many fine regulatory phenomena of biosynthesis and growth, notably those influenced by either pH or by the ATP:ADP ratio of the cell.

It is remarkable that so little effort has been made to achieve a systematic analysis of the causes of high aerobic glycolysis of tumors in the more than 50 years since this metabolic abnormality was discovered. Only recently did we undertake such a study, and in each case investigated we found that uncontrolled ATPase activity is responsible for the high glycolytic rate. It should be remembered that glycolysis, like oxidative phosphorylation, is tightly coupled. Thus ADP and inorganic phosphate must be available and must be regenerated from ATP. Many work processes, like ion transport, biosynthesis, etc., may participate in this "ATPase" activity.

Cells from several types of tumor have been analyzed in our laboratory. In one instance, the Ehrlich ascites tumor, the lesion has been identified at the Na^+K^+ ATPase site in the plasma membrane. Instead of requiring one ATP molecule for every two potassium ions moved, two to four ATP's are needed. This excessive hydrolysis of ATP provides a parallel excess of ADP and inorganic phosphate and leads to high levels of aerobic glycolysis. In another malignant cell, the polyoma virus-transformed fibroblast, the normally masked mitochondrial ATPase (F_1) is activated. This, too, leads to enhancement of aerobic glycolysis. These are but two examples of a general finding of abnormal ATPase activity. It is known that some oncogenic viruses have ATPase

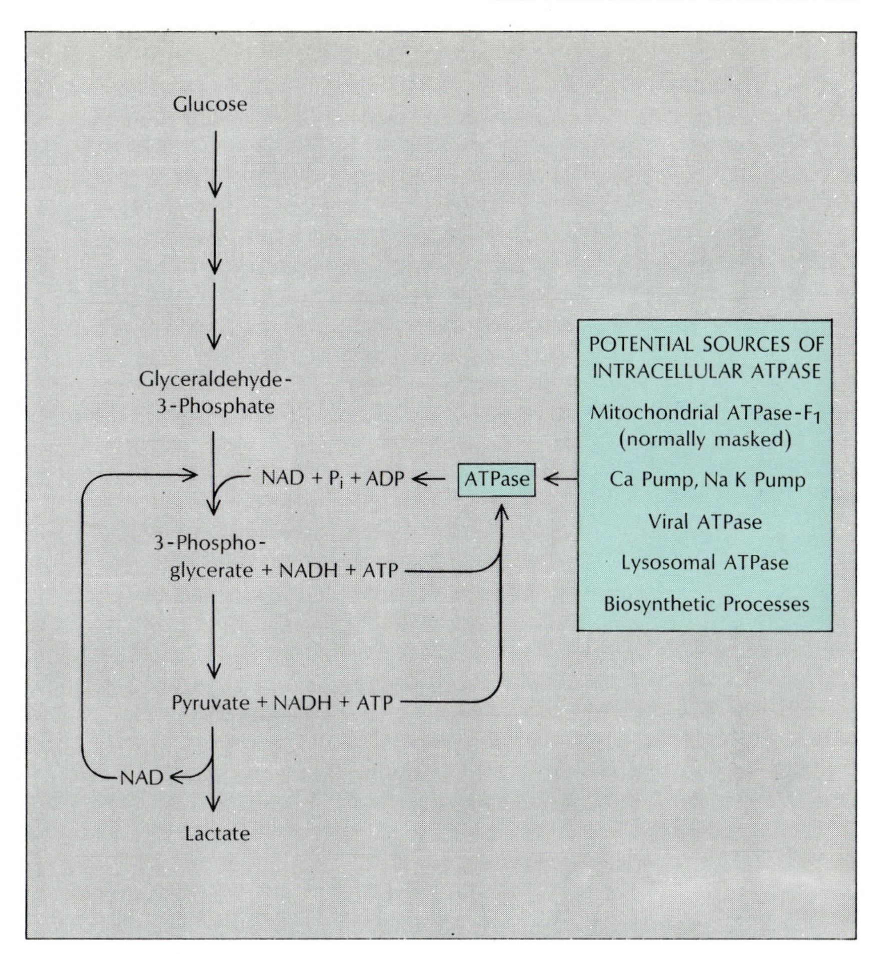

Abbreviated scheme of glycolytic pathways suggests how enhanced ATPase activity might account for the increased reliance on aerobic glycolysis demonstrated in some tumors. Possible sources of such ATPase enhancement are listed in the panel at right.

activity – though the origin and significance of this activity are unknown.

It does seem certain that an unbalanced energy metabolism caused by excessive regeneration of ADP and P_i is responsible for the high rate of lactic acid formation essential to aerobic glycolysis. While there is remarkably little known about how the various ionic pump activities in our cells are kept under control (are tightly coupled), there is reason to believe that this ignorance is a transient state. It appears that there are certain chemical compounds found in nature, the bioflavonoids, that can exert such a regulatory function. We are currently looking into the possibility that these compounds may help us to design more logically experiments that would either prove or disprove the importance of lactic acid formation in malignant growth.

If lactic acid proves essential or even helpful to rapid growth, it should be possible to design drugs that could influence aerobic glycolysis not by direct inhibition of this important pathway but by controlling the altered ATPase activity.

15

Membranes and Secretion

JAMES D. JAMIESON
Yale University

Knowing as we do that intracellular membranes play a key role in the internal economy of virtually all cells, it is obvious that (in metazoan organisms) that role will vary considerably from one type of cell to another, depending on the cells' functions in relation to the rest of the organism. This chapter is concerned with membranes in secretory cells, in particular those of the pancreas.

The historical choice by researchers of the pancreas as the archetypal mammalian secretory organ was made partly as a matter of convenience. The organ secretes its diverse digestive enzymes vigorously and, at times, even lavishly. Moreover, its secretory tissues can be obtained in relatively large quantities even from small laboratory animals (as opposed to those of such glands as the pituitary and adrenals). Without discussing in detail whether, and to what extent, the pancreatic cell is typical of secretory cells generally, we can say broadly that its mechanisms seem very similar to those of other cells possessing storage granules. These include many exocrine glands, including the parotids and various glands of the GI tract, and some of the endocrine glands. Much the same mechanisms, indeed, are involved in the secretion of lysosomal enzymes by the leukocytes.

In discussing the role of membranes in pancreatic secretion, we shall, as is usual in such matters, proceed from the known to the unknown. We begin with an overview of the secretory process, which is reasonably well understood thanks to the pioneering work of Dr. George Palade, followed by others (like myself) in his and other laboratories, and continue with a more detailed look at the role of membranes in that process, where our understanding is rather less. Finally, some consideration will be given to the biochemical mechanisms whereby the membranes perform their role – an area where our ignorance is still pronounced.

The secretory cells of the pancreas, to begin with, are grouped in roughly ellipsoidal aggregations of more than a dozen cells – the acini. The aggregation surrounds the acinar lumen, which forms the beginning of the pancreatic duct system. The individual cells are joined to one another by so-called tight junctions, desmosomes, and gap junctions (see Chapter 9, "Junctions Between Cells," by Pappas). Moreover, each cell, from the functional and structural standpoint, is highly polarized, with the manufacture of its secretory products occurring at one end (the base) and their secretion into the lumen at the other (the apex).

The secretory process begins on the membranes of the rough endoplasmic reticulum (ER), where its characteristic products – proteases such as trypsinogen and chymotrypsinogen, nucleases (RNAse and DNAse), and carbohydrate- and lipid-digesting enzymes – are synthesized. These and the other pancreatic products – there are a dozen or more – are of course enzymes or enzyme precursors and therefore proteins. Like all proteins, their synthesis occurs on the aggregations of ribosomes called polysomes, in which the genetic code for a particular protein, transcribed in the form of messenger RNA, is translated into a specific sequence of amino acids. The pancreatic enzymes (and, it is thought, secretory proteins generally) are synthesized *only* on polysomes joined to the rough ER; indeed it is the appearance of these small, attached particles under the electron microscope that accounts for the "rough" in its name. According to work done by Dr. Colvin Redman, proteins that are not manufactured for export, i.e., those that are used in the cell's own structures and metabolism, are commonly synthesized on "free" polysomes – those not bound to the ER. The vigorous secretory functions of the pancreatic cell are reflected in the fact that in it the rough ER cisternae account for a large proportion (20%) of the total cell volume. With its associated cytoplasmic matrix the ER is about 50% of cell volume and accounts for around four fifths of the surface area of all the intracellular membranes involved in processing exportable proteins. These are, of course, markedly greater proportions than are found in nonsecretory cells, where the process of protein synthesis proceeds much less actively.

The subsequent movement of the newly synthesized protein through and out of the cell can be followed quite easily by the technique known as pulse chase radioautography. Tissue pieces are cultured for a few minutes in a medium containing labeled amino acid, allowing a portion of the label to be incorporated into the secretory protein currently being synthesized; the cells are then washed free of excess unincorporated label and removed to a normal medium containing unlabeled amino acid. This chase medium serves to prevent further incorporation of label. The result is a labeled "pulse," or "batch," of protein whose progress from one organelle to the next can be tracked by removing, fixing, and radioautographing tissue samples after various times of chase. What is actually observed in the electron microscope are developed silver grains whose position on the section's surface indicates the general position of underlying labeled proteins.

From their site of synthesis in the rough ER, then, the secretory proteins move into the so-called transitional elements – portions of the ER that are devoid of polysomes, whence the term "transitional," meaning that their appearance is part "rough" and part "smooth." What appears to occur next is that the transitional elements bud off the ER as small transporting vesicles that carry their load of secretory protein toward the Golgi complex. Here we do not know exactly what happens, but it appears that the transporting vesicles probably fuse with the membranes of that organelle and in the process deliver their content to it. The Golgi complex – a receiving and packaging way station in the cell – consists of a series of flattened, smooth-surfaced saccules stacked one on top of the other, plus some associated vesicles and vacuoles. The stacks are arranged roughly in the form of a shallow cup whose convex entry side faces the rough ER basally in the cell and whose concave or exit side faces upward toward the cell apex. Precisely where the transporting vesicles fuse with the Golgi stacks is unknown, but what is clear is that all members of the stacks in the complex appear to be involved in processing the product at any one time. What is seen in most secretory cells, but to a less marked degree in the pancreas, is that the saccules nearest the concave side gradually fill with secretory material, become rounded up, and eventually bud off an immature storage granule that moves toward the center of the cup-shaped zone.

As discussed later, the secretory proteins may undergo extensive modifications during their residence in the Golgi complex. The immature storage granules, surrounded by a smooth membrane, appear to contain proteins at a concentration less than that of mature zymogen granules. Gradually, however, the concentration increases – whence the name "condensing vacuole" often applied to immature granules. The eventual result is the mature storage or zymogen granule packed with exportable proteins. Under appropriate external stimuli, these granules, which crowd the apical region of the cell, fuse with the apical plasmalemma, ejecting their contents into the acinar lumen whence they pass into the pancreatic duct and eventually into the duodenum. This process is called exocytosis.

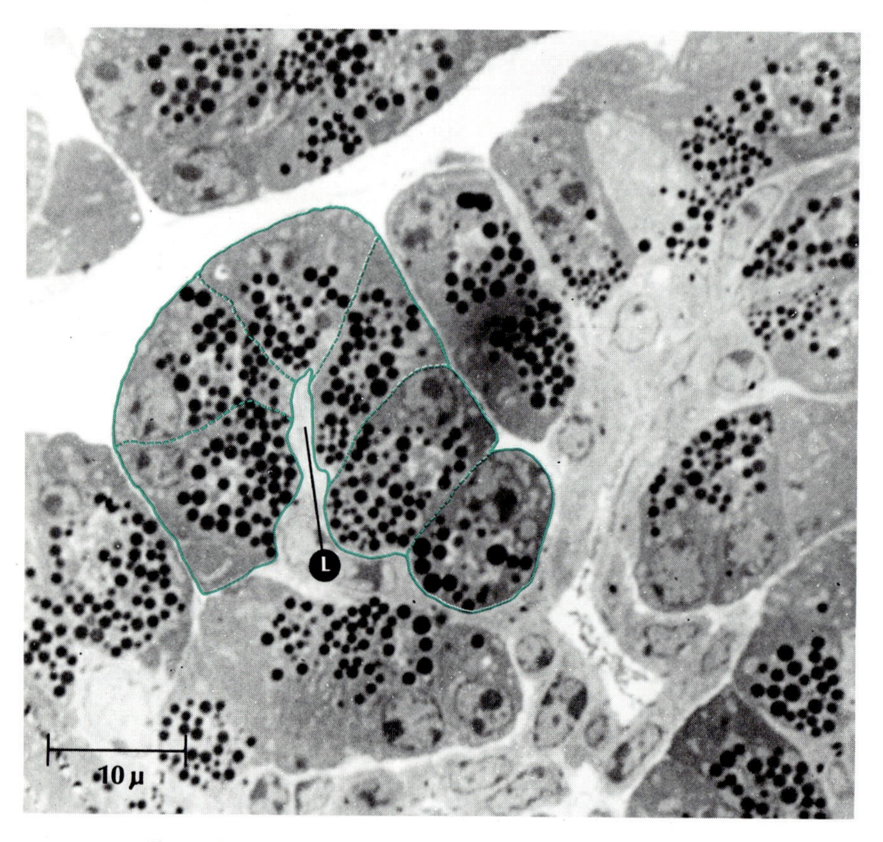

Secretory cells of the pancreas are arranged in groups of more than a dozen cells, called acini; this micrograph shows six sectioned cells in outline. Secretions pass into acinar lumen (L), and thence drain into pancreatic duct system.

Before turning back to study this secretory process in more detail, let us consider for a moment its relevance to the various types of secretory cells. As already noted, there seems little doubt that the process is typical of secretory cells possessing storage granules, which we can define as granules in the size range of 250 to 1,500 mμ. There are, however, a variety of cells that possess no such granules, yet that are clearly secretory cells – examples being the liver's parenchymal cells, the plasma cells, fibroblasts, and some endocrine gland cells. Most and perhaps all of these cells probably do possess structures apparently analogous to the storage granules, but far smaller – on the order of 50 mμ in diameter. There is evidence that the process of secretion and transport in these cells is much the same, at least in its main features, as in the pancreatic and similar cells up to the Golgi complex, but what happens thereafter is uncertain.

Certainly the secretions must somehow get from the Golgi complex to the extracellular space, and by

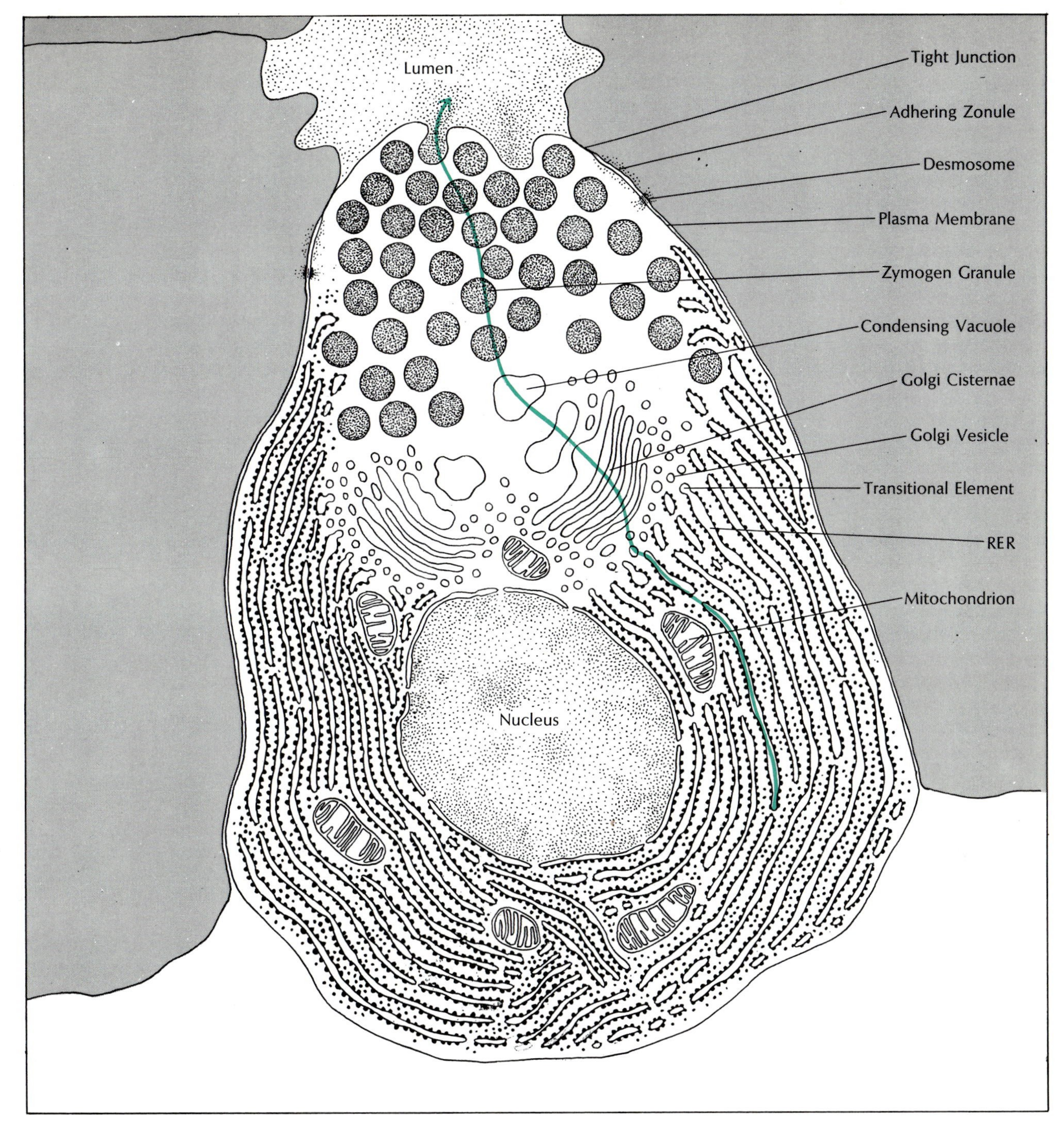

Detailed view of pancreatic exocrine cell shows structure is highly polarized. Secretory proteins are synthesized on RER in basal portion, then pass through components of Golgi complex into condensing vacuoles. These mature into zymogen (storage) granules that under stimulus eject contents into acinar lumen at apex. Note typical junctional elements joining cells laterally.

analogy, if nothing else, one would expect them to do so via the "minigranules" just cited. The problem is that the size of these bodies is well below the resolution of radioautography (about 200 mμ), so that though we may infer that the secretory products have moved into them, there is no way at present of actually tracing their progress (i.e., by labeling) beyond the Golgi complex.

However, assuming that these tiny granules are in fact analogues of the much larger storage granules in other types of secretory cells, there seems to be a clear evolutionary rationale for the difference in size. The liver and endocrine cells in question secrete more or less continuously, though of course their level of secretion can vary considerably. The pancreas and parotids, by contrast, secrete intermittently, i.e., only when appropriately stimulated. In another sense, we can say that whereas the secretions of the liver are needed by the organism at all times, those of, say, the pancreas are required only at certain times. Moreover, when the latter *are* needed, they are needed

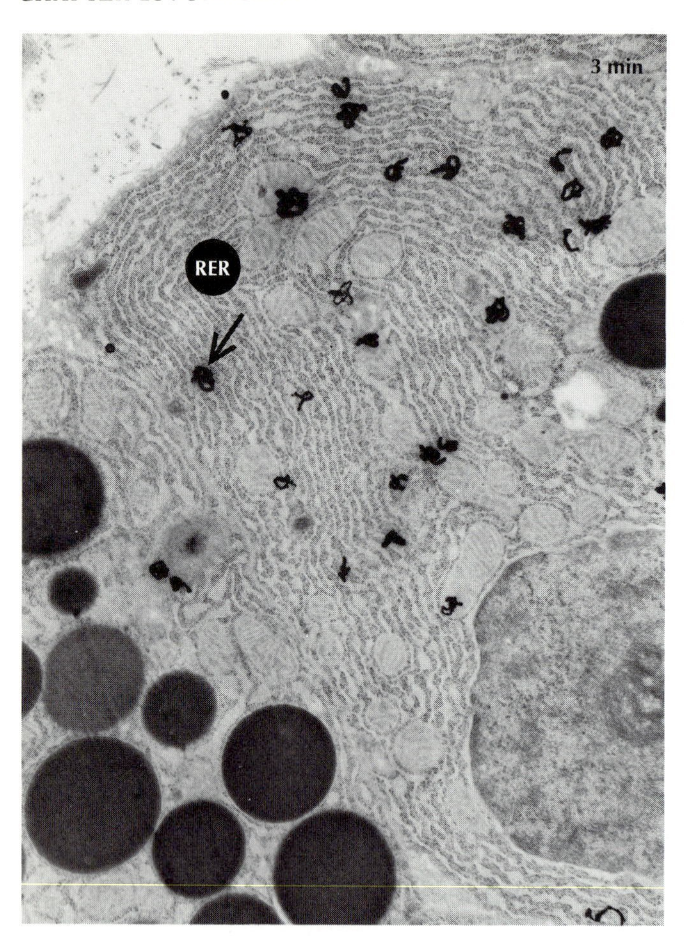

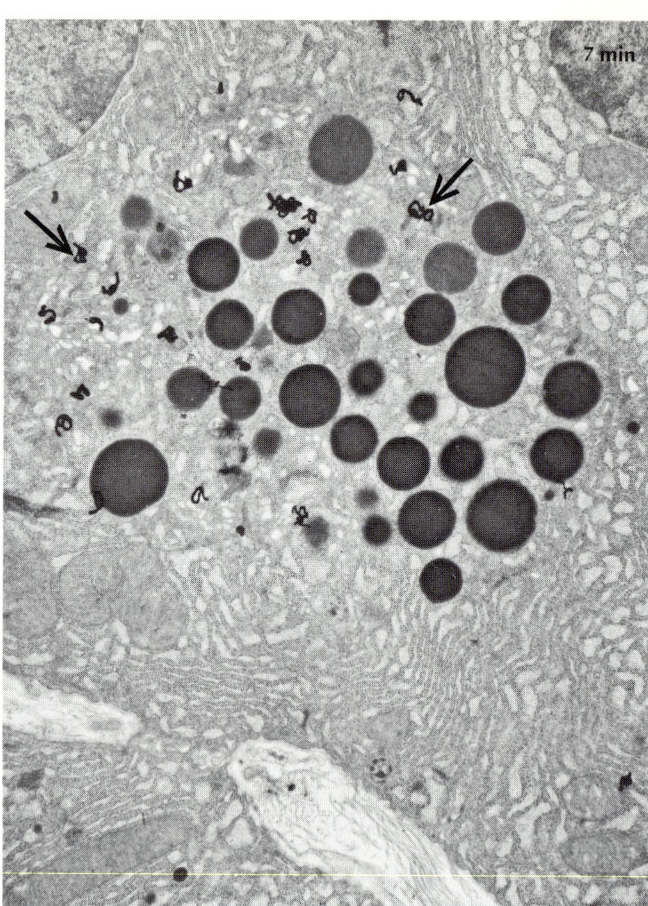

Pulse chase radioautography technique permits tracking of protein through cell compartments. Immediately following labeling (3 minutes), label (arrows) appears in RER, showing presence of newly synthesized protein. By 7 minutes, with cells in unlabeled

in relatively enormous quantities, probably exceeding the cells' capacity to meet in timely fashion through de novo synthesis the short-term demands. Hence the storage granules, whose relatively large volume provides a reserve of secretions to meet these intermittent demands, which would not be required in the more continuous and more measured activities of other types of secretory cells.

Having surveyed the main topographical features of the pancreatic cell and the migration of the secretory proteins through them, let us now turn back and examine the process in somewhat greater detail.

As noted earlier, the proteins are synthesized on the polysomes of the rough ER – specifically, the ribosomes, groups or aggregations of which are referred to as polysomes. Whether given domains in the rough ER carry polysomes programmed to synthesize a specific secretory protein or proteins is not yet known. There can be no doubt that in the later stages of the secretory process the proteins are well mixed; biochemical studies by Dr. Louis Greene have shown that their contents are of the same composition as the pancreatic juice itself, and more recently Dr. Jean-Pierre Kraehenbuhl and myself have shown by immunoelectron microscopy that from cell to cell and within a given cell the contents of all zymogen granules are qualitatively alike. Mixing likely occurs when secretory proteins reach the Golgi complex.

At this writing, only two of the pancreatic proteins have been proved to be synthesized on attached (rather than free) polysomes. However, there is every reason to believe that the same is true of the others, both because the remaining steps of the process are the same for all of them, and for other reasons that I shall cite in a moment. Assuming this to be the case, and also that it is synthesis on attached rather than free ribosomes that distinguishes exportable from nonexportable proteins, it is evident that the attached ribosomes must somehow "know" that it is their business to attach – i.e., some sort of recognition process must be operating between these ribosomes and the exterior surface of the ER membrane. It has been suggested that the ER membrane does not in fact recognize the ribosome but rather its product – the protein, or part of it, being synthesized therein, but there is as yet no evidence of this. Nonetheless, whatever the exact process, some sort of recognition must be operating.

Since so far as can be told the ribosomes are not embedded in the ER membrane but rather are "glued" to its exterior, Drs. David Sabatini and Günter Blobel have postulated another feature of the membrane: pores or channels of some sort, open either permanently or intermittently, through which the growing protein molecule can reach the interior of the ER. Similar channels have been postulated for the plasma membrane to account for

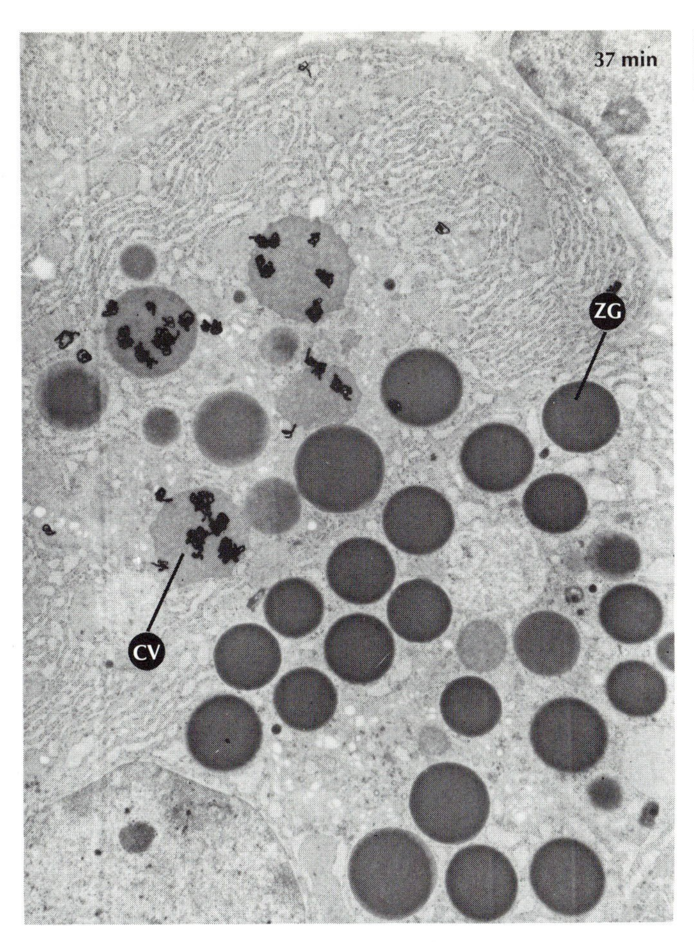

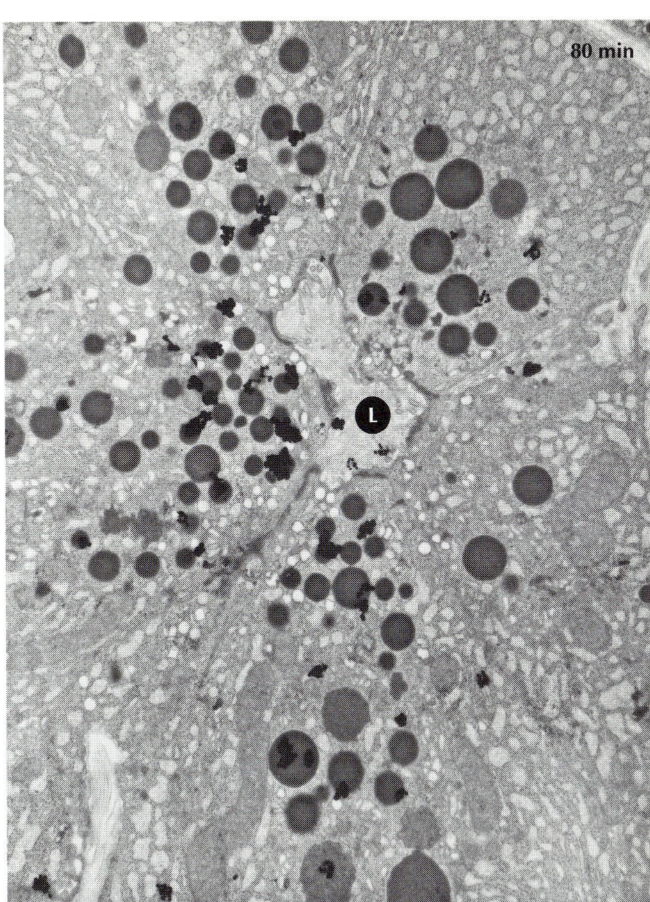

medium, "pulse" of labeled protein has moved into vesicles and cisternae of Golgi complex. At 37 minutes, most of label is in the condensing vacuoles (CV), while at 80 minutes nearly all is in darker zymogen granules (ZG) or has passed into lumen (L).

the transfer of certain small molecules into and out of the cell, and no doubt similar mechanisms – whatever they may be – are at work in the ER membrane. However, it is perhaps worth noting that in the former case, at least, the process of passage through the membrane may be passive; the protein simply slips through when its synthesis is complete, or even, if synthesis is stopped by treatment with puromycin, when it is an incomplete polypeptide.

But why, one may perhaps ask, is it necessary for the proteins to enter, and be segregated in, the ER cisternae? An obvious reason of course is that this is the "loading point" for the entire transport system, which will ultimately move the proteins where they are needed, but this is perhaps somewhat circular reasoning. More to the point is that the transport process that begins in the ER involves chemical modification as well as spatial translocation of the proteins – modifications that can presumably occur only in certain loci within the cell. It is also quite possible that were the proteins *not* segregated they might be denatured or otherwise altered by enzymes or other active substances residing in the cell sap. More certainly, we can say that without segregation the proteins would themselves exercise disruptive effects on other cell constituents.

Specifically, a portion of the pancreatic juice (small but significant in man; much larger in ruminants) consists of various DNAses and RNAses that, were they allowed free run of the cell sap, could destroy the messenger RNA and thereby bring the entire process to a halt. This has been actually demonstrated serendipitously during unsuccessful attempts to carry on cell-free protein synthesis with homogenates of pancreatic cells having substantial RNAse activity; the messenger RNA degraded almost instantly. The same problem is posed by the lipases in the pancreatic juice, which can and do disrupt the membrane lipids if they escape into the cell sap. The fact that they do not do so inside the ER (or the other organelles through which they pass) is a strong argument for asymmetry in these membranes, i.e., their inside surface must in some fashion be shielded against the activities of lipases, as their outside surfaces demonstrably are not.

The proteases that make up a large proportion of the pancreatic secretion do not seem to pose such problems because, as is well known, they are secreted in the form of inactive enzyme precursors, e.g., trypsinogen rather than trypsin, which are only converted to their active enzymatic form after leaving the cell and entering the intestine. Indeed, one of the pancreatic secretions is a rather small polypeptide that acts as a trypsin inhibitor, whose function is perhaps to neutralize any free trypsin that might accidentally get loose among the secretory proteins.

However, the interior of the ER is more than a convenient depot for on-

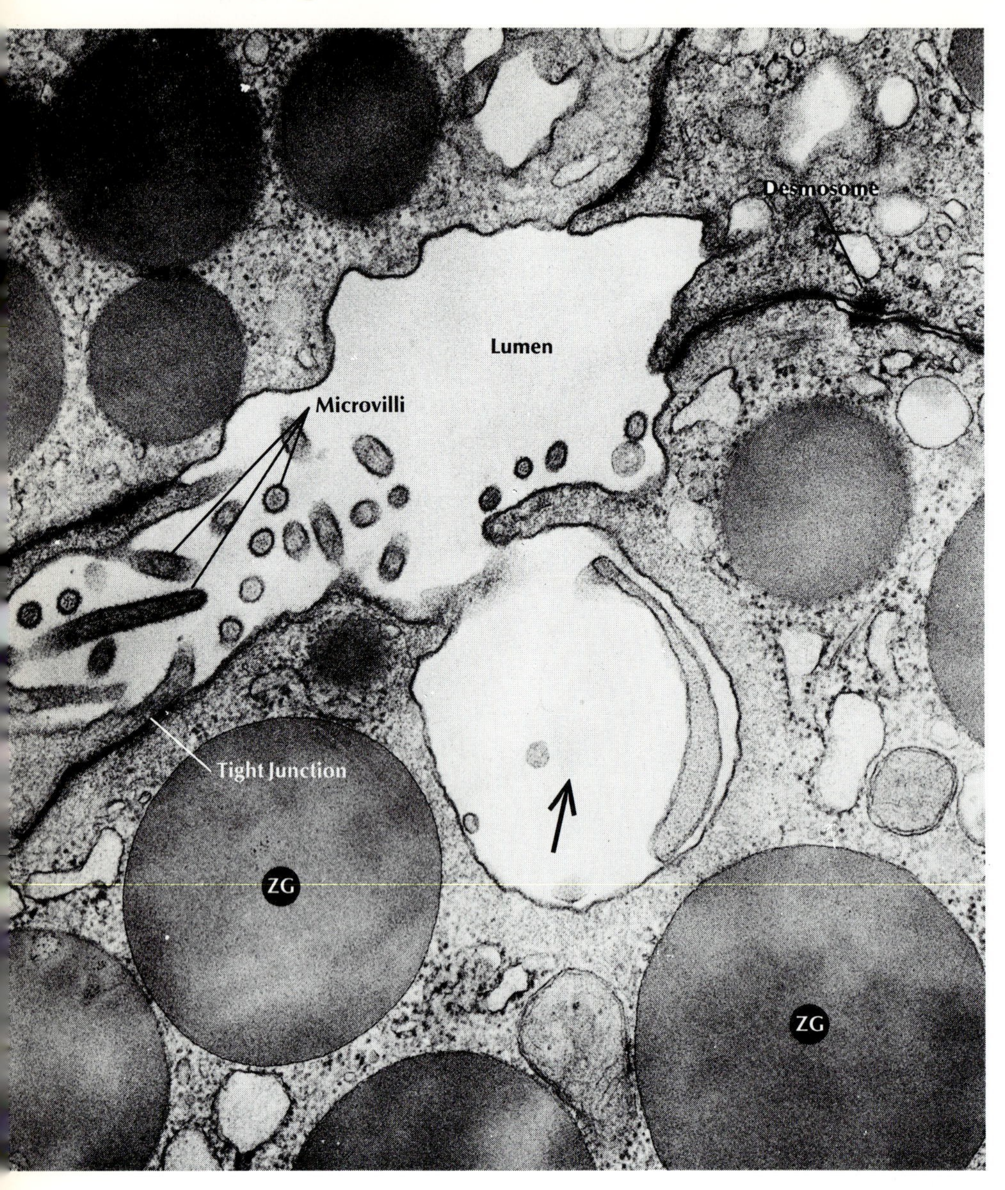

Arrow shows site of a zymogen granule that has just discharged its protein contents into acinar lumen; note also the desmosome and tight junction that link adjacent cells. Small bodies in lumen are not proteins but microvilli projecting from a cell surface.

ward shipment of the proteins; it also serves as a site for their chemical processing. In a number of secretory cells, probably including those of the pancreas, the ER contains enzymes – presumably embedded in or attached to the interior surface of its membranes – that catalyze the production of disulfide bonds. That is, to a certain extent they determine the tertiary ("three dimensional") structure of the proteins. In plasma cells, among others, the ER also includes enzymes that attach some of the proximal sugars at appropriate points on some of the protein chains destined to become secretory glycoproteins, in this case antibodies. The same may be true of pancreatic cells, since a number of their products, including amylase, lipase, DNAse, and RNAse, are glycoproteins. One should note, however, that other pancreatic secretions (e.g., the proteases) are not glycoproteins; this fact has helped to dispose of a hypothesis, formulated some time ago, that secretory proteins had to have sugars attached in order to be exported.

The next step in protein export – its movement into the transitional elements and thence into the Golgi complex – raises a number of puzzling questions. If we take appearance under the electron microscope at face value, the transitional elements seem to bud off the ER, move to the Golgi complex, and deposit their freight of proteins therein, during which process the membranes that "package" the proteins may or may not remain as part of the Golgi membranes. If the transporting vesicles indeed become part of the Golgi membranes, then with time the ER and Golgi membranes should resemble each other chemically. As against this view we have the fact that the membranes of the Golgi complex differ from those of the rough ER, not simply in their lack of ribosomes but in their actual chemical composition – i.e., the identity and proportions of the phospholipids and (especially) the protein they contain.

A possible alternative is to assume that the "buds" are in fact artifacts produced by the fixing process used for electron microscopy and that transport from ER to Golgi actually occurs through continuous tubes connecting the two regions. Such tubes have been postulated in a number of cell regions (see Chapter 13, "The Network of Intracellular Membranes," by Trump), and have actually been demonstrated, by serial sections, between the ER and Golgi complex of liver parenchymal cells. But — and there are many buts in this business — the probability that the same is true of pancreatic cells is lessened by the finding that this step in the transport process is energy dependent.

In the first place, transport is clearly not produced by a simple concentration gradient between ER and Golgi. If protein synthesis is blocked (e.g., by cycloheximide), the already synthesized proteins continue to drain into the Golgi complex at almost the normal rate. If, however, the cells are then treated with antimycin A or other substances that interfere with mitochondrial energy production, tranport to the Golgi complex ceases almost immediately. Evidently, then, the transport process requires energy, probably, since it is not hindered by substances that block glycolysis, in the form of adenosine triphosphate (ATP) generated by mitochondria.

The most plausible explanation of this drain of energy is that it is required for the pinching off of buds from the transitional elements to form transporting vesicles. As will be mentioned later, this may involve molecular rearrangements in membranes similar to those occurring during exocytosis of zymogen granules. The "tube" hypothesis would imply the existence within the tube of some sort of "valve" that would require energy to open; there is no direct evidence for such a structure. But more importantly, protein is transported always from the ER to the Golgi and does not move in the reverse direction when energy is removed from the system, as might be expected of a continuous open channel.

Within the Golgi complex the proteins undergo other chemical transformations. For instance, those proteins that are secreted as glycoproteins likely receive their polysaccharide chains here, as has been convincingly shown in other cell types exporting glycoproteins. Indeed, the enzymes involved in this modification – several types of glycosyl transferase — appear to be an integral part of the Golgi membranes and in particular are located on the inner surface of the membranes where they may be able to act in assembly-line fashion during the sequential addition of polysaccharides. We should mention in passing that the function of the polysaccharide side chains in enzyme catalysis is unknown. In cells producing acid mucopolysaccharides (including those of the pancreas), the characteristic sulfate groups of these substances are also added in the Golgi. Finally, some of the proteins themselves may undergo changes in the Golgi, as has been suggested by Dr. Donald Steiner for insulin. As is well known, this substance in its metabolically active form consists of two chains, A and B, connected by disulfide bonds. As originally synthesized in the ER, however, the two chains are joined into one, and it appears likely that in the Golgi complex they are separated by the excision of a short peptide. The reason for this process seems to be that only in the form of the continuous chain (proinsulin) can the disulfide bonds be properly positioned; if insulin is broken down in the test tube into separate A and B chains, they cannot be recombined into a functional molecule.

The passage of the secretory protein through the various cisternae of the Golgi apparatus appears to operate by simple diffusion, requiring no energy; the same may or may not be true of the formation of zymogen granules. The process by which the proteins are condensed into the relatively high concentrations of the zymogen granule is, as already noted, completed during the maturation of the condensing vacuoles, though observations in the pancreas and other cells suggest that it begins earlier in the Golgi cisternae. Rough estimates, based on the amount of label in particular compartments at particular times, suggest that the total relative concentration between ER and zymogen granule is on the order of 20 to 25 times, but we cannot yet say how much of that occurs in the condensing vacuole and how much in previous stages.

Originally it was thought that the condensation process was energy dependent, with water and other substances actually being "pumped" out of the condensing vacuole, but recent studies with various sorts of metabolic inhibitors indicate that this is true to only a limited degree. It now seems that a good deal of water loss during condensation may result from simple osmosis, caused by the low osmotic pressure of the vacuole contents. Indeed, when one calculates the osmotic pressure of the vacuoles, based on their protein content alone, vis-à-vis that of the surrounding medium (i.e., the cell sap), one finds that the proteins should act as a hypotonic solution – i.e., they should lose water "automatically." However, this is true only if one assumes that the vacuoles contain nothing but protein – i.e., no electrolytes or similar substances – and this is not in fact the case.

One way of getting around this difficulty would be to postulate the consolidation of the vacuolar proteins into still larger macromolecules that would have a still lower osmotic pressure. This might be accomplished simply by ionic interactions between proteins of different charge. In addition, the condensation vacuoles contain a small but significant proportion of acid mucopolysaccharide molecules, whose negative charges could combine with positive charges on some of the proteins, many of which are basic, thereby forming "super-molecules" in sufficient quantity suitably to lower

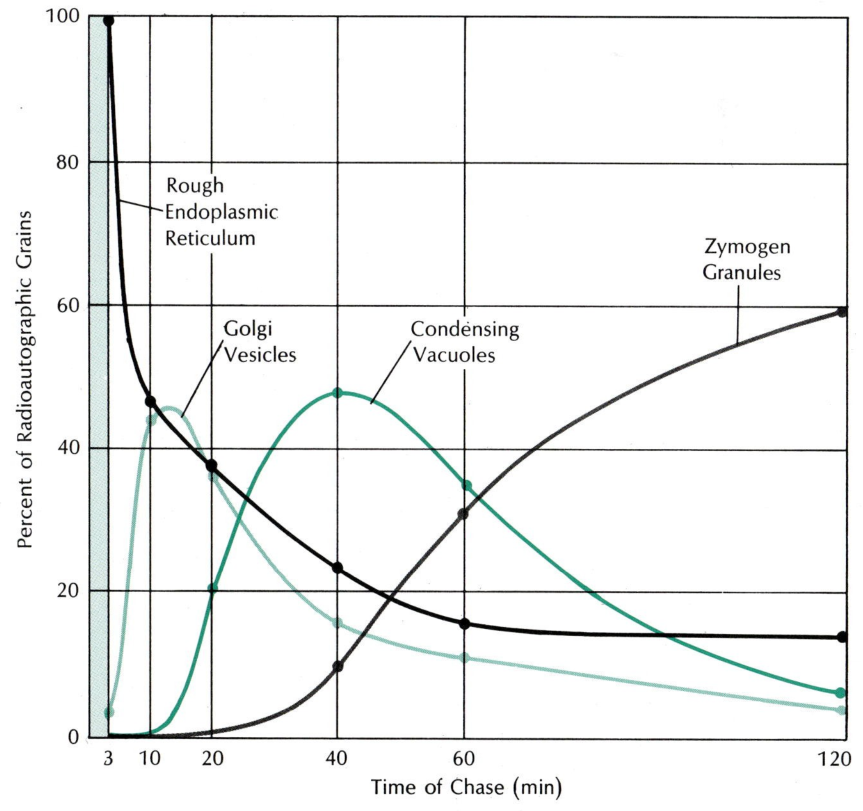

Graph of pulse-chased cells, based on counts of radioautographic grains at successive intervals after labeling period (color tint), shows "wave" of labeled protein as it passes from the RER to the Golgi complex to condensing vacuoles to zymogen granules.

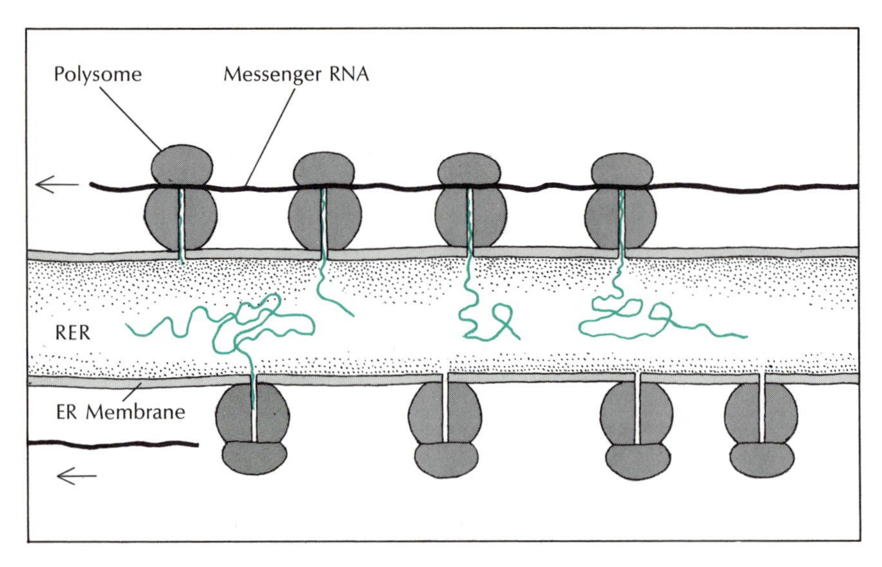

Synthesis of protein (color) occurs when a molecule of messenger RNA is "read off" by polysomes, which translate its genetic code into sequences of amino-acid residues. Presence of pores in ER membrane is inferred from the fact that during their synthesis proteins pass only into the cavity of the RER. This conclusion and the model depicted above are based on the studies of Drs. David Sabatini and Günter Blobel.

the osmotic pressure. According to this postulate, the only energy input would be that required for the sulfation of acid mucopolysaccharides in the Golgi complex, as mentioned above. Following this, condensation of secretory proteins would follow automatically. Moreover, and perhaps significantly, nobody has managed to assign any function for the mucopolysaccharides outside the cell.

The process by which the zymogen granules discharge their contents into the acinar lumen under appropriate exogenous stimulation involves two components. First, the granules must approach the apical plasma membrane, then their own membranes must merge with the plasma membrane, allowing their contents to be ejected into the lumen by exocytosis. As regards the transport of the granules to the plasma membrane, no special mechanisms need be postulated, since the distances involved – on the order of 10μ – can be covered in the known time interval by simple diffusion. The actual fusion of the granular and plasma membranes, however, is most likely to be dependent on supplies of energy – presumably in the form of ATP. This may serve to power the molecular reorganization required for the two membranes to fuse and then break apart along a new plane, but how this actually occurs we do not yet know (see Chapter 8, "The Fusion of Cell Membranes," by Lucy).

There is also a recognition process involved in exocytosis, for, though the zymogen granules frequently approach other portions of the plasma membrane very closely, they have never been seen to fuse with any but the apical portions – those adjacent to the acinar lumen. Recognition of one membrane by another is a general property of membrane systems in cells and while the molecular mechanisms involved are unknown, they probably consist of specific interactions between membrane proteins.

The mystery is deepened when we note that exogenous hormones that stimulate protein release (in the pancreas, pancreozymin) probably never reach the apical membrane through which discharge occurs. As noted much earlier, the pancreatic cells are connected by tight junctions, and while the primary function of these "fused" plasma membranes is probably to prevent leakage of the pancreatic secretions into the extracellular spaces, a secondary effect could be to block off the apical region of the cells from the circulation, through which the hormones reach their target organs. Thus, if this picture is correct, the triggering hormones must somehow effect changes in the apical membrane through their effect on the basal membrane. Whether the effect of hormones is transmitted from base to apex via cyclic AMP generated by a plasma membrane-associated adenyl cyclase, or is mediated through changes in membrane polarity with associated changes in ion flux (especially Ca^{++}, which is required for exocytosis), or is the result of these

1 Lumen 2
Plasma Membrane
Zymogen Granule
3 4

Emptying of zymogen granule begins when granule, triggered by hormonal stimuli, approaches membrane at cell apex (1). The two membranes first touch (2) and then merge (3); finally, the merged area opens to allow ejection of granule's protein contents.

and other factors is not known.

It is perhaps worth noting in passing that the discharge process does not seem to be dependent on the rate of protein synthesis in the ER, nor on the rate of transport to the storage granules; the latter continue to discharge, to the point of total depletion, even when protein synthesis is blocked. Likewise, the rate of discharge appears to have no immediate influence on protein synthesis, which continues at the same level for a considerable time whether or not discharge is occurring. Presumably, there is some sort of upper limit here, else one would expect that a prolonged cessation of discharge – e.g., in the fasting animal – would cause the cells to become choked with storage granules to the point where their functioning would be disrupted. Likewise, there seems to be some sort of control at the lower end, in that some minutes to hours after discharge protein synthesis is stepped up. But in neither case do we know the nature of the controls in question. Equally mysterious are the processes by which the polysomes are apparently "reprogrammed" to produce a different proportion of proteins in response to long-term shifts in diet (i.e., high-fat diets increase the proportions of lipases, high-protein diets of proteases, and so on) beyond the fact that these changes are measured in weeks, not hours.

There is another point about discharge that is worth noting in relation to the postulated aggregation of proteins within the storage granules: it seems to require an alkaline pH. Isolated storage granules placed in distilled water (pH about 5) will not

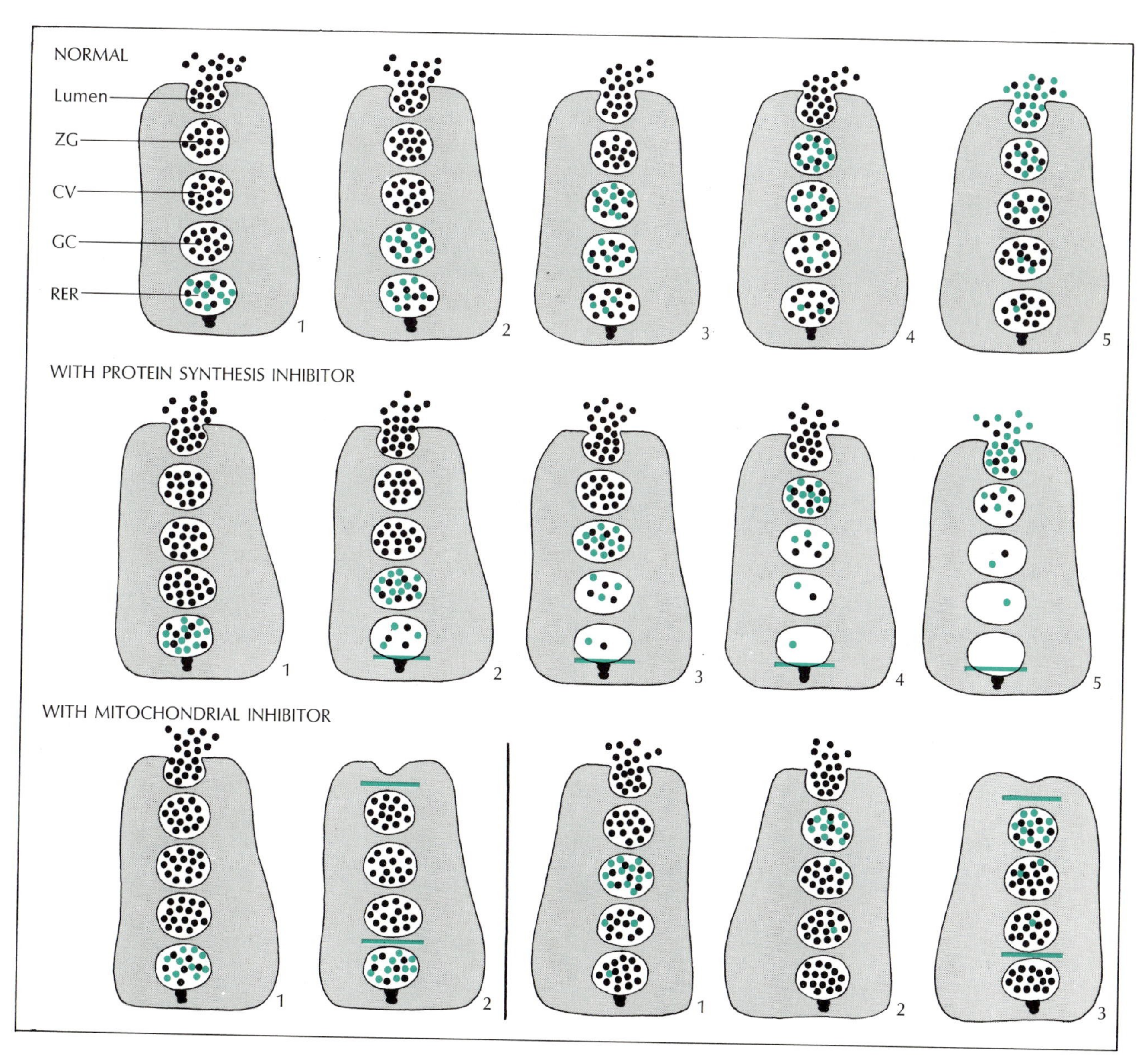

Schematic diagrams of compartmented cells show movement of proteins (labeled, color; unlabeled, black) from base to apex. Top row shows normal condition. When protein synthesis inhibitor is added (middle), movement continues as usual, showing it is not induced by "pressure" from new protein. But adding inhibitor of mitochondrial ATP production blocks passage of proteins from RER to Golgi (bottom left) or, if inhibitor is added later, from zymogen granule to lumen (bottom right). These findings indicate both these steps are energy dependent; passage from condensing vacuole to zymogen granule is not affected.

lyse, but will do so if the pH is raised toward 7. Significantly, when protein discharge is stimulated, there is also release to the acinar lumen from other pancreatic cells (probably the so-called centroacinar cells) of a water and electrolyte solution containing relatively high concentrations of bicarbonate – with a pH around 7.4. It seems likely that this alkaline medium helps facilitate the discharge process; it may break up any "super-molecules" formed from proteins and acid mucopolysaccharides, thereby returning the proteins to their "normal" soluble form.

The discharge of pancreatic enzymes and proenzymes is part of a feedback loop in which low intestinal pH (2) triggers the secretion in the intestinal wall of secretin and pancreozymin. These pass through the circulation to the pancreas, where the former stimulates the centroacinar cells to release their water and electrolytes, while the latter triggers release of secretory proteins from acinar cells. The flow of pancreatic juice into the intestine, thanks to its content of bicarbonate, raises the intestinal pH to around 7.4, leading to a cutback in secretin and pancreozymin production, and in addition providing a medium of optimal pH for the activity of the pancreas' digestive enzymes. Within the intestine, too, enterokinase initiates the sequence of reactions involved in converting pancreatic proenzymes into active enzymes. In addition, however, the pancreatic cells are also under a degree of neural control, whereby stimuli via the vagus nerve trigger protein discharge cholinergically. In fact, release can be induced in vitro by adding analogues of acetylcholine. We obviously have a great deal to learn about how these stimuli, hormonal and neural, alter the cells' plasma membranes so as to trigger release of their secretions. A later chapter in this book will discuss how cyclic nucleotides (CAMP and CGMP) influence these events (see Chapter 19, "Cyclic Nucleotides and Cell Function," by Goldberg).

Yet another mystery concerning the membranes of secretory cells, however, arises out of the fact that the transport processes within them involve not merely secretions but also their membranous containers. That is, there appears to be a continuous flow of membrane, possibly from rough ER to the Golgi apparatus, but certainly from the Golgi to the storage granules and then into the plasma membranes. The package is shipped along with its contents – but we have no accumulation of packaging materials at the "point of consumption" – i.e., exocytosis – such as notoriously occurs with man-made packaging materials. To further complicate matters, the packaging material itself undergoes qualitative changes – markedly between ER and Golgi (assuming, of course, that transport here occurs by budding rather than through tubes) and, to a limited degree, even in the later stages of the process, since storage granule membranes appear to be simplified versions of their precursors, at least so far as membrane proteins are concerned.

In the case of the ER-Golgi stage of transport, one might postulate that the transitional elements are analogous to returnable bottles, discharging their contents in the Golgi and returning to the ER for refilling. But, if this is so, no one has yet observed any signs of the "empties." And, if it *is* so, we face a situation where membrane is leaving the Golgi apparatus (in the condensing vacuoles) without more membrane coming in to replace that not removed. On the other hand, if we assume that the transitional element membranes are not thus "recycled," we are left with a problem, since on this assumption the bodies that are merging with the Golgi apparatus are considerably smaller than the condensing vacuoles that are leaving it. Because of their higher surface-to-volume ratio they will necessarily be adding more membrane to the Golgi than is subtracted by the departing condensing vacuoles.

Further loss of membrane appears to occur during the maturation of the condensing vacuoles, which begins as somewhat irregular bodies but ends as spherical zymogen granules – which, moreover, are apparently smaller than their precursors, as one would expect to be the case since their contents are more compact. Again, for geometric reasons, the shift from an irregular to a spherical configuration and the concomitant shrinkage would both necessitate a reduction in the surrounding membrane. Finally, each zymogen granule that discharges contributes its membrane to the plasma membrane, yet the latter does not expand permanently.

Here, too, some researchers have postulated a recycling of membrane, in the form of spinocytotic vesicles that return to the Golgi apparatus for reuse from the plasma membrane, but again there is no direct evidence of this. The obvious alternative, of course, is that the excess membrane at any given point is simply broken down into its constituents and new membrane is resynthesized to supply the corresponding deficiencies elsewhere in the transport process. Evidence is beginning to accumulate, however, that an intermediate situation likely exists: membrane is probably recycled in large part but, in addition, some new membrane is synthesized as part of the normal repair or turnover process.

Added to all this is the fact that the processes by which the cell's membranes are kept in balance, whether by macroscopic recycling or by resynthesis, must proceed in such a manner that the distinctive characteristics of each membrane – enzymes in the case of the ER, recognition sites for secretory granules and for exogenous hormones in the case of the plasma membrane – must remain intact so that the cell can continue to function. Clearly, researchers in the field of secretory membranes will not lack for occupation in coming years.

16

The Biogenesis of Organelles

ERIC HOLTZMAN
Columbia University

Perhaps the most fundamental trait of living organisms is their capacity to take nonliving material and transform it into living material – more specifically, into *structured* living material. Even the simplest of viruses and cells show organization and this permits them to sustain activities far more complex than are possible for unorganized mixtures of the molecules from which they are constructed. The cell lives by and through its organelles – distinctive structures with distinctive functions such as the plasma membrane, ribosomes, or mitochondria. Not the least of these functions are the continuous maintenance or replacement of the structures themselves and the assembly of new structures as required for the alteration of existing cells or the formation of new ones through cell division.

Thus the biogenesis of organelles involves a basic aspect of life processes: the transition from one level of organization, the molecular, to a higher level, the supramolecular. The mechanisms by which cells make molecules are coming to be understood fairly well; the processes by which the molecules are assembled into organelles are less well understood – but for that very reason are among the most interesting biologic problems currently under study.

The problem of organelle biogenesis can be examined from two different but complementary standpoints: conceptually, in terms of the basic principles involved, and operationally, in terms of the methods by which biogenesis is studied in the laboratory. This chapter will consider both aspects.

The central concept in organelle biogenesis is that of self-assembly. This means in essence that the macromolecules of which organelles are composed contain within them information such that, when mixed under "appropriate" conditions – a phrase we need not define in detail at this point – they will more or less spontaneously associate with one another in specific patterns to build up more complicated structures. Self-assembly is a fact, having been demonstrated in a variety of biologic systems, through successful test-tube reconstitution of structures from their components. With some important modifications the process can account for most of what we know about organelle formation and growth. But self-assembly is also, one might say, almost a philosophic necessity, for it avoids the postulation of an endless chain of structures specifying the organization of other structures. As we know, all cellular macromolecules can be traced back ultimately to the template system of the DNA molecule, whereby proteins are synthesized in accordance with the information encoded in DNA base sequences, and other molecules (e.g., lipids and polysaccharides) are synthesized in accordance with the specific enzymatic proteins that the cell produces. Unless the information, derived ultimately from DNA base sequences, made macromolecules capable of some sort of self-assembly, one would have to postulate a second, separate template system that would specify supramolecular structures as the DNA system specifies molecular ones. Then the question would arise as to how this second template system obtains *its* structure or how *it* might be duplicated from one cell generation to the next; DNA dictates its own structure, and in this sense is self-duplicating, even though the rest of the cell is involved in supporting this duplication.

Self-Assembly of Molecules and Simple Systems

The self-assembling abilities of DNA and of some proteins can be demonstrated experimentally. Thus, if one carefully heats a DNA double helix, the two complementary strands will come apart (the hydrogen bonds that hold the two strands together are relatively weak bonds that can be disrupted by moderate heating). When separate, the strands have no well-defined three-dimensional structure; they coil into varying configurations. But, if the heated DNA is slowly cooled, complementary strands will realign with respect to one another (following the base-pairing "rules" that are fundamental to DNA's role in heredity) and spontaneously reestablish the double helix.

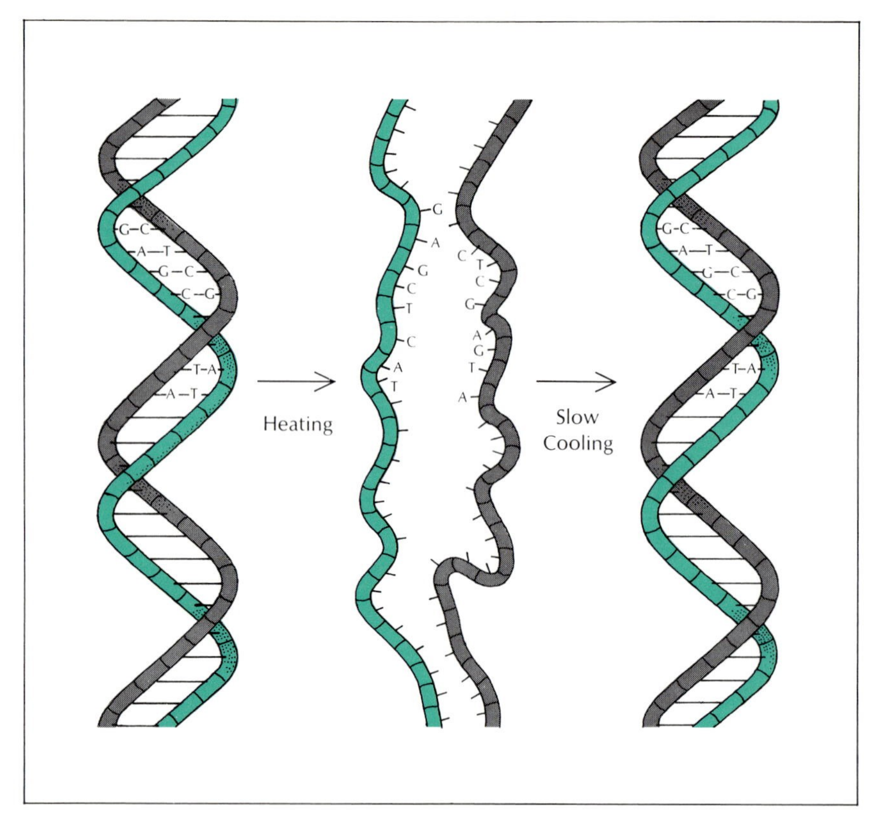

Fig. 1: Double helix structure of DNA is stabilized by the many weak (hydrogen) bonds that form between the purine and pyrimidine bases (A, T, G, C) of the two strands. Strands are related in such a way that an adenine (A) on one is always paired with a thymine (T) on the other, and a guanine (G) is always paired with a cytosine (C).

Presumably, in its normal duplication, a DNA double helix unwinds, each strand synthesizes a new complementary partner (with the aid of specific enzymes), and the new and old strands spontaneously fall into the double-helical conformation (Fig. 1).

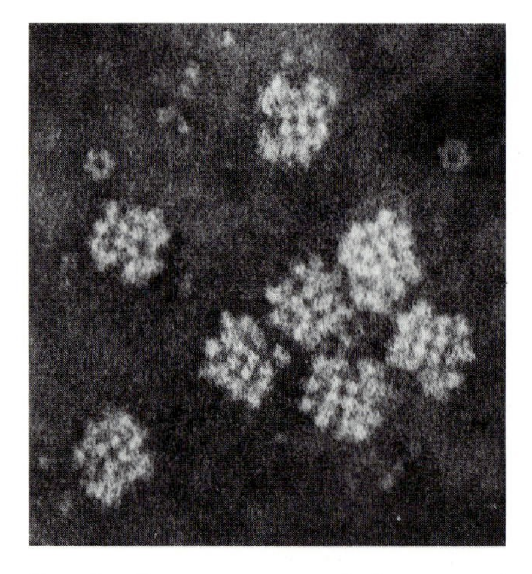

Fig. 2: Electron micrograph (courtesy of L. J. Reed, R. M. Oliver, and D. J. Cox) shows several pyruvate dehydrogenase complexes isolated from E. coli. *Each complex is composed of many small subunits, seen as light globules.*

The sequence of amino acids in a protein is determined by the sequences of bases in the DNA segment that "codes" for that protein. But proteins also have specific three-dimensional conformations that are essential for functions such as enzymatic activity. Unfolding ("denaturation") of an enzyme usually inactivates it even if the denaturation is the result of treatments such as moderate heating or exposure to urea that affect "weak" bonds (e.g., hydrogen bonds, hydrophobic bonds, ionic bonds between charged groups) but leave intact the stronger covalent links that maintain the amino acid sequence. However, some proteins (e.g., ribonuclease) are either quite resistant to denaturation or, after unfolding, can spontaneously revert to their original enzymatically active conformations when the denaturation conditions are reversed. Apparently the amino acid sequence itself contributes information that is central to the determination of the three-dimensional structure. This is most obvious when special covalent bonds such as disulfide bonds are formed between distant points along the chain of amino acids; as in ribonuclease, these bonds stabilize the folded structure. The DNA-determined locations of the amino acid cysteine that can form disulfide bonds thus are important for the folding pattern (Fig. 3, lower panel).

Other, weaker bonds also form between appropriate amino acids so that the distribution of amino acids along the chain promotes specific associations between different portions of the chain; thus, while a protein might theoretically be able to assume an almost infinite variety of conformations, one or a few will be particularly stable, and these will tend to form spontaneously through the random motions that all molecules undergo when in solution. Furthermore, for proteins such as hemoglobin, which are made of several separate subunits, the distribution at the subunit surface of amino acids with particular bonding capacities will promote correct association of the subunits; thus multisubunit proteins will often spontaneously reassemble if the subunits have been greatly separated (Fig. 3).

Implicit in this discussion is the fact that the cell can synthesize the proper amounts of requisite components at the correct time, and also can control its internal milieu in terms of ionic concentrations, pH, and so forth. Assembly processes are markedly affected by the presence or absence of particular ions, such as Ca^{++}, by temperature, and by other factors that influence the modes and rates of molecular interactions. There is no doubt that the cell can control many such factors in precise fashion, although there are large gaps in present understanding of the underlying mechanisms.

Given the existence of such controls, it is apparent that relatively straightforward self-assembly processes can contribute to the formation of fairly elaborate multimolecular arrays. For example, Figure 2 at the left shows a multienzyme complex (the "pyruvate dehydrogenase complex") isolated from the bacterium *Escherichia coli.* This structure contains a few dozen protein subunits, which comprise a number of molecules of each of three enzymes that catalyze sequential steps in oxidative metabolism. The subunits can be separated by gentle treatments, and it is found that the structure can re-form spontaneously. One can speculate that at an earlier stage of evolution the enzymes of the complex existed separately in the cell; mutations that promoted the association of the molecules with one another would be evolutionarily advantageous, since the binding of sequentially acting enzymes in a common structure can markedly enhance metabolic efficiency.

Another important self-assembly example is the fact that phospholipids and certain other lipids form bilayers spontaneously when placed in aqueous solution. As discussed in earlier chapters in this volume, such bilayers are thought to be essential features of most biologic membranes. (See: Chapter 1, "The Bilayer Hypothesis of Membrane Structure," by Danielli; Chapter 2, "Lipid Dynamics in Cell Membranes," by Chapman; and Chapter 3, "Models of Cell Membranes," by Bangham.)

Other examples of spontaneous assembly will be mentioned later. However, there are limits. One cannot simply separate all the molecules of a cell, or even of a mitochondrion, and then reconstitute the original structure merely by mixing the molecules in an appropriate ionic environment. As it contributes to the growth or formation of various cell structures, self-assembly is modified in a number of interesting ways.

Time

Instances in which gently denatured proteins do *not* resume their characteristic structure when denaturing conditions are reversed are particularly interesting because they introduce the dimension of time. In the normal course of protein synthesis the molecule is put together one amino acid at a time, starting at one end. It takes roughly one minute for a protein the length of one of the chains of hemoglobin (about 140 amino acids) to be synthesized. Therefore, completed portions of the amino acid chain may be free to fold while the remainder is still being synthesized. In the case of large and complex proteins, it is possible that the correct three-dimensional structure can manifest itself *only* under these sequential

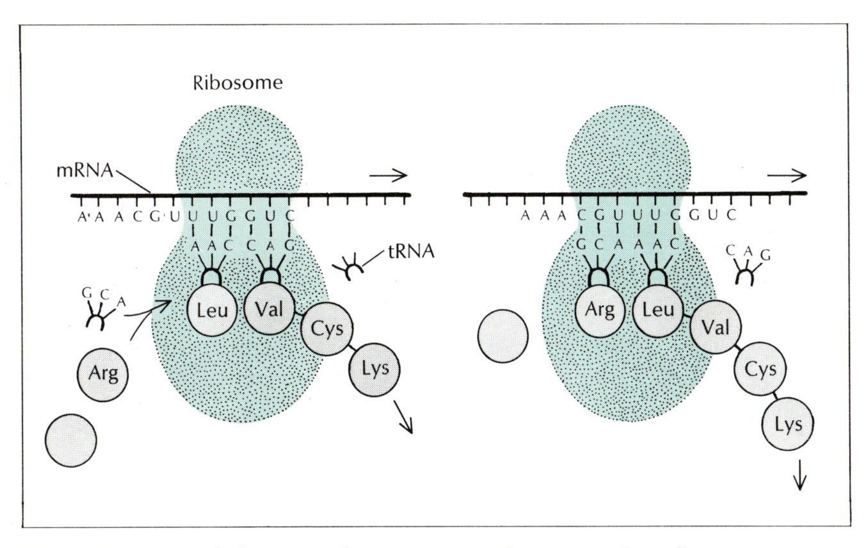

Fig. 3: As top panel shows, synthesis of a particular protein depends on a messenger RNA(mRNA) with a specific sequence of bases copied from a portion of a DNA molecule but with uracil (U) replacing thymine. Amino acids first are linked to specific transfer (t) RNAs; each tRNA includes in its base sequence a special set of three bases corresponding to the amino acid with which it can link. The amino acid–RNA complex subsequently binds to mRNA associated with a ribosome. This binding occurs only where the mRNA has a sequence of three bases with which the special tRNA set can pair according to the rule that A pairs with U, G with C. Once aligned by this pairing, the amino acid is linked to the growing end of the polypeptide chain, the fundamental polymeric structure of proteins. Below: Proteins fold and associate with one another on the basis of the distribution of specific amino acids whose side groups can form bonds, such as disulfide or ionic bonds. The sequence in which such amino acids (or any others) occur in a protein depends on the corresponding mRNA and thus ultimately on DNA.

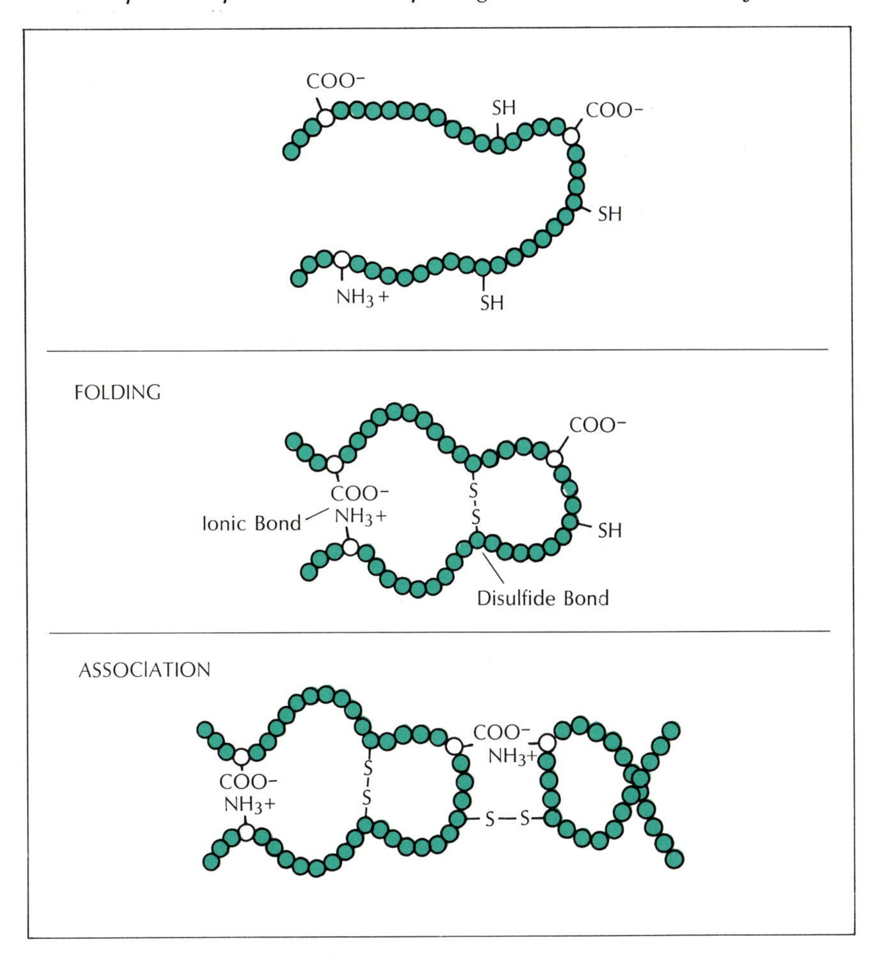

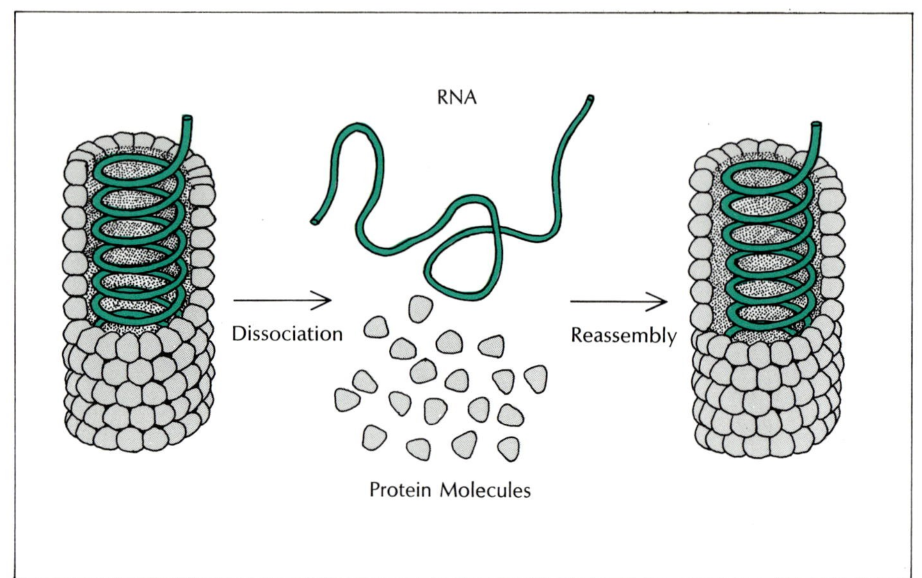

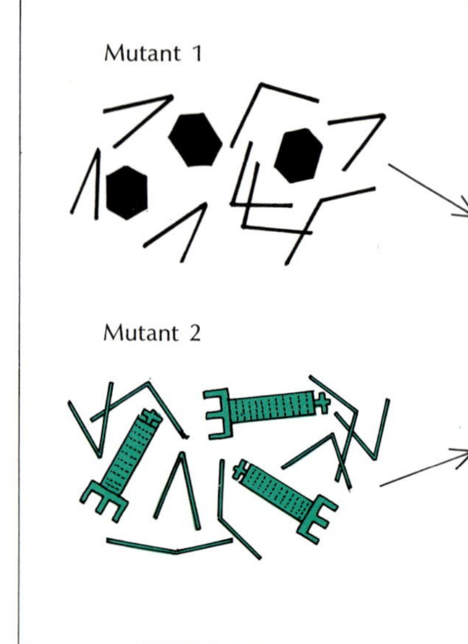

Fig. 4: As diagram at left indicates, tobacco mosaic virus (A in the photograph at far right) can be reversibly disrupted by suitable gentle biochemical procedures into separate nucleic acid and protein molecules (shown by the work of Fraenkel-Conrat and others). Above: mutants of the bacteriophage T_4 (B in photo at far right) will produce parts of viruses when allowed to grow

conditions. If the entire protein molecule is present (as in a denatured preparation), sections of it will interfere with the folding of other sections. Correct assembly of such proteins, in other words, can operate step-by-step but not simultaneously.

A similar timing problem is involved in the structuring of the insulin molecule, whose component alpha and beta chains, when separated in the test tube, cannot readily be recombined into physiologically active insulin. The two chains are synthesized by the cell as parts of a single *proinsulin* molecule, and they assume their eventual configuration while in this form; subsequently the single chain is enzymatically snipped into two, excising some amino acids but leaving the structure otherwise intact, and converting the biologically inert proinsulin into active insulin.

Illuminating self-assembly processes are seen with viruses. It has long been known, for example, that simple viruses can be reconstituted by mixing their DNA (or RNA, which substitutes for DNA in some viruses) with the protein that encapsulates the nucleic acid. With the simplest viruses, containing only one or two species of protein, this can be accomplished quite easily by careful separation and admixture of their constituents; with more elaborate viruses, phenomena have been encountered that point to the existence of additional complications in the self-assembly process. One of these involves the time-dimension, with assembly proceeding not simultaneously but stepwise. That is, different parts of the virus are assembled separately out of their appropriate subunits, following which the parts are combined into the complete virus.

Such seems to be the case, for example, with the T_4 bacteriophage, in which the heads, tails, and tail filaments of the virus are formed first out of their molecular components as separate structures, which then are assembled into complete virus particles. It has been shown that under some conditions of growth certain mutant strains of the virus produce incomplete progeny – heads and filaments only, or tails only – but if host bacteria infected with different aberrant forms and thus containing different parts of viruses are broken and their contents mixed, the parts can join together into complete and functional virus particles (Figure 4).

Enzymes and Nucleation

A second complication of T_4 assembly is the likelihood of enzymatic intervention at certain points in the process. As we have seen, many types of self-assembly, as with lipid bilayers, involve the formation merely of weak hydrophobic or ionic bonds. Others, however, evidently depend on the creation of the stronger, covalent bonds for which enzymatic action seems to be required. The structure of T_4 bacteriophage, for example, has been shown to be controlled by several dozen different genes, most of which probably specify the amino acid sequences of different proteins. However, the virus itself contains relatively few different proteins. Some of the genes presumably specify enzymes of macromolecular synthesis, but there seem also to be enzymes that preside over one or another step of the assembly process, aiding in the linking of one viral part to another.

The formation of the fibers in a blood clot proceeds in a series of steps, some of which depend upon enzymatic intervention, such as the alteration of fibrinogen molecules into fibrin molecules, which then can spontaneously form fibers based on weak bonds. Another enzyme subsequently stabilizes the fibers through the formation of covalent bonds.

I should emphasize that the existence of "assembly enzymes," like the necessity for specific environmental conditions, does no violence to the central concept of self-assembly. But enzymes do add significant elements of control and complexity. They can form stable links between specific groups on different molecules, speed

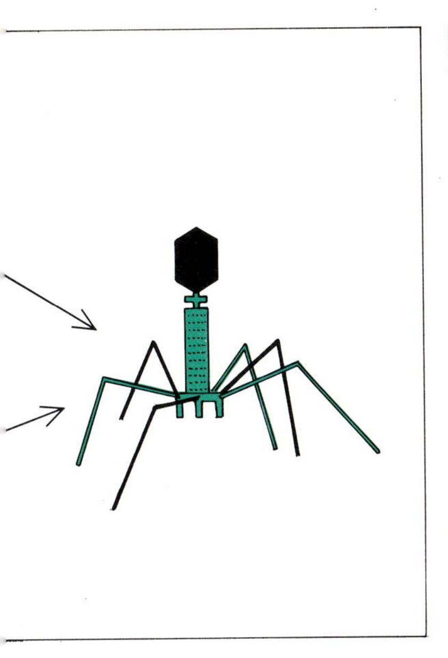

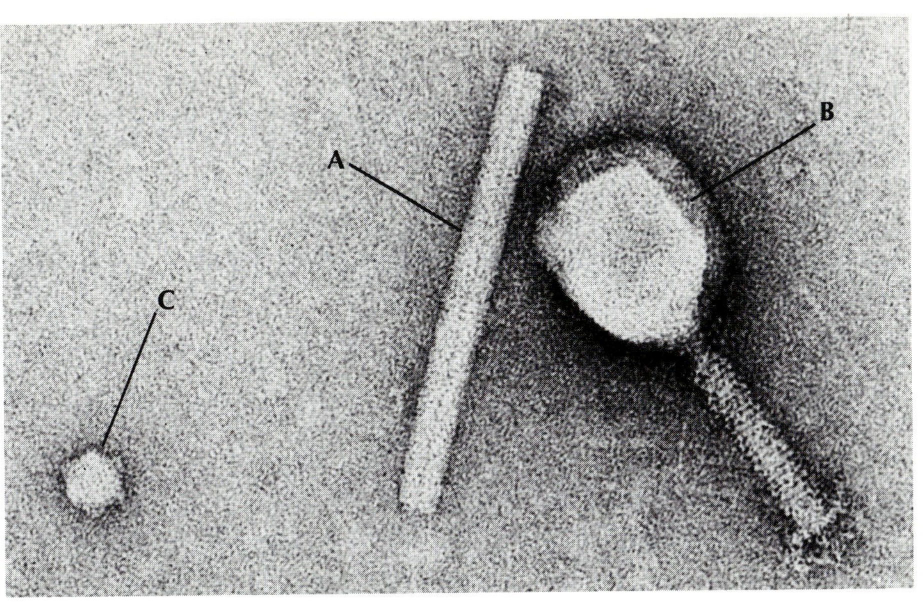

within bacteria under some conditions. If host bacteria containing different parts of viruses are broken open and their contents mixed, functional viruses form (from work by Wood, Edgar, and others). The third virus (C) in the photo is φ-X-174, which infects bacteria (as does T_4). Photo provided through courtesy of F. Eiserling, W. Wood, and R. S. Edgar.

up assembly, alter the properties of potential interactants, or otherwise modify the course of events. For instance, perhaps enzymes sometimes mediate involvement of metabolic energy in assembly processes.

Other processes of assembly include the operation of "nucleating centers." Bacterial flagella provide an example of this. These are thin filamentous structures that protrude from the surfaces of some bacteria and may play a role in motility; their structure is much simpler than that of the flagella of higher organisms to be described below. Protein molecules (flagellin) purified from bacterial flagella will assemble to form flagellum-like filaments in the test tube, but such assembly is extremely slow unless small fragments of flagella are added to the purified proteins. By analogy to the growth of inorganic crystals, it is believed that the fragments serve to "seed" or nucleate assembly by providing initial points for the oriented growth of filaments through the addition of flagellin molecules. In some cases one can show that the morphology of the newly assembled filament depends upon the morphology of the flagellum used as the seed. Growth of structures by oriented addition of subunits to a preexisting structure is probably a very common event in organelle formation; the structure that already exists can strongly influence the organization of material added to it (Figure 5).

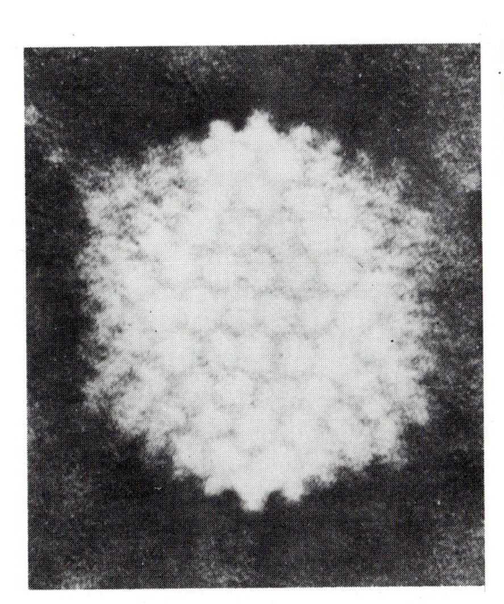

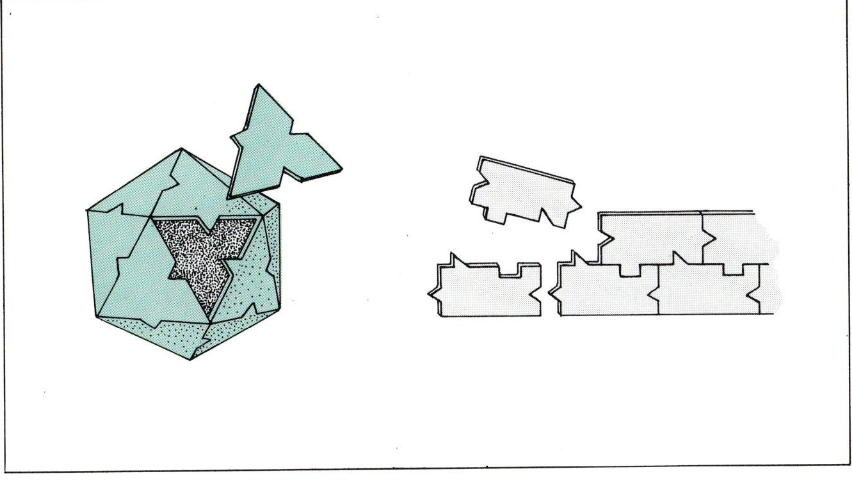

Fig. 5: In EM of a single adenovirus, the virus coat is seen to consist of many globular protein subunits arranged as the surface of a polyhedron; actually, there are 252 subunits arranged as an icosahedron (courtesy of R. A. Valentine and H. G. Pereira). Center diagram illustrates self-limited assembly, in which subunits come together to form a closed surface to which no more units can add. A similar process probably characterizes formation of polyhedral structures like some of the viruses shown on this page. Diagram at right illustrates assembly of an elongate, fiber-like structure by the oriented addition of subunits. In this case, termination of growth is not determined directly by the properties of the structure itself.

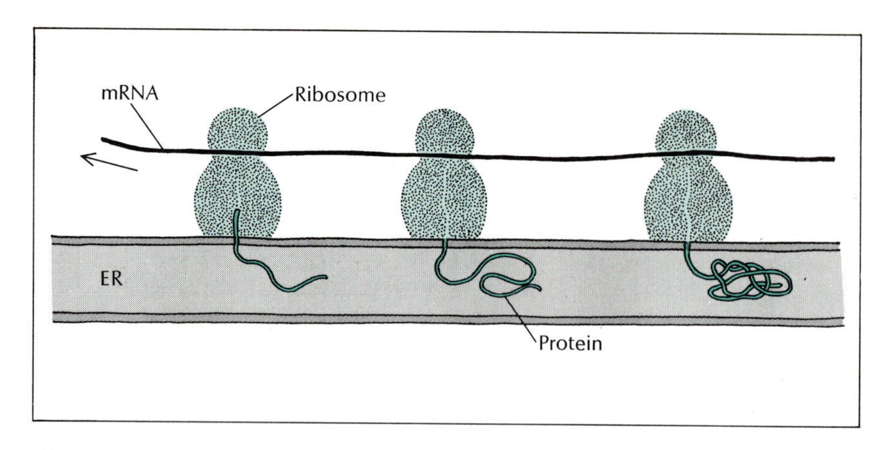

Fig. 6: A protein synthesized on a ribosome attached to a membrane of the endoplasmic reticulum may start to fold only after crossing membrane and entering the ER space.

Problems of Asymmetry and of Size

At this point it might be well to note two problem areas in the study of self-assembly – asymmetry and size. As we know, asymmetry is a fundamental property of living matter, extending down even to its constituent molecules. Many of the latter, including sugars and some amino acids, can exist in two different forms, each the mirror image of the other. Living organisms as a rule synthesize only one of the two forms; in nonenzymatic synthesis, by contrast, the normal result is a mixture containing both. Analogous problems bedevil attempts at the accurate test-tube assembly of larger structures. For example, many membranes are thought to be asymmetric, at least to the extent that they have an inside and an outside. The exterior surface of the plasma membrane probably has properties different from those of the interior surface, and the same is true of the endoplasmic reticulum, mitochondrion, and so on (see Chapter 4, "Architecture and Topography of Biologic Membranes," by Singer; Chapter 5, "Structure and Orientation of a Membrane Protein," by Marchesi). It appears that the proteins or other molecules present at one membrane surface are different from those at the other surface. When membranes are formed in the test tube from mixtures of lipid and protein molecules, it is often extremely difficult to arrange matters so that different surfaces are of different composition.

The cell makes use of devices based upon its preexisting organization to control the location of newly made components. For example, Jamieson, in Chapter 15, describes the passage of newly synthesized secretory proteins across the membranes of the endoplasmic reticulum. This passage is unidirectional in the sense that the proteins remain within membrane-delimited compartments until they are released from the cell; at no point do they pass back directly through a membrane. Probably the unidirectionality depends partly on the fact that the newly synthesized proteins do not complete their folding until they have entered the cavities of the endoplasmic reticulum (Figure 6). Thus, as they leave the ribosomes, the structures on which they are synthesized, the molecules can cross the membrane as relatively thin, unfolded chains of amino acids. Once in the cavity, they fold into a configuration of greater diameter whose ends may be buried deep within the structure and they can no longer cross membranes. It is currently considered likely by many that the formation of new membranes in cells usually—if not always—takes place by growth or by modification of preexisting membranes, so control of membrane asymmetry may depend on the fact that one almost always starts out with an "inside" and an "outside." (See Chapter 12, "Dynamics of Intracellular Membranes," by Siekevitz.)

Size also poses problems for test-tube reassembly in that the mechanisms that normally stop assembly at a particular point are often difficult to duplicate. In some cases, to be sure, the size problem does not exist either in vitro or in vivo. For more-or-less spherical structures, such as some simple viruses, subunits joining together at a determined angle will produce a sphere of fixed size. Closed structures of this type thus can exhibit self-limited assembly (Fig. 5).

Matters are more difficult for elongate cylindrical or fibrous structures. For example, it is relatively easy to polymerize molecules of the muscle protein actin to produce test-tube filaments of thickness and other properties characteristic of the corresponding type of filament found in intact muscle. But the native filaments are of quite uniform length, whereas the reconstituted filaments are extremely variable and sometimes are much longer than the native ones. Probably interaction of actin molecules with other structures or components in the cell controls the filament length. An example of this sort of phenomenon is seen with assembly of tobacco mosaic virus (TMV). The intact virus is a rod-shaped structure, 0.3μ long ($1\mu = 0.001$ mm), and composed of roughly 2,000 identical protein subunits arranged in helical fashion around a helically coiled molecule of RNA (which is a single-stranded molecule, not a double helix). The proteins and RNA can be gently separated; when separate, the RNA no longer has a helical configuration and while the proteins can assemble to form rod-like structures reminiscent of the virus, these are of very variable length and under some conditions show incorrect arrangement of the proteins. When RNA and proteins are mixed, viruses are reconstituted with proper length and proper orientation of proteins and nucleic acid. The components of the virus thus seem to interact cooperatively to ensure correct assembly; the RNA molecule, which is self-duplicating and of fixed length, provides a size control, with proteins being added to the assembling structures only until the RNA molecule is covered (Figure 4).

Turnover and Recycling

It is obvious that organelle duplication and formation of new structures must occur as part of the processes by which cells grow and divide. However, even in nondividing cells of

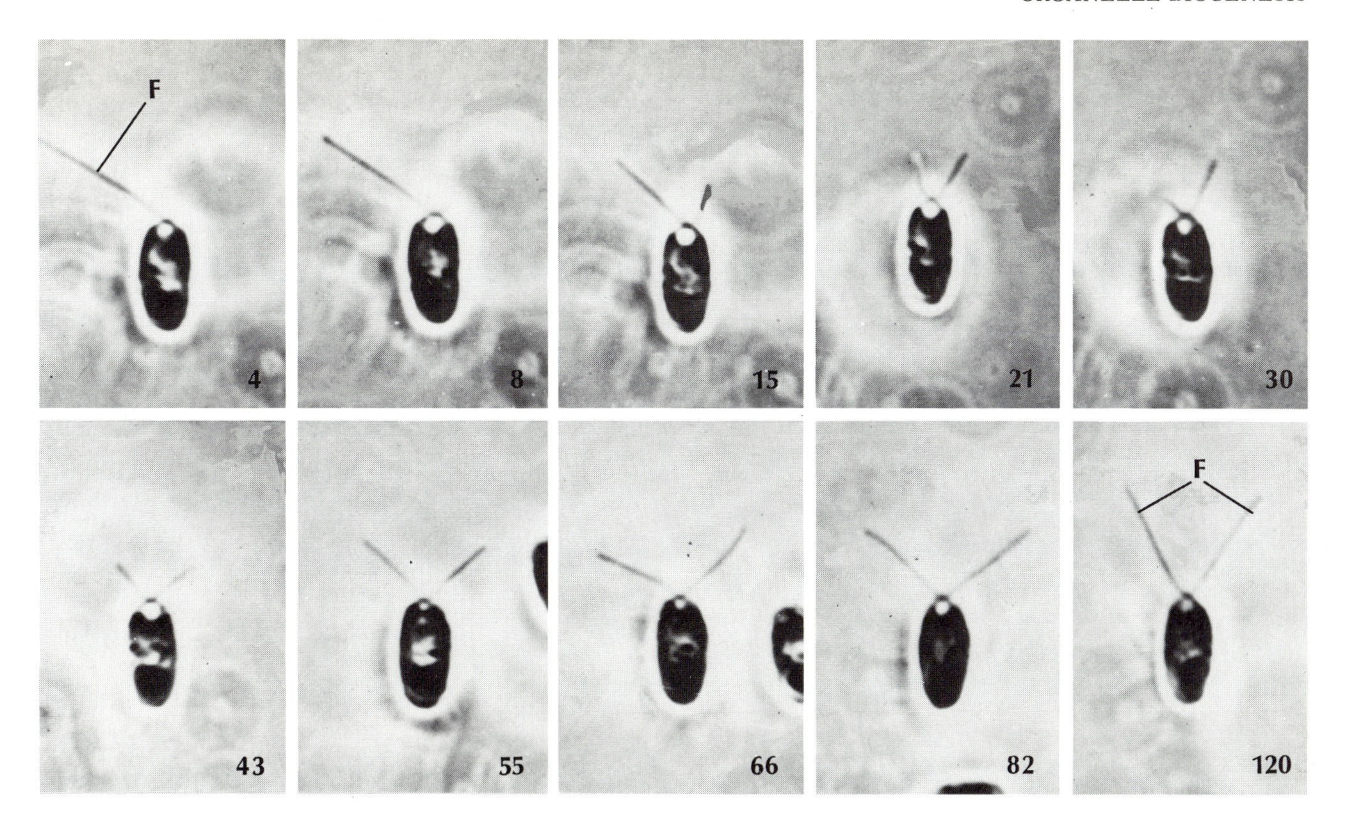

Fig. 11: This sequence of photographs (courtesy of J. L. Rosenbaum, J. L. Moulder, and D. L. Ringo) shows a unicellular organism chlamydomonas at intervals after one of its two flagella (F) was removed by mechanical agitation. (The number of minutes after the removal of the flagellum is given at the lower right of each panel.) As the second flagellum begins to regenerate, the remaining one gets shorter. After a period, the shortening stops and the two flagella both grow longer.

in association with preexisting membranes. What is membrane growth like? The assembly of lipid bilayers may occur spontaneously, as mentioned above, but how are the specific proteins added? Does formation of membrane occur at special growing sites in preexisting membranes, so that one has discrete regions of older and newer membranes? Is there more random addition of new molecules at many points along the surface? Or, are molecules added at a limited number of points but then "permitted" to diffuse throughout the membrane (see Singer chapter)? Evidence can be adduced for all these processes and perhaps they all occur, with one or another predominating under given circumstances. But definitive evaluation of such possibilities will require better techniques than are currently available.

A related area of uncertainty is the extent to which movement of membrane components from one structure to another takes place in bulk, rather than molecule by molecule. This is significant, for example, in relation to the plasma membrane, which is known to lose substance in bulk, e.g., by vacuole formation during phagocytosis or other types of endocytosis, and to gain it in bulk during exocytosis and some related processes (Figure 7).

There is also reason to believe that a proportion of plasma membrane components can be added molecule by molecule, but it is very difficult to distinguish experimentally between the two mechanisms. Certainly it is hard (though not impossible) to imagine thermodynamically plausible ways in which the distinctive proteins present at the outer plasma membrane surface (e.g., the glycoproteins involved in cell recognition) could be individually forced out from the cell interior through the lipid bilayer that forms the matrix of the membrane. Much the same uncertainty pervades the question of bulk vs molecular transport between other organelles, e.g., between the endoplasmic reticulum and mitochondria.

It should also be remembered that the membranes of different cell structures differ in their characteristic compositions and capabilities. If one structure contributes membrane to another and the two still remain distinctive from one another, then either special mechanisms must exist by which the first structure can synthesize or segregate special membranes for transport to the other, or transformation of one type of membrane into another must be possible. Jamieson in his chapter briefly touches upon such possibilities in his discussion of the interactions of endoplasmic reticulum and Golgi apparatus. Work along such lines by several laboratories can be interpreted as suggesting that membrane from the rough endoplasmic reticulum can give rise to Golgi apparatus membrane by transport in bulk. There are also those who argue that in the Golgi apparatus such membrane can be further transformed into plasma membrane and then added to the cell surface through processes resembling exocytosis.

Additional insights into membrane assembly can be expected from future studies of mitochondria and chloroplasts. Much current interest is focused on recent suggestions that the ribosomes of these organelles may be attached to the inner membranes, per-

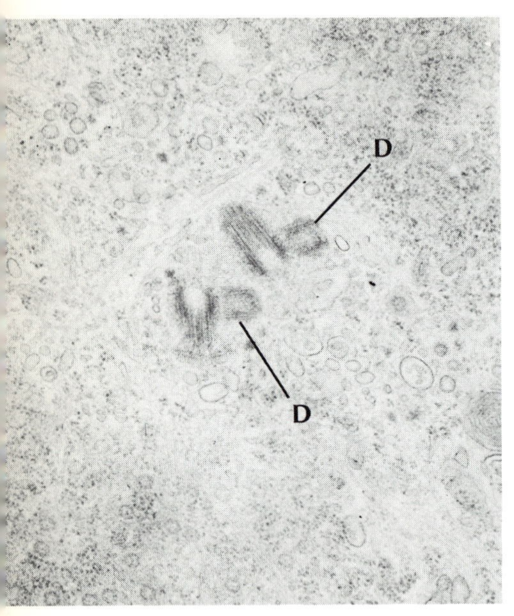

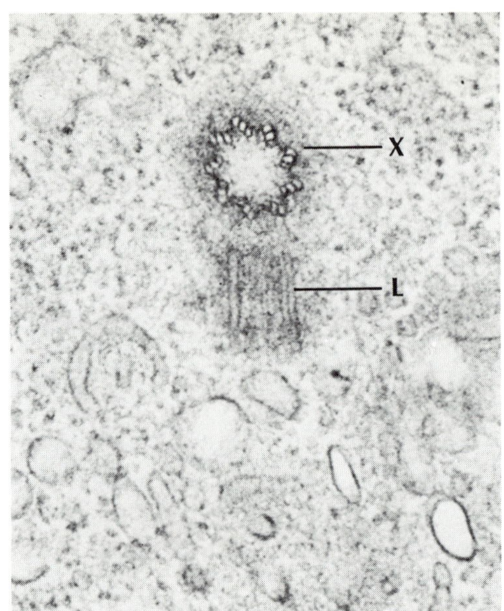

Fig. 12: Electron micrographs (courtesy of J. B. Rattner and S. G. Phillips) are of centrioles during cell division of L cells, a line of cultured mouse cells. At the left, two pairs of centrioles are seen, each consisting of a larger "parent" organelle and a smaller "daughter" centriole (D) oriented at right angles to the parent. The photograph at the right shows a pair of centrioles sectioned so as to provide a cross-sectional view of the upper one (X) and longitudinal view (L) of the lower one. The tubules of the centrioles are readily visible (cf *the basal-body diagram, Fig. 10 on page 162).*

haps in a manner similar to the attachment of ribosomes to the endoplasmic reticulum. Reconstitution approaches of various types are also being pursued. For example, the lipoprotein membranes that contain electron transport enzymes of the inner mitochondrial membrane can be separated from the small spheres that line the inner membrane and are essential for the formation of ATP via oxidative phosphorylation (see Chapter 14, "Inner Mitochondrial Membranes: Basic and Applied Aspects," by Racker). It is known that under proper conditions lipoprotein membranes and spheres will recombine spontaneously and, if a few other components are added, one can reconstruct particles capable of many of the essential steps in mitochondrial energy metabolism.

Perhaps in normal mitochondriogenesis the spheres and membranes are assembled first as separate parts and then brought together to form the final structure. More mysterious at present are the processes by which the mitochondrial membranes, as they form, come to assume the characteristic folded three-dimensional array of cristae. The cristae of mitochondria in different cell types may be tubular or plate-like, numerous or sparse, and their morphology may alter during mitochondrial function. We know virtually nothing of what underlies these shapes or changes. We know even less about what might tell a growing mitochondrion that it has gotten big enough and now should stop growing or divide into two.

Microtubules, Cilia, Flagella, and Some Related Structures

Mitochondria seem to form only by growth and division of preexisting membranous bodies – either fully developed mitochondria or in a few circumstances (e.g., yeast cells grown under specific conditions) simpler precursor structures known as promitochondria. The situation is similar for chloroplasts. We will now turn our attention to organelles that illustrate somewhat different processes.

Microtubules are ubiquitous in cells of higher organisms. The tubules are thin, elongate, hollow-looking cylinders (diameters of 0.02μ to 0.03μ) whose walls are composed of molecules of the protein tubulin arranged as illustrated in Figure 9, page 161; there probably are several closely related types of tubulin that compose different categories of microtubules.

Structures of the microtubule type play several roles in cells. For example, it is generally accepted that they participate in the establishment or the maintenance of special cell shapes; thus, the elongate axons of nerve cells contain many microtubules oriented parallel to their long axes, and this is true also, for instance, of some of the stiff modified pseudopods that protrude from the surfaces of certain types of protozoans. The tubules also are believed to take part in oriented motion of other structures within cells, either as motive elements or as guides to the direction of motion. Thus, structures are often observed to move in cells along routes that parallel the long axes of groups of microtubules. A good example is seen in cell division. Much of the essential architecture of the spindle responsible for chromosome motion is contributed by microtubules, some of which run from the poles of the spindle to the chromosomes. During division, the chromosomes move toward the poles along the paths defined by these microtubules and it is widely believed that the tubules may actually generate the force that brings about the chromosome motion.

The structure of microtubules is relatively labile; when cells are exposed to low temperature, high pressure, or drugs such as colchicine, their tubules often will break down into tubulin molecules and in consequence structures such as the spindle will disappear. These effects are reversible and when conditions are returned to normal, microtubules will re-form and spindles or other structures will reappear and function normally. This lability is reflected in the test tube; purified microtubules can be readily disaggregated into tubulin molecules and then reconstituted from the proteins. From the cell's viewpoint, the ability of tubules to break down and re-form readily seems advantageous; it provides some flexibility in the control of cell shape and also facilitates the assembly of structures such as the spindle when they are needed and the disassembly of such structures once they have done their job. (Further, one widely accepted hypothesis proposes that the mechanism for generating motion of chromosomes involves shortening of the attached microtubules by loss of subunits.) Very likely, tubulin molecules can cycle between organized structures and pools of free protein molecules many times before they are degraded.

Still unclear are most of the factors that determine where and in what pattern microtubules will form in a cell, a matter of great importance for tubule function. There is widespread consensus that microtubule formation can be controlled by nucleating centers but, with one important exception, the identity and nature of such centers are still controversial. The exception relates to the formation of cilia and flagella. These are elongate motile organelles that protrude from the surfaces of many unicellular organisms and of sperm, respiratory tract epithelia, and other cell types of metazoans. Cilia and flagella show a characteristic ultrastructure based upon the presence of microtubules arranged as illustrated in Figure 10, page 162; note the presence of nine doublets of tubules arranged in a cylinder surrounding two central tubules. The longitudinal sliding of these tubules with respect to one another is thought to be the basis of the mechanism by which the cilia or flagella themselves move.

Each cilium or flagellum forms in association with a basal body. Basal bodies are short, cylindrical structures, present at the base of cilia and showing the ultrastructure illustrated in Figure 10 on page 162; note the presence of nine triplets of tubules arranged in a cylinder. Very little is known of the chemistry of basal bodies but it is strongly suspected that they help control the formation of the tubule pattern in cilia and flagella. Each doublet of tubules in a cilium or flagellum seems to form as an extension of two of the three tubules in a basal body triplet. Growth of the doublets can take place by assembly of tubulin molecules preexisting in a cytoplasmic pool. For example, when the two flagella of the unicellular organism chlamydomonas are broken off, substantial lengths of flagella will be regenerated even by cells grown in the presence of inhibitors such as cycloheximide that prevent synthesis of new proteins. If one breaks off only one of the two flagella and maintains the cells under ordinary conditions, the second flagellum will shorten as the broken one regenerates until the two reach equal size, and then both grow out together to their original length (cf Fig. 11, page 163). This suggests that subunits can cycle from one flagellum to another.

Thus, during their regeneration (and presumably during normal growth) flagella grow by addition of new protein subunits to their tubules. This addition takes place at the tip of the growing structure not the base, as can be shown by autoradiographic studies of cells whose proteins are pulse-labeled with radioactive amino acids. Probably this means that the basal bodies initiate the formation of the flagellar tubules and then the tubules themselves continue to grow by oriented addition of subunits. This plausible proposal is also supported by the fact that in the test tube, tubules in portions of flagella can be made to grow longer by addition of purified tubulin molecules. How the tubulin molecules are transported within the cell to reach the tip of a growing flagellum requires further study, and little is presently known of the population of ribosomes that synthesize tubulin molecules.

There is much else that is still ill-understood about the relationship between the basal body and the structure growing from it. The ring of nine tubules in both of them can hardly be coincidence, but why should the basal body tubules be arranged as triplets and the others as doublets? Moreover, cilia and flagella – but not basal bodies – also contain two single tubules running up the center of the cylinder. Finally, the cilium doublets – but not the triplet basal body tubules – possess two short "arms." These are com-

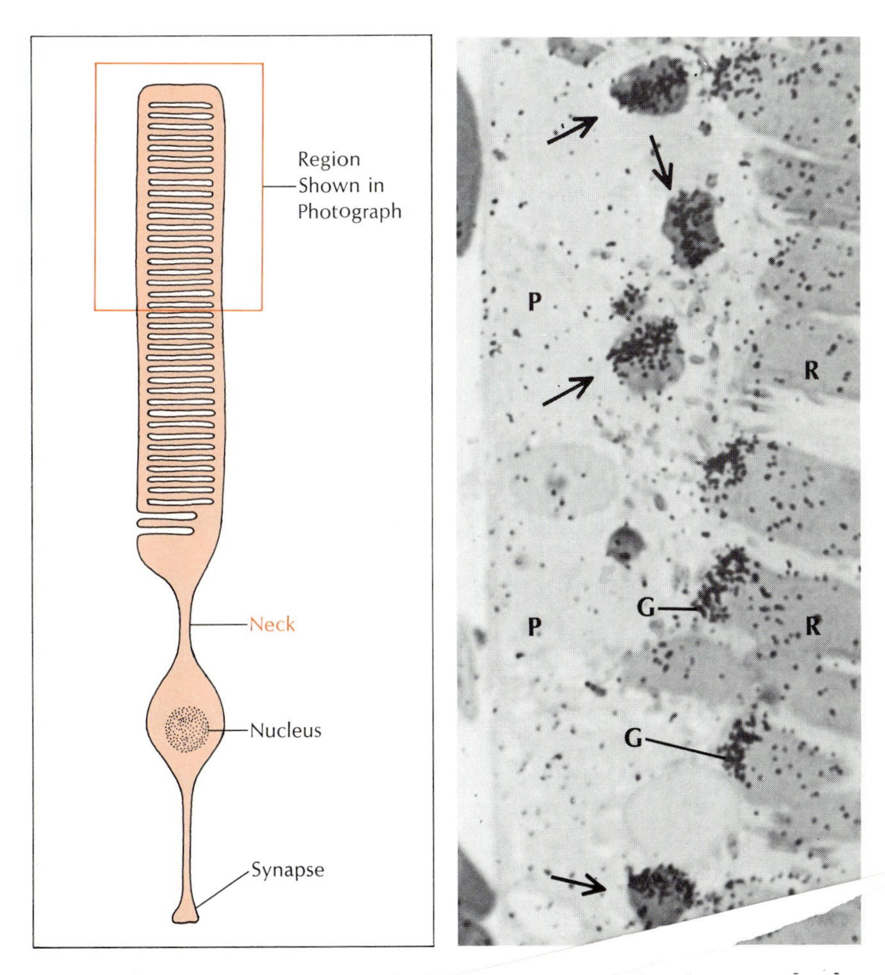

Fig. 13: At left is diagram of a retin[illegible] *one stack of discs. At right: this autoradiogram of a frog re*[illegible] *W. Young) was prepared several weeks after the cells h*[illegible] *active amino acids. The tips of several rod cells are se*[illegible] *ark grains at G correspond to radioactive proteins that have* [illegible] *rod cells as components of discs. At the arrows, grains are seen over vacuoles containing collections of discs that have already been shed by the rod cells and have been phagocytosed by cells of the pigment epithelium (P).*

posed of dynein, a protein quite different from the tubulins. By delicate chemical manipulation, the dynein can be extracted from cilia of the protozoan tetrahymena; when then added to the "armless" cilia, it recombines with them to form arms again. Functionally the dynein is important because it is an ATPase, liberating energy by breaking down ATP; almost certainly, therefore, it is an agent in the release of energy required for ciliary and flagellar motion. Apparently the tubulin of the doublets is organized so as to provide specific binding sites for the attachment of dynein.

What other nucleating centers control microtubule distribution? There are several candidates but little hard evidence. For example, structures essentially identical to basal bodies are present at the poles of the spindle in cell division of most animal cells and of some plant cells. These structures are referred to as centrioles, and microtubules are very closely associated with them, often seeming to end in small masses of amorphous material that are present near the centrioles. Microtubules also seem to show association with centrioles in cells that are not dividing. However, since there are many plant cells that have neither basal bodies nor centrioles but do have microtubules, it is obvious that the former structures are not essential for tubule formation. Some investigators believe that even in these plant cells nucleating centers are present in the form of small masses of amorphous material too nondescript to attract the immediate attention of microscopists. It also has been proposed that the special region of the chromosome (the kinetochore) responsible for attachment of the chromosome to the spindle may be capable of organizing microtubules, at least during spindle formation.

What is known of the controls and mechanisms that influence the distribution and number of potential nucleating centers? Again, far too little. In many cases centrioles and basal bodies show signs of the capacity for some kind of modified form of self-duplication. For example, during the cell division cycle of an animal cell, new centrioles appear in association with the old ones; a small disc-like procentriole arises near each older organelle, and this then grows to form a cylinder with the proper ultrastructure, oriented at right angles to the older centriole (Fig. 12). The synthesis and assembly processes underlying such phenomena are unknown. Further, there are many situations in which centrioles or basal bodies form in very different ways from this. Most difficult to explain are the cases where organelles with the characteristic basal body - centriole structure appear in cells that previously had no such organelles. Does this mean that the organelles can form "de novo" by spontaneous assembly of subunits? Or is there a special self-duplicating nucleating or orienting device always present but lacking a distinctive enough appearance to be recognized by microscopists? Future studies of the chemistry of centrioles or basal bodies may help clear up such uncertainties by revealing what, if any, hereditary information they carry and what metabolic capacities they possess.

Additional work is also needed on such problems as the controls that determine the pattern of basal bodies (and hence of cilia or flagella) present in a given cell. In ciliated protozoa such patterns are known to be precisely inherited from one cell generation to the next, and in cells of higher organisms the distribution of cilia or flagella characteristically differs for different cell types. How the cell "establishes" where cilia or flagella will form and how long they will become or how the plasma membrane grows out, as it must to bound a forming cilium or flagellum, are aspects of questions of great interest for students of cellular morphogenesis.

A fascinating relative of the systems just discussed is the rod cell of the vertebrate retina. These cells each consist of a more-or-less conventional nerve cell body, which is attached to an elongate rod-shaped portion by means of a short neck. The rod-like segment is bounded by a plasma membrane and is filled with a pile of discs, much like a tall stack of poker chips. The "chips," however, are flat membranous sacs whose surfaces are rich in the visual protein rhodopsin. The discs provide an enormous surface area for the reception of light by an organized array of molecules. The rod segment as a whole is thought to be a highly modified cilium; the neck contains the typical ring of nine doublet tubules (though not the two single, central tubules), and a basal body is present (Figure 13). New discs form continually as infoldings of the plasma membrane just above the neck. By pulse-labeling their proteins with radioactive amino acids, it has been shown that after separating from the cell surface, individual discs maintain their original set of macromolecules as they migrate up the length of the rod (or are pushed up by the formation of new discs below), and after a period of one to several weeks (depending on the species of animal) reach the tip. There they are shed from the cell as units, and are then phagocytosed by cells of the retinal pigment epithelium, within which they are broken down by lysosomes. Future studies of disc formation, migration, and degradation should contribute much to understanding of membrane assembly and turnover.

Perspective

Twenty years ago, the molecular bases of heredity and of macromolecular synthesis were largely unknown. This certainly is no longer the case. And now, as I hope this chapter has demonstrated, we are well along the path to understanding how molecules get put together to form structures. Waiting in the wings is the next set of problems – those concerning the mechanisms by which macromolecular synthesis and assembly are controlled and coordinated to achieve orderly growth and division of cells and the formation and maintenance of specialized cell types.

17

Membranes in Synaptic Function

VICTOR P. WHITTAKER
Max Planck Institute of Biophysical Chemistry

As the extracellular, "open" component of nerve communication, the synapse may be thought of as the weak link in the chain through which the nervous system controls the body. Most vulnerable to toxic and pathologic processes, the synaptic region is by the same token most accessible to experimental manipulation and accordingly has long been a prime focus of neurologic research.

In recent years, synaptic research has been stimulated by the development of techniques whereby the neuron's presynaptic region – the terminal portion of the axon – can be isolated as a sealed structure. These bodies, called synaptosomes, can be kept alive, or at any rate functional under appropriate conditions, as a species of "cell in miniature." They have thus served as a sort of halfway house between the whole cell and the completely artificial lipid vesicle described earlier in this volume (see Chapter 3, "Models of Cell Membranes," by Bangham). As such, synaptosomes have been used in a number of research laboratories to investigate various carrier-mediated uptake processes, involving such substances as K^+ and Ca^{++} ions, neurotransmitter compounds (e.g., norepinephrine, acetylcholine), and certain amino acids and their metabolic precursors.

In addition, synaptosomes can be further broken down into their constituents – cytoplasm and its soluble components, intraterminal mitochondria, synaptic vesicles (of which more later), and external membrane. The last named not only represents a relatively pure preparation of synaptic membrane but also, we have reason to believe, is fairly representative of the neuron's external membrane in general.

The light that these and related studies are throwing on neuronal and (especially) synaptic function is of course of great interest for physiologists, but it is also not without implications for the clinician. A number of potent toxins (botulinin is one) are known to act by disrupting synaptic function in one way or another, and the same is true of certain neurologic disorders (e.g., myasthenia gravis, parkinsonism). In addition, there is increasing evidence that disorders of the synapse may play at least a contributory role in a number of diseases hitherto generally considered purely psychologic in their etiology, notably schizophrenia and manic-depressive psychosis. Thus there is good cause to expect that enhanced understanding of synaptic function will improve our understanding of, and ability to control, these very varied pathologies.

The synaptic region – a roughly club-shaped enlargement of the narrow axonal filament – can be fairly described as the business end of the axon. It is here that the nerve signal carried by the neuron achieves its *raison d'être* – triggering the discharge of another, distal neuron or activating a muscle or gland cell. In a few varieties of synapse, the nerve impulse is transmitted electrically through the synapse. The presynaptic membrane of the afferent neuron is linked with the postsynaptic membrane of the target cell through one of the so-called tight or gap junctions (see Chapter 9, "Junctions Between Cells," by Pappas). In effect, so far as their electrical properties are concerned, the two cells are one; therefore an impulse can pass from one to the other almost instantaneously. In certain fish, we find such electronic synapses in anatomic locations where speed is crucial – as in controlling muscles involved in escape from an enemy.

But electrical neurotransmission, though speedy, is also unselective. It is an all-or-nothing affair, incapable of being quantitatively modified or integrated with other afferent impulses. Accordingly, the vast majority of synapses, in both vertebrates and invertebrates, employ chemical rather than electrical transmission, which is not subject to these limitations. In the chemically mediated synapse, presynaptic and postsynaptic membranes are separated by a distinct cleft or gap, across which molecules of the transmitter substance – one of several different compounds – diffuse. On the other side of the gap, the transmitter molecules activate the postsynaptic membrane

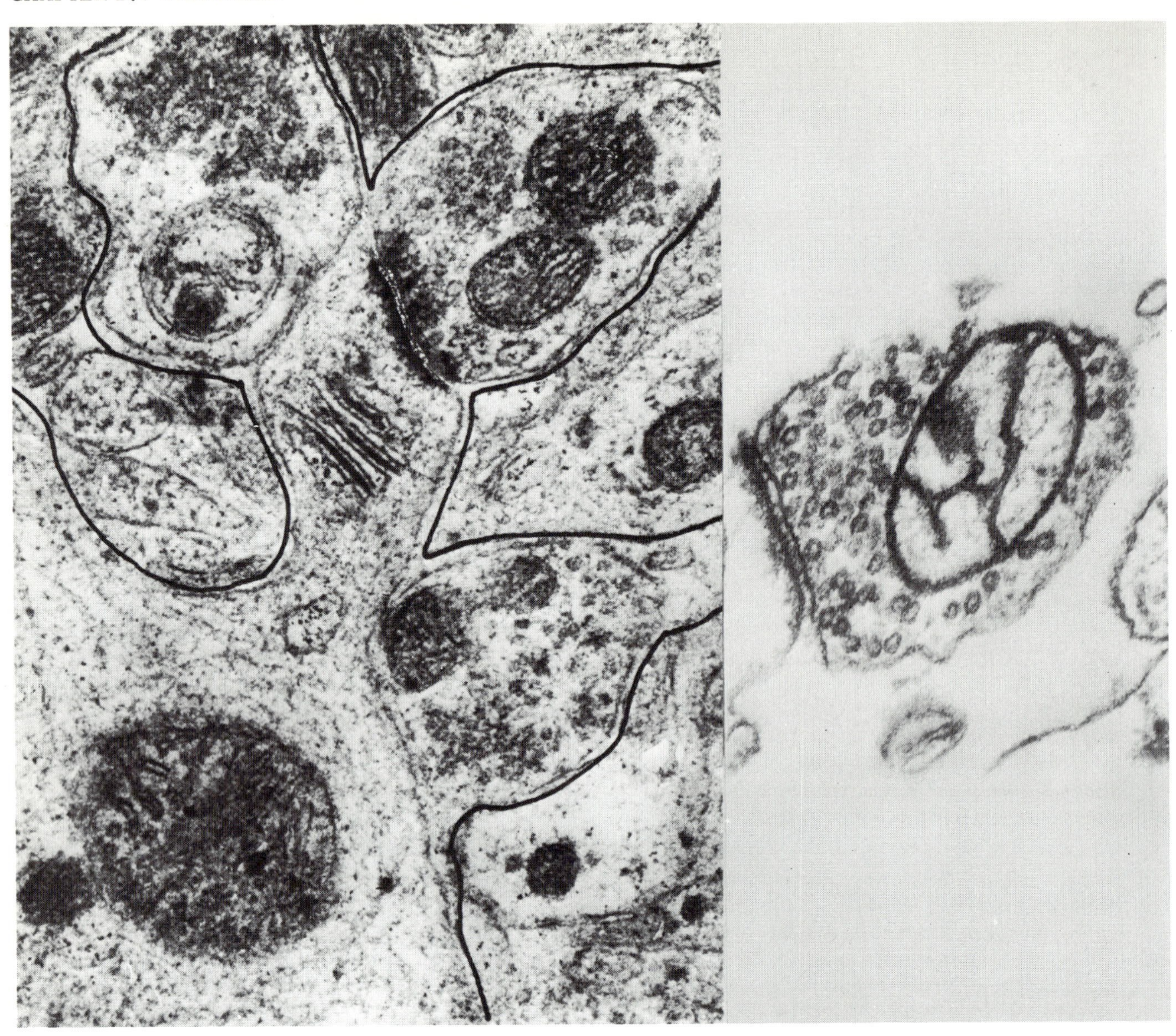

of the target cell, either depolarizing it – leading to excitation of the cell – or hyperpolarizing it – making the cell less susceptible to excitatory impulses from other synapses.

All this is fairly basic but its repetition is important for understanding what follows.

My own work on synaptic function began with an interest in what has been called bound acetylcholine in the brain. It has long been known that acetylcholine – which is, of course, one of the neurotransmitter compounds just discussed – is found in brain tissue. Yet even though it is a highly soluble substance, aqueous media such as Ringer's solution will extract only a very small fraction of it. More drastic treatment, with acidic reagents or organic solvents, or both, will liberate the remaining acetylcholine, which in its natural state is bound to some form of particulate matter; the binding serves to inactivate it pharmacologically, and also to safeguard it from the action of destructive enzymes such as cholinesterase. I was interested in finding out more about the nature of bound acetylcholine, since it promised to throw light on how the neurohumor is stored in neuron terminals and released from them.

At just about this time (1956), an important clue turned up through the work of Blaschko at Oxford and Hillarp in Sweden, in the shape of the chromaffin granules of the adrenal medulla. The two investigators had succeeded in isolating the granules, by means of density gradient centrifugation in sucrose of homogenized medullary tissue, and had found that they contained up to 10% of their weight in epinephrine. However, injection of the intact granules into an animal showed that the epinephrine had very little pharmacologic activity; only if the granules were disrupted, releasing their freight of neurohumor, was the characteristic epinephrine response observed.

All this, of course, looked very much like "bound" epinephrine, meaning that the techniques Blaschko and Hillarp had used might well be applicable to the study of bound acetylcholine. And in fact simple centrifugation of gently homogenized brain tissue revealed that the acetylcholine, like the medullary epinephrine, remained bound, being associated with the so-called crude mitochondrial fraction.

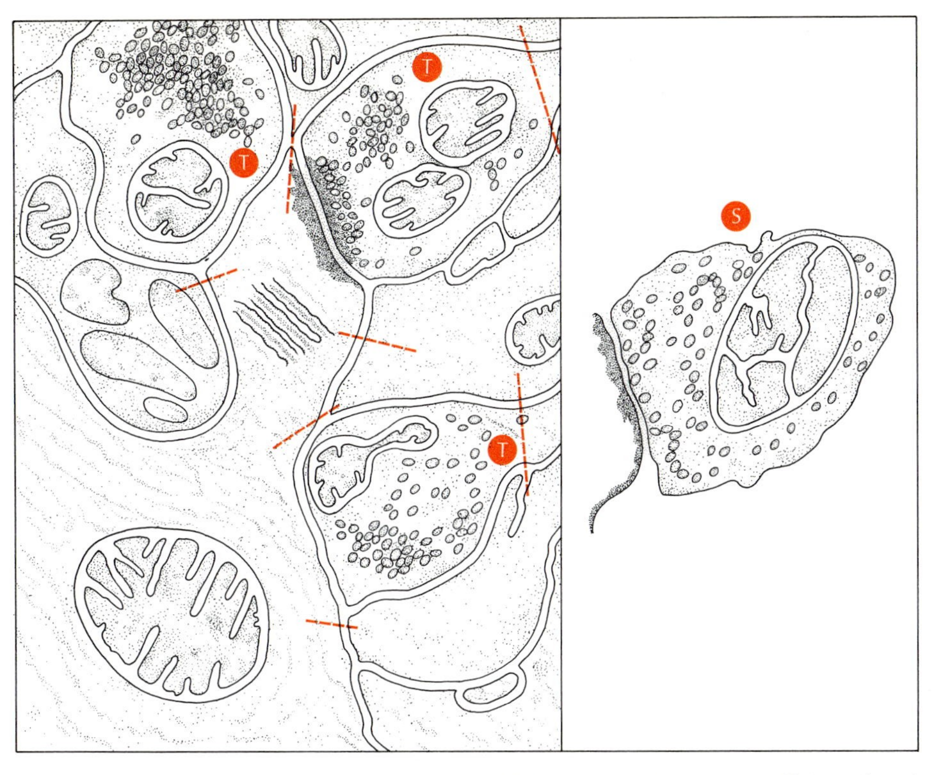

Presynaptic neuron terminals and associated dendrites of their target cell (from brain) are seen in electron micrographs at left and in drawing above. Terminals (T) are rich in vesicles containing neurohumors such as acetylcholine, which, when released into narrow synaptic cleft, stimulate postsynaptic receptors on target cell. Homogenization detaches terminal and postsynaptic region at points shown by broken lines, forming synaptosomes (S), whose composition and behavior can be studied separately.

More refined fractionation by the density gradient technique then showed that the bound acetylcholine was not in fact associated with the mitochondria – which is to say, with enzymes such as succinate dehydrogenase, which are "markers" for mitochondrial membrane. Rather, the acetylcholine was found in an intermediate layer consisting of small particles. When George Gray and I examined these particles under the electron microscope, we found to our surprise and pleasure that they were detached presynaptic nerve terminals. Evidently they had been formed by the breaking of the thin, axonal stalks (which are only about 0.1 to 0.2 μ across), after which the external membrane of the detached terminals spontaneously "resealed" itself at the point of breakage. This "synaptosomal" fraction of our preparation also contained other neurotransmitter substances, such as norepinephrine and serotonin, along with acetylcholine.

Among the features revealed by the electron microscope were a host of tiny, membranous vesicles, which filled the synaptosomes as they do the terminal region of intact neurons. It seemed not unlikely that the bound acetylcholine was contained in the vesicles, whose membranous walls would account for its pharmacologic inactivity. This possibility was strengthened by some recently reported experiments that had demonstrated, by complex but convincing methods, that the amounts of acetylcholine released in the synapse by nerve discharge were discontinuous – or "quantized." That is, the amount of acetylcholine released in a single discharge could not be random but had to have a value corresponding roughly to multiples of a lowest common factor – just as the atomic mass of any pure isotope represents a multiple of the atomic weight of hydrogen. It seemed very likely that the observed quanta of acetylcholine corresponded to the vesicles. In other words, the neurotransmitter did not simply diffuse through the presynaptic membrane during discharge, a process that could presumably involve any number of molecules whatever, but was discharged into the synaptic gap through the emptying of one or more vacuoles, the contents of each one corresponding to one quantum or "packet" of the neurotransmitter.

However, though these conjectures were very plausible, they were bound to remain of merely theoretical interest unless it could be demonstrated that the vesicles did in fact contain acetylcholine. Accordingly, we set about devising a method of isolating the vesicles from the remainder of the synaptosome, in order to see whether they – and only they – contained the neurotransmitter. The technique we finally arrived at (it took us more than two years) involved gently disrupting synaptosomes by exposing them to hyperosmotic sucrose, then subjecting the disrupted preparation to further fractionation by the density gradient method. The result was an almost pure vesicle fraction; just as the initial separation process had isolated the terminal region (i.e., the synaptosome) from the rest of the neuron, so this second-stage purification isolated the vesicles from the cytoplasm and external membrane of the synaptosome.

Neither the cytoplasmic nor membranous fractions showed any significant level of acetylcholine – nor did the vesicular fraction, so long as the vesicles remained intact. By suitable treatment, however, it could be released from what was evidently its bound form and detected in concentrations that accounted for up to half of the acetylcholine content of the synaptosomes – or of intact neurons, for that matter. The origin of the lost acetylcholine is uncertain: it may be transmitter released (and destroyed) during the isolation of the vesicles.

The final step in linking acetylcholine quanta with the vesicles was to estimate the amount of acetylcholine per vesicle. Bernard Katz, who had done the original work on the quanta, had estimated that each quantum must contain somewhere between 2,000 and 10,000 acetylcholine molecules. The question we confronted was whether the vesicles also contained an amount of acetylcholine in this range.

We of course knew the amount of acetylcholine present in a given quantity of our vesicular fraction; what we did not know was how many vesicles

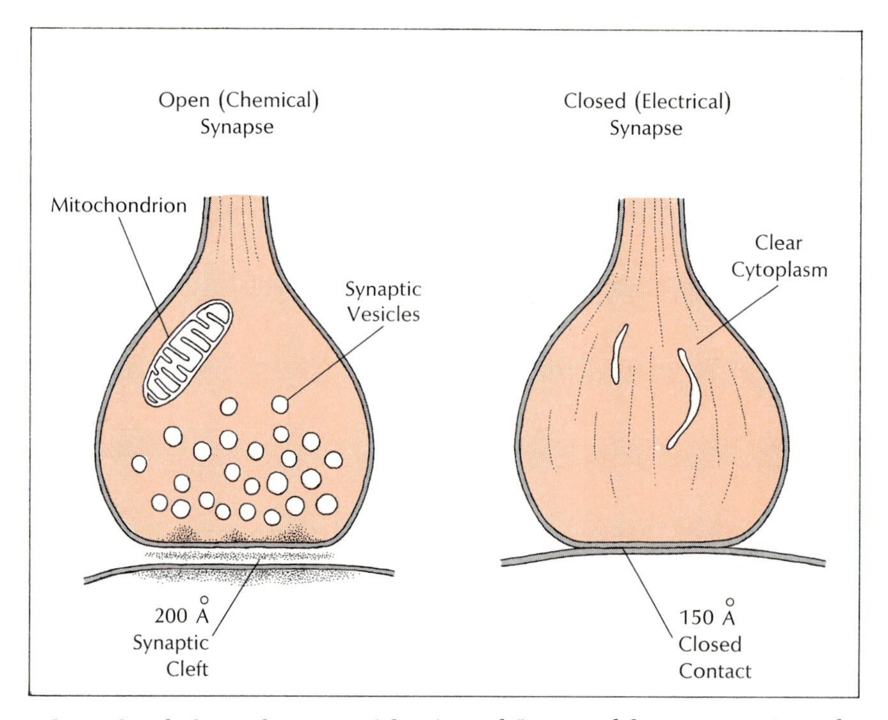

Chemical and electrical synapses (above) are differentiated by presence of vesicles containing neurohumor in the former and absence of synaptic cleft in latter, in which pre- and postsynaptic membranes are fused. Schematic diagram of chemical synapse (below) shows steps in its action. Acetylcholine is synthesized from its precursors (1) by enzymatic action (2), then incorporated into vesicle, which under appropriate stimulus ejects it into synaptic cleft (3). The neurohumor stimulates a receptor on the target cell (4) and is then enzymatically broken down to inactivate it (5). Breakdown products eventually return to cell (black lines) where they are recycled.

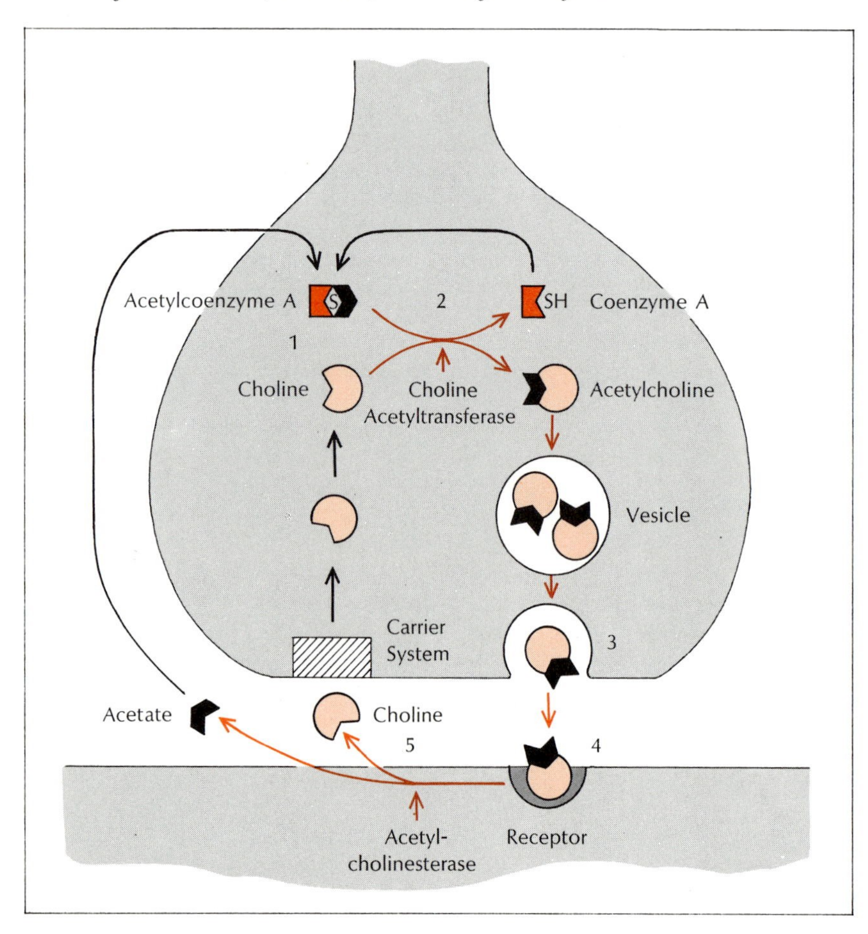

were included in the same volume. They could be seen and counted under the electron microscope, but there was no way of telling precisely how large a volume was being viewed and counted. To solve this difficulty, we adapted a method used to count virus particles: adding to the preparation a milky suspension containing a known quantity of tiny, submicroscopic polystyrene beads. The number of beads visible in the electron microscope field would then give us an accurate estimate of the volume of fluid being viewed. To illustrate with some arbitrary figures: if the concentration of beads was known to be 10,000 per ml, a field in which 10 beads were visible would contain about 1/1,000 ml of fluid. And if this same field contained, say, four vesicles (which have a quite different appearance from the beads), clearly there would be about 4,000 vesicles per ml.

Given figures of this sort, plus the known concentration of acetylcholine per ml, it should have been easy to calculate the number of acetylcholine molecules per vesicle. Unfortunately, there was one major snag: we had every reason to believe that the original brain tissue from which the axon terminals (i.e., the synaptosomes) were obtained included both cholinergic and noncholinergic terminals. That meant that only a proportion of the vesicles contained acetylcholine as opposed to other neurohumors. Correcting our calculations for this mixed population of vesicles involved a good deal of guesswork; the final figure we emerged with was approximately 2,000 acetylcholine molecules per vesicle.

This figure was encouraging in that it lay within the postulated limits of molecules per vesicle; had we come up with a figure of 500 or 50,000, our whole hypothesis would have been thrown into question. However, it still embodied more assumptions and guesswork than we cared for. Accordingly, we have most recently been examining another type of tissue, obtained from the torpedo or electric ray. The electric organs of the biggest of these remarkable fish – they are related to the sharks and skates – can deliver a shock of up to several hundred volts, involving a total energy expenditure of between 1 and 2 kw – comparable to the shock obtained from an ordinary American 15-amp household electric line, which will deliver up to about 1,650 watts before the fuse goes out.

For our purposes, however, the interesting thing about these organs is not their remarkable structure, whereby the microcurrents of many thousands of cells are combined into discharges that can stun or even kill prey, but the nature of the synapses that activate them. The torpedo's electrocytes are thought to be modified muscle cells and, like all muscle cells, are activated by acetylcholine; there is no evidence that they respond to any other neurotransmitter. Unlike other muscle cells, however, the postsynaptic region – the end-plate – is not limited to a small area, but almost completely covers one side of the cell, and this enlarged postsynaptic area is paralleled by a similarly enlarged presynaptic terminal of the activating neuron. What we have here, then, is a type of tissue in which the terminals (and of course their contained vesicles) are presumably of purely cholinergic type. No other neurohumors are present to complicate the picture. In addition, the greatly enlarged synaptic area implies that the electric organs must be much richer in acetylcholine than ordinary muscle tissue – at a guess, somewhere between 200 and 500 times as great.

However, isolating vesicles from the electric organs has turned out to be rather difficult. The constituent cells are held together by collagen, a very slippery, gelatinous material, so that although the tissue can be homogenized the mechanical conditions for synaptosome formation are not present. We have, therefore, had to devise a rather different technique – we think rather a neat one – for extracting the vesicles directly.

We begin by freezing the tissue in liquid nitrogen or Freon-12, which makes it intensely hard and brittle; it can then be reduced to a coarse powder by pounding in an ordinary ceramic mortar. When one looks at the surface of these particles under the electron microscope, they seem at first glance to be no different from normal electric tissue. Examined more closely, however, it turns out that though the electrocytes, including their postsynaptic membranes, are not much damaged, the mem-

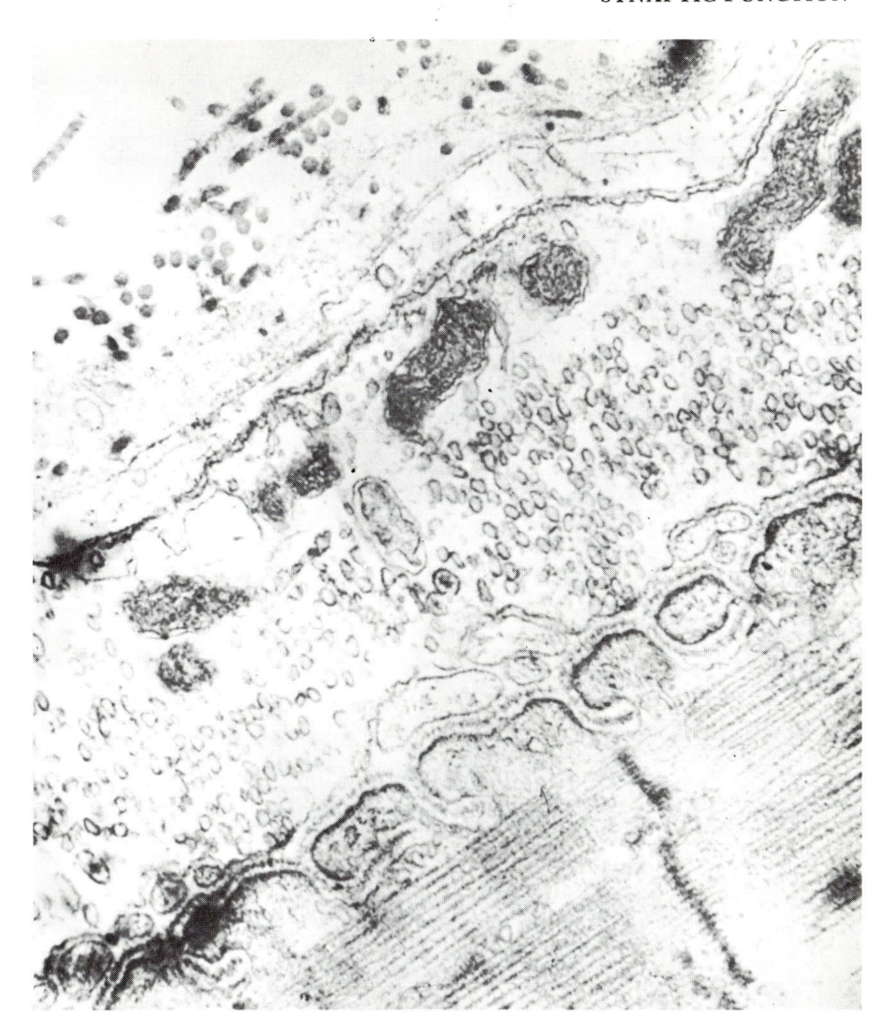

Axon terminal activating a muscle cell is seen here as a diagonal band across the center of the electron micrograph, distinguishable by its characteristic population of neurohumoral vesicles. Striations at lower right are muscle fibrils; between them and the terminal lies the end-plate region containing receptors for the neurohumor.

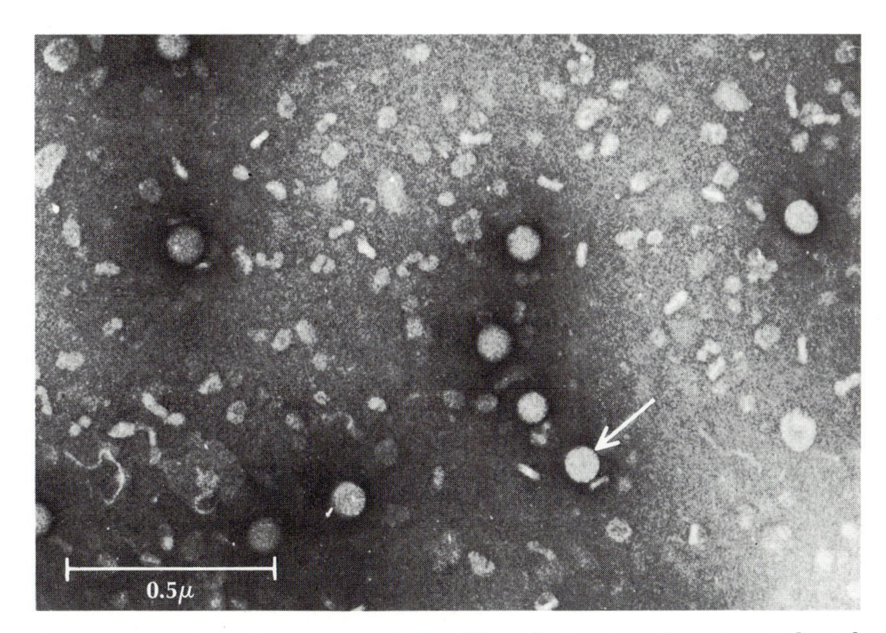

Concentration of vesicles is estimated by adding plastic microspheres (one indicated by arrow) in known concentration to preparation of isolated vesicles. Count of spheres defines volume in field; count of vesicles in field then establishes their concentration.

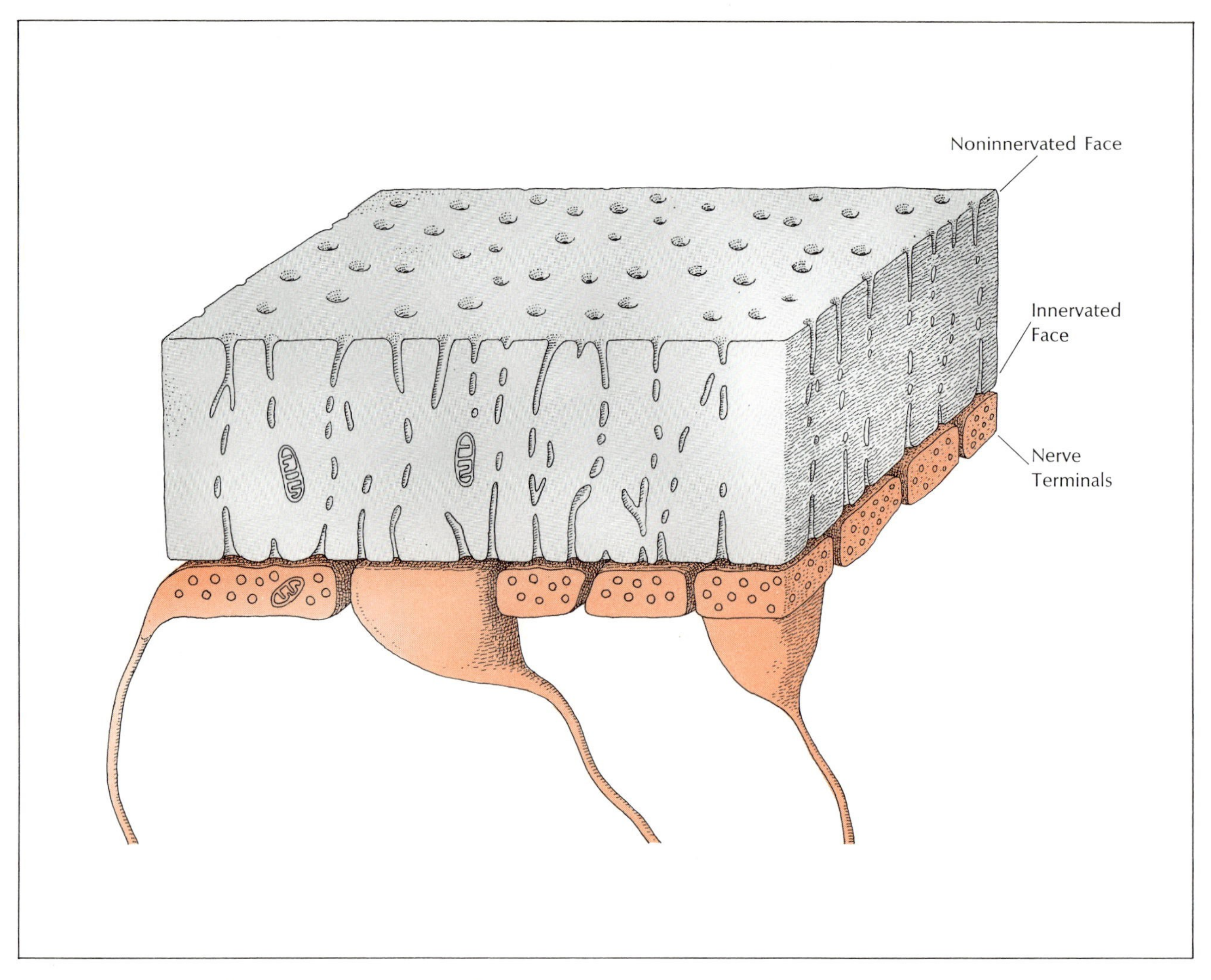

Drawing of part of an electrocyte from a torpedo's electric organ shows the extraordinary innervation of these cells; one side is entirely covered by axon terminals. This rich innervation enables cell to discharge totally in a few microseconds.

branes of the axon terminals are missing over large areas. Specifically, the terminal membrane facing the synaptic cleft remains attached to the electrocyte, while the rest of the synaptic membrane is ripped away with the rest of the axon.

What we have, therefore, is a sort of shallow, open cup of terminal membrane, filled with frozen cytoplasm – which is rich in acetylcholine vesicles. By successive centrifugations, the cytoplasm and vesicles can be separated from the coarse particles of electrocyte and axon and then further fractionated to give soluble cytoplasmic proteins, and an almost pure preparation of vesicles. From such preparations, which, as already indicated, contain only acetylcholine vesicles, we have been able to recompute more accurately the quantity of acetylcholine per vesicle. It turns out that the average vesicle contains about 70,000 molecules of the neurotransmitter: this is quite a lot more than estimated for cholinergic vesicles from brain, but these vesicles from torpedo are considerably larger, with volumes up to 10 times that of vesicles from brain.

We have also been using these preparations to study the structure of the vesicles themselves, as well as the mechanisms by which they take up and release their load of acetylcholine. Under the electron microscope, the vesicles look like tiny, spherical bubbles, enclosed – like so many intracellular vesicles and vacuoles – in a simple "unit membrane" consisting primarily of a lipid bilayer. Of the total wet weight of the vesicles, acetylcholine accounts for about 4%. This may seem small but in terms of the usual physiologic concentration of the compound it is relatively enormous.

Further research has established that the vesicles are not simple bubbles containing an aqueous solution of acetylcholine, comparable in structure to the gelatine capsules in which some liquid drugs are administered. A considerable portion of the vesicles' mass is accounted for by an acidic protein whose molecules occupy much of the space inside the vesicular membrane. And there is reason to believe that acetylcholine is stored within the vesicles as an actual salt of this protein, in which the positive charge of the acetylcholine molecule is electrostatically attracted to the protein's negative charge.

This picture is based in part on a certain amount of direct evidence, but in part on analogy with the chromaffin granules that store epinephrine in the adrenal medulla. These are half a micron in diameter, making them more than 10 times the size of the vesicles, but there are grounds for thinking that similar storage mechanisms are involved in both cases. The chromaffin granules also contain an acidic protein or proteins, as do also the synaptic vesicles that store norepinephrine; indeed, some of the proteins seem identical. Thus it seems likely enough that the acetylcholine-containing synaptic vesicles involve similar storage mechanisms.

The questions we are currently tackling concern how the acetylcholine gets into the vesicles and (especially) how it gets out of them during synaptic discharge. And here, unfortunately, we must still rely to a great extent on analogy (i.e., with the much better understood chromaffin granules) and plain guesswork.

Acetylcholine, in the first place, is manufactured in the neuronal cytoplasm by an enzyme that transfers acetate from acetyl-coenzyme A to choline. The enzyme itself is apparently manufactured in the body of the neuron, but migrates down the axon to the terminal region, where its concentration is far higher than elsewhere in the cell. This suggests that much or most of the cell's acetylcholine synthesis occurs in the terminal region; however, it does not take place in the vesicles, since the synthesizing enzyme is not found there. The vesicles, like the enzyme, are probably also made in the cell body, whence they pass down the axon into the terminal region. These formed vesicles presumably include both the protein and its enclosing lipid membrane, since there seems no very plausible way in which the protein could get inside the membrane after its formation.

There is some reason to believe the vesicles begin taking up acetylcholine on their journey down the axon to the terminal region, but if, as we suspect, the bulk of the acetylcholine is manufactured in the latter site, most of the acetylcholine uptake would presumably occur there as well. One could speak more definitely on this point if it were possible to isolate populations of vesicles from different regions in the cell and determine their acetylcholine content. Something of the sort had been done with noradrenergic vesicles: by constricting the axon one can obtain an accumulation of vesicles on the proximal side of the constriction, and these do contain norepinephrine. But whether, and to what extent, the same is true of cholinergic vesicles remains to be determined.

There can be little doubt that the acetylcholine is discharged into the synaptic cleft by the very common cellular process of exocytosis, but many of the details are still obscure. At some point, obviously, the vesicular membrane must fuse with the external, presynaptic membrane and form an opening through which the acetylcholine, at least, can escape. And evidently it does this as a concomitant of neuron activation – i.e., in response to the action potential of

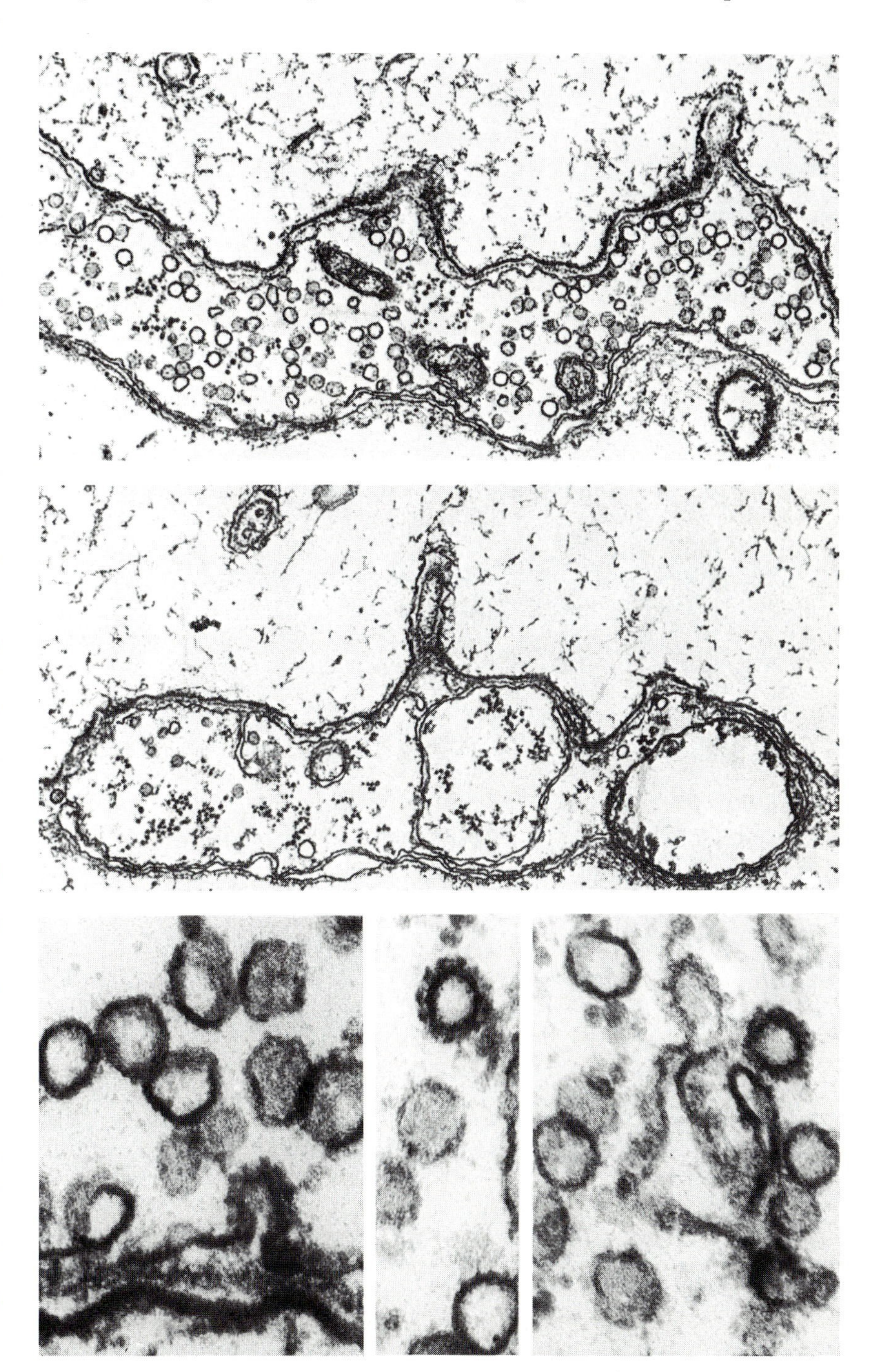

Under "resting" conditions (top), an axon terminal innervating a torpedo electrocyte is rich in acetylcholine-containing vesicles; after stimulation (center), vesicles have almost completely disappeared. Enlargements (bottom) show some vesicles merging with external membrane of the terminal. Electron micrographs are by Dr. H. Zimmermann.

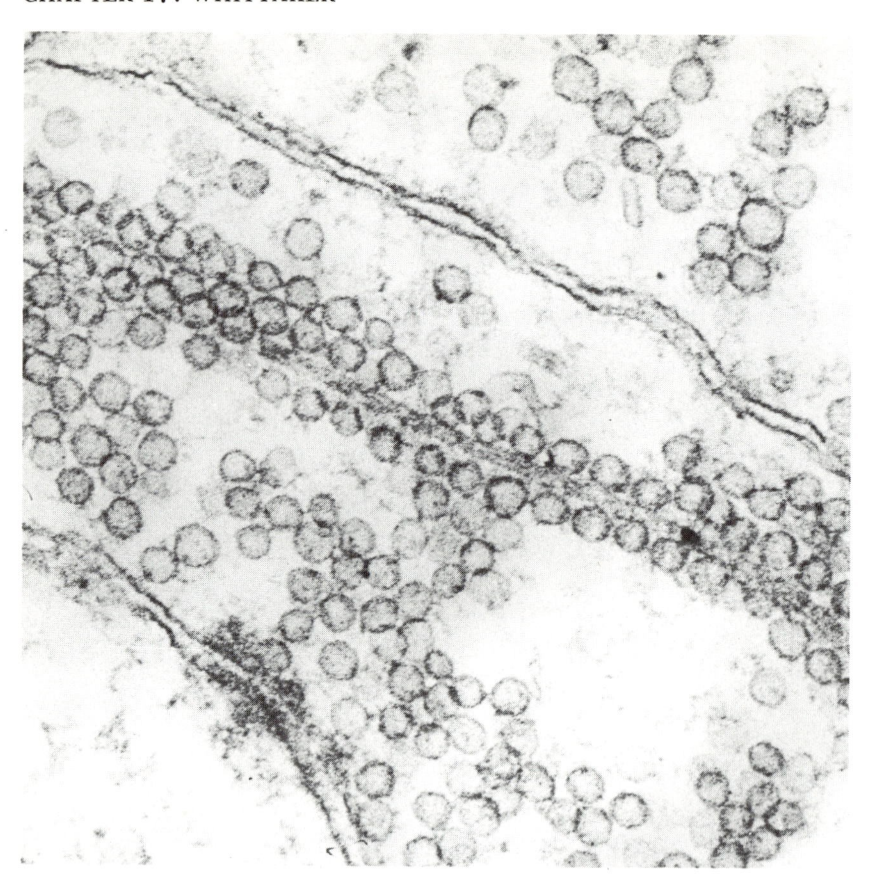

Spontaneous "lineup" of neurohumoral vesicles in axon terminal suggests the possibility that a group of them may fuse into a single structure that discharges as a unit (electron micrograph by Dr. H. Smith).

the cell. In this respect, vesicular discharge would resemble the discharge of any other type of storage granule, already described in this volume (see Chapter 15, "Membrane and Secretion," by Jamieson).

But with the vesicles – as with other storage granules – we still do not know much about how vesicles that are a considerable distance away from the membrane are induced to discharge through it. With ordinary storage granules, indeed, this may not pose much of a conceptual problem. Given the distances involved, simple random diffusion would probably maintain a sufficient population of granules directly adjacent to the external membrane to maintain an adequate flow of their contents under appropriate stimuli. The conditions surrounding synaptic transmission, however, are rather more demanding. It has long been known that there is a threshold value of acetylcholine release below which the target cell will not respond, and it is now apparent that this value corresponds to the contents, not of one but of several hundred vesicles. Moreover, we know that acetylcholine does not accumulate on or in the postsynaptic membrane but is broken down almost immediately, i.e., its effects cannot sum up over any but the very briefest time interval.

What this implies is that for nerve transmission to occur a sizable number of vesicles must discharge simultaneously, or almost so; the cell probably cannot "wait" for diffusion to bring the vesicles one by one to the presynaptic membrane. Accordingly, a number of investigators have suggested that the vesicles interact with one another, so that as a vesicle adjacent to the external membrane fuses with it, it also fuses with a more distant vesicle on the other side, and that one with a third vesicle, and so on. The result would be a sort of tunnel into the cell, through which all the component vesicles could discharge more or less simultaneously.

There is a certain amount of evidence supporting this "tunnel" hypothesis. Simply by warming synaptosomes, one can produce intracellular tunnels or tubules by vesicle fusion. Similar formations have been observed in nerve cells treated by the neurotoxin of the black widow spider, the effect of which is greatly to accelerate acetylcholine discharge, to the point where the terminal becomes completely depleted of vesicles. Even in normal terminals, the vesicles, under the electron microscope, often seem to be lined up in rows of some sort, and somewhat the same thing has been observed in preparations of isolated vesicles: they form chains rather like poppet beads. Refined staining techniques show what appear to be connections among chains of vesicles, albeit tenuous ones.

Given the exocytotic discharge of a vesicle (or chain of vesicles), an obvious question is whether they discharge their entire contents (i.e., both acetylcholine and protein) or the neurotransmitter alone. Here we have as yet no direct evidence; our assay techniques are insufficiently sensitive to determine whether or not the protein is present in the synaptic cleft. In the case of the chromaffin granule, however, there is rather good evidence that the entire contents of the granule, including the protein, is discharged. In the case of the norepinephrine vesicle, on the other hand, it appears that while the norepinephrine is discharged into the synaptic cleft (along with other readily diffusible substances), most of the protein (about 85%) remains imbedded in the presynaptic membrane. But which (if either) model applies to the cholinergic vesicle remains to be seen.

This question is closely bound up with another, the fate of the emptied (or partly emptied) vesicle. Some researchers have suggested that it may be recycled: reabsorbed as a more or less intact structure into the interior of the axon terminal and there refilled with acetylcholine. In favor of this hypothesis are the so-called vesicles in baskets, first observed by some Japanese workers using isolated vesicles and unorthodox staining techniques. Subsequently, similar findings were reported by Prof. George Gray in London, using intact tissue sections. These consist of basket-like structures that enclose some vesicles attached to the interior of the presynaptic membrane, later dropping off into the cytoplasm. Gray has obtained some rather remarkable stereo micrographs from thick sections that appear to show all stages of this process, from which he has concluded that the vesicles are not, as was thought, formed in the neuron body, but rather by invagination of the terminal membrane, stimulated in some fashion by the "basket."

As against this possibility, we have the fact that the vesicle membranes, to the extent they can be chemically characterized, seem to be quite different in composition from the exterior terminal membrane. The latter contains high concentrations of both cholesterol and ganglioside – constituents thought to be characteristic of all external membranes. The vesicle membrane, on the other hand, contains much less of both these components. Chemically, it is quite similar to the other intracellular membranes such as the microsomes, which would suggest that vesicles are formed, not by invagination of the external membrane but by budding from some internal membrane (probably that of the Golgi apparatus), as is known to be true of other storage granules.

However, the two hypotheses are not mutually exclusive. It is conceivable that some or all of the vesicles could be recycled as described in Gray's basket hypothesis, but they would lose some of their structural integrity at each recycling, becoming unusable after a few cycles and being replaced by new vesicles from the neuron interior.

Perhaps the least understood aspect of synaptic function is the process by which vesicles are (or are not) discharged. On the face of it, the process must clearly be triggered by the action potential of the discharging neuron, but precisely how this stimulates the mobilization and discharge of the vesicles is not known. A fuller understanding of this process should contribute to our understanding of the many toxic and pathologic processes that affect synaptic function. Botulinus toxin, for example, in some fashion blocks release of cholinergic granules, while black widow toxin raises the rate of release of these granules to damagingly high levels. Paralleling the action of these toxins on the neuron terminal, we have others that affect the target cell, such as curare, which desensitizes the endplate to the action of acetylcholine, and the cholinesterase inhibitors (such as the nerve gases), which prevent the breakdown of the neurotransmitter after it has done its job.

Less certainly, but very plausibly, we have such conditions as schizophrenia and manic-depressive psychosis, where there is at least a suspicion that some synaptic disorder may be involved. While reports of the presence of abnormal metabolites in the spinal fluid of schizoid and manic-depressive patients are still somewhat controversial, it seems increasingly certain that these conditions tend to "run in families" in a manner suggesting the existence of some organic, biochemical component in their etiology. As further evidence, we have the existence of the many psychoactive drugs that are analogues of neurotransmitters. Thus LSD, for example, resembles serotonin, while such compounds as Ditran (sometimes used as an antidepressant) are structurally analogous to acetylcholine. Many of these substances, in the proper (or improper) dosage, can induce hallucinations comparable to those found in "ordinary" psychotics. To the extent that we can better understand normal synaptic function, we should be able to confirm whether, and to what extent, synaptic disorders are involved in these troublesome conditions, and, if so, perhaps how to correct them.

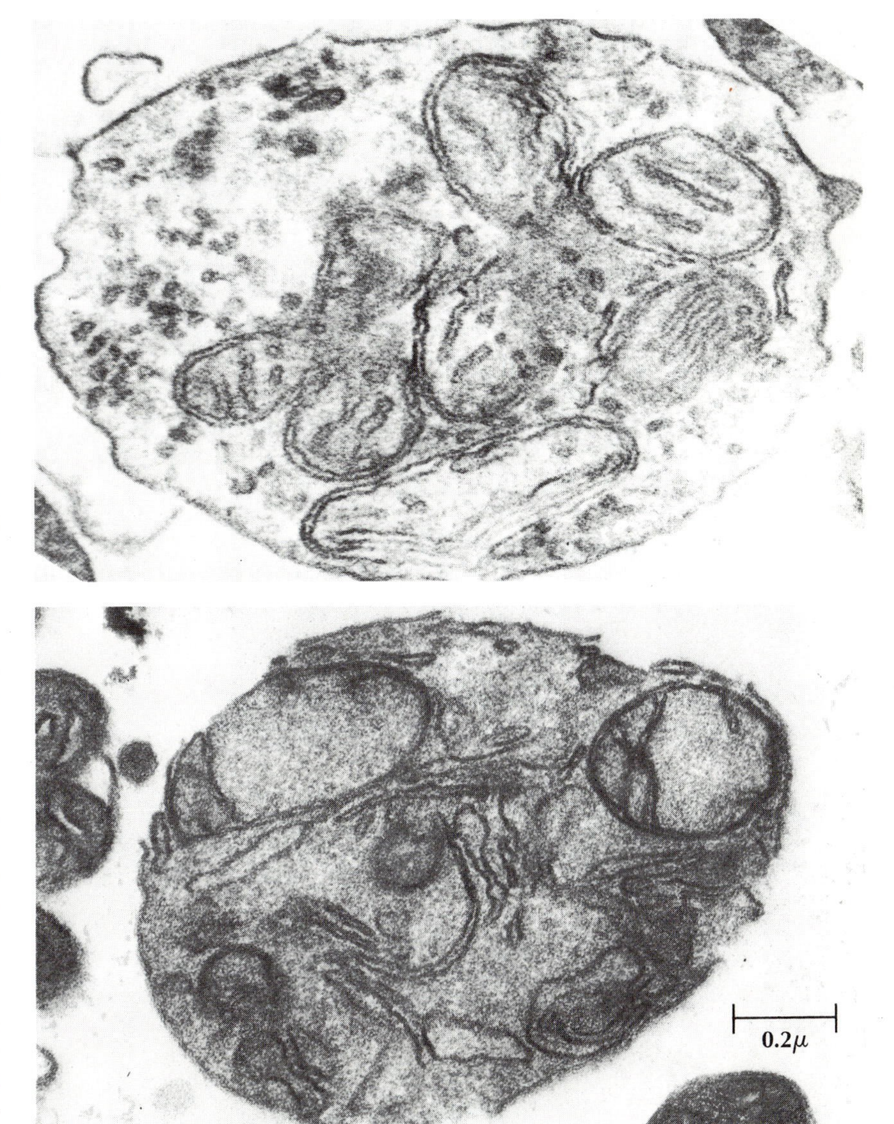

Hypothesis of vesicular fusion is strengthened by experiments in which discrete vesicles of normal synaptosome (top) can be induced to fuse by incubation at 10° C (bottom).

18

Hormone-Receptor Interactions And the Plasma Membrane

PEDRO CUATRECASAS

Johns Hopkins University

Earlier chapters in this book have discussed in much detail the intracellular processes whereby hormones are thought to modulate cell functions. My own concern has been with a more proximal aspect of hormone action: the processes whereby hormone molecules "recognize" and attach themselves to receptor molecules on the exterior of the cell's plasma membrane and, having done so, induce changes in (or around) the membrane that set in train the intracellular processes just alluded to. I shall focus on the relationship between insulin and its receptors, mainly because it is by all odds the best understood among hormone-receptor interactions.

Conceptually, most investigators in this area see the hormone-receptor relationship as involving two distinct and (at least in principle) separable processes: recognition-binding and activation. In part, this view rests on analogy with enzyme action, in which recognition-binding between enzyme and substrate is clearly separable from the former's catalytic action on the latter. To be sure, catalysis obviously cannot occur without binding, but binding can clearly occur without catalysis. For example, enzyme-inhibiting substances typically bind to the enzyme – often, indeed, more firmly than the normal substrate – but cannot then launch catalysis. Instead, they tend to remain attached to the enzyme, thereby preventing it from reacting with its normal substrate. Similarly, it can be shown that in certain enzyme defects the normal substrate binds to the abnormal enzyme, but is not catalyzed by it.

As regards the hormone-receptor relationship, the situation is less clear-cut. While "hormone inhibitors" (analogues) that can attach themselves to a receptor without activating it are known for certain hormone systems such as glucagon, angiotensin, and ACTH, none are known for insulin. The insulin molecule can be modified in a number of different ways such that its biologic activity is markedly reduced, but in all such instances the change in activity can be shown to be a function of a change in the molecule's binding capacity. There is no known instance of normal binding capacity coupled with subnormal activation. Nonetheless, it is important to distinguish the two processes, if only for expository purposes. Moreover, as we shall see later, there is considerable indirect evidence indicating that binding and activation are indeed separable functions.

Before considering in detail how insulin, our hormonal prototype, binds to its cellular receptors, it is perhaps worth reminding ourselves that its biologic actions are considerably more diverse than is sometimes appreciated. Thanks to the clinical significance of insulin in diabetes, everyone is well aware that it stimulates cellular uptake of glucose and the conversion of glucose to glycogen in liver and muscle tissue. More broadly, however, insulin can be described as an overall anabolic hormone. It stimulates cellular uptake of amino acids, and their incorporation into protein, in addition to inhibiting protein degradation; likewise, it stimulates uptake of nucleotides and their incorporation into DNA and RNA. Finally, it steps up lipogenesis, which is, of course, another mode of energy storage, paralleling the conversion of glucose into glycogen.

The cellular sites at which insulin triggers these diverse activities were long a subject for debate; nowadays, however, everyone agrees that they are located on the cell surface. The evidence for this view is voluminous, so that I will note merely a few of the experiments involved. It has been shown, for example, that insulin bound to plastic beads of various types is fully capable of activating cells in the normal manner, though the beads cannot and do not enter the cell (some of them, indeed, are a great

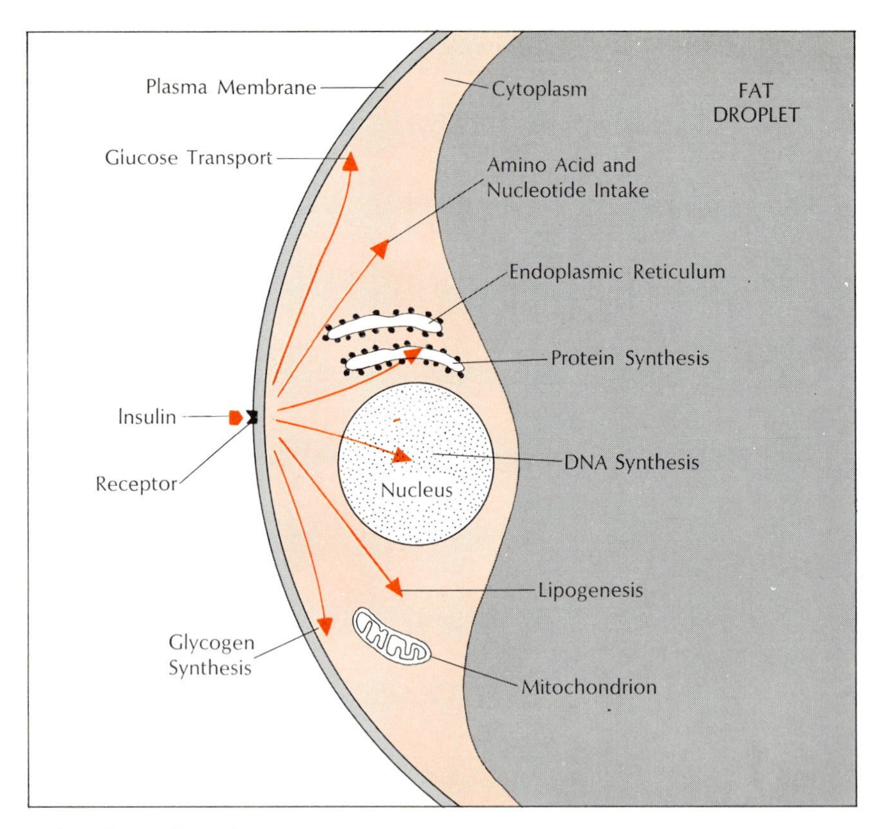

When the insulin molecule activates the receptor in the plasma membrane, various anabolic effects ensue, involving changes in both the membrane and the organelles. It is estimated that an average fat cell, schematized above, has some 10,000 such receptors.

deal larger than the cell). Another type of experiment involves the production of "ghosts" – empty cell membranes – of, for example, fat cells, which bind insulin as do the intact cells themselves. By certain procedures, however, it is possible to prepare the ghosts in an everted, inside-out form – which does not bind the hormone. Moreover, if insulin is added to the preparation before the ghosts are everted, it becomes trapped inside them – i.e., it is released to the medium far more slowly than insulin bound to normal ghosts or intact cells. From these and many other experiments, there can be no doubt that the cell's insulin receptors are located in the plasma membrane (rather than in the cytoplasm), and on only one side of the membrane – its exterior.

The binding of insulin to these receptors was at one time thought to be an essentially irreversible process, involving a covalent bond – perhaps produced by the opening of one of the disulfide bonds that hold the insulin molecule in its characteristic configuration and the formation of new disulfide bonds with sulfur atoms in the receptors. Homeostasis would then require that the insulin at some point be removed from the receptor; otherwise the receptors would rapidly become saturated with the hormone, producing the cellular equivalent of insulin shock. Thus the covalent bond theory implied that the attached insulin molecules would be degraded, either by being absorbed into the cell interior or in some other fashion.

Further research, however, has shown that this view is far from the truth. Labeling experiments, for example, indicate that insulin is in fact bound rather loosely to the receptors. The rates at which its molecules associate with and dissociate from the receptor molecules show the kinetic characteristics of simple dissociable processes, being governed by concentration (and also temperature). For example, labeled insulin bound to cells will pass into the surrounding medium more or less rapidly, depending on whether insulin levels in the medium are low or high – which would not be the case if a significant fraction of the bound insulin were being degraded. Moreover, and even more important so far as degradation is concerned, those molecules that are bound to receptors and that dissociate into the medium spontaneously are capable of binding again to receptors – and of activating biologic responses.

Experiments with various analogues of insulin, involving modifications of its molecule, indicate that their binding strength is unrelated to their degradability. For example, proinsulin (the hormone's biologic precursor) shows only about 1/20 the activity of normal insulin, and this can be entirely accounted for in terms of its reduced binding capacity. Indeed, proinsulin (at binding capacity and activity equal to that of insulin) is *less* easily activated by liver or its membranes. Rapid degradation of proinsulin does occur in vitro, but under conditions that probably reflect the presence of contaminating protease enzymes rather than interactions with the receptor. Finally, intact cells (e.g., lipocytes) appear to have no effect on the insulin that they bind, since its biologic activity remains unimpaired after release.

This is not to say, of course, that insulin is not degraded in the body; like all body constituents, its molecules are sooner or later broken down, to be replaced by newly synthesized molecules from the pancreas. The site of its breakdown, like that of so many pharmacologically active molecules, is the liver, but very little is yet known about how, how fast, or even where in the liver cell this occurs. It is possible that insulin may also be broken down by other types of cells; certainly it is vulnerable to a great many homogenized cell and tissue preparations. However, this may reflect merely the presence in these fragmented preparations of generalized proteolytic enzymes. Some of these reside in intracellular membranes, some of which always contaminate plasma membrane preparations; others are soluble proteases from the cytoplasm, which during homogenization may adsorb strongly to the membranes. Proteases, unfortunately, are ubiquitous, and can give spurious results in almost any in vitro study; in vivo, however, they are under very fine control as regards localization, compartmentalization, and so on.

As authors of this book have frequently pointed out, the plasma membrane to which insulin attaches itself is composed of two main types of molecules: amphipathic lipids (chiefly phospholipids), and proteins of various types, so that it is among these types that the receptor molecules must be sought. The phospholipids can be ruled out en bloc: when cell membranes are treated with phospholipases, far from becoming less "receptive" to insulin, their binding capacity increases dramatically. (The increase can be partially reversed by adding exogenous phospholipid to the mixture.) These findings not only demonstrate that the receptors are not phospholipids but suggest that the cell contains a sizable number of "occult" receptor sites that normally are blocked by the ionically active heads of the surrounding phospholipid molecules. Further evidence of this is provided by the finding that the same increase in binding capacity can be achieved simply by adding salts (such as NaCl) to the medium. Evidently the salt ions react in some fashion with the phospholipid heads to reduce or eliminate their blocking effect; so long as the salts are present, phospholipases produce no further increase in binding capacity.

Presumably, then, the receptors are protein molecules of some sort. And in fact when cells are treated with proteolytic enzymes such as trypsin, their response to insulin drops. A number of quite elegant experiments along these lines, in which trypsin is combined with phospholipase and other reagents, have provided important clues to the nature and topology of the receptor sites.

First, mild trypsin treatment alone produces no reduction in the cell's capacity to bind insulin, though it does reduce its affinity for insulin. In other words, the cell's binding capacity for insulin (as measured by labeling experiments) and its response to insulin (as measured, say, by glucose uptake) are not impaired by mild trypsin digestion, but both maximal binding and maximal response require considerably higher concentrations of the hormone. Evidently, then, mild trypsin treatment does not destroy receptor sites, but it does alter them

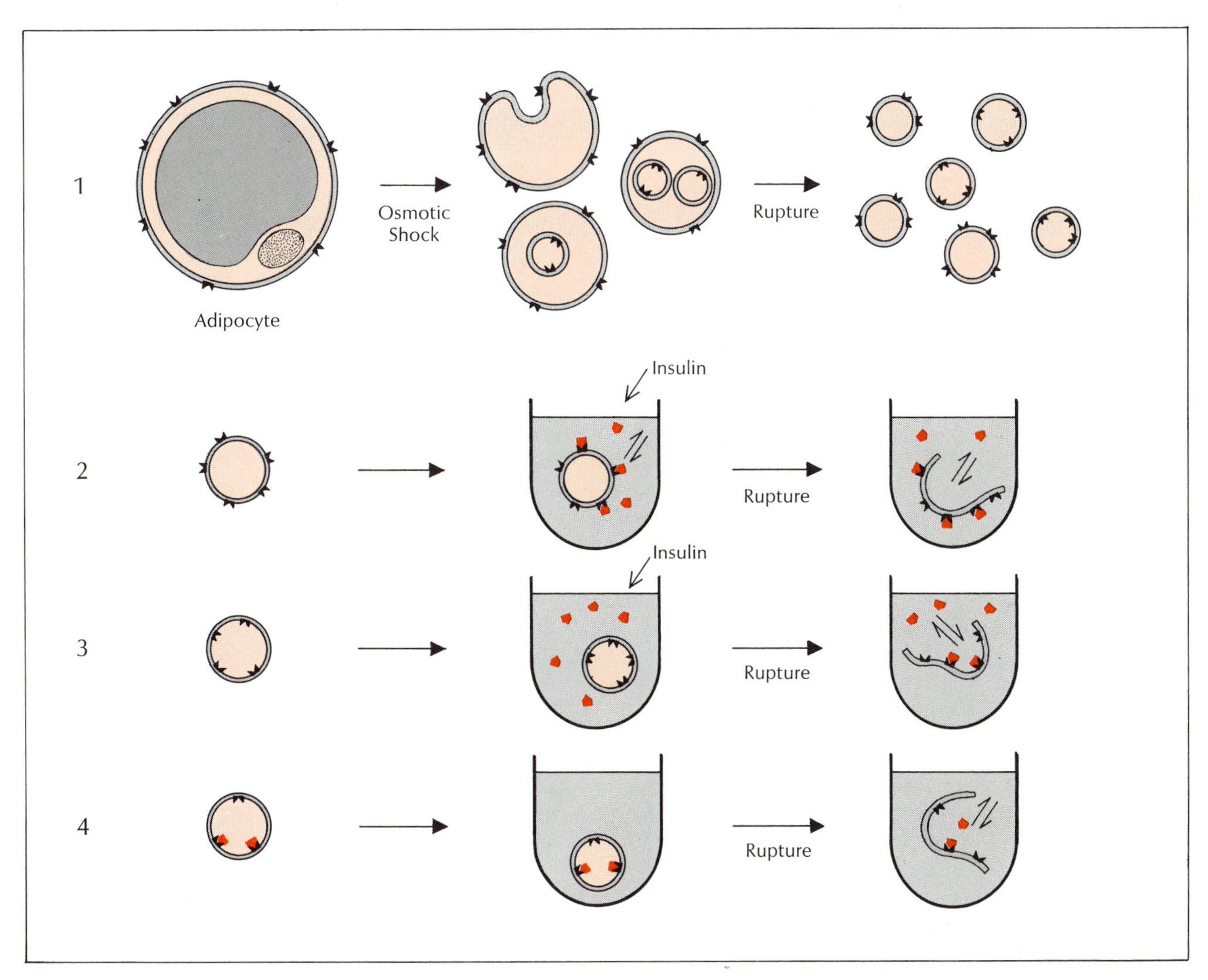

By suitable treatment, "ghost" membranes of fat cells can be converted into small, closed vesicles, some of which are inside out (1). Right-side out or normal vesicles, like cells, bind and release insulin in equilibrium with the surrounding medium, whether they are intact or broken (2). Inside-out vesicles do not bind the hormone unless they are broken (3). If inside-out vesicles are prepared in a medium containing insulin, it becomes trapped inside them and is released only if vesicle is broken (4). These experiments, along with many others, have established that insulin receptor lies only on exterior side of the membrane.

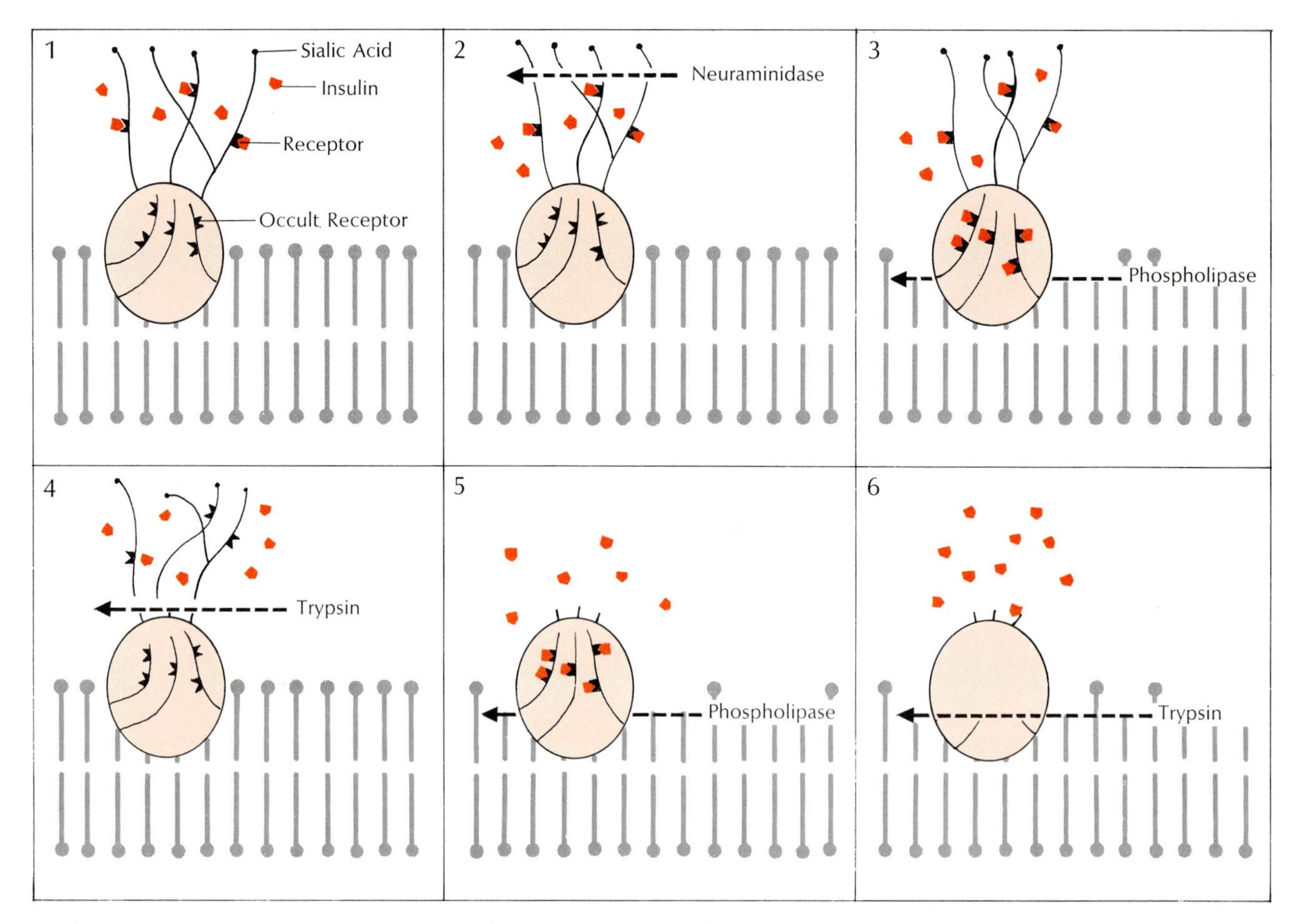

Insulin receptor molecule (1) is thought to include oligosaccharide chains tipped with sialic acid, which, however, is not part of binding site, since its removal by neuraminidase (2) does not impair binding capacity. Treatment with phospholipase steps up binding capacity, exposing "occult" binding sites (3). Trypsin destroys normal binding sites (4), but subsequent phospholipase treatment more than replaces these with occult sites (5). If phospholipase is followed by trypsin, all sites are destroyed (6).

enough to loosen the attachment of the insulin molecule – presumably by impairing its fit with the receptor molecule.

By contrast, drastic digestion with much higher concentrations of trypsin produces a marked loss in the membrane's insulin-binding ability, regardless of concentration (and of course a comparable loss in its insulin response), implying that some binding sites have been destroyed. Significantly, however, the loss can be more than restored by subsequent treatment with phospholipase, which evidently compensates for the destroyed normal binding sites by uncovering an even larger number of occult binding sites. Some binding sites are nonetheless permanently destroyed by the trypsin treatment. This is shown by the fact that the maximum binding levels achieved, though higher than in the untreated membrane, are only two thirds the level of those in membranes treated with phospholipase alone. In the latter situation the uncovered occult sites do not replace destroyed normal sites but add to their number.

Finally, if phospholipase-treated cells are subjected to even mild trypsin treatment, they undergo a marked – and, again, permanent – loss in binding capacity. Evidently, the phospholipase "opens up" binding sites to the point where even mild trypsin digestion can destroy them, presumably by attacking them at a more critical point in their structure. Significantly, the most drastic loss of binding sites occurs after treatment with reagents such as tetranitromethane, whose molecules, far smaller than those of trypsin, can evidently gain access to the vulnerable portions of the binding sites, both normal and occult, as larger bodies cannot. After such treatment, phospholipase cannot restore binding capacity, since there are evidently no binding sites left to be uncovered by its action.

Quantitative studies of binding capacity under various conditions indicate that the occult sites are nearly three times as numerous as the ordinary sites. Why these sites, which cannot bind insulin under normal conditions, should exist at all, let alone in such numbers, is unknown. It is possible, of course, that under normal conditions they serve to bind some other molecule (or molecules), becoming accessible to insulin only when uncovered by the phospholipase treatment. However, no one has yet found evidence that they can bind anything but insulin. Nor is there evidence that they differ in any other respect, once they are uncovered, from normal sites. Conceivably, they could be evolutionary vestiges, remnants of a stage of evolution in which more sites were needed for some reason, but there is no evidence for this either.

The fact is that we do not know why they exist – but they do.

Even more interesting are the effects of various enzymes that detach sugar residues from proteins or from polysaccharide chains. For example, treatment of cells with the enzyme neuraminidase, which detaches the amino-sugar sialic acid, will (if the experimental conditions are just right) abolish the cell's response to insulin, in terms of glucose transport and inhibition of lipolysis. However, it does *not* abolish, or even impair, the cell's capacity to bind insulin. What are we to make of this? The most likely explanation is that the insulin receptor includes one or more sialic acid residues but that these are not part of the receptor site. Rather, they serve in some fashion to communicate the "message" of the hormone-receptor combination to other molecules in the membrane, which in turn set into motion the insulin-activated processes of glucose transport and lipolysis inhibition. In this case, at least, binding and activation would appear to be truly separable processes, since the latter requires the presence of sialic acid on the receptor while the former does not.

That the sialic acid in question *is* on the receptor and not on some molecule activated by it is suggested by other experiments. If, for example, the normal cell is treated with a mild preparation of β-galactosidase, nothing happens. If, however, it is first treated with neuraminidase (or with both enzymes together), the effect is profound: the cell loses not only its responsiveness to insulin (which the neuraminidase alone would accomplish) but also its capacity to bind insulin. The probable situation is that galactose forms part (and evidently a pretty essential part) of the actual receptor site, while sialic acid, though not part of the receptor site as such, is attached to the galactose in such a way as to shield it from the action of β-galactosidase; one visualizes a chain of sugars, with galactose as the next-to-last link and sialic acid as the last.

This evidence alone would strongly suggest that the insulin receptor is one of the class of molecules known as glycoproteins, amino acid chains with short chains of sugar residues attached at various points along their length. Glycoproteins are known to subserve receptor and antigenic functions of many sorts; earlier in this book (see Chapter 5, "The Structure and Orientation of a Membrane Protein"), Marchesi described erythrocyte glycophorin, which carries blood group determinants on red cells and is a receptor for viruses, the reagent phytohemagglutinin, and so on. Further experiments have demonstrated that the insulin receptor is indeed one of this group.

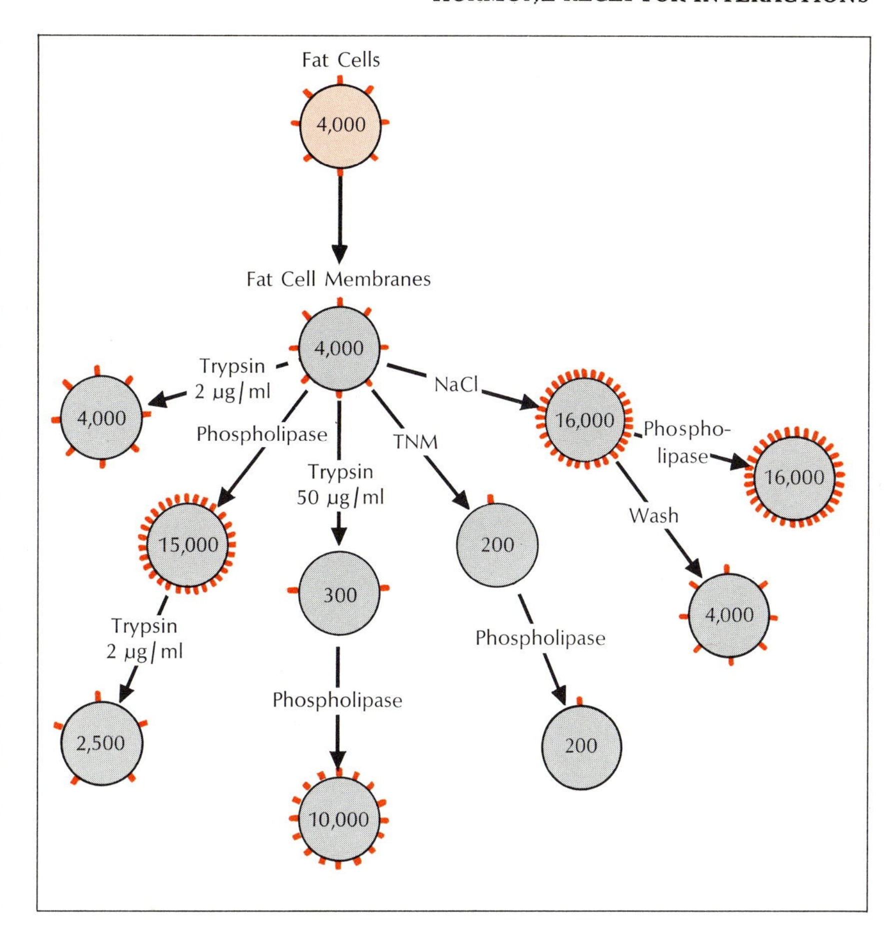

Changes in cellular binding capacity (numbers indicate counts per minute of labeled insulin) under various manipulations have elucidated the presumed structure of the insulin receptor, depicted in previous illustrations. NaCl steps up binding capacity, presumably because its ions neutralize charges on phospholipid heads; TNM, a small molecule, can penetrate phospholipid layer to destroy all sites.

We have found, for example, that the plant lectins concanavalin A (ConA) and wheat-germ agglutinin (WGA), in low concentrations, produce insulin-like effects on fat cells; in higher concentrations, they inhibit insulin binding to the cells. The insulin-like effect of these substances is interesting in its own right, as I shall explain shortly. In terms of the character of the receptors, however, the important point is that both these substances are known to bind to a variety of glycoproteins. They can be displaced by various simple sugars, with which they combine preferentially – and displacement occurs also with their binding to the insulin receptors.

It is conceivable, of course, that the lectin binding might not involve the receptor but rather some nearby molecule. However, this possibility has been eliminated by experiments with preparations of isolated receptors, obtained by a rather complex process known as affinity chromatography.

The first step is to obtain isolated membranes from fat cells, from which the protein component can be removed by various detergents; as one would expect, the deproteinized membranes no longer bind insulin. The soluble detergent fraction, separated from the membranes by centrifugation, is a very crude preparation, since it contains a large proportion of the membrane proteins, among which the insulin receptor accounts for less than

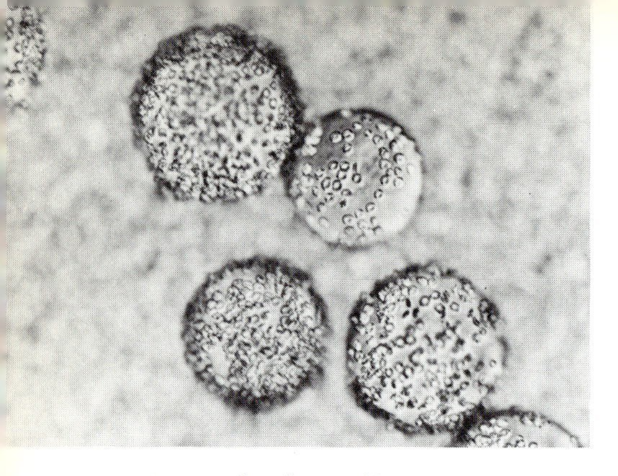

Agarose beads, used to purify receptors (see below), will also bind erythrocytes as shown here if treated with phytohemagglutinin.

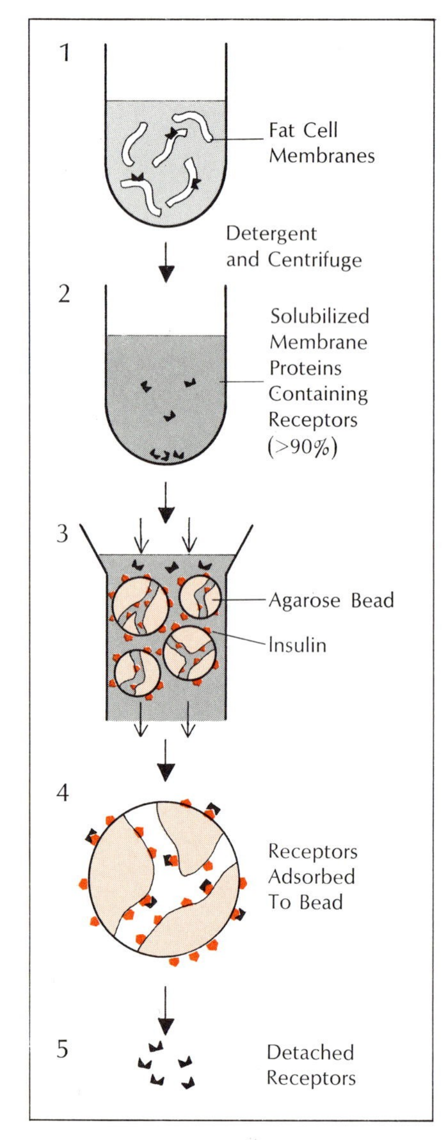

Relatively pure preparations of receptor sites are produced by affinity chromatography, in which protein fraction of membranes is passed through column of beads pretreated with insulin. Beads bind the receptor molecules while allowing the >99.9% of nonreceptor protein to pass through. After other proteins are washed off, receptors are detached from beads.

1%. After some preliminary purification steps, the mixture is passed through a column of agarose beads – highly porous particles with a correspondingly large surface area. These have been previously prepared by reacting with insulin and certain other substances in such a manner that the insulin is firmly bound to the bead column. The insulin receptors attach themselves to the bead-bound insulin, while most of the rest of the protein passes on through. By delicate chemical manipulation, one can then gently detach the receptor from the bound insulin, thereby obtaining a quite concentrated solution of receptor molecules (we estimate a purity of about 50%).

These purified preparations have enabled us to estimate the size of the receptor molecule (its molecular weight is about 300,000) and also to compare its properties with those of ordinary, membrane-bound receptors. It turns out that the isolated receptors bind insulin in a wholly normal manner, and also bind ConA and WGA precisely as do ordinary receptors. Thus there is no doubt in our mind that the receptors are in fact glycoproteins. What remains is to estimate the quantitative importance of the sugar component, as compared with that of other glycoproteins (such as erythrocyte glycophorin), but we have not yet managed to do this because of the very small quantities of purified receptor thus far obtained.

Through quantitative studies of these purified preparations and other means, we have estimated that the average lipocyte membrane includes some 10,000 receptors (this of course refers to normal, not occult, receptors). How they are distributed in the membrane is something of a puzzle. On the one hand, there is reason to believe that under normal conditions their distribution is essentially random (indeed a certain amount of evidence suggests this is the case). On the other hand, in variants of the original experiments using fat cells and insulin bound to large plastic beads, one can juggle the proportion of beads to cells in such a way that one can be certain that each cell is in contact with no more than a single bead, a situation in which, for simple geometric reasons, only a small part of the cell surface can be reacting with the bead-bound insulin. Yet under these apparently restrictive conditions the cells still show a full or almost full response to the hormone. One is forced to conclude that if the receptors are indeed randomly distributed on the cell surface (as seems likely), they still retain considerable mobility. This would enable them fairly quickly to locate themselves in that area of the surface in contact with the bead, and, binding to the insulin there, to remain clumped in one place. This seems not unreasonable, since similar clumping has been observed in other receptors, such as those for lectin and virus particles.

Our work on the insulin receptor has naturally led us to wonder about its implications for the hormone's mode of action. A full discussion of this question would take us well beyond the bounds of this chapter and into the area of cyclic nucleotides, a topic that will be examined extensively in Chapter 19. Our thinking in this matter began with the fact that insulin's actions, as noted earlier, are very diverse, so that any theory would have to include some mechanism whereby activation of a single type of receptor could trigger a great variety of intracellular processes. Another central fact was the finding that insulin inhibits the action of the membrane-bound enzyme adenylate cyclase, reducing cellular concentrations of cyclic AMP.

One possibility, then, would be that the insulin receptor lies in close proximity to some other membrane molecule containing adenylate cyclase. Activation of the receptor molecule would induce some change in this adjacent molecule, inhibiting the enzyme; the resulting drop in cAMP would then trigger the secondary processes such as lipolysis inhibition, glucose transport, and so on. In favor of this hypothesis is the fact that many effects of insulin seem to be associated with a drop in cAMP levels. Some, to be sure, have not been shown to be so associated, but this may merely reflect our present ignorance. However, one must also note that association is not in itself proof of causal relationship.

A second possibility is that the insulin receptor does not of itself directly influence the activity of adenylate cyclase. Rather, it might achieve its effects by releasing some substance that inhibits the enzyme and also activates other insulin-modulated processes such as glucose transport. In favor of this possibility is the fact that insulin has been shown, in our laboratory and elsewhere, to sharply increase intracellular levels of a substance that inhibits some of the effects of cAMP. This is the "opposite" cyclic nucleotide, cGMP; insulin evidently increases its concentration by activating the enzyme guanylate cyclase, which is to cGMP as adenylate cyclase is to cAMP, but the exact mechanism by which this occurs is not known. Whether insulin's capacity to stimulate production of cGMP is sufficient to account for its diverse physiologic activities remains to be determined.

A third possibility – not wholly inconsistent with the other two – is that activation of the insulin receptor molecule, by changing its shape or charge, induces some major change in membrane conformation. This presumably could modulate the activity of enzymatic processes (e.g., cAMP production) and other membrane activities (e.g., glucose transport) even though the molecules mediating these processes were not physically contiguous to the receptor. It has been shown, for example, that minute concentrations of growth hormone can induce major changes in the overall structure and conformation of erythrocyte membranes, and the same may be true of other peptide hormones. Present knowledge does not permit us to choose among these hypotheses.

Another poorly understood, but very interesting, aspect of the insulin receptor is its possible involvement in growth processes, both normal and abnormal. The facts bearing on this area, however, are still very patchy, albeit provocative.

First, it is well known that the plant lectins, such as ConA and WGA, stimulate cell mitosis, and conceivably this may be related to their insulin-like properties in cells, which in turn may relate to their capacity to react with insulin receptors. It should be noted that these substances are mitogenic in cells that demonstrably do not possess insulin receptors, such as circulating lymphocytes. However, it is conceivable that such cells possess "incomplete" receptors that will respond to an unnatural stimulus such as a lectin but not to the hormone itself. (There is evidence that such receptors exist in immature mammary gland cells.) Alternately, the biologic insulin-like properties of these plant lectins may reflect the ability of these molecules to interact with and thus perturb molecules distal to the insulin receptor but that would under normal circumstances be in the insulin-receptor train of reactions. (The lectins may thus bypass the receptor by stimulating other membrane structures.) Another interesting fact is that when lymphocytes are transformed by lectins – which can be done by varying the experimental conditions – they acquire full-fledged insulin receptors. These also appear in the "naturally" transformed lymphocytes from patients who have acute lymphocytic leukemia. A final point is that the substance somatomedin (sometimes called "sulfation factor"), which is thought to mediate the action of growth hormone, can react directly with insulin receptors in several types of cells, showing marked insulin-like properties.

It seems reasonable, therefore, that

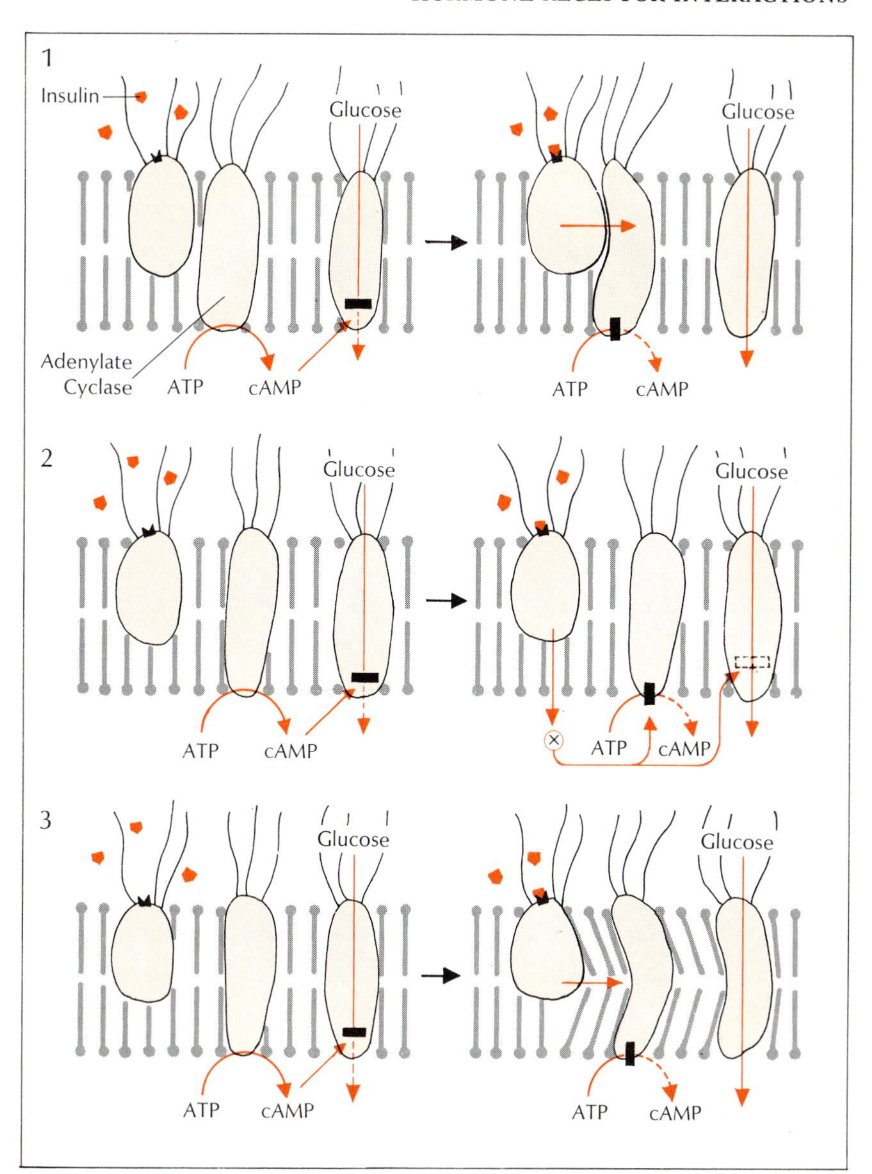

Three possible modes of insulin action are shown schematically. Activation of receptor molecule by hormone is known to block cAMP production, which in turn may activate other insulin-dependent processes, typified by glucose transport (1). Or, activation may release "X" substance (perhaps cGMP or a related nucleotide), which both blocks cAMP production and triggers other processes (2). Finally, activation may produce cellular responses by physically deranging membrane structure (3).

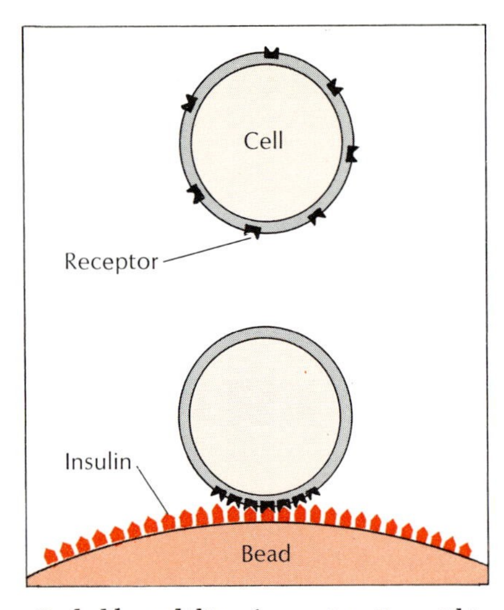

Probable mobility of receptor sites within plasma membrane is shown by cell's response when it is "treated" with insulin-agarose bead. Cellular response is close to normal, indicating that nearly all its receptor sites must be activated.

the receptors for insulin, which as we have seen is a "broad spectrum" anabolic hormone, should be involved at some point in growth processes. But the exact nature of that involvement remains to be clarified.

A natural question for anyone dealing with insulin receptors is their role in insulin-resistant diabetic states. Thus far, the only studies have been with animals. We have found that in rats that have been made insulin-resistant by starvation or by treatment with prednisone or streptozotocin, the hormone receptors appear to be normal both in number and in their affinity for insulin. This suggests that the disorder involves processes further along in the chain of events. However, in mice suffering from the obese-hyperglycemic syndrome (a recessively inherited trait characterized by insulin resistance), the liver cells show a marked decrease in their capacity to bind insulin, apparently caused by a sharp drop in number of receptor sites.

Let me conclude this account with a very summary description of recent findings concerning membrane receptors for other hormones. Glucagon, the "opposite number" of insulin because it steps up glycogenolysis rather than glycogen synthesis, appears to be bound by its receptors in a rather similar manner. As with insulin, binding does not seem to be related to inactivation of the hormone, though here too protease contamination has given some misleading results. But the binding of glucagon to its receptor and its physiologic effect (the activation of adenylate cyclase) are more clearly separable processes than with insulin. For example, analogues of glucagon exist that in suitable concentrations displace the hormone from its receptors by preferential binding, yet show no physiologic activity. Also, the binding and/or activity of glucagon is dependent on phospholipids in a quite different manner than with insulin. Digestion of membranes with phospholipases does not increase response to glucagon but rather decreases it markedly. This finding led to the suggestion that the receptors themselves may be phospholipid molecules, but subsequent work indicates instead that the loss in response involves the activation rather than the binding phase of the process. The glucagon receptors, it seems likely, are glycoproteins of the same general type as the insulin receptor.

The catecholamines, epinephrine and norepinephrine, also bind to membranes, but under experimental conditions, at least, do so in quantities several orders of magnitude larger than insulin – and far larger than seems consistent with their physiologic effects. Furthermore, stereoisomers of these substances show similar binding activity, though their biologic activity is markedly different. Because of this apparently nonspecific binding, specific receptors for these hormones – if they exist – have not yet been identified.

Acetylcholine receptors have been identified in muscle end-plates, and have been purified from preparations of the electric tissues of fish, though not yet from mammals. Like the insulin receptors (and, indeed, like all membrane hormone receptors thus far characterized) acetylcholine receptors are basically lipoproteins, but whether they are lipoglycoproteins has not been determined. Perhaps the most interesting thing about them is that they are far more densely concentrated than are insulin receptors. Physiologically speaking, this seems reasonable enough. Insulin action probably resembles that of glucagon, which involves an enzymatic cascade whereby the effect of a single hormone molecule is amplified by many orders of magnitude. For this very reason, however, the speed of the response is measured in seconds or tens of seconds. The muscle's response to acetylcholine, however, is necessarily much faster, meaning that there is no time for any amplificatory cascade to operate. Thus an adequate physiologic response would evidently involve the simultaneous action of a great many acetylcholine molecules – meaning a great many more variable receptors.

Of the other polypeptide hormones (such as thyrotropin, ACTH, calcitonin) all that can usefully be said in this brief account is that specific receptors for them appear to exist in their target cells, but that little is as yet known about the nature of the receptor molecules.

In summary, then, while the intricate operations of the insulin receptor are probably in some sense a paradigm of those governing other hormone receptors, it will take a great deal of work before we can be certain just how close the resemblances are, and the extent and nature of the differences. It would be gratifying if all hormones operated according to a single, simple model – but it does not now appear that nature has made things that easy for us.

19

Cyclic Nucleotides And Cell Function

NELSON D. GOLDBERG
University of Minnesota

In order to survive, every living organism must accommodate to changes constantly occurring in the environment. This is accomplished by readjusting or calling into play the specialized functions contributed by each of the cells of which it is comprised. Hormones can be viewed as the body's chemical messengers that are liberated in response to environmental changes, which are represented by chemical or neurogenic stimuli. The hormone messenger molecules percolate through the body by way of the bloodstream, and program the organism to meet the new requirements by interacting with the appropriate target cells and modulating their function accordingly. That this process occurs is certain, yet our understanding of how cellular function is regulated has long been marked by a serious conceptual and factual gap. Most hormones – the steroids are a notable exception – do not enter the interior of the target cells whose functions they modulate. Instead, they interact with the exterior of the plasma membrane, which can be thought of as the switchboard, or signal receiver, of the cell, while a so-called second messenger internalizes the hormone message so the proper changes may be induced in the functional apparatus within the cell.

There is now good reason to believe that two second messengers have been identified in the form of two cyclic nucleotides, cyclic 3′, 5′-adenosine monophosphate (cAMP) and 3′, 5′-guanosine monophosphate (cGMP). These two compounds, chemically similar but physiologically quite distinct, have been implicated in a host of hormonally mediated processes, and even in nonhormonally induced changes in function.

In all frankness, there is much we still do not understand about these remarkable messenger substances. Present knowledge of them can be compared to a half-completed jigsaw puzzle, in which areas showing clear and meaningful patterns are interspersed with blank regions, as well as with isolated pieces that cannot be fitted in anywhere at present. Yet the pattern as it is emerging is one of extraordinary interest. On the one hand, the mechanisms by which the cyclic nucleotides operate – insofar as they are understood – appear to be infinitely complex; on the other, the overall relationship of the cyclic nucleotides to the regulation of cell function seems to be one of almost exquisite simplicity, since it appears that these two substances, acting in opposition to one another, can modulate dozens and quite possibly hundreds of diverse cellular processes in most if not all living cells.

Much of the early work on cAMP – the first of the two nucleotides to be discovered and still much the better understood – was done in the laboratory of the late Earl W. Sutherland Jr., at Case Western Reserve University and at Vanderbilt University, who in 1971 received the Nobel Prize for his discovery of cAMP and the elucidation of its role in the expression of hormone action. Around 1957, Sutherland and his associate, Theodore Rall, had begun investigating the action of such polypeptide hormones as glucagon and of the catecholamine neurohumor epinephrine. Since these compounds, like other polypeptide hormones (e.g., insulin) and neurohumors (e.g., acetylcholine), were thought to find their final site of action outside their target cells, it was evident that they must be activating receptor molecules on the cell surface, which in some fashion triggered the appropriate changes within the cell.

Sutherland and Rall had already shown that epinephrine and glucagon, both of which increase serum glucose levels, do so by activating the enzyme glycogen phosphorylase, which converts glycogen – the form in which liver cells store glucose – into glucose-1-phosphate, which in turn is converted into glucose, which is released into the blood. But the activation – first demonstrated with liver slices – clearly involved intermediates in the cell membrane, because the enzyme could not be activated in a membrane-free preparation made from homogenized cells. However, if the hormones were added to the membrane pellet remaining after centrifugation and the mixture was then added to the membrane-free supernatant, activation occurred.

Further experiments showed that the activation involved a stable substance deriving from the membrane that remained after the membrane itself had been denatured by heat. This was identified as CAMP, chemically related to the ubiquitous energy-yielding compound adenosine triphosphate (ATP) – subsequently identified as its chemical precursor – and even more closely to ordinary adenosine 5′-monophosphate (AMP), a component of ribonucleic acid. Like all nucleotides, CAMP consists of a base – in this case, adenine – joined to a molecule of the five-carbon sugar ribose. The phosphate group is attached, not to a single position on the ribose group (as is the case with ordinary varieties of AMP) but at two different points, the 3′ and 5′ positions, producing a ring that accounts for the term "cyclic."

In a series of experiments, Sutherland and his associates, namely Rall, Butcher, and Robison, were able to nail down the connection between glucagon action and CAMP in four different ways which were eventually adopted as criteria for establishing the nucleotide's role as a mediator of hormone action in most other systems. First, as already noted, they demonstrated that glucagon did, in fact, elevate CAMP levels in liver cells. Second, they showed that CAMP is formed by the plasma membrane through the action of the enzyme adenylate cyclase (AC) (formerly called adenyl cyclase), and the hormone can activate that enzyme in broken membrane-containing fractions obtained from homogenized liver cells. Third, they established that adding CAMP to a liver cell preparation mimics the effects of the hormone; subsequently it has been found that a more potent effect can be obtained by using certain derivatives of the nucleotide that are believed to pass through the plasma membrane more easily and/or undergo a less rapid hydrolytic degradation. Finally, Sutherland and associates demonstrated that CAMP is degraded to 5′-AMP by the enzyme phosphodiesterase, which they found to be inhibited by caffeine or theophylline, and that if an inhibitor of that enzyme is added to a cell preparation, the hormone action is potentiated – i.e., by maintaining cell CAMP at artificially high levels.

By satisfying these criteria, CAMP has been identified as the messenger substance that mediates the actions of a number of hormones and other biologically active substances that serve as signals in a variety of cells; broadly speaking, it appears to play a role in regulation rather than "maintenance" of cellular activities such as ATP synthesis in the mitochondria; the maintenance activities are more or less self-regulatory through internal controlling mechanisms involving "feedback" systems in existing metabolic pathways.

Superimposed upon the maintenance activities is the influence of hormones and other biologic signals that reflect environmental changes. Through the generation of CAMP they modulate the specialized cell functions that subserve the fluctuating needs of the organism for survival in a changing environment. In the liver, as already noted, CAMP mediates glucagon action by modulating carbohydrate metabolism – stimulating glycogen breakdown, inhibiting glycogen synthesis, and promoting the formation of glucose from metabolites such as lactate and amino acids. In fat cells, it mediates the action of several hormones (epinephrine, glucagon, TSH, etc.) that stimulate lipolysis and inhibit lipogenesis. In pancreatic beta cells, it stimulates the secretion of insulin in response to glucagon and perhaps even glucose and other substances signaling for the release of this hormone. Similar processes involving CAMP are thought to govern many if not all other types of secretory cells. For example, the actions of most trophic hormones (ACTH, TSH, FSH, etc.) have been shown to be mediated through CAMP. Furthermore, the secretion of anterior pituitary trophic hormones stimulated by specific hypothalamic releasing factors can also be shown to be brought about through CAMP. In kidney tubule cells, CAMP stimulates the reabsorption of water in response to antidiuretic hormone; in the bones it mediates the action of parathyroid hormone in stimulating bone resorption and thereby regulates serum calcium. In the myocardium, it enhances both the rate and the force of contraction in response to epinephrine.

In addition, CAMP appears to be implicated in a number of processes of cell regulation (or, in one case, malregulation) in which hormones may not be involved. It is concerned in the phenomenon sometimes called "contact inhibition" of cell growth in that it "turns off" the proliferation of cells in culture under the appropriate conditions (a phenomenon we shall discuss in more detail later). There is also evidence from Mark Bitensky's laboratory at Yale that retinal cells' response to light involves CAMP (but perhaps CGMP more directly). Finally, it is now generally accepted that cholera toxin exerts its noxious effects on the intestinal mucosa by activating intestinal adenylate cyclase and thereby raising CAMP levels in this tissue. An interesting aspect of this finding is that here – in contrast to other situations – the activation of AC appears to be relatively permanent, which would of course explain the very severe and prolonged intestinal derangements in the disease.

An obvious and fascinating question is how a single molecule can manage to do so many different things. Here, unfortunately, we come into one of the less discernible areas in the jigsaw puzzle already alluded to. The detailed mechanisms of hormone-CAMP action have been worked out for only one system – the epinephrine- and glucagon-induced regulation of glycogen breakdown and synthesis. However, taking this system as a prototype – which seems reasonably safe at this point – we can deduce a good deal about the probable mechanisms by which CAMP serves as a mediator in so many diverse processes.

The action of glucagon on a liver cell or epinephrine on a muscle cell begins when a molecule of the hormone interacts with a receptor molecule on the cell surface. The receptor, when occupied by the hormone, in turn activates adenylate cyclase – but not, it appears, directly. There is consider-

The cyclic nucleotides consist of a purine base combined with ribose (a five-carbon sugar) and a phosphate group. Cyclic AMP is derived from ATP through loss of two phosphate groups and relocation of the third to form a ring attached to the ribose. Cyclic GMP, which differs from cAMP only in a portion of the purine group (shaded), is similarly derived from GTP (guanosine triphosphate).

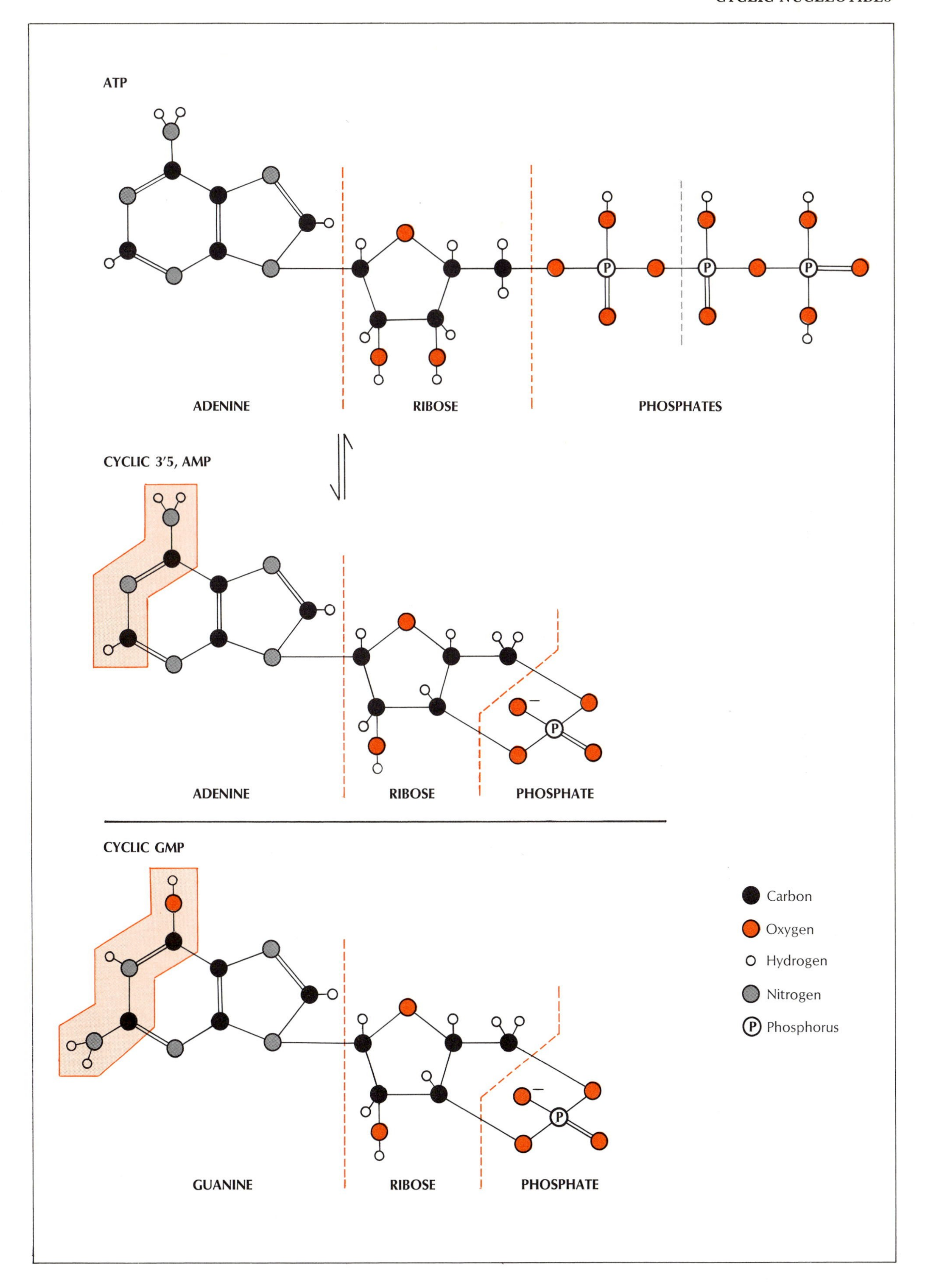

ATP
P
ADENINE
RIBOSE
PHOSPHATES
CYCLIC 3′5, AMP
ADENINE
RIBOSE
PHOSPHATE
CYCLIC GMP
GUANINE
RIBOSE
PHOSPHATE
Carbon
Oxygen
Hydrogen
Nitrogen
Phosphorus

able evidence from Martin Rodbell's laboratory at the National Institutes of Health that the activation involves an additional component situated within the membrane, of unknown nature, interposed between receptor and AC, which has been christened the transducer. Moreover, the transducer's activation of AC may not be (as one might perhaps expect) an all-or-nothing process, but is modulated by a number of substances. Chief among these is guanosine triphosphate (GTP), a chemical analogue of ATP and also – to get a bit ahead of our story – the chemical precursor of cyclic GMP. In both the liver membrane and some other systems, GTP in micromolar concentrations somehow sensitizes the receptor-transducer-AC system to hormone action; likewise, depleting the membrane of GTP minimizes activation of the enzyme. But how this action of GTP occurs – whether by affecting the transducer molecule or the enzyme itself – is not known.

Once activated, adenylate cyclase catalyzes the synthesis of CAMP from ATP. And CAMP, in turn, activates another enzyme – actually, a family of similar enzymes – known as protein kinase, which was discovered in the laboratory of Edwin Krebs at the University of California, Davis. This enzyme, Krebs and his colleagues have shown, consists of at least two subunits, one catalytic, the other regulatory or inhibitory; when the two are complexed together, the enzyme is inactive. Activation occurs when CAMP complexes with the inhibitory subunit, allowing the catalytic subunit to get about its business: the phosphorylation of certain proteins. The phosphorylation reaction involves the linking of phosphate groups (contributed by a molecule of ATP) to amino acids – namely, serine – of the protein. In muscle and liver cells, one of the proteins that undergoes phosphorylation is another enzyme, phosphorylase kinase, which is involved in glycogen metabolism (and perhaps the control of other cellular processes according to some recent discoveries by the Krebs group). When phosphorylated, phosphorylase kinase becomes activated and, in turn, catalyzes the phosphorylation of the final enzyme in the sequence, glycogen phosphorylase, which catalyzes the phosphorylation of the glycosyl units comprising glycogen into glucose-1-phosphate, which in the liver eventually becomes transformed into glucose. I should emphasize that it is the phosphorylation of phosphorylase kinase and glycogen phosphorylase that results in their activation, and when activated, each of these enzymes can promote another phosphorylation in which the "new" phosphate group is contributed by ATP. The process by which phosphorylase kinase and glycogen phosphorylase are inactivated involves the removal of the phosphate groups from the enzymes, a process catalyzed by enzymes known as phosphoprotein phosphatases. The protein kinase, activated as a result of the interaction with CAMP, is inactivated when the enzyme's inhibitor subunit recombines with its catalytic subunit. This occurs when the concentration of CAMP is no longer great enough to support complex formation with the inhibitor subunit. It is also possible that some other component may be generated that prevents CAMP from forming the complex.

All this may sound rather like a surrealist biochemical version of "The House That Jack Built"– this is the hormone that triggered the receptor that stimulated the transducer that activated the adenylate cyclase that formed the CAMP that activated the protein kinase that phosphorylated the phosphorylase kinase that phosphorylated the glycogen phosphorylase that phosphorylated the glycogen that yielded the glucose that ended up in the bloodstream. However, there seem to be at least two very important reasons for this elaborate chain of events: biochemical amplification and alternative regulatory input. An enzyme is a catalytic molecule and it seems to transform into product not just one molecule of its substrate, but many, referred to as the enzyme turnover number. Each of the enzymes involved in the glucagon-CAMP system can conservatively be estimated to turn over 100 molecules of substrate to product in a matter of only seconds. Thus the action of glucagon initiates a "cascade" in which *the effect of the original hormonal stimulus is amplified at least 100 times at each of the four enzymatic steps in the process*, for a total amplification of 10^8 or more; thus a single molecule of glucagon can theoretically trigger the release of about 100 million glucose molecules!

Hormones have long been known as extraordinarily potent substances; in the case of glucagon and epinephrine, at least, we may now know why. The intricate cascade of events also endows the system with the ability to respond to other messages that may be signalling – perhaps for different reasons than glucagon – for a change in the rate at which glycogen should be metabolized. For example, phosphorylase kinase, as we have seen, can be activated by protein kinase–catalyzed phosphorylation. Alternatively it can be activated by calcium in certain tissues. Neurogenic stimulation of skeletal muscle contraction results in an influx of cellular calcium. It is conceivable, in this case, that the energy-utilizing contractile event is coordinated with an enhanced availability of energy-producing glucose-1-phosphate through a calcium-induced activation of phosphorylase kinase that promotes a greater rate of glycogen breakdown. Therefore, activation of glycogen phosphorylase can be triggered at the second step in the cascade by a nonhormonal signal that employs calcium instead of CAMP as the key effector molecule.

As noted above, it is impossible to say at this point how closely the other CAMP-dependent processes in vertebrate systems resemble this model. All of them involve CAMP, and some of them, at least, have been shown to involve protein kinase as well, which is known to phosphorylate many different proteins, whether directly or through a secondary action of phosphorylase kinase or some similar enzyme. Paul Greengard of Yale has proposed that all effects of CAMP are expressed as a result of its activation of a protein kinase. In the liver cell,

Formation of glucose in response to glucagon or epinephrine involves a "cascade" of enzymatic steps, beginning with the activation of adenylate cyclase. Biochemical "amplification" (sequence at right) occurs at four of the steps, through the capacity of a single enzyme molecule to transform many molecules of substrate; amplification at each step has been estimated to be at least 10^2. Through this process, a single molecule of hormone can release at least 10^8 molecules of glucose.

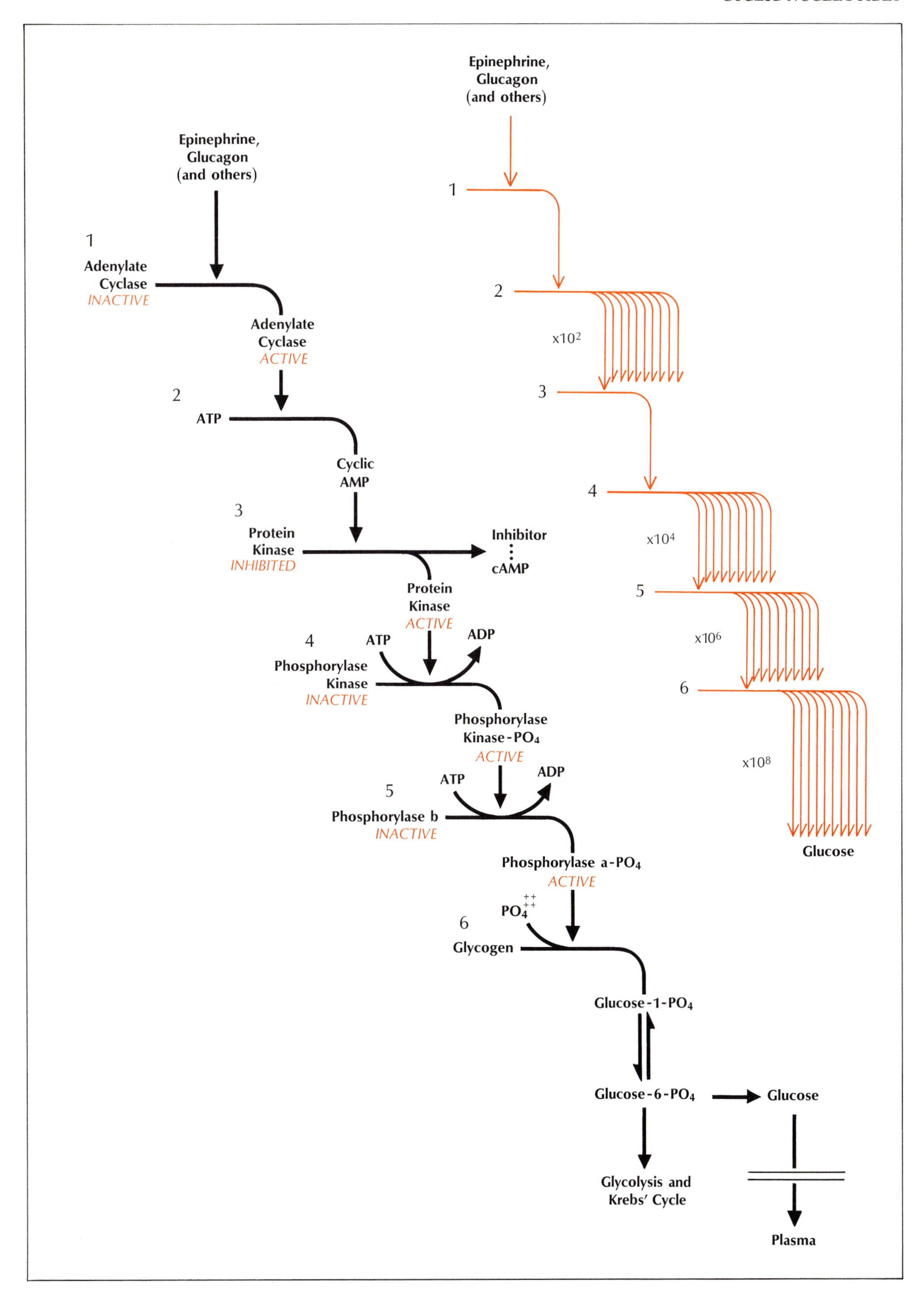
Epinephrine,
Glucagon
(and others)
1
Adenylate
Cyclase
INACTIVE
Adenylate
Cyclase
ACTIVE
2
ATP
Cyclic
AMP
3
Protein
Kinase
INHIBITED
Inhibitor
cAMP
Protein
Kinase
ACTIVE
4
ATP
ADP
Phosphorylase
Kinase
INACTIVE
Phosphorylase
Kinase-PO4
ACTIVE
ATP
ADP
5
Phosphorylase b
INACTIVE
Phosphorylase a-PO4
ACTIVE
PO4++
6
Glycogen
Glucose-1-PO4
Glucose-6-PO4
Glucose
Glycolysis and
Krebs' Cycle
Plasma
Epinephrine,
Glucagon
(and others)
1
2
x10²
3
4
x10⁴
5
x10⁶
6
x10⁸
Glucose

in fact, the protein kinase system also phosphorylates glycogen synthetase that catalyzes formation of glycogen rather than its breakdown. This appears contradictory, until we note that whereas phosphorylation *activates* glycogen phosphorylase it *inactivates* glycogen synthetase, i.e., the latter enzyme is more catalytically active in its dephosphorylated form. Thus the CAMP-stimulable protein kinase when activated simultaneously stimulates glycogen breakdown and inhibits its formation. This reciprocal or dual type of control on opposing, unidirectional metabolic steps is necessary, since otherwise the likely effect would be only an enhanced synthesis to keep pace with the accelerated breakdown —which would constitute a so-called metabolic short circuit.

As already intimated, there appears to be not one but a whole family of protein kinases – sometimes there are different species of protein kinase within a single cell. One distinguishing feature of the apparently different forms is that they exhibit strikingly different catalytic ability with different protein substrates. How different they are functionally is not yet clear. It has been shown that a protein kinase from one type of cell will phosphorylate proteins from other types of cells. Cross-reactivity can even be demonstrated with mammalian protein kinases and proteins from bacterial cells. It may be, therefore, that protein kinase somewhat resembles many other proteins (e.g., hemoglobin, myoglobin) in showing minor differences from one species to another, but also striking phylogenetic similarities. The question of their functional differences – even for the different forms within a given cell – remains to be resolved.

Functionally alike or not, the majority of the protein kinases described to date are activated by CAMP – specifically by formation (already mentioned) of a complex between the nucleotide and the kinase's inhibitory subunit, detaching the latter from its catalytic "partner." (A few exceptions have been uncovered where CAMP is inhibitory and some species are preferentially activated by CGMP). Once activated, the kinase(s) can phosphorylate certain plasma membrane proteins, including those from kidney tubules concerned perhaps in the transport of electrolytes and water. Kinase has been shown to phosphorylate a protein found in microtubules, structures involved in cell proliferation, secretion, and in the formation of certain organelles (see Chapter 16, "The Biogenesis of Organelles," by Holtzman), though what effect (if any) this may have on microtubule assembly or disassembly is still controversial. Protein kinase will phosphorylate microsomal proteins, though here again the functional significance of the process (indeed, of the protein itself) is not known. Krebs and coworkers have also shown contractile protein of skeletal and cardiac muscle to be phosphorylated – interestingly, more specifically by phosphorylase kinase than by protein kinase. Finally, protein kinase will phosphorylate the intranuclear acidic proteins and basic proteins called histones, whose functions are also somewhat controversial, though there seems little doubt that they interact with nuclear DNA and are intimately involved in the progression of events associated with the proliferative as well as the differentiative processes.

In the case of the histones, at least, it seems all species of protein kinase are not equal; indeed it is possible to differentiate them on the basis of which histone (or histones) they will phosphorylate. In addition, the process has not been shown to be always CAMP-dependent in vitro; that is, isolated protein kinases may phosphorylate the histone at the same rate in the presence or absence of CAMP. Here, however, it is likely that in at least some cases what has been isolated is only the active portion of the kinase molecule – i.e., it has presumably already been activated by CAMP. Thomas A. Langan at the University of Colorado has shown that injection of glucagon, epinephrine, or CAMP into an animal not only raises blood sugar (and activates phosphorylase kinase and phosphorylase) but also induces the phosphorylation of a specific histone, which, he proposes, prevents it from binding to and "masking" a site on DNA involved in the synthesis of messenger RNA (mRNA), which may code for enzyme proteins induced by glucagon (i.e., tyrosine transaminase). The CAMP–protein kinase pathway seems by all odds the most likely mechanism by which these phosphorylations are accomplished in vivo.

The combination of CAMP and protein kinase would help explain how the nucleotide – which in mammalian tissues is not yet known to interact directly with any substance but protein kinase – can produce so many diverse effects. But these of course depend in the first place on the distinctive characteristics of the cell itself: whether it is or is not sensitive to a particular hormone (i.e., whether its plasma membrane is equipped with the appropriate receptors), and, at the other end of the process, what proteins are present that can be phosphorylated by one or another of the protein kinases. In effect, CAMP could conceivably simply transmit the "message" it has received, via the receptor, from the hormone. Whether it receives a message at all depends on what receptors are present in the plasma membrane and what hormones impinge on them; where the message it "forwards," via protein kinase, eventually ends up depends on who is "listening" – on what proteins are available in the cell for phosphorylation. But the content of the message, so far as CAMP itself is concerned, appears to be much the same in most cases.

Another possibility worth considering at this time (with the knowledge that several different species of CAMP-stimulable protein kinases may exist within a given cell type) is that the particular species of the enzyme with which CAMP will interact – and therefore, the specialized protein that will undergo phosphorylation, which will, in turn, determine the cellular event affected – may be dependent upon other factors or specific combinations of factors. The latter determinants may be represented by such substances as the steroids, thyroid hormones, and prostaglandins – which are known to enter the cell interior readily – and, perhaps, the prevailing levels of certain electrolytes such as sodium, potassium, and, most importantly, calcium. Another interpretation of the view just expressed is that a combination of signals received by the cell may be integrated in such a way as to direct specific events that would not occur in response to any one of the signals alone or a different combination. The participation of a cyclic nucleotide could be envisaged in a variety of these combinations and, as stated above, through effects on specific species of protein kinase.

The emphasis on protein kinase as the cellular component with which cAMP may interact to express the action of hormones intracellularly is certainly justified from the reports that have appeared, especially with regard to mammalian tissues. It should be understood, however, that cAMP is present in the cells of lower phylogenetic forms, including bacteria, plants, slime mold, etc. The biologic events, in which cAMP seems to play a regulatory role in these phyla, are truly fascinating but will not be dealt with in this chapter. One feature of cAMP uncovered through the work conducted with these lower phyla, which should be pointed out because of the relevance it may ultimately have in mammalian systems, is the way cAMP functions in certain bacteria. This role has been beautifully developed by Ira Pastan and Robert Perlman at the NIH and by Geoffrey Zubay at Columbia University. Very briefly summarized, cAMP appears to act as a controlling factor that allows for the expression of specific genetic information in *Escherichia coli*. In this scheme its regulatory function does not involve a protein kinase, but instead a specific binding protein that, when complexed with cAMP, binds to a specific site on DNA. The interaction of the cAMP complex with DNA allows mRNA, which codes for the synthesis of specific enzyme proteins, to be synthesized by RNA polymerase. From this information, it may be concluded that in *E. coli* the actions of cAMP are not mediated through activation of a protein kinase – and this conclusion raises the possibility

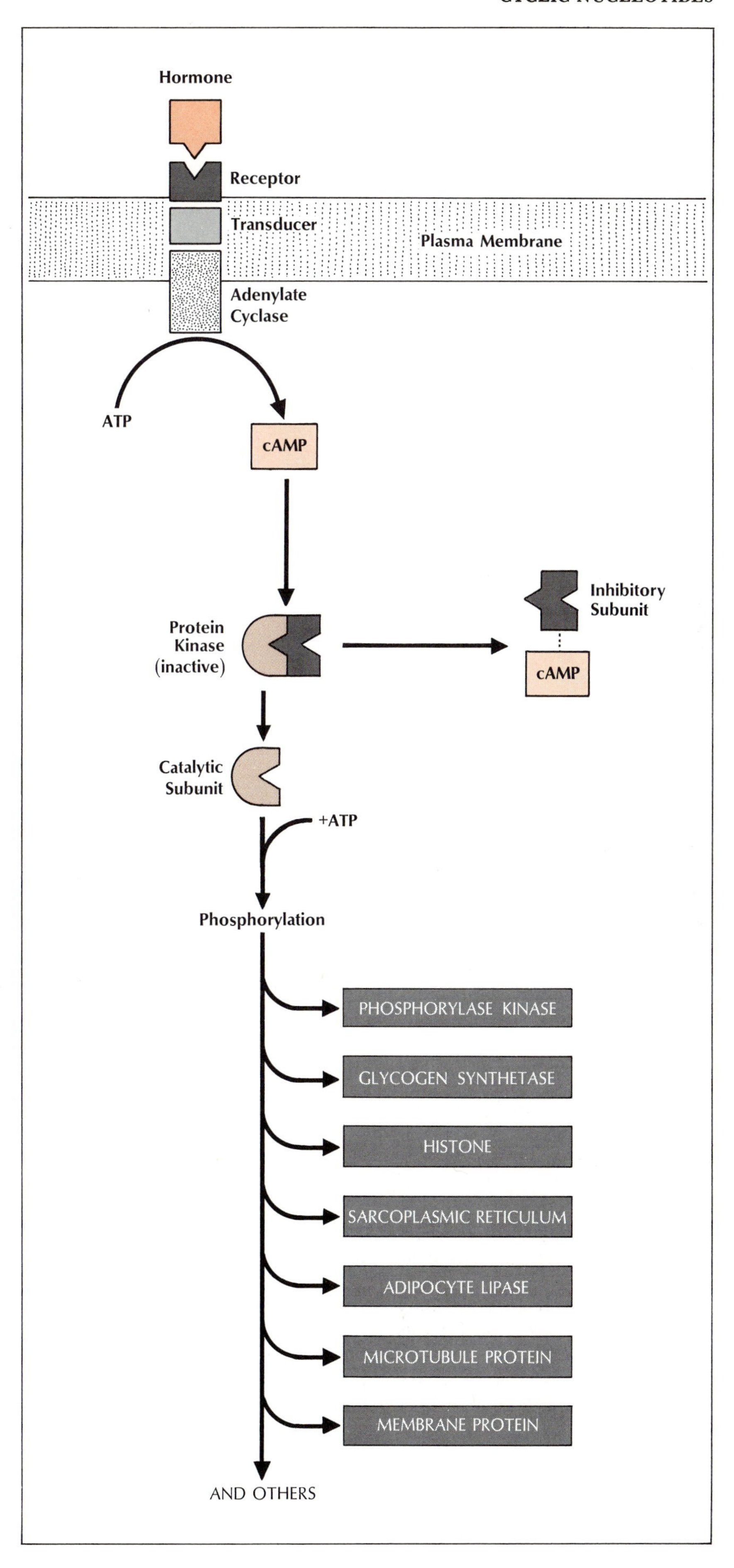

"Messenger" function of cAMP in a number of systems is thought to depend on its capacity to activate the class of enzymes known as protein kinase, by dissociating an inhibitory portion of the enzyme molecule. The activated kinase, in turn, activates or inativates various enzymes or other cellular proteins by phosphorylating them. Which proteins are thus transformed is believed to depend both on which are present and also on what particular kind (or kinds) of protein kinase is found in the particular cell.

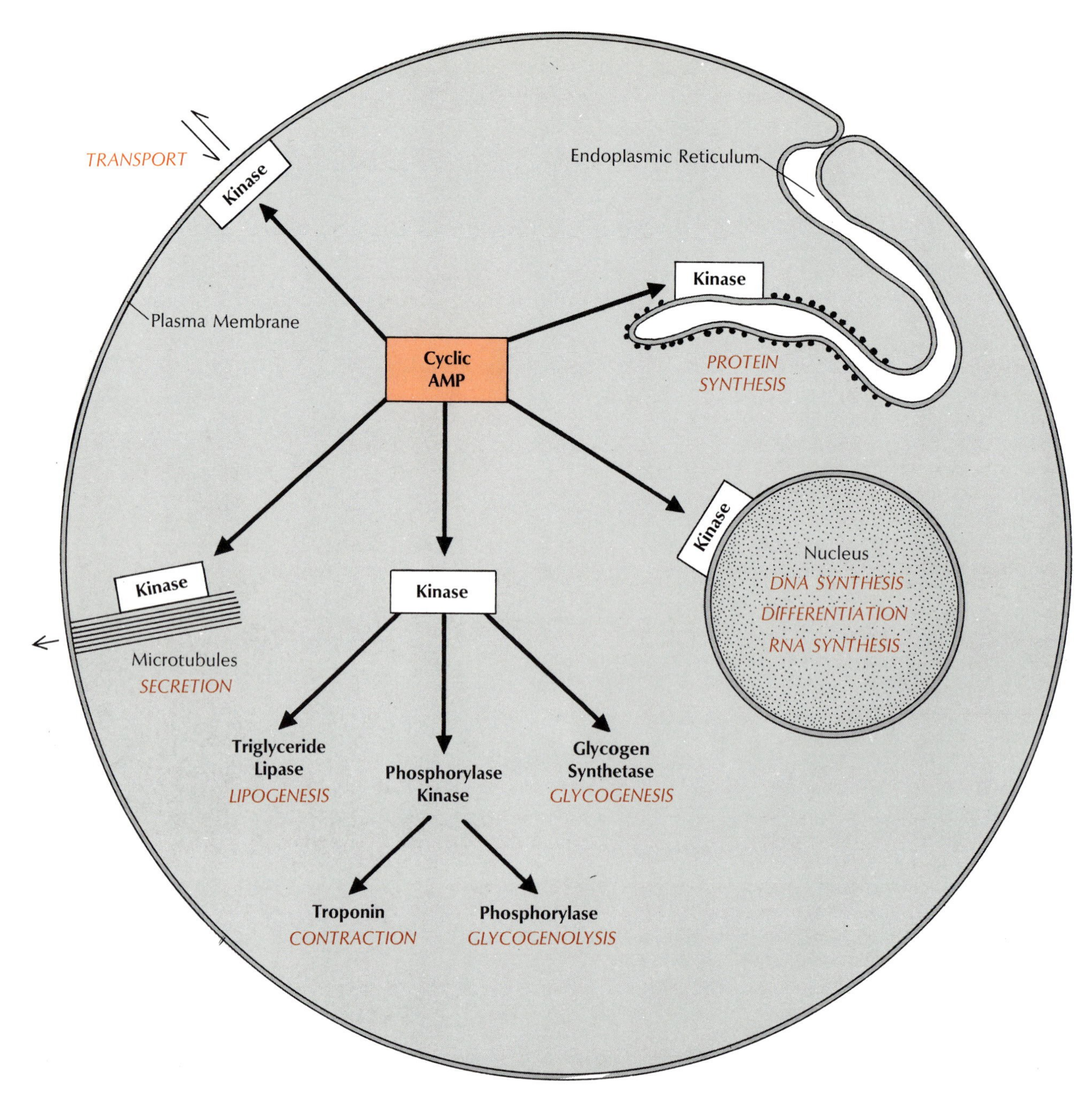

Protean quality of cAMP, resulting from its capacity to activate one or a variety of protein kinase, is thought to involve it in modulating processes in many different parts of cell, as schematically suggested here. (Only a few such processes will be going on in any given type of cell.) Most of these cAMP-kinase activities are based on inference; details are known for very few.

that in mammalian systems certain of its actions may also involve a component other than protein kinase.

There can be no doubt that cAMP acts as the second messenger for some hormones. But does it serve in this capacity for all hormones that require a cell-membrane-to-interior component? It is well known that hormones frequently occur in pairs that act in opposing ways – e.g., parathyroid hormone and calcitonin, or (to take an even more familiar example) glucagon and insulin, the first of which breaks down glycogen, while the latter builds it up. More specifically, it has been shown that whereas glucagon action brings about an activation of glycogen phosphorylase and inhibition of glycogen synthetase by stimulating the phosphorylation of both molecules, insulin action leads to the inactivation of the phosphorylase and the activation of the synthetase – conceivably by dephosphorylating both molecules. Since glucagon produces its effects by increasing cellular levels of cAMP, it would be easy to conclude – as many investigators have – that insulin produces its effects by *lowering* cAMP levels, either by inactivating adenylate cyclase, which catalyzes cAMP formation, or by activating phosphodiesterase, the enzyme that breaks it down.

This unitary hypothesis of "bidirectional" control through changes in cAMP levels alone was troublesome to me and my associates because it implied among other things that some critical level of cAMP must be necessary for normal or basal cell function.

And this did not square very well with the generally accepted view that hormones serve to modulate rather than maintain cell function, considering that CAMP was depicted as a mediator of hormone action (though we now know that this is probably too narrow a definition of the biologic roles played by the cyclic nucleotides). On testing the hypothesis with Joseph Larner in 1967, we found that in both muscle and liver cells one could obtain full expression of insulin action either without any lowering of CAMP levels or before any lowering was detectable. Our natural conclusion was to suspect that some other messenger-like substance might be involved, the most logical candidate being another cyclic nucleotide. Our first guess, based on earlier knowledge, was that it might be cyclic 3′,5′-uridine monophosphate (CUMP). But almost a year of experiments turned up no evidence that this substance was present in mammalian tissues – nor have others subsequently identified CUMP or two other nucleotides, cyclic 3′,5′ inosine monophosphate (CIMP) and 3′,5′ thymidine monophosphate (CTMP).

Our attention was directed next to cyclic 3′,5′-guanosine monophosphate. This substance, which as noted is similar to CAMP with the exception that the purine base guanine replaces the adenine ring of the latter, had been shown to exist in nature in 1963, when a group at Columbia headed by Price found it in rat urine. However, it was not known where the urinary CGMP originated, and there was no notion about what its importance, if any, might be. Thus, whereas Sutherland's original work might be thought of as starting off with a particular function (stimulated glycogenolysis) in search of a compound, the CGMP story began with a compound in search of a function.

Our attempts to elucidate its possible biologic importance began by devising an analytic system sufficiently sensitive and specific to detect and quantitate the presence of CGMP, whereby we could determine if it was present in mammalian tissues, and if so, in which tissues. This was accomplished in 1969, when we reported that CGMP is a naturally occurring component of several mammalian tissues, including brain, liver, and kidney. This was soon confirmed by Sutherland's laboratory, which also reported that CGMP was present in insect tissues and identified an enzyme, guanylate cyclase (GC), in mammalian tissues that promoted CGMP synthesis from GTP. Joel Hardman with Sutherland at Vanderbilt and Arnold White, then at NIH, showed that GC was distinguishable from adenylate cyclase on the basis of cellular distribution (AC was present only in membrane fractions, but GC in both membrane and cytoplasmic fractions), ionic requirements, and responsiveness to the action of modifiers such as hormones (GC activity was not stimulated by the hormones that stimulated AC activity).

These early experiments also revealed what might be considered to be one of the major roadblocks in studying CGMP: its extremely low levels in tissues. In urine, as a result of about a 100-fold concentration by the kidneys and other as yet unknown factors, the CGMP level is in the same range as that of CAMP (about 1 μmole/liter), but in tissues the concentrations are usually less than those of CAMP by a factor of up to 100. Monitoring changes in CAMP levels has always been very difficult, since it requires measurements accurate to something like 10^{-13} moles (most analytic procedures for tissue metabolites will do no better than about 10^{-9} moles). With CGMP, it has proved necessary to increase the sensitivity into the range of 10^{-15} moles if the substance is to be detected in a reasonably sized sample of material – and the analytic difficulties increase almost at the exponential rate of the sensitivity. A big step forward was recently made with the development of a radioimmunoassay for CGMP, by Alton Steiner and his coworkers at Washington University in St. Louis.

We were able to show at the outset that hormones such as glucagon or epinephrine, which elevate CAMP levels in liver and some other cells, had little if any effect on the cellular concentrations of CGMP. But all that we could conclude at this point was that CGMP probably had a biologic role much different from that of CAMP. Our initial attempts to involve CGMP in the expression of insulin action were unsuccessful because of a failure to appreciate the subtleties that now appear to characterize the cellular functions of CGMP.

Some of these subtleties did start to become appreciated by my associates and me quite early in the game, when reports began to appear from other laboratories describing the effects of CGMP added to whole-cell preparations or tissue homogenates to ascertain whether the cyclic nucleotide would mimic the effects of any hormone. It was found, in fact, to produce either no effect at all or to duplicate the effects of CAMP – it was merely a poor substitute for it. Indeed, at this writing reports continue to suggest that the effects of 10^{-3} molar CGMP in certain systems are similar or identical to those produced by CAMP – which we consider to be an unfortunately misleading representation of CGMP.

These findings, while certainly unpromising, were not wholly unexpected so far as we were concerned. From earlier work with CGMP we were aware that it was far less soluble than CAMP, both in lipid and in aqueous media (there is a good teleologic reason for this as will be seen in a moment). Accordingly, the chance of effective amounts of CGMP passing through the plasma membrane of a number of cell types seemed unlikely. In cell-free systems, Keith Schlender (then working at the University of Minnesota) was able to show that CGMP could indeed mimic the effects of CAMP – specifically, by activating an enzyme then known as glycogen synthetase-I kinase by the Larner group that was subsequently found to be identical with protein kinase. Considering the chemical kinship of the two substances, this was not very surprising, but the important fact that emerged from these studies was that something like 200 times as much CGMP was required to mimic CAMP action – and this was more than 2,000 times as much CGMP as was actually found in cells. We concluded, therefore, that the CAMP-mimicking effect of CGMP in this system was not representative of its "normal" function, and that the reported CAMP-like effects of CGMP probably resulted from a nonspecific effect, when employed at the extremely high, artificial concentrations reported by others.

Having gained an appreciation of

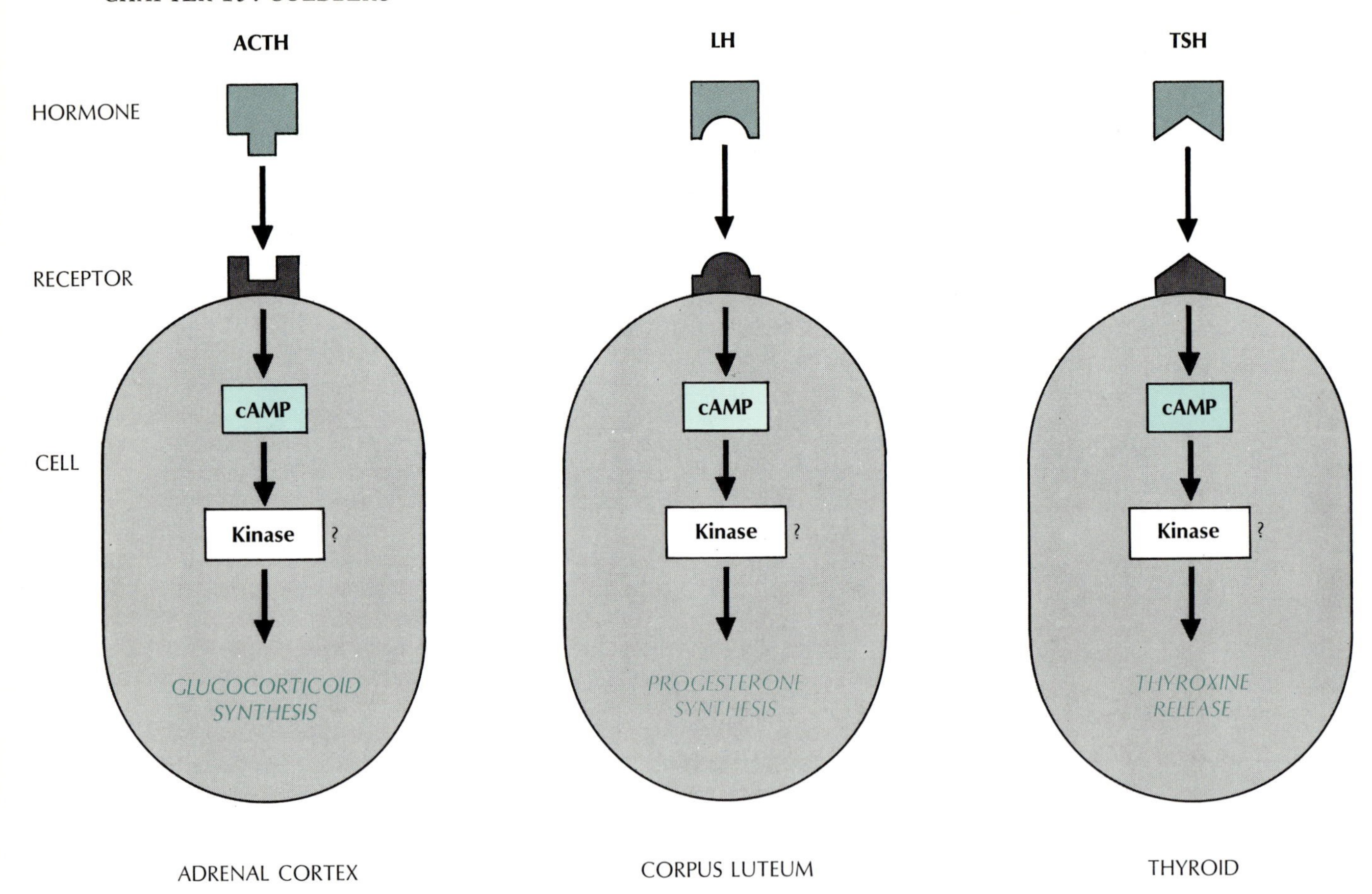

Differential response to hormones of various types of cells depends in the first place on what hormonal receptors are present on the cell surface; each hormone is thought to be able to trigger only its "own" receptor. Cyclic AMP then serves as a "common pathway" to effectuate many of the subsequent cellular reactions. In some cases (e.g., PTH), the same hormone-receptor combina-

what the pitfalls might be in elucidating the biologic role of cGMP by "dumping" it into intact or cell-free systems, we returned to the more tedious approach of attempting to find a biologically active substance that would stimulate the cellular accumulation of cGMP in association with a definable alteration in cellular function. After more than a year of failure we finally succeeded in identifying such an agent – acetylcholine.

In many physiologic systems this substance acts oppositely to epinephrine (whose beta-adrenergic actions had already been shown to be mediated by cAMP); thus in the myocardium, for example, acetylcholine diminishes both the rate and the force of contraction while epinephrine increases them. We managed to show that acetylcholine, in physiologic concentrations, produced an almost instantaneous rise in the cGMP of something like 300% in the rat heart – coincident with its characteristic depression of cardiac function. This work was accomplished with the help of two postdoctoral fellows, William George and James Polson. We then showed that isoproterenol, an analogue of epinephrine, depresses myocardial cGMP levels at the same time as it elevates cAMP concentrations. Finally, we found that acetylcholine, along with elevating cGMP, will depress somewhat the levels of cAMP.

These findings revived our original notion, which had been put aside after our initial failure: the theory that cGMP functions as the regulatory counterpart of cAMP, but in the opposite sense – as we put it in a later paper, the Yin and the Yang of the biologic world. In pursuit of this notion we tested a whole series of other systems. I might add at this point that I was extremely fortunate in acquiring the help of an unusually talented colleague, Mari K. Haddox, who came to our laboratory in 1968 at the age of 17 as a Minnesota Heart Association Summer Fellow. During the ensuing six years, she has been instrumental in refining the analytic procedures, helping to determine the course of the investigations, and pursuing them to their logical conclusion.

Tests of other systems established that the association between cholinergic stimulation and enhanced cellular accumulation of cGMP was by no means unique to myocardial tissue; the same turned out to be true when we examined uterine smooth muscle, lung, brain, and human peripheral blood lymphocytes. Next, we determined that not just acetylcholine but a number of other hormones can raise cGMP levels; these include oxytocin, calcitonin, and insulin, all polypeptide hormones; two other neurohumors, serotonin and histamine, as well as epinephrine (i.e., the alpha-adrenergic variety) and prostaglandin F_2 alpha. In all cases, the effect of the cGMP-linked hormone is opposite to that of one linked to cAMP, as in the acetylcholine-epinephrine contrast already mentioned. A number of these results have been confirmed while the obser-

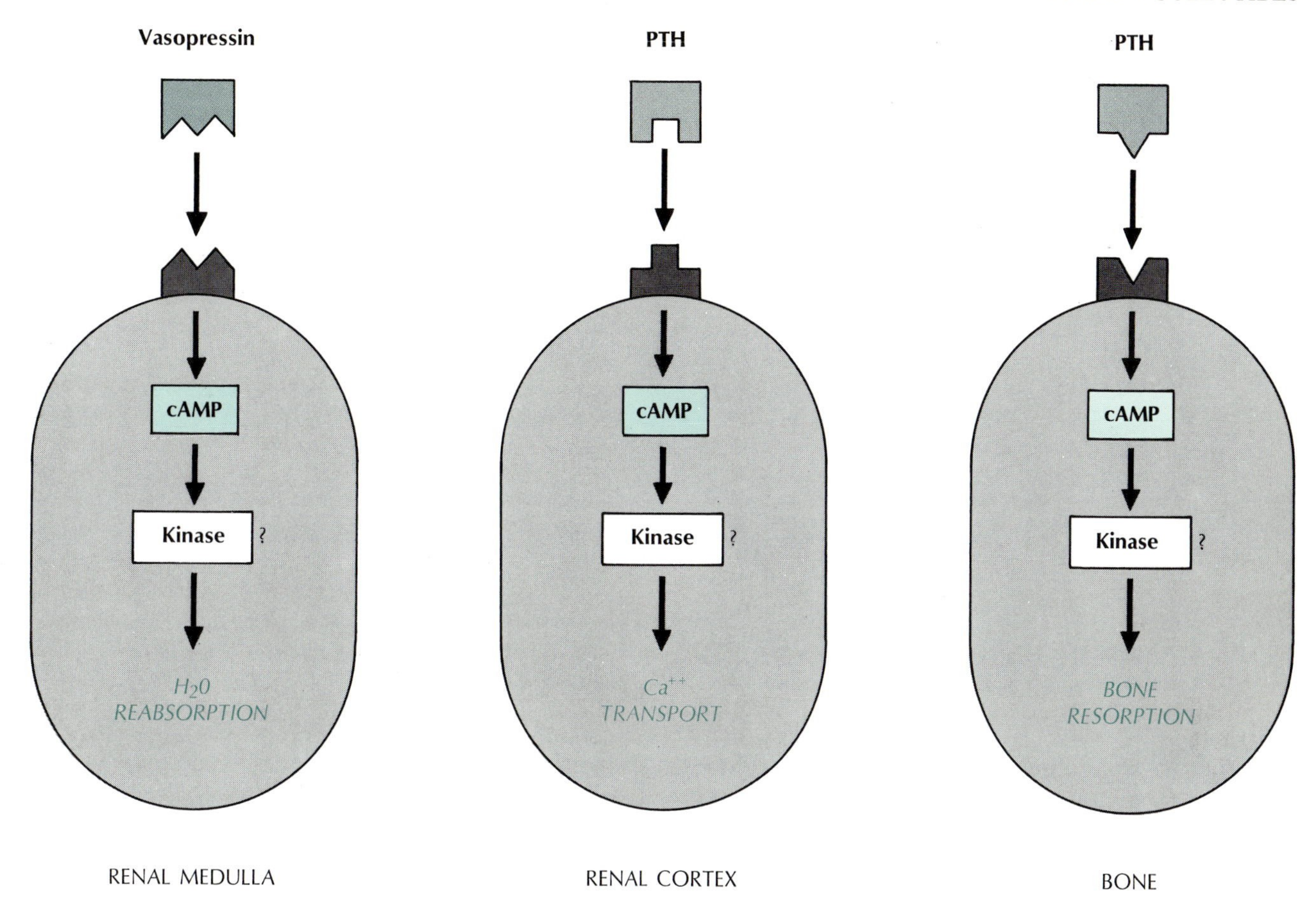

tion may trigger different reactions in different types of cell, though the mechanism is not known. In a number of the processes shown, the hormone also induces Ca^{++} influx; in others, such as smooth muscle contraction, Ca^{++} efflux accompanies the hormone action. In some processes also, cyclic AMP itself is believed to modulate Ca^{++} influx and efflux.

vations have been extended to include the actions of hormones such as angiotensin and cholecystokinin.

Of particular interest is the demonstration by Guenther Schultz and Joel Hardman at Vanderbilt, and later by our group, that epinephrine action of the alpha-adrenergic type is associated with an increase in cellular cGMP. This for the first time establishes a plausible rationale for the two different kinds of adrenergic action. It has been known for quite some time that there are two kinds of adrenergic receptors in the body, producing opposite types of cellular responses, called alpha and beta. Sutherland had already shown that the beta receptors (e.g., in myocardial cells) are linked to cAMP generation; it now appears that actions deriving from alpha-adrenergic receptor stimulation are linked to cGMP. Currently we are hoping to demonstrate a similar contrast between the two types of histamine receptors, called H-1 and H-2. In addition, we have hopes of linking the two nucleotides to substances that produce one response in some types of tissue but an opposite response in others. In this area we have achieved partial success by showing that prostaglandin E_2, which can *relax* vascular smooth muscle in association with an increase in cAMP and a lowering of cGMP, stimulates *contraction* in association with an increase in cGMP.

Now that it has been demonstrated that many hormones can raise intracellular cGMP levels, a number of researchers, including some in my laboratory, have again tackled the problem of mimicking hormone actions in cells by adding exogenous cGMP – and with a good deal more success than in the past; in fact, many of the hormone actions that have been shown to involve increases in endogenous cGMP can now be shown to be mimicked with the exogenous nucleotide. Obviously this raises the question of why the earlier experiments were so unsuccessful.

One reason seems to be that cGMP is effective only within a relatively narrow range of concentrations – in a number of cases, extraordinarily low ones. The earlier experiments involving the exogenous nucleotide employed concentrations in the 10^{-3} molar range; the later ones used levels generally below 10^{-5} molar and often as low as 10^{-10} or even 10^{-11} molar. For example, John W. Hadden and Leslie Johnson, at Sloan-Kettering Memorial Hospital, have shown that cGMP in the range of 10^{-8} to 10^{-11} molar in the presence of calcium increases DNA-dependent RNA polymerase I activity in isolated lymphocyte nuclei. Whitfield and McManus at the Canadian National Research Institute were among the first to demonstrate effects of cGMP in the 10^{-9} to 10^{-11} molar range as a promoter of DNA synthesis in intact thymic lymphocytes. Albert Wollenberg in East Germany demonstrated acetylcholine-like effects of cGMP on embryonic heart cells in culture, and Frank Austen's group and Terry Strom at Massachusetts General Hospital showed consistent effects of relatively low cGMP concentrations (or a less readily metabolized derivative) in mimicking the action of acetylcholine and other hormones that enhance mast-cell release of histamine or increase the cytotoxic action of lymphocytes. Gerald Weissmann at NYU has shown that cGMP and acetylcholine enhance the microtubule-dependent release of lysosomal enzymes from human leukocytes: an effect opposite to that of cAMP or beta-adrenergic agents. Harry Hill and Richard Estenson in my group have shown that cGMP at concentrations between 10^{-6} and 10^{-9} molar stimulate human leukocyte chemotaxis, while at concentrations above 10^{-4} molar it produces inhibition of cell motility – as occurs with cAMP at any concentration. The role that these cyclic nucleotides play in the central nervous system has long been a major area of study, because both of them occur in relatively high concentrations in brain tissue. A major step in clarifying this elusive question may have been taken by the laboratory of Paul Greengard, where it was demonstrated that cAMP can hyperpolarize the postsynaptic membrane in superior cervical ganglia while cGMP depolarizes it.

To sum up: in all of the systems just mentioned, in which cGMP can be shown to have an effect on cell function similar or identical to the effect of a hormone that can elevate cellular concentrations of the nucleotide, it has also been shown that cAMP (or hormones that stimulate its generation) has an opposite effect. Another point to note is that in most of the intact cell systems in which opposing effects of the two cyclic nucleotides have been demonstrated, the process affected must first be initiated by a "primary promoting substance" – such as a chemotactic factor in the case of leukotaxis, complement in the case of lysosomal enzyme release, etc. The cyclic nucleotide under these conditions seems, therefore, to serve as a modulator of the cellular process set in motion by a nonhormonal factor. This is not an unexpected role for the cyclic nucleotides if they are truly acting as mediators of the action of hormones that are only to modify cellular function. Whether the cGMP and the cAMP normally present within the cell also participate in setting the machinery in motion is not yet known and will be very difficult to ascertain in the future.

While the use of lower concentrations of exogenous cGMP seems to be an important factor in the success of the experiments just described, the reason why it is often ineffective is not yet understood. One explanation, which was pointed out earlier, may involve the impermeability of the plasma membrane to this nucleotide. Another view recently offered – which I will discuss more fully later on – is that many if not most of the actions of cGMP require an influx of cellular calcium; thus adding cGMP alone, without providing this additional companion component to the cell interior, would preclude the expression of the nucleotide's action. As of this writing, however, there is no really satisfactory answer to these questions; all we can say is that cGMP works – when it does work – often at improbably minute concentrations.

One of the most fascinating aspects of the cyclic nucleotides is their apparently antagonistic regulatory influence on cell proliferation. It has been known for a number of years, thanks to the work of such researchers as Ira Pastan, his coworkers at NIH, and John Sheppard at the University of Minnesota, that the dibutyryl derivative of cAMP, or agents that elevate cAMP levels, inhibit cell proliferation. For example, cultures of malignant cells, exhibiting no contact inhibition, will acquire such inhibition when treated with dibutyryl-cAMP or with prostaglandin E_1, which stimulates endogenous cAMP production in these cells. It has also been shown that normal cells transformed, for example, by viral infection have diminished cAMP levels.

If one views cGMP, as we do, as the physiologic opposite of cAMP, one would expect it to be involved in stimulating cell proliferation, and the evidence is mounting that this may indeed be the case. In one series of experiments, for example, John Hadden and I treated human peripheral blood lymphocytes with phytohemagglutinin or with concanavalin A, both well known as promoters of lymphocyte proliferation, and found that intracellular cGMP concentrations increased by 10 to 50 times. Larger increases have recently been reported to occur with rat lymphocytes, and qualitatively similar effects on cGMP in comparable cells from mice upon exposure to concanavalin A. Both experiments incidentally showed little or no change in cAMP levels, which suggests that cGMP under certain con-

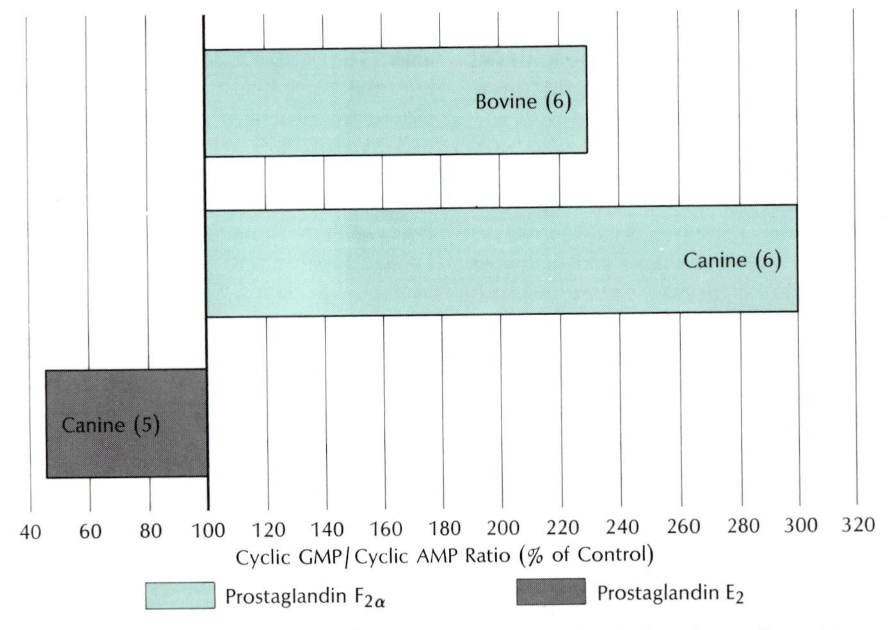

Changes in levels of the two cyclic nucleotides correlate with physiologic effect of two types of prostaglandin (PG) on vein smooth muscle. $PGF_{2\alpha}$ constricts the veins and also raises cGMP: cAMP ratio; PGE_2 relaxes veins and reduces the ratio. Numbers on the bars denote the number of animals in each experiment.

ditions may represent a component in at least some cell types that serves to trigger the cellular events involved in the proliferative process.

It is worth noting that this trigger effect is not inconsistent with the notion of the cyclic nucleotides as mediators between the cell and its environment; in the case of lymphocytes, at least, we know that their proliferation occurs when they receive the proper external signal, which is an antigen. Whether the nucleotides are involved in "normal" cell proliferation – e.g., the regular replacement of epithelial cells in the intestinal mucosa – is another question, and one that as yet cannot be answered.

Similar involvement of cGMP with growth has been demonstrated in other systems. For example, in collaboration with Carlos Lopez and John Hadden, we have shown that insulin in high doses can act as a mitogenic agent – and when it does so (e.g., in cultured fibroblasts) produces a 10- to 40-fold rise in cellular cGMP. Rapidly proliferating fibroblasts in culture exhibit markedly higher levels of cGMP than the same cells that have become contact-inhibited and have ceased to proliferate. The addition of fresh serum to the confluent, contact-inhibited cells stimulates their proliferation and causes a concomitant increase of several-fold in cGMP levels while lowering levels of cAMP. Phorbol myristate acetate (the active principle of croton oil) is a cocarcinogen (i.e., requires an "inducer" substance like benzpyrene to produce carcinogenesis), but will also stimulate a round of division in fibroblasts by itself; Richard Estenson has shown that it produces a several-fold rise in cGMP. A University of Michigan researcher, Raymond Ruddon, has shown that benzpyrene-induced two-stage carcinogenesis is associated with the induction of a microsomal enzyme called arylhydrocarbon hydroxylase, but induction occurs only in the presence of active cell proliferation or of an inducer such as phorbol. Interestingly, very low concentrations of cGMP (10^{-8} molar) have been shown to substitute for whatever the requirement is that derives from rapid proliferation in this system.

The relationship between cGMP and cell proliferation has also shown up in psoriasis, a pathologic state that involves excessive epidermal cell proliferation. In collaboration with the University of Michigan dermatologist John Voorhees, we have shown that psoriatic cells contain at least twice as much cGMP as nonpsoriatic cells from the same patients. Dr. Voorhees has also reported that there is a significant deficiency in cAMP in cells from psoriatic lesions.

The relationship between cGMP and proliferation is not limited to mammalian cells. In at least two strains of bacteria, for example, we have found that cGMP levels are high during rapid (i.e., log-phase) growth; when the growth rate diminishes – because of a depletion of nutrient in the medium – the nucleotide levels fall to about 10% to 20% of the original level, while those of cAMP rise, if that nucleotide is present. (In the case of bacilli, no cAMP has been detected by anyone who has attempted to measure it.)

We have observed similar relationships in plants. In pea and bean seedlings, for example, there are regions in which cells are actively proliferating, e.g., at the tips of roots and shoots (called meristematic regions) and regions in which cells are no longer dividing but are merely undergoing elongation. We have found that cGMP levels are four to 50 times higher in the tips of both roots and shoots than in the elongating cells.

Finally, in collaboration with Fred Kuehl of the Merck Institute, Constance Sanford and Susan Nicol in my laboratory have demonstrated that steroid hormone–induced proliferation may also involve cGMP. Estrogens, for example, are well-known stimulators of proliferation in uterine endometrium, and we have found that when ovarectomized rats are treated with either estradiol or diethylstilbestrol, cGMP levels in the endometrial cells increase markedly while cAMP concentrations drop. Moreover, in monitoring cyclic nucleotide levels at each stage of the estrous cycle of the rat, uterine cGMP levels were found to be highest during proestrus, when plasma estrogen levels are also at a peak, while cAMP levels are minimal at this time. Progesterone, when coadministered with estradiol or clomiphene (which is antiestrogenic), blocks both estrogen-induced proliferation and the enhanced accumulation of cGMP.

These findings may seem to be somewhat inconsistent with some of the introductory remarks depicting cyclic nucleotides as intracellular messengers of agents that are not internalized by the cell but interact with receptors on the cell membrane. The observations should, however, be viewed as an extension of Sutherland's original hypothesis, for they indicate that cyclic nucleotides may be

Enhanced Cell Proliferation Associated with Elevated Levels of Cyclic GMP

Cell Type	*Mitogen or Condition*
Lymphocyte	Phytohemagglutinin Concanavalin A
Fibroblasts	Insulin Phorbol Serum
Epidermis	Histamine Phorbol Psoriasis
Uterine Endometrium	Diethylstilbestrol Estradiol-17β
Escherichia coli Bacillus licheniformis	Log Growth
Plant Roots and Stems	Meristematic Regions
Slime Mold (Amebae)	?

involved in the expression of at least some of the actions of agents such as steroids, which enter the cell interior. This conclusion would not be inconsistent with the fact that guanylate cyclase is located both in membranous and cytoplasmic fractions of broken cell preparations. If cGMP can be implicated in mediating some of the actions of steroids, is there evidence that cAMP may also be involved in mediating certain actions of this class of hormones, and are the actions elicited by steroids that appear to involve cAMP opposite to those that involve cGMP? From the evidence now accumulating, the answer would appear to be affirmative on both counts.

Steroid Action

First, there have been several reports that adenylate cyclase is not (as originally thought) confined to only the plasma membrane but is also an identifiable component of intracellular organelles such as the nucleus and endoplasmic reticulum. Furthermore, in a report by Coffey and Middleton of the Children's Asthma Research Institute and Hospital in Denver it was shown that anti-inflammatory steroids elevate lymphocyte cAMP several-fold – and we and others have found them also to prevent phytohemagglutinin-induced increases in cGMP.

These observations are of particular interest in the context of regulating cell proliferation as well as the mechanism of steroid hormone action, since anti-inflammatory steroids exhibit potent antimitotic effects on lymphocytes. Recalling that increasing cellular cAMP can be inhibitory to the proliferative process and to the release of lysosomal enzymes, it may be difficult in the future to avoid implicating the cyclic nucleotides even in steroid action. At the risk of belaboring the point, I should also mention that an opposite effect of an anti-inflammatory steroid – stimulating proliferation in a selected line of cultured cells – has been reported by Gordon Sato's group at the University of California to be associated with marked increases in cGMP levels.

At this point, however, we reach a large area of terra incognita concerning cGMP. If hormones can increase cytoplasmic levels of this nucleotide, then one would expect that the same hormones would stimulate the enzyme, guanylate cyclase, that catalyzes its formation – as is known to be the case with adenylate cyclase. This has indeed been difficult to demonstrate although some progress has recently been achieved. Rudland, Seifert, and Gospodarowicz at the Salk Institute have shown that the hormone, fibroblast growth factor (FGF), which Gospodarowicz has isolated from pituitary and brain, produces a marked activation of guanylate cyclase in membranes isolated from mouse fibroblasts. This observation is consistent with the demonstration by this group that FGF produces greater than a tenfold increase in fibroblast cGMP concentrations in association with stimulating these cells to proliferate.

Probably the first bona fide activator of guanylate cyclase thus far identified (by Schultz and Hardman) is Ca^{++} ion, which also tends to inactivate adenylate cyclase in some systems; this finding has led researchers to suggest that Ca^{++} may be the "real" second messenger, especially in the cGMP system, operating by modulating the enzymes involved in the formation and degradation of cGMP and cAMP. A concept involving Ca^{++} as a messenger molecule is discussed extensively in Chapter 20 by Dr. Howard Rasmussen, so that my attention to it at this point will be minimal. However, I feel a few comments regarding calcium and the cyclic nucleotides are in order at this time, at least to establish a point of view held by cyclic nucleotide researchers. First of all, I should like to make clear my belief that there is indeed a very intimate relationship between cyclic nucleotides and calcium in the modulation of cellular function induced by hormones or other biologically active substances. There may be instances where calcium or a cyclic nucleotide alone may appear to be sufficient to promote the event, but I believe that many more instances will be found where the two act in concert.

The latter situation is best illustrated by the cellular events "turned on" by hormones that also promote cGMP accumulation. From our present knowledge, these would include smooth muscle contraction (uterine, vascular, and intestinal), platelet aggregation, cell proliferation, histamine release, lysosomal enzyme release, cytotoxic response of lymphocytes, leukotaxis, and nuclear RNA synthesis. All of these processes are not only associated with enhanced cGMP but also require the presence of calcium; moreover, in the few cases examined by Schultz and Hardman, cellular cGMP levels cannot be elevated, and in some cases basal cellular levels of the nucleotide decline markedly following removal of calcium from the medium. These observations would indicate that calcium is required for maintenance of cGMP levels – and therefore, probably, for its formation. This is consistent with the observation cited earlier that Ca^{++} ion can be shown to stimulate guanylate cyclase. It also seems likely, however, that at least in the case of smooth muscle, calcium is required not merely for maintaining cGMP levels but also for the expression of cGMP action (i.e., contraction). Or cGMP may be necessary for the expression of calcium action.

In a very simplistic model to explain how cGMP may be involved in promoting an event requiring calcium – such as smooth muscle contraction – cGMP could be envisaged to act by promoting a greater passive influx of calcium through an action on the cell membrane and by promoting its release (or preventing its active uptake) by intracellular storage structures, the net effect of cGMP in this type of system being to enlarge the intracellular pool of free Ca^{++}. Because cAMP has the opposite influence on smooth muscle contraction (i.e., relaxes it), its influence on Ca^{++} flux would be opposite to that of cGMP – it would bring about a suppression of extracellular Ca^{++} influx and promote its active uptake by intracellular storage structures. In secretory systems where cAMP in concert with Ca^{++} promote the process – the influences of cAMP and cGMP on Ca^{++} flux described above would be reversed. Some credit should be given for at least the cAMP half of this model to Dr. Michael Berridge of Cambridge University who has a keen insight into the working relationship between cAMP and Ca^{++} and with whom I have had fruitful discussions.

It might also be worth considering

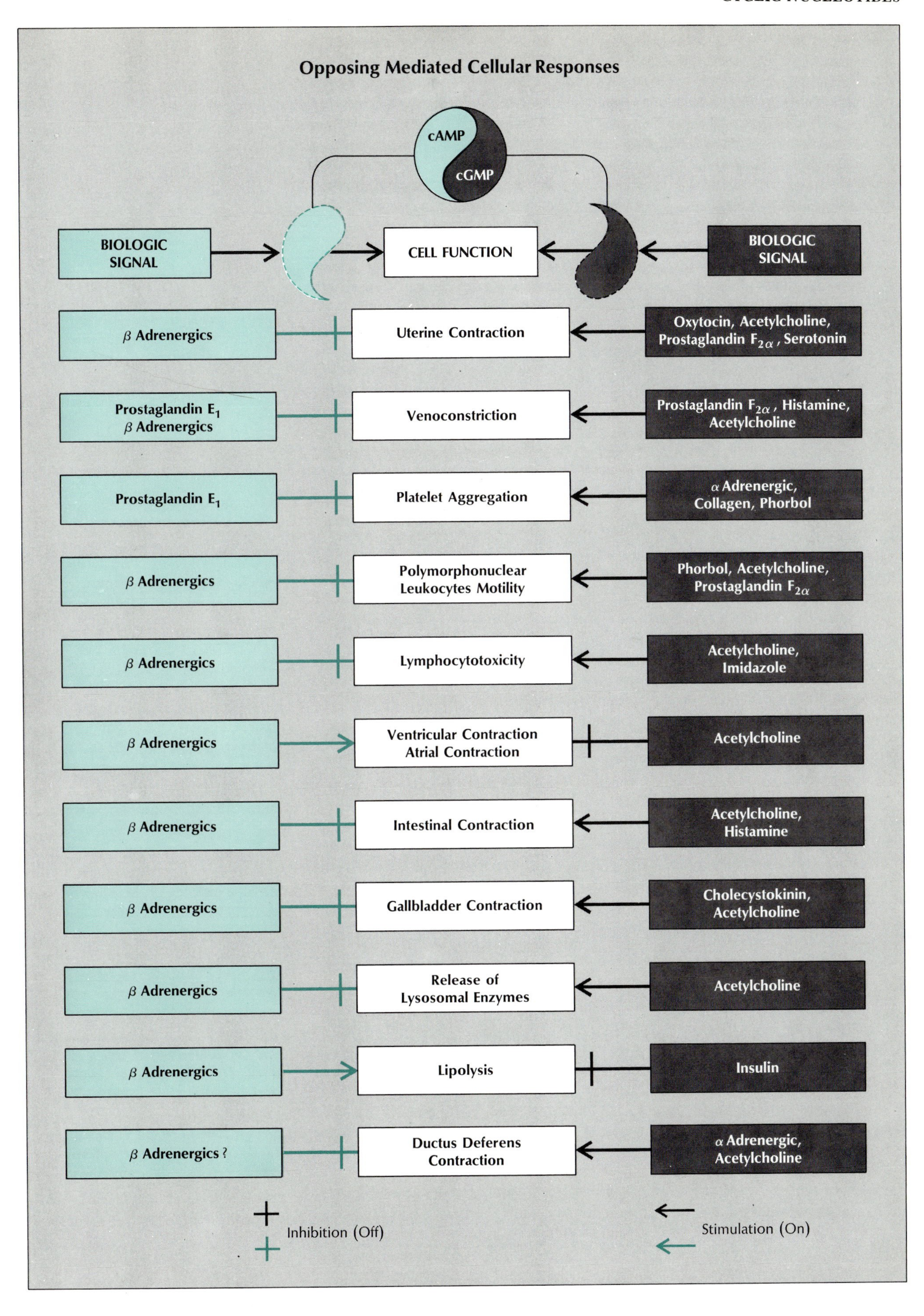

Opposing Mediated Cellular Responses
cAMP
cGMP
BIOLOGIC SIGNAL
CELL FUNCTION
BIOLOGIC SIGNAL
β Adrenergics
Uterine Contraction
Oxytocin, Acetylcholine, Prostaglandin $F_{2\alpha}$, Serotonin
Prostaglandin E_1 β Adrenergics
Venoconstriction
Prostaglandin $F_{2\alpha}$, Histamine, Acetylcholine
Prostaglandin E_1
Platelet Aggregation
α Adrenergic, Collagen, Phorbol
β Adrenergics
Polymorphonuclear Leukocytes Motility
Phorbol, Acetylcholine, Prostaglandin $F_{2\alpha}$
β Adrenergics
Lymphocytotoxicity
Acetylcholine, Imidazole
β Adrenergics
Ventricular Contraction Atrial Contraction
Acetylcholine
β Adrenergics
Intestinal Contraction
Acetylcholine, Histamine
β Adrenergics
Gallbladder Contraction
Cholecystokinin, Acetylcholine
β Adrenergics
Release of Lysosomal Enzymes
Acetylcholine
β Adrenergics
Lipolysis
Insulin
β Adrenergics ?
Ductus Deferens Contraction
α Adrenergic, Acetylcholine
Inhibition (Off)
Stimulation (On)

Opposing Effects of cGMP and cAMP or Their Derivatives

Cell or Tissue	Response	Cyclic GMP	Cyclic AMP
Mast cell	Histamine release	↑	↓
Polymorphonuclear leukocytes	Release of lysosomal enzymes	↑	↓
	Leukotaxis	↑	↓
Lymphocytes (T cells)	Cytotoxicity	↑	↓
	Binding of sheep RBC's	↑	↓
	PHA-induced proliferation	↑	↓
Embryonic heart cells	Contractile rate	↓	↑
Gastric fundus	Contraction	↑	↓
Postsynaptic membrane	Polarization	↓	↑

↑ =Stimulation of the process ↓ =Inhibition of the process

that the detection of an increase in cellular cGMP may not be the criterion on which to infer participation of the nucleotide. A biochemical component in any system does not necessarily have to increase in concentration in order to participate in an event; it may already be, as it were, hanging around waiting for its metabolic partner (i.e., Ca^{++}) to arrive. For the same reasons, it would be equally unwarranted to conclude that simply because certain in vitro systems can be "turned on" by the addition of exogenous cGMP, calcium does not therefore participate in the response.

The point I am trying to make is that both of these substances do indeed exist in the cell, and it is doubtful that evolution would have produced this arrangement and permitted its survival if both were not of some utility as effectors of the events in question. I anticipate that further research will reveal many relationships between cyclic nucleotides and calcium, and between them and steroids, sodium, potassium, etc.; it seems likely that many systems will be catalogued on the basis of whether calcium, in the presence of a particular cyclic nucleotide, would exercise a facilitatory or inhibitory effect on the cell.

The mechanism by which cGMP acts in (or on) the cell is, for the most part, only a matter of conjecture at present. In the glucagon–liver cell system, as noted earlier, we know that cAMP achieves its effect by setting in motion the train of events that phosphorylates both glycogen phosphorylase and glycogen synthetase, thus activating the former and inactivating the latter. Obviously, cGMP, whose effects are opposite to those of cAMP in this system, "ought" to operate by *de*phosphorylating both these enzymes – but no one has managed to show that it in fact does so. Paul Greengard and coworkers have shown that certain protein kinases found in invertebrate tissues (e.g., lobster tails and moth larvae) can be activated more selectively by cGMP than by cAMP, and evidence that similar cGMP-stimulated protein kinases exist in mammalian tissues has begun to appear, but the relevance of these findings to cGMP action remains obscure. Because cGMP appears to be involved primarily in promoting anabolic processes (as opposed to catabolic by cAMP) and since such events are usually of a highly complex and integrated nature, it would seem theoretically that cGMP might be expected to interact selectively with several cellular components. This notion is beginning to receive some support from recent findings. Green and Martin, at the University of California in San Francisco, have shown that phosphoribosyl pyrophosphate synthetase is activated, selectively, by extremely low concentrations (10^{-9m}) of cGMP. This enzyme catalyzes what could be considered the first step toward the biosynthesis of nucleotides – an action that would be expected to be accelerated upon stimulating the proliferative process. We have also found several relatively low molecular weight proteins that bind cGMP exclusively (and not cAMP) and do not exhibit any protein kinase or phospho-protein phosphatase activities. One interesting characteristic shown by some of these proteins is a dependence on cations, especially calcium, for interaction with cGMP.

The first unequivocal demonstration of an action of cGMP on an enzymatic reaction was shown by Joe Beavo in Hardman's laboratory to be an activating influence on phosphodiesterase, the enzyme that breaks down cAMP specifically–which is, of course, perfectly consistent with the fact that this nucleotide can fall to basal levels in systems when cGMP increases.

It is important to emphasize once again, however, that though a mechanism may exist whereby cGMP can reduce cAMP levels, this effect alone probably does not account for its physiologic effects. We have already noted the fact that insulin can stimulate glycogen synthesis by liver cells *without* lowering cAMP, assuming the nucleotide is at basal levels to begin with. It has also been shown that in liver cells both glucagon and cAMP can bring about induction of the enzyme tyrosine transaminase, and that addition of insulin, curiously enough, also increases induction of the enzyme additively in combination with glucagon. Clearly, if glucagon brings about induction by raising cAMP levels, which seems likely, insulin could hardly bring about further induction by lowering these levels.

Furthermore, in certain types of salivary gland cells both alpha- and beta-adrenergic receptors are fully functional, and the latter are known to stimulate cAMP synthesis when occupied by epinephrine. If the former worked by lowering cAMP, instead of (as we believe) through independent effects of cGMP, it would clearly be impossible to evoke both alpha and beta activity at the same time, since one set of receptors would simply cancel out the effect of the other. But in fact the cells can respond in both ways simultaneously, pointing to the existence of two essentially independent mechanisms.

In conclusion, we might note that the probable role of cyclic nucleotides in pathologic processes such as proliferative diseases (i.e., psoriasis, cancer), smooth muscle disorders (i.e., hypertension, asthma), inflammatory states, immunologic disturbances, and so on, suggests that substances facilitating or inhibiting their synthesis may one day find application in clinical medicine. I think this is likely enough; indeed, I would go further and say that the next few years may well show that a great number and variety of pathologic states involve disturbances of cyclic nucleotide metabolism and therefore an imbalance in the amount of one in respect to the other – and may be correctable by normalizing the imbalance. But this may be quite a long way off. Before we can even think about moving from the laboratory into the clinic, we must fill in the blank spaces in our jigsaw puzzle by expanding and amplifying our understanding of these extraordinary protean substances.

20

Ions as 'Second Messengers'

HOWARD RASMUSSEN

University of Pennsylvania

The problem of the "second messenger" – the substance (or substances) that transmits the instructions of hormones or other small molecules at the cell membrane to produce physiologic changes within the cell – has already been discussed at length by Dr. Goldberg in the previous chapter. There seems to be no doubt that one messenger is the compound, cyclic adenosine monophosphate (CAMP), which evidently serves as a mediator in a wide variety of hormonally triggered processes. But there is growing evidence that CAMP is only one of several second messengers.

A number of laboratories have been investigating the possible messenger role of another nucleotide, cyclic guanosine monophosphate (CGMP), and have in fact developed considerable evidence implicating this compound in certain hormonally induced intracellular changes. However, the chain of evidence as currently described seems to me to contain at least one serious gap: the lack of a plausible enzymatic mechanism for the manufacture of CGMP in, or at, the plasma membrane.

Cyclic AMP is known to be synthesized from adenosine triphosphate (ATP) by the enzyme adenylate (or adenyl) cyclase, and this enzyme has been found in membrane fractions from a variety of cells, *in a form that can be activated by the appropriate hormones.* Similarly, CGMP is synthesized by the enzyme guanylate cyclase, which is known to be present in some plasma membranes (as well as in the cell cytoplasm). But no one has yet shown that the membrane-bound enzyme can in fact be activated by any hormone. If, then, the existence of hormonally activated adenylate cyclase in plasma membrane is one of the strongest bits of evidence for the messenger role of CAMP (which I believe to be the case), the apparent absence of hormone-activated guanylate cyclase in plasma membrane seriously weakens the case for CGMP's messenger role. It remains a conspicuously unfilled area of the cyclic nucleotide "jigsaw puzzle" described by Goldberg.

For this and other reasons, my associates and I have focused our researches on the possible messenger role of a quite different class of substances, the metallic ions (cations) – in particular, Ca^{++}. We were led in this direction by a number of aspects of hormone action that seemed to indicate that the second messenger hypothesis (associated with the late Earl Sutherland) was not an adequate description of the cellular events.

First, the hypothesis as originally stated described a unidirectional process – what the systems engineers call an "open loop." The hormone interacted with a receptor on the cell surface, which activated adenylate cyclase, which synthesized CAMP, which somehow caused the specific cellular response. There was no suggestion of any "closed-loop," feedback process whereby the results of this series of events could in any way influence the antecedent steps. It was clear enough how CAMP production could be turned on, but far from clear how it was turned off, once the cell had responded appropriately.

Now on the supracellular level, at least, feedback loops are the rule rather than the exception. In the endocrine system in particular, almost everyone is aware that just as increased production of a trophic hormone (e.g., ACTH) by the pituitary stimulates production of the appropriate "target" hormone (e.g., cortisol), so rising serum levels of the target hormone cut back production of the trophic hormone, thereby holding the whole process within rather narrow limits. It seemed strange to us that, given such precise controls (actually far more elaborate than described here) at the manufacturing end of the hormone system, there should be no evidence of similar feedback at the action end.

A second problem was that, in the case of insulin, the peptide hormone most extensively studied (largely because of its obvious pathologic and therapeutic significance), all the evidence indicates that its interaction with its receptor on the plasma membrane produces a number of quite different changes that seem to be largely or entirely inde-

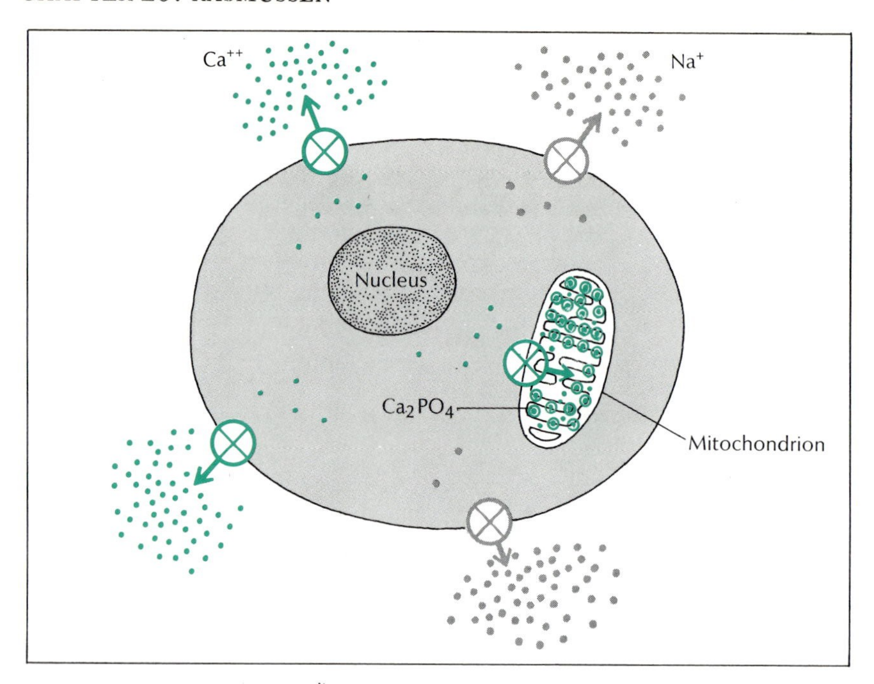

Ionic imbalances within cell and between cell and environment are maintained by energy-dependent "pumps" in plasma and other (e.g., mitochondrial) membranes. These pumping reactions account for some 25% of the cell's basal energy expenditure.

pendent. Insulin increases cellular uptake of glucose, of amino acids, and of monovalent cations (e.g., K^+). In some systems, at least, insulin appears to *inhibit* the activity of adenylate cyclase – that is, it reduces intracellular levels of CAMP.

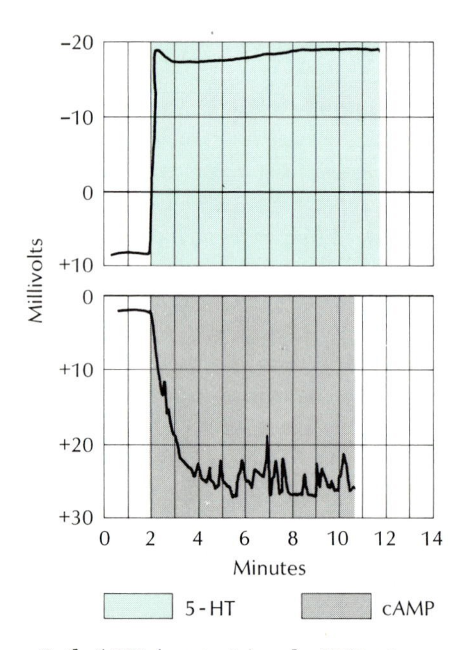

Both 5-HT (serotonin) and cAMP trigger secretion from blowfly salivary gland, but electrically the effects of the two substances are opposite. Transepithelial potential becomes more negative with 5-HT, more positive with cAMP.

None of these actions of insulin can be shown to be primary. One can, for example, stimulate cellular uptake of glucose with insulin in a medium lacking amino acids, and vice versa, so that neither process is dependent on the other. Even more interesting is the fact that simply by manipulating the ionic environment of the cell one can induce certain insulin-like changes – e.g., increased glucose uptake. Hypothetically, then, one can conceive that insulin produces its effects by a redistribution of ions across the membrane – but this is extremely difficult to prove. Eventually, of course, both glucose uptake and amino acid uptake will probably be shown to depend on some change in the membrane structure, but at present we do not know the nature of this change. Nevertheless, this single postulated change may lead to not one but to several second messengers.

Before developing this line of thought further, I should now say something about the general role of ions in cells. Virtually all animal cell membranes are actively involved in ion transport. These ion transport processes lead to an asymmetric distribution of ions across the membrane. For example, in most mammalian cells the cytoplasm is considerably higher in K^+ and lower in Na^+ than the extracellular fluid surrounding the cell because of the ion transport activity of the plasma membrane. Less widely appreciated, perhaps, is the fact that within the cell the ionic "pumping" of the mitochondrial membrane produces an asymmetric distribution of Ca^{++} and Mg^{++} between cell cytosol and the interior of the mitochondria. Maintenance of these ionic asymmetries requires the consumption of approximately 25% of the cell's basal energy production.

It seems hardly conceivable that processes this important in the cell's energy budget would not possess comparable importance for cell function, and in fact we know that they do. In neurons, for example, the ionic gradients confer on the plasma membrane the property of excitability, whereby a disturbance in ionic flow across the membrane can be propagated from one end of the cell to the other. In cells generally, ionic pumping also subserves the important function of regulating cell volume by controlling the inflow and outflow of water. Finally, there is increasing evidence that ionic asymmetry is one way of coupling events that occur on the cell surface with what goes on inside (see Chapter 11, by Loewenstein).

Of key importance in this area is the Ca^{++} ion. The distribution of this divalent cation in and around the cell is rather different from that of the monovalent cations just cited, in that the bulk of the cell's calcium is concentrated within the mitochondria. Taking a typical mammalian cell, for example, the concentration of calcium outside the cell will be around 10^{-3} molar, that within the cytoplasm perhaps 10^{-7} molar (or about 1/10,000 of the external concentration), but within the mitochondria it will be about 10^{-2} molar, or much higher than outside the cell. The situation is somewhat complicated by the fact that most of the mitochondrial calcium (some 99%) is not ionized but rather is stored in the form of a phosphate salt. Nonetheless, this still leaves a concentration of Ca^{++} ion of about 10^{-4} molar within the mitochondrial matrix space, making for a sharp calcium ion gradient across the mitochondrial membrane, and an even sharper one across the plasma membrane separating cytoplasm from extracellular fluid. Both gradients, of course, are maintained by energy expenditure; indeed the uptake of calcium has actually been demonstrated in isolated mitochondria, in an energy-dependent process that is (as one would expect) facilitated by the presence of phosphate.

The energy for mitochondrial calcium uptake is provided by the same process that provides energy for the manufacture of ATPase described by Racker in Chapter 14. The intramitochondrial calcium-transport and storage system, in fact, is an arrangement for buffering calcium, phosphate, and H^+ ion within the cell (the latter being generated when the Ca^{++} and phosphate combine within the mitochondrion). The H^+ ion, in turn, couples this system with the bicarbonate-CO_2 system that serves primarily to maintain intracellular pH. Finally, at the level of the plasma membrane, the movement of Ca^{++} is coupled with that of Na^+. The simplest way of describing these interrelated systems is as an ionic network, with each compartment having its specific ionic composition, maintained at the expense of cell energy, and in which perturbing events in either the mitochondrial or the plasma membrane can change ionic distribution throughout the cell. Thus, the ionic net can act to propagate and amplify an original signal perceived by a membrane receptor (see Chapter 18, "Hormone Receptor Interactions and the Plasma Membrane," by Cuatrecasas).

The second general fact about ions that is important to this discussion is that a great many enzymes are regulated or activated by them. A number of enzymes, for example, are activated by Mg^{++}, and by changing the concentration of this ion one can change the "effective" concentration of an activator or inhibitor of the enzyme. Pyruvate kinase, to take a specific example, is known to be activated both by Mg^{++} and by monovalent cations, especially K^+. And many enzymes can be activated or inhibited by Ca^{++}. All of which amounts to saying that changes in ionic distribution, mediated by the "ionic net," are one way in which cell function can be altered.

Returning now to the second messenger problem, we note that in many of the systems in which cAMP has been implicated as a second messenger, Ca^{++} ions are also involved – and seem to play a role, moreover, apparently quite as important as that of the nucleotide. The detailed evidence – which we shall consider in a moment – indicates that, as one would expect, the specific role of calcium varies with the particular system under investigation, and also that in some systems, at least, the ion and the nucleotide act in an interrelated manner, so that in a sense neither of them is "the" second messenger; rather, that role is filled by a cAMP-Ca^{++} system.

One of the first biologic systems in which the interrelated role of Ca^{++} and cAMP was demonstrated was in the salivary gland of the blowfly,

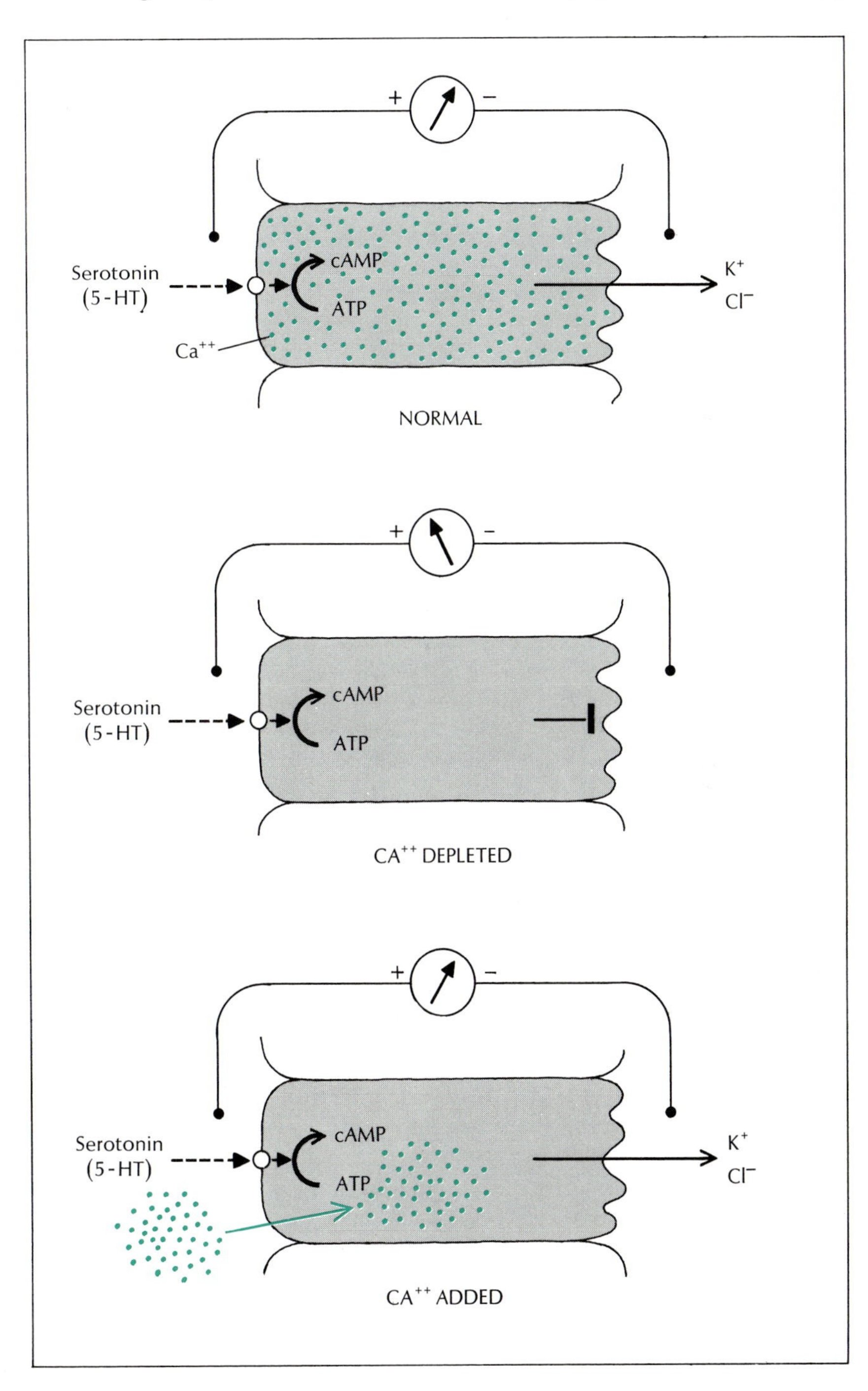

Serotonin produces increased cAMP levels and characteristic (negative) change in the transepithelial potential in the normal (top) and a positive change in potential in the calcium-depleted (middle) salivary cells of blowfly, but triggers secretion only in the former. Addition of exogenous calcium to medium surrounding depleted cell allows secretion to proceed normally and the potential to become negative.

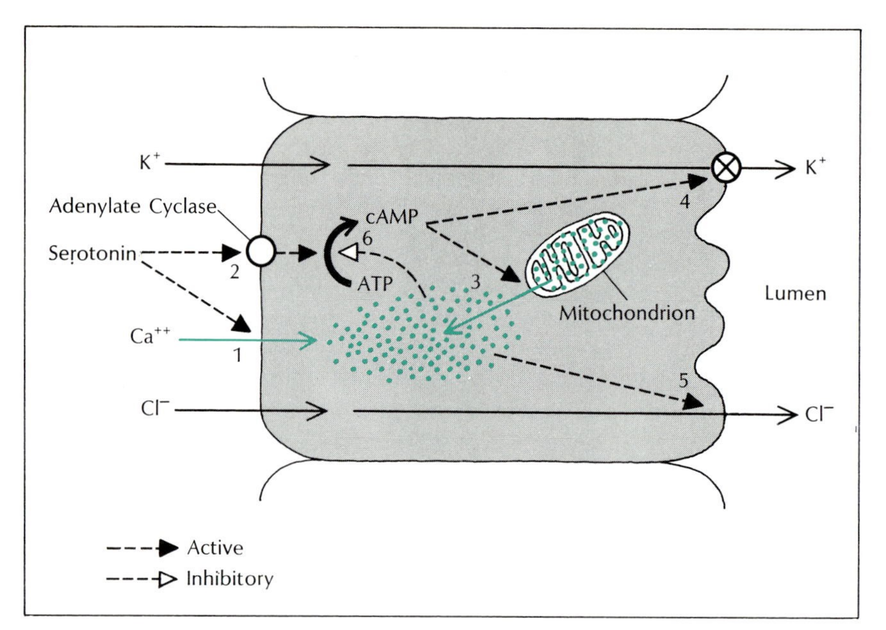

Serotonin-triggered secretion from fly salivary gland requires both calcium and cAMP. Serotonin produces calcium influx into cell (1) and also activates adenylate cyclase (2), leading to rise in cAMP, which in turn triggers calcium efflux from mitochondria (3). The nucleotide also activates "pump" that expels K^+ ion from cell (4); meanwhile, the calcium alters permeability of the luminal membrane to Cl^- ion (5). However, the calcium also blocks further production of cAMP, thereby serving as a feedback control (6).

through a series of researches that I carried out with William Prince and Michael Berridge at Cambridge University over a period of several years. The abdominal portion of this organ is convenient to study, both because it is of very simple structure and because it can quite easily be removed intact from the insect. Its function is to secrete a fluid – mainly isotonic potassium chloride – that liquefies the insect's food (such as rotting meat), enabling it to be lapped up (the fly possesses no teeth). The gland is not neurally innervated, as is the case with the human salivary gland, but is wholly controlled by the hormone serotonin, or 5-hydroxytryptamine (5-HT), which, as in vitro studies have shown, can rapidly raise the rate of secretion from near zero to a significant figure.

The control of the gland's secretion by 5-HT clearly involves CAMP, and indeed fulfills the classic criteria originally set up by Sutherland and his coworkers for demonstrating the involvement of that substance. For example, the hormone's action is potentiated by theophylline, a drug that blocks the enzyme phosphodiesterase, which breaks down CAMP; it thereby augments the rise in CAMP concentration resulting from a submaximal dose of the hormone. Again, the hormone's action can be mimicked by adding exogenous CAMP to the system, albeit at relatively high concentrations. It is presumed that high concentrations are necessary because only a small proportion of the exogenous CAMP actually gets into the cell. Finally, serotonin increases CAMP levels within the cell. This was difficult to demonstrate because of the small quantity of tissue and very low concentrations (around 10^{-8} molar) of CAMP involved, but a very refined method developed by Dr. Alfred Gilman, presently at the University of Virginia, has established these increments.

By all these criteria, then, serotonin apparently activated the salivary cell by stimulating CAMP production. Other studies, however, have shown that in certain respects the two substances (hormone and cyclic nucleotide) are by no means equivalents but may indeed exert opposing actions. For example, if one measures the transepithelial potential of the salivary cells by placing a microelectrode on the outside of the gland and another in the secreted fluid within the lumen, it turns out that whereas serotonin makes the potential more negative, CAMP makes it more positive – both changes being quite significant. Clearly, these differing changes in membrane potential argue some change in ion flow – and a difference in ion flow in response to CAMP vis-à-vis serotonin.

On the basis of our earlier work with other systems, it seemed to us that Ca^{++} might well be the source of these ionic differences. Accordingly, we depleted cells of Ca^{++} by incubating them for some time with a chelating agent that sequesters the ion, and found that under these circumstances serotonin produced no increase in secretion – yet it did increase CAMP concentration within the cells. A second messenger was being produced, but the message was not being read. However, when we then added exogenous calcium, secretion began immediately, but only in cells that had been already "primed" with serotonin, or in which the effects of serotonin had been mimicked by adding exogenous CAMP. Evidently, then, both the ion and the nucleotide were necessary for secretion.

We next set about determining precisely why they were both necessary, and how they acted jointly (if in fact they did) to induce secretion. Further experiments revealed some additional, and suggestive, differences between serotonin and CAMP in their effect on calcium. By different sorts of labeling experiments, we found that serotonin increases the uptake of calcium whereas exogenous CAMP does not. On the other hand, both exogenous CAMP and serotonin increase the efflux of calcium from the cells of glands previously incubated with radioactive Ca^{++}, so as to label the cell calcium pools.

In order to explain these rather puzzling observations – which on their face seemed to imply that serotonin was doing two contradictory things simultaneously – we set up a conceptual model. According to this hypothesis, when serotonin reacts with its receptor on the plasma membrane, there are at least two different events: first, activation of adenylate cyclase, with a consequent increase in cellular CAMP, and, second, an increase in the plasma membrane permeability to Ca^{++}. And since the extracellular Ca^{++} concentration is, as noted, much higher than the cytoplasmic, the result is a rapid flow of ion into the cell.

However, given the active calcium

uptake and buffering action of the mitochondria already described, this alone would be expected to have little or no effect on cytoplasmic Ca^{++} concentrations. Accordingly, we believe that just as 5-HT increases plasma membrane permeability to Ca^{++}, so CAMP increases the ionic permeability of the mitochondrial membrane. Since calcium concentration within these organelles is, as noted earlier, far higher than in the cytoplasm, the result is a flow of Ca^{++} into the cytoplasm from *both* the mitochondria and the extracellular fluid. That such a change in mitochondrial membrane permeability does occur is indicated both by our own experiments and by some more recent ones, involving isolated mitochondria, performed by Andre Borle at the University of Pittsburgh.

Finally, there is evidence, from these and other experiments, that as the level of calcium in the cytoplasm rises, the generation of additional CAMP is inhibited – presumably by the inactivation of adenylate cyclase. Blockage of CAMP synthesis by rising cytoplasmic levels of calcium of course provides the feedback loop we had postulated as needed in the system: the CAMP regulates calcium release from the mitochondria, and the calcium then regulates CAMP production. The net effect is a new steady state, with both CAMP and Ca^{++} within the cell cytosol held at a somewhat higher-than-normal level.

But in the salivary gland system, at least, calcium is more than a feedback regulator; it also acts as a second messenger itself, along with CAMP. Specifically, the latter apparently activates the "pump" that passes potassium ion (K^+) out of the cell into its lumen, while the Ca^{++} similarly takes care of the "other half" of the secretory process, by increasing the permeability of the luminal membrane to chloride (Cl^-). We have inferred this on the basis of a number of different studies.

First, a change in transepithelial potential occurs in response to either exogenous CAMP or 5-HT (presumably mediated by endogenous CAMP) even in cells that have been depleted of calcium; this we interpret as reflecting activity of the K^+ pump. However, no secretion occurs under these conditions – which is what one would expect, given the electrogenic nature of the process. If a cation (K^+) is being pumped out, its charge must be balanced either by entrance of another cation (e.g., Na^+), or by the more or less simultaneous exit of an anion (in this case Cl^-). But in the absence of calcium, the latter process cannot occur owing to the impermeability of the luminal membrane of the cell to Cl^-. Operation of the pump thus is blocked by the ionic gradient it must work against, even though CAMP concentrations in the cell are rather higher than normal. (Given the absence of the calcium

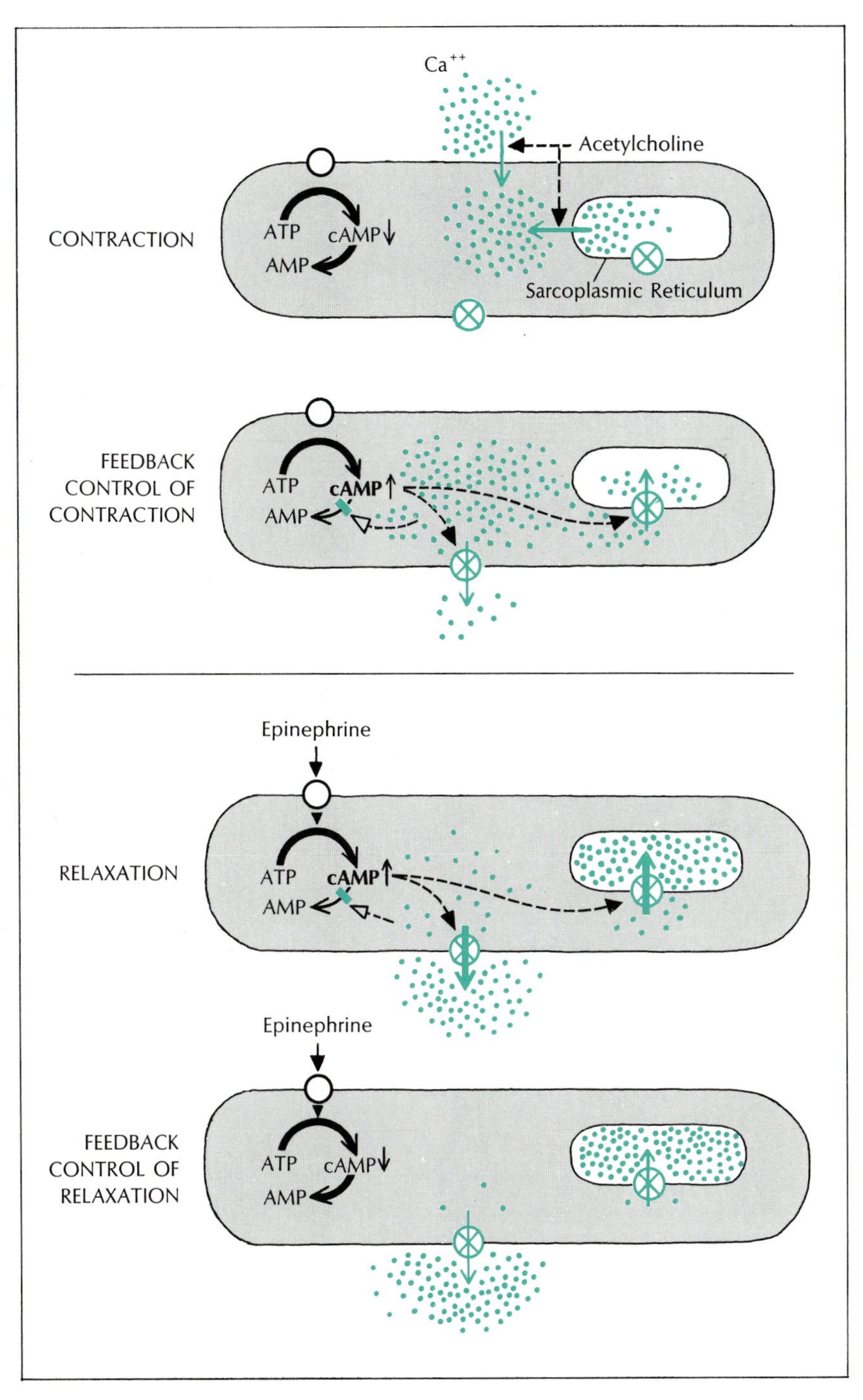

Acetylcholine produces contraction of smooth muscle by triggering calcium influx into cytoplasm from the medium and sarcoplasmic reticulum (SR). Calcium also serves as feedback control, blocking the breakdown of cAMP by phosphodiesterase; the increased cAMP levels then produce reaccumulation of calcium in the SR. Epinephrine acts in opposite sense, triggering rise in cAMP that causes calcium reaccumulation in SR; drop in cytoplasmic calcium produces feedback control by unblocking breakdown of cAMP.

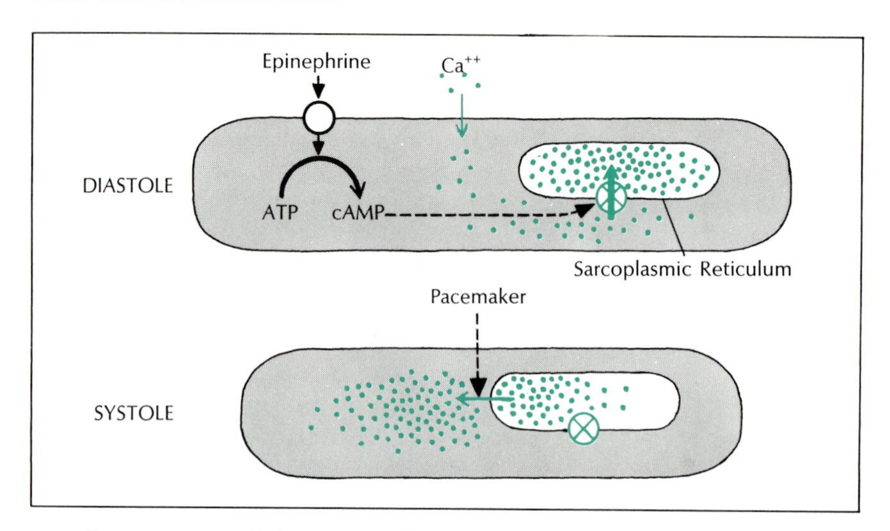

Stimulation of myocardial contraction by epinephrine is thought to occur through increased accumulation of calcium in sarcoplasmic reticulum, which is mediated by cAMP, as in smooth muscle. Cell contracts when calcium is released by pacemaker impulse.

feedback control this is what one would expect.) However, if calcium is now added to the system, chloride permeability increases and secretion proceeds normally.

One can also perform the opposite type of experiment, in which calcium levels in the cell are increased without increasing cAMP. This can be done by means of compounds called ionophores, one of which (A23187, prepared by Eli Lilly and Company) can change the permeability of some membranes to calcium. Adding this substance to the medium surrounding the salivary gland produces an uptake of calcium like that produced by 5-HT, a change in membrane potential in the same direction as that produced by the hormone, and an increase in secretion, albeit to submaximal (about 60% maximal) levels – all occurring *without* any rise in cAMP concentration within the cell. Thus in this system, Ca^{++} alone acts much like 5-HT. We interpret this finding to mean that while a certain amount of cAMP is being produced at all times, even without stimulation by 5-HT, only in the presence of Ca^{++}, which allows the outflow of Cl^{-}, can it stimulate the pumping of K^{+} out of the cell. Thus the amount of Ca^{++} is evidently the limiting factor in the secretory process. Nonetheless, in response to the natural hormone, it is evident that both Ca^{++} and cAMP are second messengers in this gland.

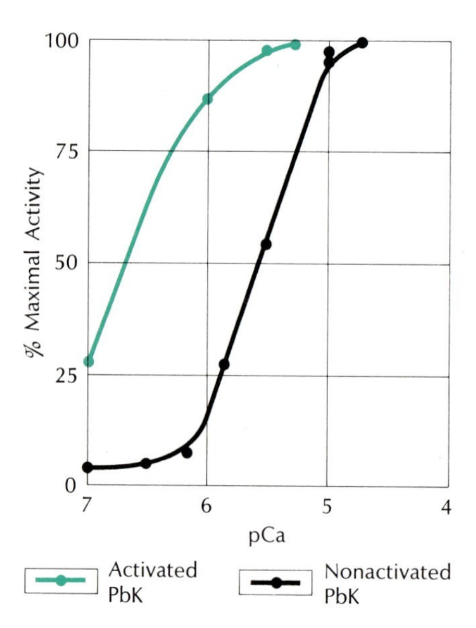

Calcium ion produces activity in both activated (phosphorylated) and nonactivated enzyme (PbK from skeletal muscle) but affects the latter only in much higher concentrations.

One naturally wonders whether the relationship between cAMP and Ca^{++} as joint second messengers of hormone action also obtains in mammalian cells. Much evidence accumulated in the past five years shows that it certainly does. The systems in which this is the case include the effect of ACTH on the adrenal cortex, the effect of parathyroid hormone on the isolated renal tubule, the effect of the same hormone on the bone cell, and the effect of melanocyte-stimulating hormone on melanophores. In all of these situations, the evidence – though less complete in some than in others – suggests that the same basic relationship obtains between cAMP and Ca^{++} as found in the fly salivary gland, including both the feedback effect and the need for both to be present if the hormone is to act on its target. Thus it appears that the cAMP-Ca^{++} system is not specific to the action of one or two hormones, but is one of the general biochemical mechanisms developed early during evolution for activating cells. In this view, hormone specificity depends, first, on the specific hormone receptors present on the cell surface and, second, on the cell's unique enzymatic makeup rather than on a unique means of turning on the cell. Stated in another way, calcium ion and cAMP appear to be two of the few general second messengers involved in cell activation, and these two often operate in a cooperative and stereotyped way.

However, the specific way in which the two messengers work together is likely to show considerable variation as a consequence of evolutionary adaptations. We have already noted one general relationship exemplified by the fly salivary gland and certain mammalian cells. A good contrasting example is the hormonal regulation of longitudinal intestinal smooth muscle in mammals.

As is well known, the basic physiology of this tissue is simple: acetylcholine (or one of its analogues) causes the muscle to contract; epinephrine (or any other beta-adrenergic agent) causes it to relax. So far as contraction is concerned, everything we know about muscle, smooth or striated, tells us that calcium is the prime contender for the title of second messenger. One would therefore expect that acetylcholine would produce increased entry of calcium into the cell and/or its release into the cytoplasm from some intracellular compartment – and this is precisely what one finds. Likewise, one would expect epinephrine to reduce calcium concentrations in the cytoplasm and to stimulate its reaccumulation in some intracellular compartment, and again this is what one finds. In this type of tissue, however, the intracellular compartment involved in calcium storage, as it turns out, is not the mitochondria, but a specialization of the microsomes called the sarcoplasmic reticulum. In muscle cells, and perhaps some other specialized types, these bodies appear to perform the calcium-buffering role taken in most other types of cell by the mitochondria – but apparently in an opposite direction. That is, whereas CAMP induces the *release* of calcium from mitochondria, it seems to stimulate calcium *uptake* by the sarcoplasmic reticulum.

So far, so good. If, however, we now measure the effect of the two "opposing" hormones, not on calcium but on CAMP, it appears that they produce not opposite but essentially identical results: as has been shown by Anderssen and his coworkers in Sweden, both increase CAMP. This paradox can be resolved if we make the assumption that calcium in this system inactivates phosphodiesterase instead of activating it; the result would be that any CAMP manufactured would gradually build up because phosphodiesterase would not be available to break it down.

This may seem like a rather daring assumption, but there is a good deal of evidence, albeit indirect, to support it. First, it is definitely known that phosphodiesterases from different types of cells are not all of a piece; some are calcium activated, others, calcium inhibited. The second bit of evidence involves measurements of the differing patterns of CAMP increases produced by acetylcholine and epinephrine respectively. Acetylcholine, it will be recalled, produces an influx of calcium into the cytoplasm, which is followed by a relatively gradual rise in CAMP – according to hypothesis – because the rising calcium is inhibiting the phosphodiesterase. By contrast, epinephrine produces an immediate rise in CAMP, presumably through direct activation of adenylate cyclase, producing a fall in calcium. This is followed by a gradual fall of CAMP as calcium levels in the cytoplasm drop and – by hypothesis – phosphodiesterase becomes activated. Thus it would appear that in the muscle's response to acetylcholine, it is the calcium buildup that causes both the contraction and rise in CAMP concentration. The CAMP, in turn, by stimulating reuptake of calcium into the sarcoplasmic reticulum, acts as a feedback loop to prevent "overshoot." In the epinephrine response, on the other hand, relaxation is produced by CAMP, which stimulates the membranous uptake of calcium. The resulting drop in calcium, by activating phosphodiesterase and thereby breaking down CAMP, serves as the feedback loop.

Yet another system in which calcium and CAMP work together is in the action of epinephrine on heart muscle. As is well known, there are at least two important components to this: an increase in the force (and also the rate) of contraction, and the breakdown of stored glycogen into glucose. The first of these effects, on its face, seems to be opposite to that resulting from the action of epinephrine on intestinal smooth muscle, which (as just noted) is to produce relaxation. Is there any way in which we can resolve this apparent contradiction without multiplying hypotheses indefinitely? I believe there is.

In the first place, we should note that the contradiction is not quite as real as it looks. Whereas in intestinal muscle, epinephrine triggers relaxation, it does not trigger its opposite, contraction, in cardiac muscle. Cardiac contraction occurs, of course, in response to a neural trigger, the pacemaker system, not a hormonal one. A second difference is that heart (as opposed to the intestinal) muscle is an intrinsically unstable system, in the sense that it contracts and relaxes repetitively – which evidently involves a similarly repetitive release and reaccumulation of calcium from the sarcoplasmic reticulum. And it must be obvious that there is some sort of relationship between the force of contraction and the amount of calcium released during that particular beat.

All this still does not clarify the role of epinephrine, which in intestinal muscle, as we have seen, produces reaccumulation of calcium rather than its release. However, it is evident that just as the force of contraction depends on the amount of calcium released, so the amount released will depend on the amount accumulated during the previous diastole: the more accumulated, the more available for release. Thus epinephrine, by stimulating CAMP production, will cause a more vigorous calcium uptake by the sarcoplasmic reticulum, making for a more vigorous contraction when that

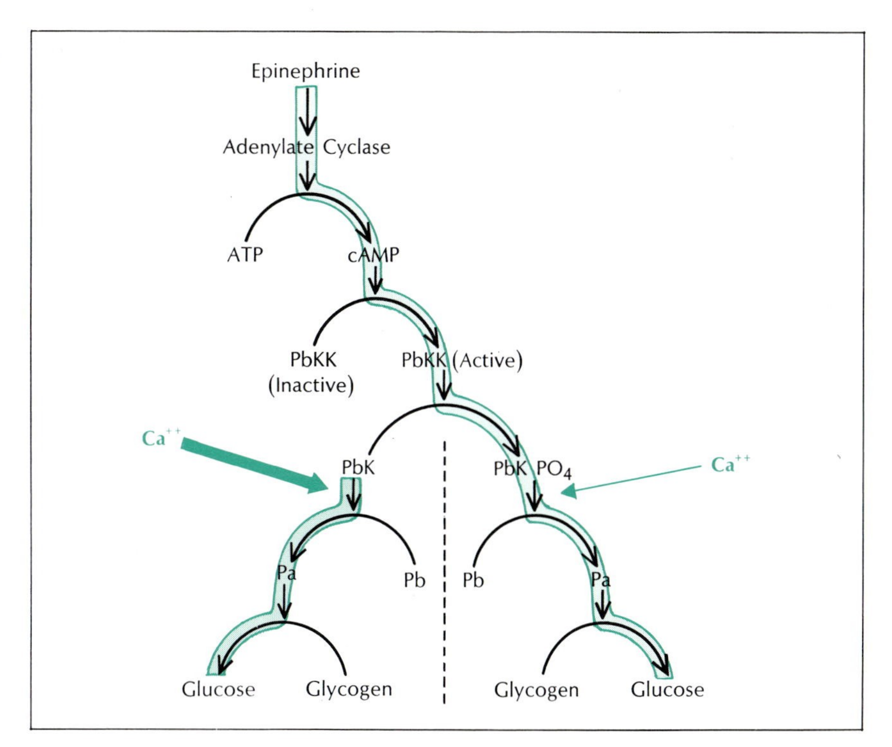

Glycogen breakdown in striated muscle can occur via two pathways, as schematized above. High calcium concentration, produced by repeated contraction, activates nonphosphorylated PbK (left), releasing glucose to supply energy for muscle itself. Epinephrine triggers phosphorylation of PbK, which sensitizes it to low calcium concentration (right). The glucose that results is eventually released from the muscle as lactate and serves to increase the energy supply to the entire body.

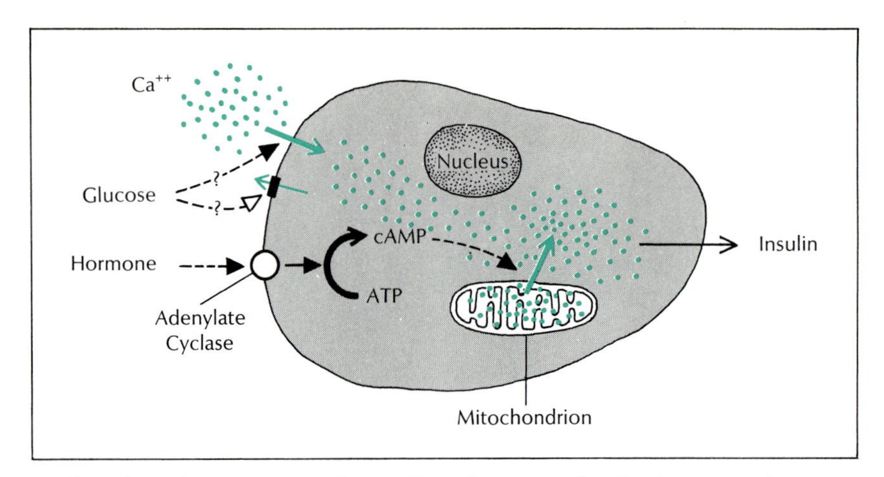

Insulin release from pancreatic beta cell can be triggered either by various hormones or by high concentrations of serum glucose. The hormonal actions are mediated by cAMP, which triggers calcium release from mitochondria, producing insulin secretion. Glucose is thought to act via a different pathway, increasing calcium directly by facilitating its influx into the cell and/or blocking its efflux.

calcium is released in response to the nonhormonal pacemaker stimulus. At the same time (and there is independent evidence for this) the inflow of exogenous calcium into the myocardial cells is also stimulated, so that overall calcium concentration rises.

The epinephrine-induced breakdown of glycogen in the myocardium involves another type of mechanism, again involving both CAMP and calcium, as shown by the work of many investigators including Edwin Krebs of the University of California at Davis and Steven Mayer of the University of California at San Diego. The process here is essentially the same as that for glycogen breakdown in the liver, with the trigger in this case epinephrine. Epinephrine activates adenylate cyclase, which catalyzes the synthesis of CAMP, which activates protein kinase, which, in turn, activates phosphorylase kinase by catalyzing its phosphorylation. The active form of phosphorylase kinase catalyzes the transformation of phosphorlylase b into its active form, phosphorylase a, which breaks down glycogen into glucose-l-phosphate, which serves as an energy source for the heart.

This enzymatic cascade, however, though it is set off by CAMP, is also calcium dependent, in the sense that one of the enzymes in the chain – phosphorylase kinase – requires Ca^{++} for its full catalytic action. By increasing or decreasing concentrations of the ion over the range of 5×10^{-7} to 5×10^{-6} molar, one can proportionally increase or decrease the enzyme's activity. Here, then, CAMP and Ca^{++} are operating at two different, sequential steps in the process: CAMP to set off the cascade, Ca^{++} to modulate its flow.

In skeletal muscle, we have a variant of this relationship between CAMP and calcium in regulating glycogen-glucose metabolism, reflecting the functional differences in that process between skeletal muscle and myocardium. In the heart, glycogen serves essentially as an emergency energy reserve for the organ itself, for which reason its breakdown is closely coupled to the organ's energy demands, through the mechanisms whereby epinephrine modulates both contraction (demand) and glycogen breakdown (supply). In skeletal muscle, on the other hand, glycogen may serve local needs (e.g., in an anoxic or rapidly working muscle), but also acts as an energy reserve for the entire body (as, of course, does liver glycogen); when the organism is challenged, epinephrine is released and stimulates the breakdown of muscle glycogen, ultimately into lactate, which is carried to the liver, where it is recombined into glucose, which is then processed to yield energy in the ordinary way.

Now evidently these two modes of glycogen breakdown in skeletal muscle are – must be – independent processes. The local need of an overdriven muscle for reserve energy has no necessary connection with the body's total energy needs; similarly, the rise in blood glucose triggered by epinephrine-induced glycogen breakdown when the organism is challenged has no necessary connection with whether muscles are actively contracting. Indeed, as is well known, the epinephrine response is an essential part of the organism's "alarm reaction" that *prepares* it for possible muscular exertion – which, however, may or may not subsequently take place.

Given these functionally independent processes, what one would expect to find is two independent control mechanisms governing glycogen breakdown – and this in fact seems to be the case. When a skeletal muscle contracts in response to an electrical stimulus to its motor nerve, we find no evidence of any increase in CAMP, though if the contraction is repeated often enough we do get glycogen breakdown. Evidently this process is, under these circumstances, CAMP independent; it is triggered instead by a rise in calcium (probably combined with other changes) which, as in the myocardium, increases the activity of phosphorylase kinase, one of the enzymes in the glycogenolytic cascade. If, however, the muscle is stimulated by epinephrine, calcium levels do not change (which would produce a probably inappropriate contraction or relaxation of the muscle in question), but CAMP levels go up – as does glycogenolysis.

This dual control system forces us to take a closer look at the enzymatic cascade through which it acts. It turns out that the key step involves phosphorylase kinase. Hitherto we have spoken of this enzyme as either active (phosphorylated) or inactive (dephosphorylated), but the reality turns out to be more complicated. It appears that *both* forms are active – in skeletal muscle but under different conditions. Both are activated by Ca^{++}, but the phosphorylated form is much more sensitive to the ion – specifically, in concentrations as low as 10^{-7} molar, which is around the normal level in the resting muscle. The dephosphorylated form, in contrast, will not respond to calcium in this dilution, but will do so if the concentration is raised by a factor of around 10, i.e., *to levels found in the contracting muscle.* The amount and activity of the two different forms of phosphorylase kinase has actually been measured by stimulating muscle in various ways and then freezing it in liquid nitrogen, which at the same time "freezes" its enzymatic composition so that the latter can be assayed.

Thus the rise in Ca^{++} consequent on muscular contraction will activate phosphorylase kinase in the dephosphorylated form – and thereby increase glycogen breakdown to meet the metabolic needs of that contraction. The rise in cAMP consequent on epinephrine stimulation will phosphorylate phosphorylase kinase, sensitizing it to the resting calcium levels – thereby increasing glycogen breakdown to meet the general metabolic needs of the challenged organism.

One final example showing the variety and complexity of the cAMP-Ca^{++} control systems in the body concerns the release of insulin by pancreatic beta cells. The basic trigger here is of course a rise in serum glucose that, functionally speaking, releases insulin so that the latter will stimulate the conversion of glucose into glycogen by the liver or its uptake into muscle or adipose tissue cells. The consequent reduction in serum glucose then cuts back insulin release, thereby completing the feedback loop. But insulin release can also be induced by a number of other substances – certain hormones, amino acids, and so on. An important point, however, is that many or most of these substances are glucose dependent; in effect, they seem to make the beta cell more sensitive to the sugar – whose presence, nonetheless, is required if insulin release is to occur.

What is the role of calcium and cAMP in these processes? There can be no doubt that the cell's response to glucose is mediated by calcium; this can easily be demonstrated in vitro and also in vivo, in both animals and human beings suffering from severe hypocalcemia. However, it has been shown more recently that insulin release can also be stimulated by theophylline (which the reader will recall inhibits the breakdown of cAMP), and by exogenous dibutyryl cAMP (a modification of cAMP that is thought to enter cells more easily). This, in turn, has led some investigators to suggest that glucose acts by activating adenylate cyclase, thereby raising intracellular cAMP levels. However, nearly all the evidence we have indicates that this is not the case – e.g., glucose by itself does not raise cAMP levels.

But if calcium is the prime second messenger in this system, as appears to be the case, one would expect that the evident capacity of cAMP to stimulate insulin release would itself be mediated by calcium. And, indeed, Dr. W. Malaisse in Brussels has shown quite clearly that in the beta cell cAMP has the same effect as in the fly salivary gland: it mobilizes intracellular calcium by releasing it from some special pool or compartment (quite possibly the mitochondria). Glucose, on the other hand, either stimulates calcium influx or inhibits calcium efflux across the cell membrane. Thus, we have here two means of regulating the cytoplasmic calcium concentration: one controlled by a direct effect of glucose on the plasma membrane of the cell and the other (mediated by hormones, etc.) functioning through activation of adenylate cyclase, with a consequent rise in intracellular cAMP that mobilizes calcium from an intracellular store. Unlike the case of the fly salivary gland, these controls operate in an independent, mutually supplementary fashion, rather than in a mutually dependent fashion.

The foregoing discussion of the cAMP-calcium interactions, together with Dr. Goldberg's description of the relationship between cAMP and cyclic guanosine monophosphate (cGMP) in the previous chapter, should suggest to the reader that there must be a fundamental relationship between Ca^{++} and cGMP. This is certainly so, but the details of this association have not been as fully explored as those of Ca^{++} and cAMP.

Nevertheless, within the past year, some exciting new insights into the Ca^{++}-cGMP association have emerged. The calcium ionophore, A23187, which we used so effectively to explore the Ca^{++}-cAMP relationship in the mechanism of fluid excretion in the fly salivary gland, has now been used to study the role of calcium as a second messenger in a host of other systems. These include: the release of granules from mast cells; stimulation of smooth muscle contraction; activation of the unfertilized egg; and transformation of peripheral human lymphocytes. In all of these systems, previous evidence had strongly suggested that calcium ions had some messenger function, but a relationship of this calcium involvement to changes in cAMP concentration could not be established.

The ionophore experiments have shown that increasing the uptake of calcium into the cell and/or causing its release from an internal store by treatment of the particular cell with an appropriate concentration of A23187 cause an activation of the cell to either, respectively, initiate the early events of fertilization, the contraction of the smooth muscle, the extrusion of the granules from the mast cell, or the stimulation of transformation in human peripheral lymphocytes.

The most interesting aspect of these results is that, in systems where it has been studied, A23187, in addition to causing an increased entry of calcium into the cell, causes a calcium-dependent increase in cGMP concentration in the cell. That is to say, if A23187 is added to the cell in the absence of external calcium, there is no increase in cGMP content of the cell; but if calcium is now added, there is a prompt rise in cGMP concentration. Likewise, when the plant mitogen, phytohemagglutinin, is added to human peripheral lymphocytes, it causes a calcium-dependent rise in cGMP concentration early in the response. Thus, in the case of at least some cell systems, the process of cell activation first appears to involve the interaction of the extracellular messenger with a surface receptor, which causes a change in the plasma membrane. This change then leads to an increase in the permeability of the membrane to calcium ions; and the subsequent rise in Ca^{++} concentration within the cell cytosol activates guanylate cyclase and leads to a rise in cGMP within the cell. Hence, in such a system, cGMP is not really a second but a third messenger in the hierarchy of information transfer in the control of cell function.

This general model is supported by the findings of Hardman and Shultz of Vanderbilt University showing that isolated guanylate cyclase from smooth muscle is activated by calcium ions. What we don't yet know is whether this is the only way in which the Ca^{++} and cGMP systems interact or how, in a system such as lymphocyte transformation, the rises in Ca^{++} and cGMP are integrated into the control of cellular response. In addition, we have not yet learned the nature of the feedback signals in this control system that determine the magnitude of this response; but it is of interest that agents that cause a rise in cellular cAMP in this system, in general, also cause an inhibition of transformation.

Perhaps the most exciting aspect of these studies is the utility of A23187 as a tool to dissect out the role of calcium as an intracellular messenger. A23187, in its own way, is to this field of research what theophylline has been to the study of cyclic nucleotides. With its use and the use of drugs that block calcium channels in membranes, we can expect rapid and significant advances in our understanding of the role of calcium in cell activation and its relationship to the cyclic nucleotides, cAMP and cGMP.

Section Three

Pathology

21

Surface Properties Of Neoplastic Cells

MAX M. BURGER
University of Basel

To readers of earlier chapters in this book, the fact that malignant cells show surface properties differing from those of normal cells is well established. However, though such phenomena as the loss of contact inhibition by tumor cells have been cited as part of the rationale for studying cell membranes, and correctly so, the phenomena themselves have not been discussed at any length. Such a discussion is clearly relevant to our understanding of the malignant process and to the improvement of our techniques for diagnosing and treating it. In the course of the discussion it will become apparent that the process of examining the anomalous surfaces of malignant cells leads to certain conclusions concerning the surfaces of normal ones.

The first clear demonstration of surface differences between normal and neoplastic cells dates from the 1940's, when D. R. Coman measured the adhesiveness of the two types. He did this by inserting thin glass needles into two adjacent, adherent cells and measuring the pull on the needles required to separate them. For tumor cells – he used a tissue sample from a cervical carcinoma – the force was much less; they were, so to speak, less "sticky" than normal cells. Not much has been done since to follow up this observation, but it is of obvious significance for the phenomenon of metastasis. Clearly, the less adhesive a malignant cell, the more easily it can break loose from the main tumor mass and become lodged elsewhere.

Even before Coman's demonstration, however, many qualitative observations, both clinical and pathologic, had led to the strong presumption that malignant cells have anomalous surfaces. Metastasis itself implies that the vagrant tumor cell is not recognized as "foreign" by the tissues in which it implants itself; if it were, we should expect it to be destroyed by leukocytes. Admittedly, not all types of tumor cells metastasize. But of those that do not, most show invasive growth into adjacent tissues —i.e., they do not simply damage these tissues passively by pressure of their expanding mass but actively intrude "arms" of tumor growth into them. Such invasion, in turn, implies either that the invaded tissue does not recognize the invader, or vice versa—or both. In any case, the failure of recognition can be most simply explained by surface alterations in the malignant cell that modify its effect on other cells, or theirs on it.

Closely allied to these recognition failures is the anomaly of the surface of tumor cells most frequently cited: loss of contact inhibition. Before discussing this in detail we should emphasize that contact inhibition actually involves two distinct phenomena: inhibition of movement and inhibition of growth. What one actually sees in cultures of *normal* cells is first multiplication of cells together with movement of individual cells up to the point at which they form a continuous monolayer. At this point both movement and growth decrease dramatically in the cultures of normal cells, but we are unable, thus far, to say whether inhibition of multiplication or inhibition of movement (if either) plays the key role in contact inhibition. In the cultures of *tumor* cells, by contrast, the cells continue to divide and also to move over each other, so that they can pile up, sometimes many layers deep. Much recent work suggests that loss of the inhibition of multiplication, rather than loss of the inhibition of movement, seems to correlate more reliably with the malignant state; for this reason we shall limit our discussion to the former phenomenon.

To study contact inhibition one begins with a bit of normal or tumor tissue, minces it up with dissecting scissors, and then dissociates the individual cells by treatment with proteases (primarily trypsin). The resulting suspension of single cells can then be plated and overlaid with the necessary nutrients and the growth observed. A standard technique is to add to the culture a given amount of nutrient medium with a low content of calf serum (10%), which is not renewed, and wait until the number of cells stops increasing, i.e., has reached its saturation density. With normal cells, as already noted, the saturation den-

Photomicrograph at left shows 3T3 mouse fibroblast cells grown to confluency; if injected into a mouse of the same strain, these cells do not give rise to tumors. Note their regular growth pattern. At right: Polyoma virus–transformed 3T3 mouse fibroblast cells have an irregular growth pattern and quite often they grow on top of each other. If injected into a mouse, these cells give rise to tumors. (Photos were taken by P. Turner with Nomarski optics; magnification 200x).

sity reaches a plateau as soon as the cells have formed a continuous monolayer, while tumor cells, given the same quantity of medium, pile up atop one another to reach a saturation density up to 10- to 20-fold greater.

The same high density occurs with "transformed" cells (and in fact was historically first observed with them). These are normal cells that have been treated with certain viruses – both DNA- and RNA-containing – or with chemical carcinogens or with x-rays. In addition, in most tissue cultures a few cells transform spontaneously, with no exogenous manipulation. We should perhaps clarify the word "transformation," which is a somewhat ambiguous concept. Sometimes it is defined simply as loss of contact inhibition, a definition not very suitable for our purposes since it involves a certain amount of circular reasoning. Alternatively, it may be defined as tumorigenicity – the capacity to give rise to neoplastic growth when injected into an experimental animal. However, though most transformed cells that have been tested show this capacity, not all do (for a variety of reasons, not all of them known); moreover, some cannot be tested at all (e.g., transformed human cells would be expected to grow primarily in a human subject – but such a test would be unlikely or impossible). In general, therefore, transformation is defined on the basis of certain specific changes in cell morphology (especially that of the nucleus) and stainability. Since it is often more convenient to use transformed rather than "natural" tumor cells, many of the experiments to be described have been performed with these. While the resulting findings are presumed to apply to natural tumor cells as well as to transformed cells, actual proof of this assumption will obviously require repetition of the experiments with natural tumor cells, and more than correlations only.

The difference in contact inhibition – i.e., in saturation density reached – between normal and transformed cells could conceivably be due to some nutritional deficiency affecting the former, so that in effect they run out of culture medium. However, this appears unlikely, since both cultures are started with the same amount of medium; indeed, the same difference occurs even if the two types of cells are cultured together in the same dish. The clinching evidence, however, came from experiments of R. Dulbecco a few years ago, in which a monolayer culture of confluent, normal cells was "wounded" by scraping an open line across the center. The cells on both sides of the wound immediately began dividing again and continued to do so until the wound was closed. It seems pretty clear that the normal cell possesses some capacity to recognize its fellows. Transformed cells have lost this capacity either to receive or to transmit the information of a crowded environment.

Theoretically, the difference might still be due to some "microenvironmental" effect – for example, a difference in permeability of the trans-

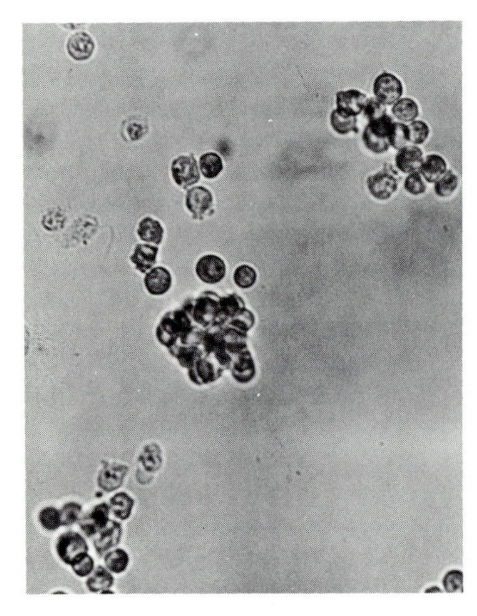

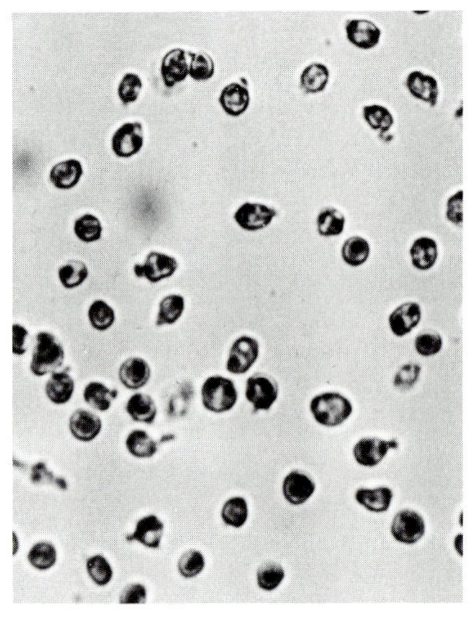

Agglutination is present five minutes after 50 μg/ml of wheat germ agglutinin (WGA) has been added to polyoma virus–transformed 3T3 mouse cells at room temperature (left, magnification 125x). In contrast, agglutination is not seen in normal 3T3 mouse fibroblasts under the same experimental conditions.

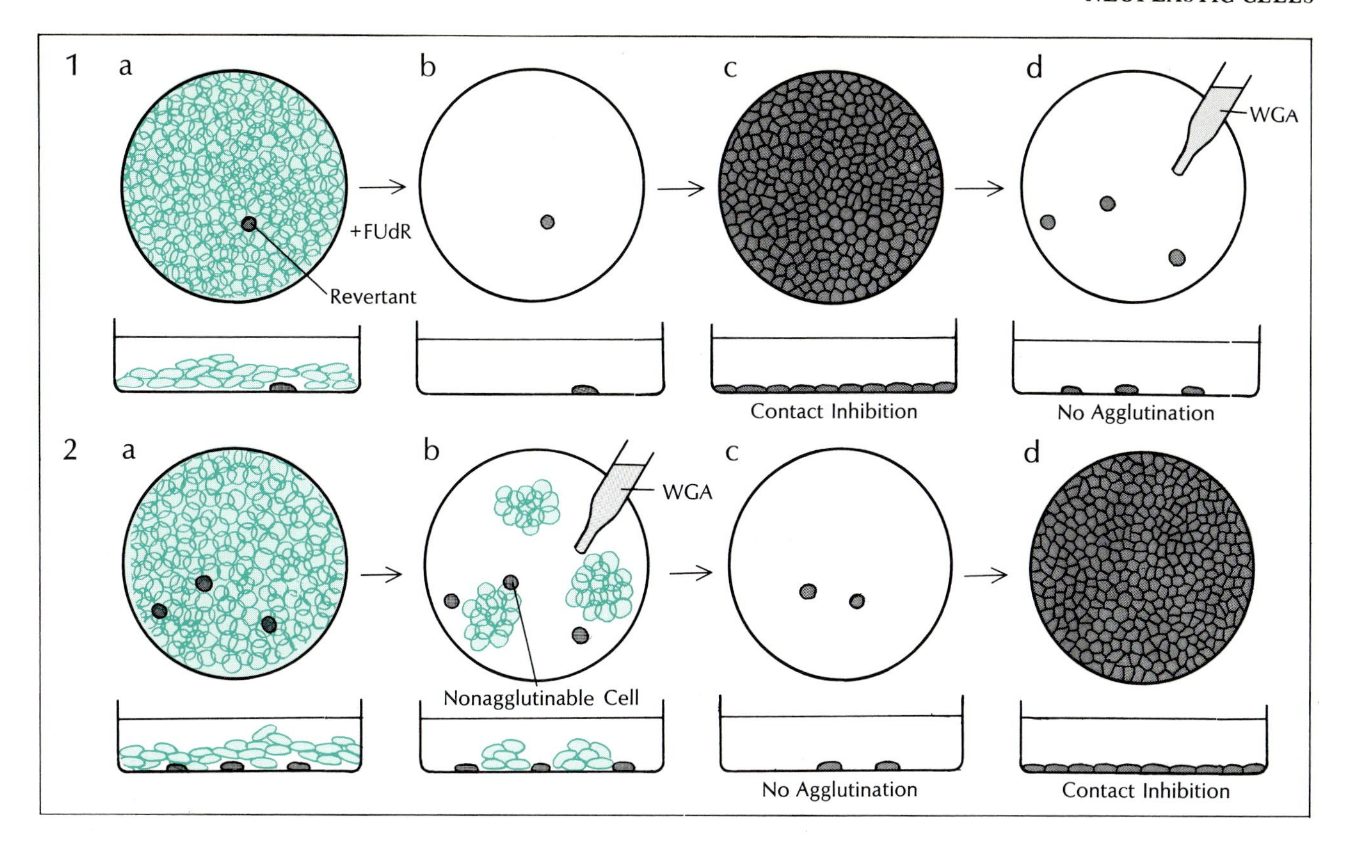

Culture of transformed cells shows pileup of cells because of loss of contact inhibition of growth (1a). Occasionally a cell (gray) spontaneously reverts to normal growth behavior and morphology; if isolated by killing all the growing transformed cells with FUdR (1b) and allowed to multiply, it shows contact inhibition (1c) and failure of agglutination if WGA is added (1d). If WGA is added to a population of transformed cells, those that bind WGA, and are killed, aggregate and float off from the culture dish while a few that are not killed by WGA (which always seem to be present) remain attached (2a, 2b). Transferred to another dish, such cells not only agglutinate poorly (2c) but regain some of the contact inhibition typical of normal cells (2d).

formed membrane that enabled it to take up nutrients more effectively at close quarters – and one must concede that this possibility has not yet been totally excluded. However, when I was at Princeton, my associates and I – along with most other researchers in the field—had made the assumption that the difference was due to some change in the transformed cell membrane that reduces its capacity to respond to crowding in the culture. This assumption has, I think, proved a very productive one, but I should stress that it is still just an assumption.

Any recognition system obviously involves two subsystems, one for sending, the other for receiving, and a failure of the system may be due to interference with either (or both) of the subsystems. Thus we can fail to recognize an acquaintance by sight if he is wearing a mask (signal interference) and also if we are wearing a blindfold (receiver interference). Some experiments by M. G. P. Stoker suggest that the loss of contact inhibition is due to derangement of the "sender," not the "receiver." He has found that in certain systems at least, tumor cells recognize normal cells surrounding them and stop growing – but are not inhibited by contact with other tumor cells. At present, the broader significance of this finding remains uncertain, but we can say, at any rate, that many experimental data have established clear and consistent surface differences between transformed and normal cells.

At the time (the mid-1960's) when my associates and I set about trying to identify these surface differences, other researchers had been investigating them for years by chemical means – breaking up the membranes and analyzing their constituents. However, since this approach had not singled out any clear-cut differences, we decided to work with intact cells, utilizing immunochemical rather than analytic techniques. We considered this approach the more promising in that immunochemistry not only can identify the presence of a given substance in the membrane, as does the analytic approach, but can also give some clues to the location of the substance, and therefore to the membrane's molecular architecture.

For our immunologic material we chose certain plant proteins called lectins rather than the more common globulin antibodies; the name comes from the Latin *legere*, "to select," and was coined because some of these substances can distinguish among the different human blood groups. We chose lectins rather than globulins because the former are much more stable – an important consideration inasmuch as we anticipated having to conjugate them with various markers such as ferritin and some aggressive radioactive labels, and we wished to minimize the likelihood of their being denatured by this treatment.

The discovery of the wheat germ lectin is worth a short digression, since it involves not one but two examples of scientific serendipity. In the late 1950's, two English investigators, Ambrose and Easty, were attempting to find an enzyme that would selectively inhibit the growth of transformed cells. They tested a whole series, including protease and lipase. At the time, there were two types of lipase in the reagent catalogues, one obtained from pancreatic tissue, the other from wheat germ. Perhaps since the wheat germ enzyme was cheaper, they tested wheat germ lipase, and this turned out to be the enzyme that would in fact inhibit the transformed cell cultures much better than the normal cells. That was the first bit of serendipity.

Joseph Aub of Massachusetts General Hospital, seeking to repeat these experiments, found that the material also selectively agglutinated transformed cells; he tentatively identified it as a mucopolysaccharide. When we took up the work, we questioned this conclusion, since mucopolysaccharides do not show the sharp chemical specificity found in proteins (e.g., gamma globulins and enzymes). When we purified Aub's "wheat germ agglutinin" (WGA), we found that in fact it was not a mucopolysaccharide, nor was it lipase, but rather an impurity of the lipase preparation; this was the second bit of serendipity. The purified WGA turns out to be a glycoprotein with a molecular weight of about 17,000 daltons. Originally, we hypothesized a structure with active sites at both ends of the molecule, enabling it to form a bridge between complementary sites on transformed cell membranes and thereby agglutinate the cells. It seems now that the wheat germ agglutinin in its most active form consists of two such molecules providing possibly four active sites, and the exact mechanism for the agglutination is still not clear.

Before proceeding further along these lines, however, we needed to be quite certain that WGA's capacity to agglutinate transformed cells reflected not merely an irrelevant surface difference between them and normal cells but a significant difference – one clearly related to tumorigenicity. WGA has, indeed, been shown to aggregate more than 60 different lines of tumor cells – every one we have tested so far – but it also reacts with some types of normal cells, notably erythrocytes. An occasional exception among the tumor cells (so far two reported by other laboratories) should be tested for one of the many other lectins that in the meantime have been found to preferentially agglutinate tumor cells. Since lectin agglutinability of tumor cells seems to become a general phenomenon for several lectins, the tumor cell that does not agglutinate better than its parental normal cell with at least one of the many lectins is probably an extremely rare tumor cell and will first have to be found.

WGA, as it turns out, not only reacts with natural tumor cells but also with transformed cells, regardless of the method of transformation – chemical, x-ray, virus, or spontaneous. Even more significantly, if the transforming virus is such that it will alter the cells only under certain conditions, WGA reacts with the cells only under those same conditions. This was shown in an experiment with a virus strain developed by Dulbecco and Eckhart at the Salk Institute, a temperature-sensitive mutant of the polyoma virus called Ts-3. This virus, as its discoverers had already found, will transform certain cells (such as those from kidneys of baby hamsters) at 32° C (the "permissive" temperature), but the transformed cells revert to normal morphology and near-normal growth patterns if the temperature is then raised to a "nonpermissive" 39°. By lowering and raising the temperature the shift can be induced as often as one likes. We found that the change from permissive to nonpermissive temperatures always followed a shift from agglutinability to nonagglutinability, and vice versa. Similar results have been obtained with a temperature-sensitive mutant of the Rous sarcoma virus (an RNA rather than a DNA virus), and also with another type of temperature-sensitive mutation affecting not the

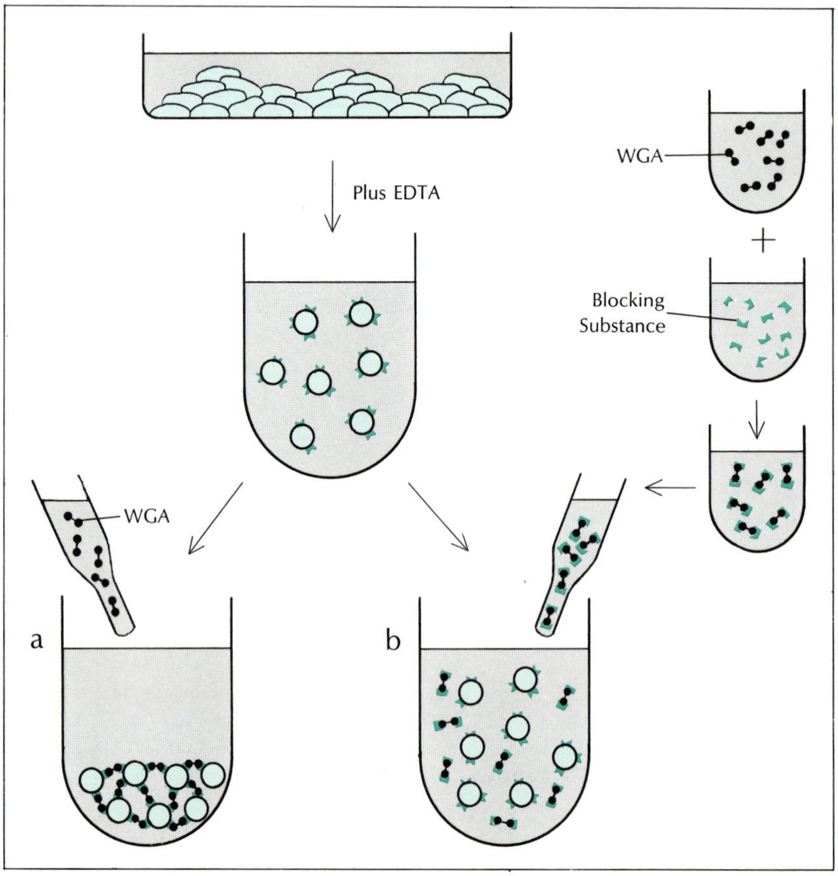

WGA is thought to agglutinate cells by binding together receptor sites on adjacent cells; if these are removed with EDTA from the culture dish and WGA added, cells agglutinate (a). Adding a particular saccharide that binds to the active sites on WGA (upper right) blocks this action and agglutination does not occur (b), demonstrating that the cell receptor sites contain the same inhibiting substance, a disaccharide.

transforming virus but the host cell itself (discovered by Basilico and Renger at New York University).

Together with Robert Pollack at Cold Spring Harbor we were able to relate tumorigenicity to agglutinability in another way. Just as in a population of normal cells an occasional one will spontaneously transform, so in a population of tumor cells an occasional one will spontaneously revert to normal. Pollack succeeded in isolating several such variant cell lines, which had regained their contact inhibition, and in all cases they were shown to have also lost their agglutinability. The reverse has also been shown by several laboratories. By treating a transformed cell culture with WGA or with concanavalin A (another lectin found to behave like WGA by Inbar and Sachs at the Weizmann Institute in Israel), one can isolate the mass of agglutinable cells from the few that have spontaneously *lost* their agglutinability – and it turns out that the latter have at the same time regained part (though not all) of their normal capacity for contact inhibition. Thus, by several criteria, agglutinability correlates with the transformed state and loss of contact inhibition of growth (always excepting the normal erythrocyte), whereas nonagglutinability correlates with the presence of contact inhibition of growth. A straight correlation between the tumorigenic state and the agglutinable state could be shown when comparing fibroblast cells from isogenic BALB/c mice with the polyoma virus–transformed derivatives of those cells.

Having determined that transformed, tumorigenic cells possessed some distinctive surface substance that bound WGA, we then set about identifying the substance. It seemed likely that it was a sugar, since sugars attached to erythrocyte membrane proteins had already been identified as the blood-group antigens with which some other lectins reacted. We therefore reasoned that if we added an appropriate sugar to the cultures along with WGA, it would bind to the WGA's active sites and prevent these from interacting with the cell surfaces (that is, from subsequently agglutinating the cells). By testing a number of sugars, we found that WGA could indeed be inactivated by – and only by – the sugar *N*-acetylglucosamine, most conspicuously by its dimer, di-*N*-acetylchitobiose.

Tom Shier of the Salk Institute then carried our reasoning a step further. If the WGA receptor on tumor-cell surfaces was in fact di-*N*-acetylchitobiose, as our experiments seemed to indicate, then immunizing an animal against that disaccharide should at the same time immunize it against tumor cells. He thereupon bound the di-*N*-acetylchitobiose to a carrier, polyaspartate (an artificial polypeptide composed entirely of aspartic acid residues), injected the substance into mice, and indeed found that, under certain conditions, the animals were protected against subsequent challenge with tumor (myeloma) cells. Furthermore, Shier has attempted to elaborate these results in a direction more relevant to possible clinical situations (since outside the laboratory, animals are seldom exposed to "infection" by actual tumor cells). He has sought to determine whether immunizing an animal against artificial agglutinin receptor sites will protect it against the effects of *chemical* carcinogens – through immunologic destruction of cells transformed by the chemical. These experiments, however, still permit several interpretations as to the mechanism involved in protecting the animal, and for my part I wonder whether, even if successful, they are likely to prove of clinical value. Since lectins react with normal erythrocytes as well as with transformed cells, immunization against lectin receptors on tumor cells might well involve simultaneous immunization against the presumably similar receptors on erythrocytes or other normal cells, leading to clotting or some form of autoimmune disease. Considering their relevance for oncology, they should nevertheless be pursued.

Our own continuing effort has been to study in further detail the nature of the surface changes that render the virally transformed cell capable of being agglutinated by lectins. There are basically two possibilities. First, either the normal cell membrane possesses no agglutinin receptors or too few for agglutination, whereas in the transformed cell the virus induces or accelerates synthesis of the receptor molecule. A second possibility is that the normal cell has the same number of receptors as the transformed cell,

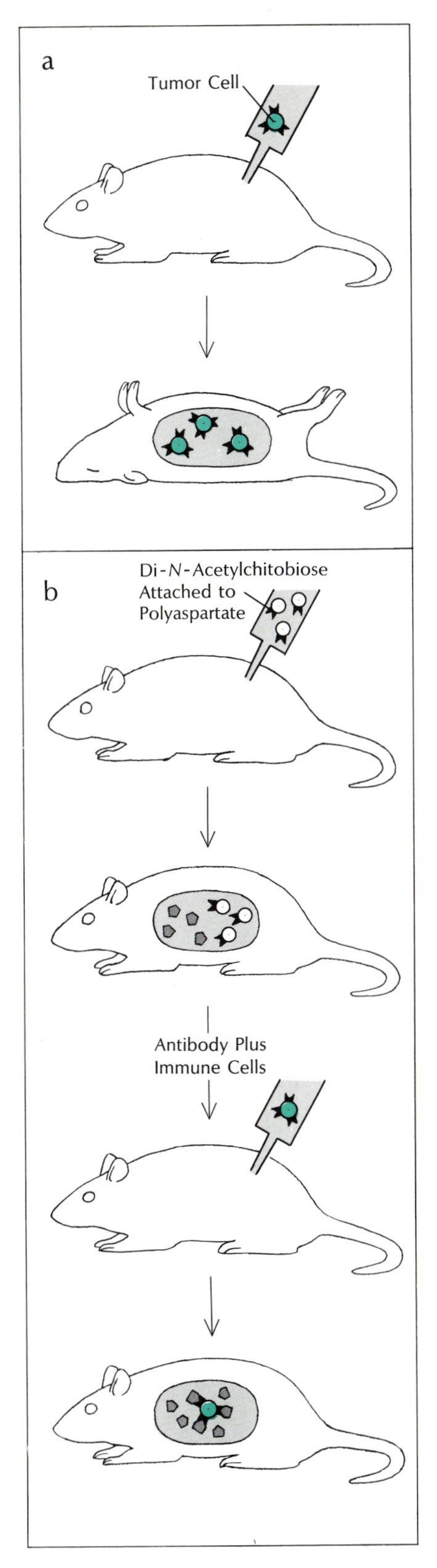

Normal mouse injected with tumor cells dies when they multiply (a). However, if animal is first injected with "receptor site" disaccharide bound to artificial peptide carrier (b), it forms antibodies and immune lymphocytes against the disaccharide; these then bind to receptor sites on injected tumor cells, inactivating or killing them.

but they are so arranged in or on the membrane that they are nonfunctional; in the transformed cell the virus changes the membrane structure so as to activate what may be called their cryptic capability of being agglutinated. The first possibility can be ruled out for many reasons, particularly since a semiquantitative analysis of normal and transformed membranes shows equal amounts of the "receptor substance," i.e., the *N*-acetylglucosamine containing glycoproteins.

The second possibility, of cryptic capability, could come about in a number of ways; the most obvious is that in the normal cell the agglutination sites are physically buried within the membrane. Other explanations were recently suggested, such as increased fluidity of the receptor sites in transformed cells and thereby cluster formation by the multivalent lectin of the receptor sites. This model is primarily based on similar amounts of overall binding sites that are measured with isotopically labeled lectins. Until the relevant binding sites can be sorted out from the multitude of lectin-binding glycoproteins and glycolipids, which most probably have different lectin affinities, comparable binding studies cannot be performed and conclusions about the cause for the differential agglutinability of untransformed and transformed cells will have to remain ambiguous.

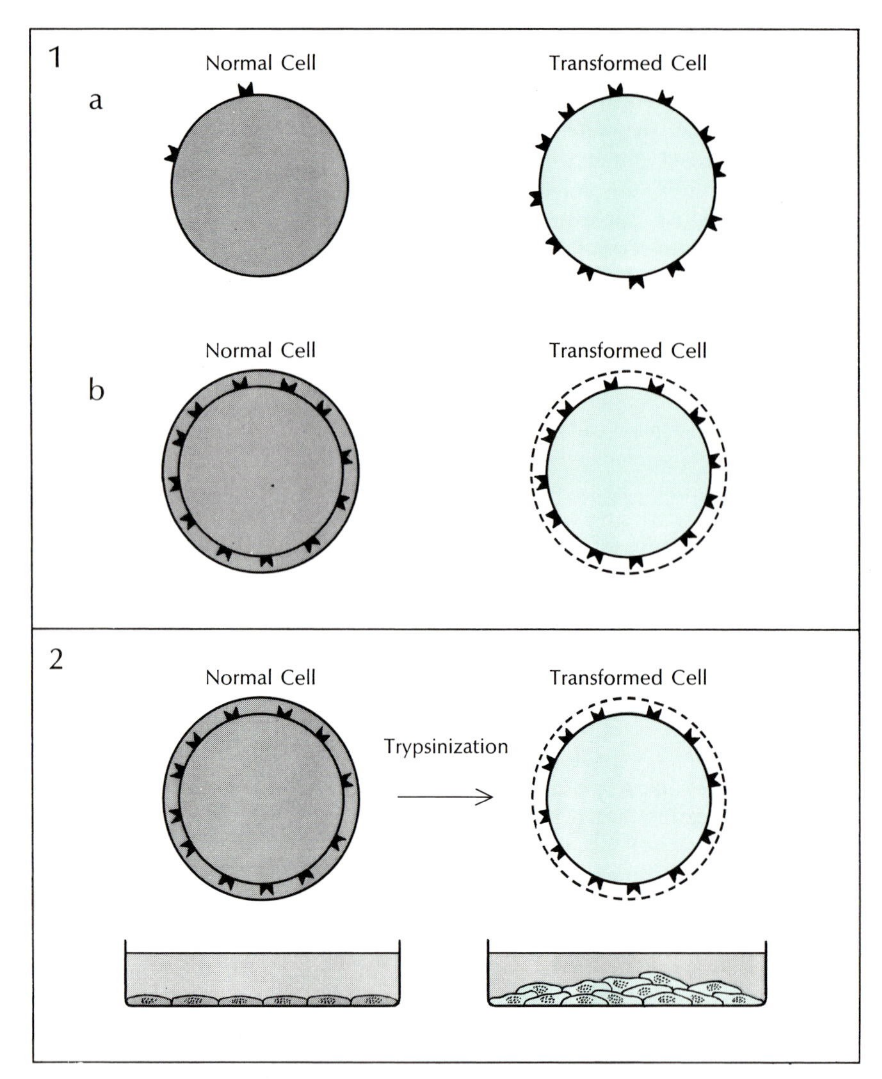

Surface differences between normal and transformed cells can be explained in two hypothetical ways. Normal cells may have too few receptors for agglutination, or none at all (1a); transformation triggers synthesis of new or more receptor sites. Alternatively, normal cells may contain the full complement of receptor sites, which, however, are covered by an exterior layer or are poorly accessible for some other reasons (1b). Second hypothesis is strengthened by an experiment in which normal cell, with normal contact inhibition of growth (left), is temporarily changed by treatment with trypsin to a "transformed" cell with no contact inhibition. Trypsinization may, for instance, remove the outer coating that was blocking the receptor sites. Such treated cells will, however, repair themselves and revert to the normal nonagglutinable state quite rapidly, as true transformed cells will not. While confirming the appearance of increased agglutinability after protease treatment of untransformed cells, recent studies indicated at the same time that growth stimulation of 3T3 fibroblasts was not a general phenomenon (R. D. Glynn, C. R. Thrash, and D.D. Cunningham, Proc Natl Acad Sci USA *70:9: 2676, 1973). Such protease stimulations could in principle, however, be seen with other cells like chick embryo fibroblasts (B. M. Sefton and H. Rubin,* Nature *[London] 227: 843, 1970).*

To us, this immediately suggested an experiment in which we treated normal cells for short periods with minute amounts of trypsin to remove such a hypothetical outermost layer of the normal cell surface – or at any rate the protein portion of that layer – without destroying the cell. We found that trypsin-treated cells did indeed become agglutinable and, even more important, that they temporarily lost their capacity for contact inhibition, dividing and growing like transformed cells. The loss of inhibition is temporary (about one cell generation), since, as many types of experiments have shown, normal cells can rapidly repair many kinds of cell damage, including the kind inflicted by our mild protease treatment.

Of course this experiment by itself did not prove that the trypsin was actually uncovering occult agglutination sites; conceivably it might have deranged the cell in some other way by passing through the plasma membrane into the cytoplasm. We were able to exclude the latter possibility by binding the trypsin to plastic microspheres too large – about 1/10 of a cell diameter – to pass through the membrane (and observation confirmed that they did *not* enter). The results were the same – and assay afterward showed that the trypsin did not leak off and enter in soluble form but was still attached to the beads.

We do not yet know whether the protease treatment removed a covering protein, thereby unveiling cryptic sites, or whether it simply split some peptide bonds that in some yet unknown fashion also rendered the cells agglutinable and capable of multiplication. Whatever the mechanism, the fact that protease treatment could induce growth immediately suggested to us a mirror-image experiment. If one could cover the agglutinating sites on transformed cells, would they then behave like normal cells? An obvious way of testing this would be to add lectins to the culture, which would cover the receptor sites by binding to them. However, we found – not unexpectedly, since the observation had been made previously by Leo Sachs' group in Israel – that the lectins tended to immobilize, agglutinate, and eventually kill the cells. What was needed was a modified lectin that would not agglutinate the cells yet would still remain able to bind to them. We obtained this by treating the lectin with a proteolytic enzyme, chymotrypsin, producing a preparation that would not agglutinate, and that would, moreover, prevent agglutination when normal lectin was added, evidently by blocking the binding sites on the cells.

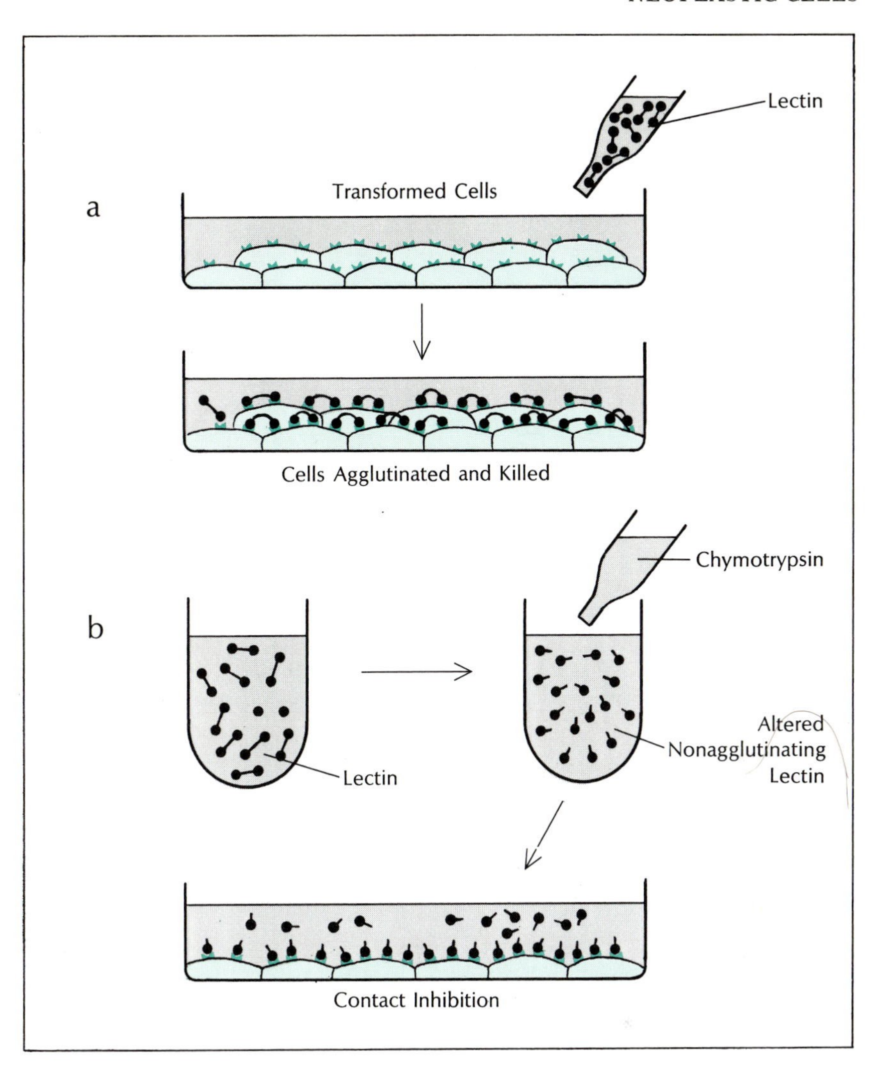

*Experiment in blocking receptor sites with lectin failed, as expected, when transformed cells were agglutinated and killed (a). However, when lectin molecules were broken apart with protease, they could block receptor sites without agglutinating cells; covering of the cell surface receptors with such a protease-treated lectin fraction induced normal contact inhibition of growth in the transformed cells (b).**

Hypothetically, we had fragmented the normal lectin molecule to the point where each fragment had no more than one active site capable of binding to the receptor, whereas to bridge between receptors (or cells) would obviously require at least two active sites. The exact molecular mechanism that led to the desired lectin preparation is not yet known. But hypothesis aside, this modified lectin did prevent tumor-like growth of the treated transformed cells. It did not stop growth as such; the cells, like normal cells, would grow up to the point of confluency with other cells – but at that point, like normal cells, they would stop. Moreover, the phenomenon was reversible. When we added to the medium the specific carbohydrate reactor substance, which competed for the active sites of the modified lectin, thereby uncovering the agglutination sites, the cells lost contact inhibition again and reverted back to the growth type seen prior to the addition of the modified lectin.

The finding that the exogenous protease-treated trypsin temporarily removed contact inhibition in normal cells suggested the possibility that a derangement in endogenous proteases of the cell surface might be responsible for the loss of inhibition in transformed and tumor cells. Assays of transformed cell membranes with presently available nonspecific and not very sensitive protease assays reveal no general increase in proteolytic enzymes, which is to say that the derangement, if it exists, is not likely to be a quantitative one. Indeed, we had not expected it to be. Rather, we believe that the derangement affects only certain membrane proteases, either increasing their total content or changing their specificity. We have found, in fact, with H. P. Schnebli, that when protease inhibitors such as TLCK, TAME, or ovomucoid are added to transformed and normal cultures they inhibit the growth of the transformed cells at much lower doses

**Although the proteolytic preparation of nonagglutinating, but still binding, derivatives of lectins seems to be of variable quality, such preparations have been reported recently again (M. S. Steinberg and I. A. Gepner,* Nature New Biology *241: 249, 1973; P. Evans and B. Jones,* Exp Cell Res *88: 56, 1974). A better defined and similar preparation can be obtained by succinylation of concanavalin A, which at low doses does not inhibit growth of transformed cells (I. Trowbridge and D. Hilborn,* Nature *250:304, 1974), probably since it binds to serum glycoproteins and is inactivated; while at higher doses, inhibition can be observed (R. Mannino, unpublished).*

than they do the normal ones. Recently Edward Reich and Dan Rifkin at Rockefeller activated this field of protease alterations in transformed cells. They found that transformed cells carried a very specific protease which could activate another inactive protease in the serum containing medium (plasminogen) to its active state (plasmin). Cells isolated from normal tissues did have very little activity of this plasminogen activating protease. Although such clot-dissolving enzymes could have potential tumorigenic relevance for the establishment of metastases, no such clinical relevance has been established so far.

The difference in agglutinability between tumor and normal cells suggested to us the possibility of using this phenomenon as a tool to diagnose neoplasia. If, for example, one could label the agglutinin with a fluorescing molecule, tumor cells should fluoresce brightly and normal cells should not, thereby permitting easy and quick identification of the former. More remotely, one could imagine conjugating the agglutinin with some antimetabolite or other antitumor drug, thereby ensuring that the drug would attack only tumor cells, rather than, as is now too often the case, damaging normal cells as well.

A pilot experiment with fluorescent-labeled agglutinin and transformed cells established that there were indeed differences in fluorescence between these and normal cells, but these were not of a sufficient magnitude to permit exploitation in clinical diagnosis. This does not mean the idea is invalid, merely that it did not show clinical promise in this system, using WGA and cells of a particular type. However, the control experiments turned up the interesting observation that *normal* cell populations always contained a few strongly fluorescent cells. Initially, we thought that these were probably dead cells, in which the breakdown of permeability barriers had accelerated the uptake of fluorescein. Further investigation, however, established that these highly fluorescent cells were by no means dead; rather, they were actively dividing. In fact, normal cells during mitosis turn out to undergo temporary surface changes very similar to, if not identical with, the permanent changes we find in transformed and tumor cells. Thus it is not strictly accurate to assert that normal cells never agglutinate; they *do* become agglutinable, but only for about one hour out of their (say) 20-hour growth cycle.

F. Jacob, in experiments with bacterial cells, had already postulated an intimate connection between cell surface and cell division; changes in the cell membrane, he suggested, might "tell" the cell to replicate its genetic apparatus; the apparatus, having gone through replication, then "tells" the cell membrane to undertake the changes necessary for cell division. From our experiments, it appears that the same process may operate in animal cells. It has in fact been known for some time that completion of the so-called S-phase of replication, in which DNA is synthesized, somehow triggers mitosis; it seems not unlikely that surface changes at the completion of mitosis in turn might be required to initiate the next S-phase, thereby establishing a complete, closed-loop feedback system.

As yet we have not secured any evidence that mitotic change triggers DNA synthesis in this manner, but we have obtained some results that suggest how it might do so. These grew out of findings at several laboratories that addition of cyclic AMP to the culture medium can change the morphology of tumor cells to something approximating that of normal cells, and the further finding that normal cells contain more CAMP (at least twice as much) than tumor cells. Why these differences exist is not known, but it is a fact.

Combining these results with our own findings, we speculated that the low levels of cyclic AMP in tumor cells somehow constitute a "message" in the cell's interior to continue in its cell cycle (CAMP is known to act as a "messenger" in many other cell processes). If the mitotic surface alteration really had signaling function, as we proposed earlier, one would expect to find similar low levels of CAMP in *normal* cells during mitosis – and that is just what we did find. Moreover, Sheppard in Minneapolis and our laboratory found that the same low levels could be induced in contact-inhibited normal cells by trypsinization – which, the reader will recall, had earlier been shown to render these cells temporarily like transformed cells, both with respect to growth and agglutinability. Finally, we found that when we added CAMP to these trypsinized cells, even in very low doses, we could prevent them from growing, just as CAMP had previously been shown to "normalize" the morphology of tumor cells.

All this, of course, does not prove that the drop in CAMP levels actually triggers DNA synthesis, still less that the surface change triggers the drop in CAMP levels. What all these experiments do add up to is a series of less than conclusive but highly suggestive correspondences: active growth corresponds with agglutinability, which corresponds with protease activation, which corresponds with low CAMP levels; inhibited growth corresponds with nonagglutinability, which corresponds with protease inactivation, which corresponds with high CAMP levels. And there, for the moment, the problem stands. It seems inconceivable to me that these elaborate correspondences do not signify something of great import concerning the relationship between cell surface and cell growth, in both normal and malignant cells, but exactly what remains to be determined.

22

The Membranes of Lymphocytes

JONATHAN W. UHR
University of Texas

The plasma membranes of lymphocytes are inevitably of special interest to both the biochemist and the cytologist. The basic physiologic function of these cells – defending the body against invading organisms – involves the presence on their surfaces of receptors that can detect "foreign" antigens in the serum and thereby set the cells' defense mechanisms into action. In addition, the process by which immature bone marrow cells differentiate into fully immunocompetent lymphocytes is known to involve the sequential appearance on (and also disappearance from) their surfaces of various antigenically distinctive molecules. And though the functions of these "marker" substances, like the differentiation process itself, remain poorly understood, it seems likely that the better we can identify and characterize them, the better we shall comprehend the development of immunocompetence itself – with all that this implies in such clinically important areas as infection, transplantation, and neoplasia.

As background to some of the recent findings in lymphocyte membrane research, I should like to review some of the fundamental facts about these cells. They are, to begin with, of two basic types. Both originate as "stem" cells in the bone marrow, but on passing into the system take divergent paths. One group migrates to an organ known in chickens as the bursa of Fabricius; its analogue in mammals is not known with certainty but bursal functions seem to be distributed between the bone marrow itself and the spleen. Once in these sites, the stem cells differentiate into B lymphocytes possessing immunocompetent surface receptors capable of recognizing and responding to foreign antigens. On stimulation with an appropriate antigen and with the help of a second cell type, to be discussed below, a B cell differentiates further into a much larger plasma cell, which then fulfills its defensive function: the release into the serum of antibody that can react with the antigen.

The second type of lymphocyte, the T cell, matures not in the bursa or its mammalian analogue but in the thymus, where it differentiates into a cell also possessing immunocompetent receptors, though these, as we shall see, may not be chemically identical with B cell receptors. When its receptors are appropriately stimulated by antigen, the T cell reacts in two ways. First, either it or some substance that it releases interacts with B cells, helping to trigger their transformation into plasma cells. In most cases, indeed, it appears that unless the B cell receives *both* stimuli – the antigen plus one from the antigenically stimulated T cell – it will not mature into a plasmocyte. This helper function of the T cell appears to depend in large measure on the T and B cell sharing certain products of the major histocompatibility locus, e.g., T and B cells of different mouse strains may not cooperate. But the T cell plays a direct as well as an auxiliary role in the body's defenses: antigenic stimulation promotes its development into a "killer cell," which can attack and destroy many types of foreign cells, including those of transplanted tissues.

From the biochemist's standpoint, there are three specially interesting aspects of lymphocyte surfaces. First is the nature of the immune receptors of both B and T cells, along with their sites of manufacture within the cell and their regulation on the cell surface. Second is the nature of the products of the major histocompatibility locus that play a major role in T and B cell cooperation, as summarized above. And third is the nature and significance of the surface antigens, or markers, that appear and disappear on lymphocytes as they mature. Recent studies have begun to throw light on each of these areas – though, as in most aspects of immunology, not nearly as much light as would be required for complete illumination.

Formerly, lymphocyte surface antigens were studied by treating intact cells with specific antibody. For example, one could treat the cells with a specific antibody possessing cytotoxic properties; if the cell lysed, this would point to the presence of the corresponding antigen on the cell surface. One could also employ an antibody labeled with a fluorescent dye; the appearance of fluorescence on the cell surface under the microscope would again indicate the presence of the corresponding antigen.

These techniques were useful in dealing with intact cells, but they could not deal effectively with homogenized cells or cell fractions. For this purpose biochemical techniques were needed.

The biochemical study of lymphocyte surfaces was long blocked by two major hurdles. First, the molecules of interest were present in extraordinarily small quantities. Second, they were mixed with large amounts of other molecules from which they could be separated only with great difficulty and poor efficiency. It has been estimated that one would need to process gallons of cells to obtain an adequate and reasonably pure sample of any one of the molecules in question. Thus, the only practicable way in which they could be characterized was to label the molecule with an isotope and to isolate it with a specific antibody.

In 1971, my associates (Bauer, Schenkein, Vitetta, and Sherr) and I described a method for selectively labeling surface molecules on lymphocytes and for separating various classes of them for more precise characterization. This technique is based on an earlier observation that the enzyme lactoperoxidase (a constituent of raw milk) could attach iodine to the surface proteins of intact erythrocytes. By using radioiodine we found that the labeled surface proteins could be easily distinguished from those of the cell interior. Moreover – and an important point for analytic studies – the labeling could be quite heavy; the iodine is attached to tyrosine residues, which are quite common in most types of protein.

Our own method begins with the radioiodination of a lymphocyte suspension. The cells are then lysed, following which they are treated with an antiserum to the particular molecule we desire to identify – for example, an immunoglobulin (Ig) of a particular class, such as IgM, IgA, and so on. The antiserum, of course, precipitates the surface molecules with that antigenic specificity, along with any similar (but unlabeled) molecules from the cell interior. For

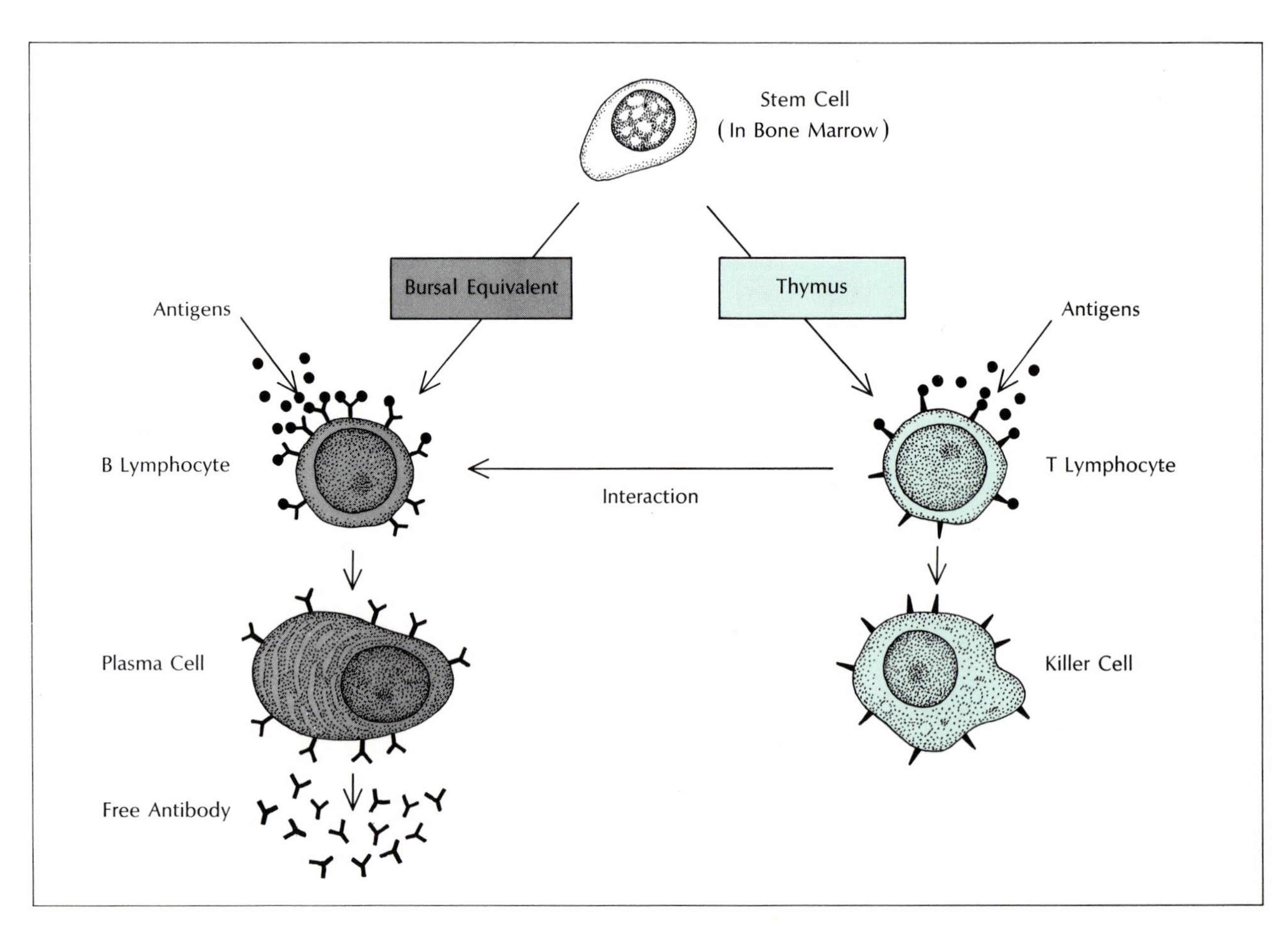

Stepwise differentiation of stem cells into B and T lymphocytes includes changes in cell surface such as development of receptors. Interaction of T and B cells with antigen results in further differentiation. B cell surface receptors are immunoglobulins resembling free antibody, which the cells later secrete, while the composition of T cell receptors is presently unknown.

our purposes, however, it does not matter whether a particular molecule is or is not present within the cell, so long as we can be certain (by means of the labeling) that it is or is not present on the cell surface. To use a homely metaphor, we want to know whether there is butter on the bread; provided we can answer that question, the question of whether there is butter *in* the bread (i.e., incorporated into the dough) is irrelevant.

Some of our early findings concerned the identity of the surface immunoglobulins on B lymphocytes. Before citing them, however, a little more background may be in order. As most people know, immunoglobulin molecules consist of four polypeptide chains: two linked "heavy" chains, each of which is joined to a "light" chain that lies more or less parallel to it. One end of the Ig molecule, including about half the length of the light chains and a fourth or fifth of the heavy chains, gives it immunologic specificity – the capacity to react with one or another specific antigen. Considered from this standpoint, any individual possesses tens or hundreds of thousands of different Ig's, each keyed to its own antigen.

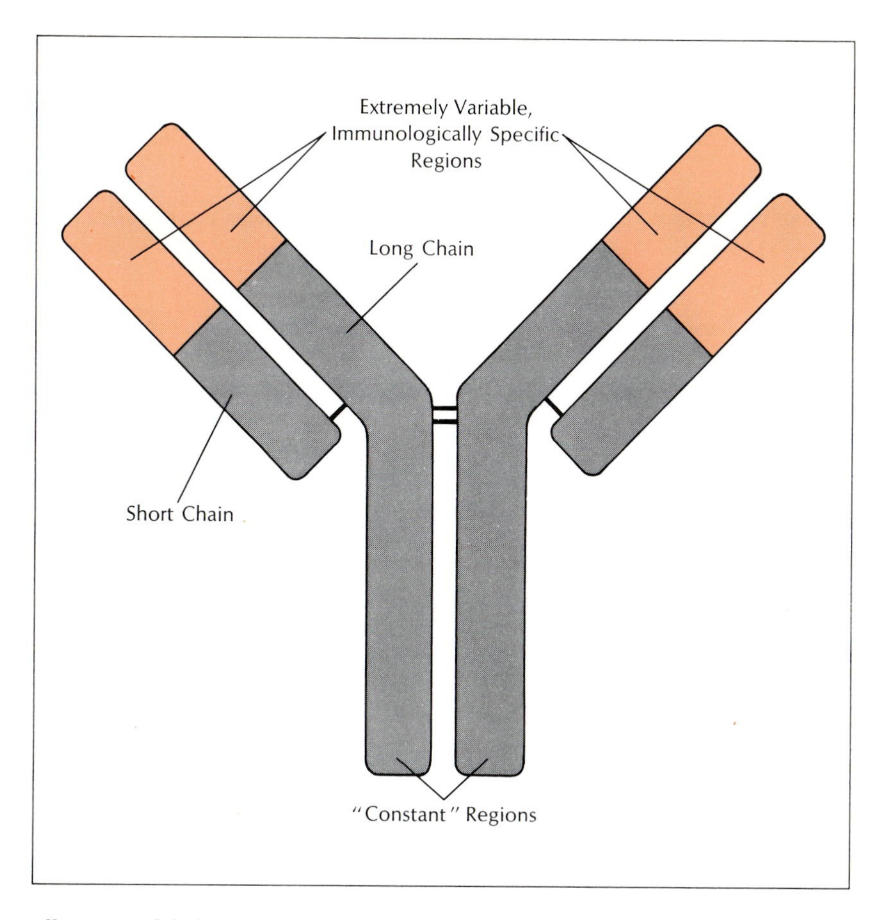

All immunoglobulins, bound or free, are of basically the same structure; two long protein chains linked to two short ones. Colored portions of both chains exist in thousands of variations, each keyed to a specific antigen. Gray sections of the long chains are of only five known classes (A, M, G, D, E), and of the short chains of two types, each differing in amino acid residue sequence.

The remainder of the Ig molecule also varies, both in molecular weight and in the sequence of amino acid residues that compose it, but the variability is far narrower. Thus far, indeed, only five classes of heavy chain are known, and they have been labeled IgA, M, G, D, and E. Their distinctive properties are thought to reflect different roles that they play in the body's defenses. Broadly speaking, one can say that whereas the specific end of an Ig molecule determines the antigen it will recognize and bind to, the nonspecific "type" end determines what it can do after binding.

Earlier experiments had already provided strong evidence that immunoglobulin of some type was present on the surfaces of B lymphocytes, where it – at least in part – served as the receptor that was stimulated by exogenous antigen to transform the cells into plasmocytes. Through our radioiodination technique, we were able to identify this immunoglobulin as IgM, with a molecular weight of about 180,000. This Ig had long been known to be present in serum, but in pentameric form, with a molecular weight around 900,000. The lymphocyte surface IgM, however, was clearly the monomer, a difference presumably reflecting its different function in that locus.

Very recently, in collaboration with a team of NYU scientists, we have identified another Ig on the surfaces of B lymphocytes. Its molecular weight is almost identical with that of IgM – making it difficult to distinguish from the latter by most methods of electrophoresis — but the two differ in certain other respects. Specifically, both are technically what are called glycoproteins, with chains of sugar residues attached to their basic protein framework (see Chapter 6, "Sugars of the Cell Membrane," by Roseman). They differ, however, in that the "new" Ig appears to contain less protein (a shorter amino acid chain) than IgM, but a larger proportion of sugar. Thus the two can be separated by modifying the electrophoresis technique so that the sugar moieties contribute less to the molecules' mobility.

This "high sugar" Ig from mouse lymphocytes has many of the properties assigned to human IgD, and we have tentatively concluded that it is in fact the murine counterpart of that immunoglobulin. Even more remarkable is our finding that in some mouse lymphocytes – those from peripheral lymph nodes, and from some areas of the gut and thoracic duct – this putative IgD makes up something like 90% of the identifiable surface Ig.

Several experiments then provided evidence to place surface IgD in the development of B lymphocytes. It

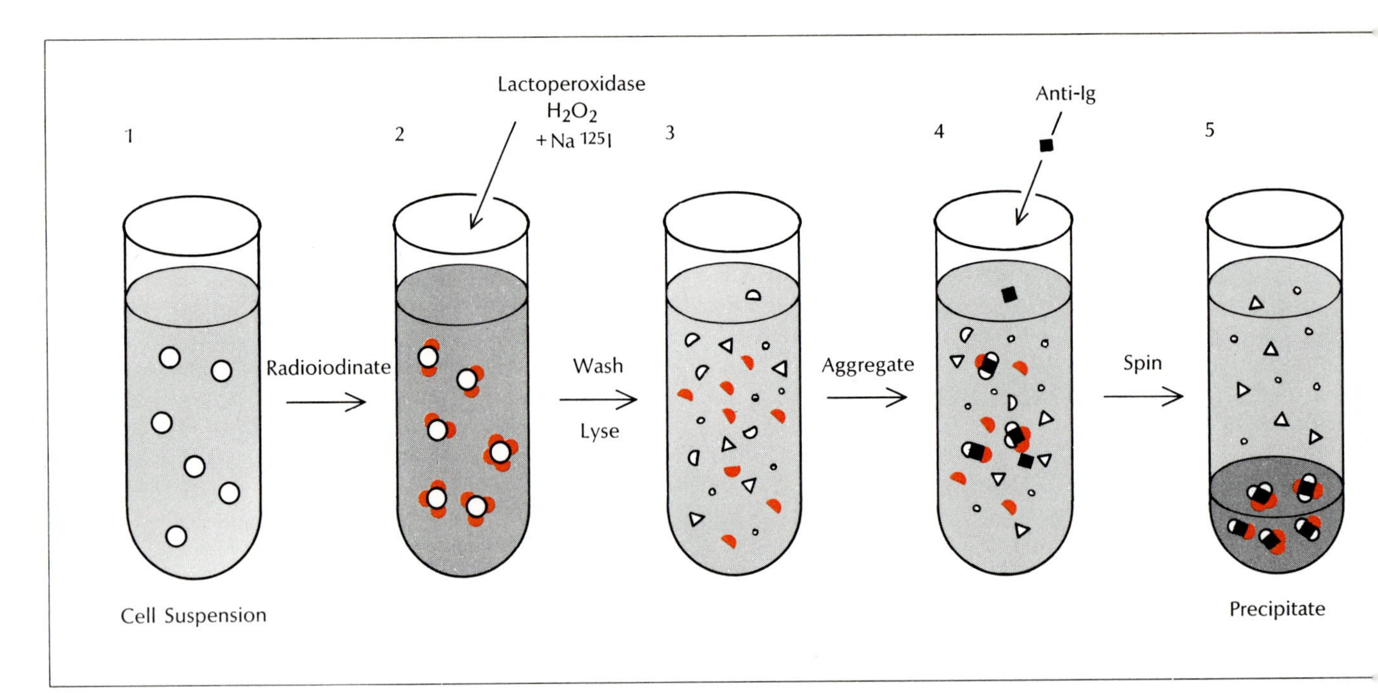

Identification of lymphocyte surface molecules begins with enzymatic iodination of intact cells (1,2). After lysis (3), preparation is treated with antibody specific to suspected surface molecule (4), which precipitates this substance with any cyto-

was shown that it was not present on spleen cells of newborn mice; these cells did have cell-surface IgM. IgD first appeared at about two weeks of age. This time of appearance was not altered in germ-free animals, suggesting that the acquisition of IgD did not depend upon antigenic stimulation of the cells. IgD also appeared in conventional amounts in mice lacking a thymus gland, indicating that T cells were not essential to its development. Finally, we have found only IgM on the surfaces of cells from bone marrow, which are of course the precursors of B lymphocytes.

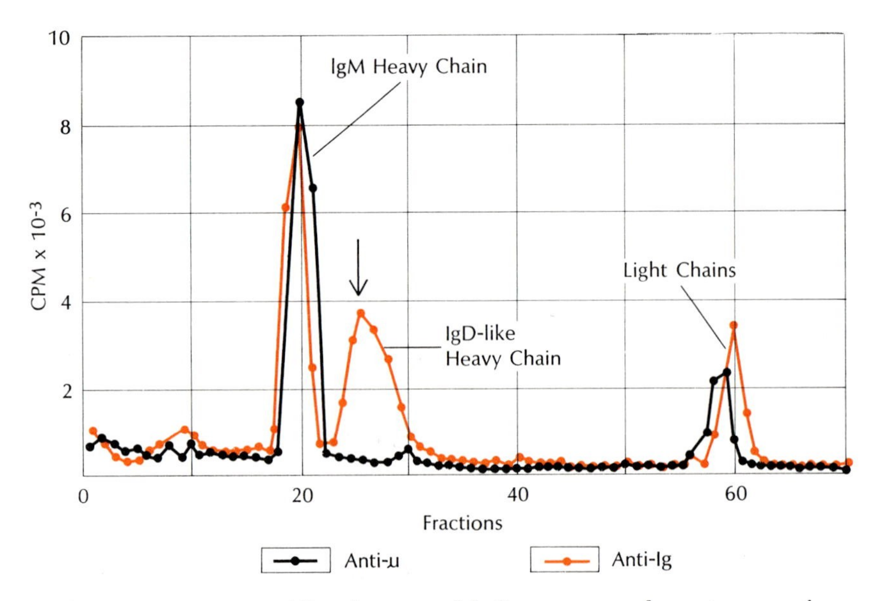

Evidence for appearance of "new" immunoglobulin on mouse splenocytes comes from electrophoretic studies in which radioiodinated cell membranes are precipitated with anti-μ (against Ig heavy chains) and anti-Ig antibodies. Anti-Ig peak, which appears to right of IgM peaks, is analagous to human IgD heavy chain.

According to our hypothesis, therefore, the B cells start out with IgM on their surfaces, but during the differentiation process in the spleen they acquire IgD. We believe they then migrate to peripheral lymphoid tissues as small lymphocytes now fully competent immunologically to respond to antigen.

After stimulation by antigen, B cells mature into plasma cells, which secrete IgM and, later in the immune response, IgG and IgA. (It may be noted that IgA predominates in the Ig secreted by mucous-membrane-associated plasma cells, e.g., gut, whereas IgG appears to be secreted by the plasma cells of most lymphoid organs.) Yet one would have to assume, from concepts of immunology and from some experiments of others, that the antigenic specificity of the Ig is not changed, meaning that while the back end of the molecule is shifted from M to D to G or A type, the front end remains unaltered. One could explain this by assuming that the front end and the back end are controlled by different genes, one of which is switched off and replaced by another during maturation. This certainly appears possible, although much more evidence will be needed

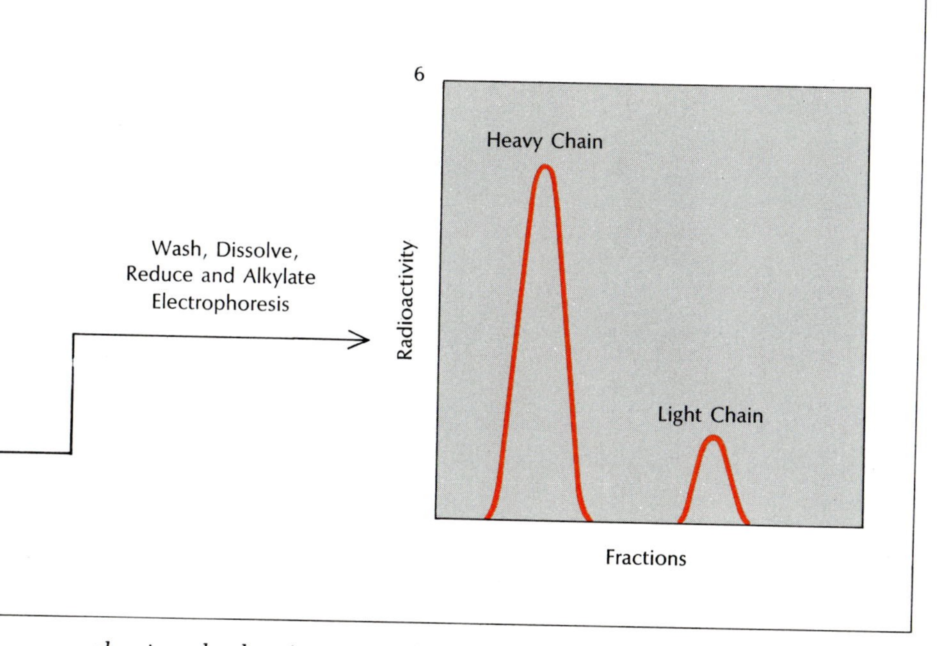

plasmic molecules of same specificity (5). After electrophoresis, label clearly distinguishes band including surface molecule from those with only cytoplasmic ones (6).

before it can be considered certain.

As to why IgD appears at all, we have no data. Since IgD is not secreted in significant amounts, it must function mainly as a receptor in contrast to the other classes of immunoglobulin. We speculate that the IgD receptor facilitates "triggering" of lymphocytes by antigen.

Another question that has concerned us is where surface Ig is manufactured in the lymphocyte, and how it gets to the surface. Tracer experiments, using amino acids labeled with tritium, have shown – not very surprisingly – that the site of manufacture is the same as for other proteins: the endoplasmic reticulum. The Ig is then transported into the Golgi complex, where its sugar chains are added. By this time it is already "outside" the cell, in the sense that it is separated from the cytoplasm by the Golgi membranes. The molecule could then be transported to the exterior through the formation of vesicles that would migrate to and merge with the plasma membrane, in the same manner observed with such secretory proteins as those of the pancreas. We have not, indeed, been able to demonstrate the existence of such vesicles in lymphocytes under the electron microscope; they may be too small to show up. One should note in this connection that the large size and consequent high visibility of transport vesicles such as those in the pancreas are due to their serving for storage as well as transport. There is no reason to suppose that lymphocytes store membrane immunoglobulin.

One may then inquire why surface Ig remains bound to the exterior of the cell while secreted Ig does not – even when (as with IgM) they are of the same basic type. The most likely explanation is that the surface molecules are bound to the Golgi membranes from the beginning and remain attached when these membranes are exteriorized. The secreted Ig, on the other hand, presumably floats free within the Golgi complex, and is then free to pass into the serum when transported to the exterior. This would imply that there is some difference between membrane IgM and serum IgM, and we already know that the former is a monomer and the latter a pentamer. However, this does not seem to explain why one is bound and the other is free; we know that in certain human pathologic conditions, and normally in certain groups of animals such as the primitive bony fishes, IgM is secreted in monomeric form. It seems necessary to postulate, therefore, that the two forms of IgM differ slightly in some portion of their molecules. For example, the membrane-bound form may possess an extra short hydrophobic portion at its base that could intercalate with the membrane lipids. The sequence of amino acids of secreted IgM is known, and it does not contain any lengthy hydrophobic portions such as

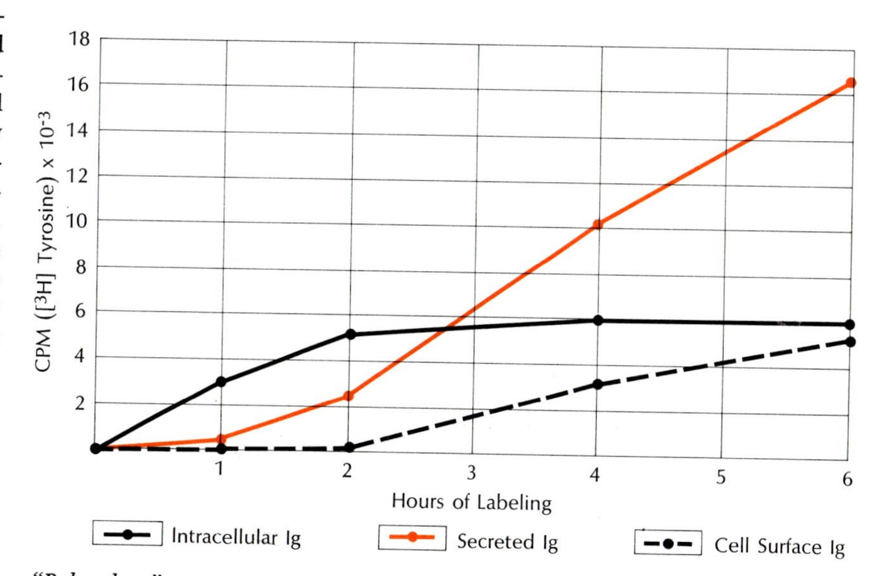

"Pulse chase" experiment employing labeled amino acids shows that label appears first in intracellular Ig, indicating that there is a time-consuming transport step between synthesis and appearance of Ig at the cell surface or in the incubation medium. Slow appearance of label in membrane-bound Ig reflects its slower metabolic turnover.

have been found in the erythrocyte membrane protein glycophorin.

Another interesting fact about membrane-bound Ig is that under certain conditions it is released quite rapidly from the membrane into the medium. Closer investigation, however, has established that the molecule is not precisely released; rather, it is shed into the medium while still attached to a portion of the membrane. This can be shown in a number of ways. For example, unless the preparation is treated with a detergent, the Ig is precipitated as part of a heterogeneous mixture of membrane proteins. With the detergent it can be precipitated separately. One therefore visualizes the shedding as pinching off of a bit of membrane with a variety of proteins embedded in or attached to it, one being Ig – with protein separation becoming possible only if the lipid portion of the membrane fragment is dissolved. Whether an equivalent process occurs under physiologic conditions is not yet known.

While it is generally agreed among immunologists that the antigen receptors of B cells are immunoglobulins, there has been much debate over the nature of T cell receptors. It was therefore interesting to find that when we employed our radioiodination technique on T cells we were not able to identify surface Ig of any type. As a check, we ran an experiment in which B cells were mixed with T cells in the proportion 1:200, and found that Ig was easily detectable. This and other control experiments convinced us that the "fault" was not in the technique but in the cells: surface Ig was not detected because it was either not there or was present in only trace amounts.

But if the T cell receptor is not an Ig like the B cell receptor, what is it? Our own suspicions have fallen upon the antigens coded by the major histocompatibility gene complex because of the work of Benacerraf and McDevitt. They have shown that an animal's capacity to recognize particular antigens at the level of T cells is controlled by genes within this complex and that T and B cell cooperation is also affected by this gene complex.

There are many genes in this complex and only a few of the antigens that they code have been identified. Foremost are the H-2 alloantigens, which we are currently studying. These are the substances responsible for graft rejection in mice; their counterparts in man are the HL-A antigens. The H-2 alloantigens are not restricted to lymphocytes but occur on the surfaces of all nucleated cells. The question one would like to answer is what function they serve there – graft rejection can hardly be it. Thus far, no general answer has been obtained.

Nathenson has shown that H-2 antigens, like the immunoglobulins, are proteins, with attached carbohydrate moieties. They resemble Ig in another way in being composed of a long and a short chain. The resemblance is even closer when we note that the H-2 short chain is similar if not identical to that of what is called beta-2-microglobulin – a cell component in humans that has an amino acid sequence that bears a great resemblance to that of Ig.

The H-2 antigens also resemble Ig in other respects. Through tracer experiments, we have shown that H-2 antigens, like Ig, are synthesized and contained within membranes until these antigens reach the surface of the cell. In contrast to Ig, the H-2 antigens are not secreted and, after reaching the surface, stay there, i.e., they do not appear to be shed. We do

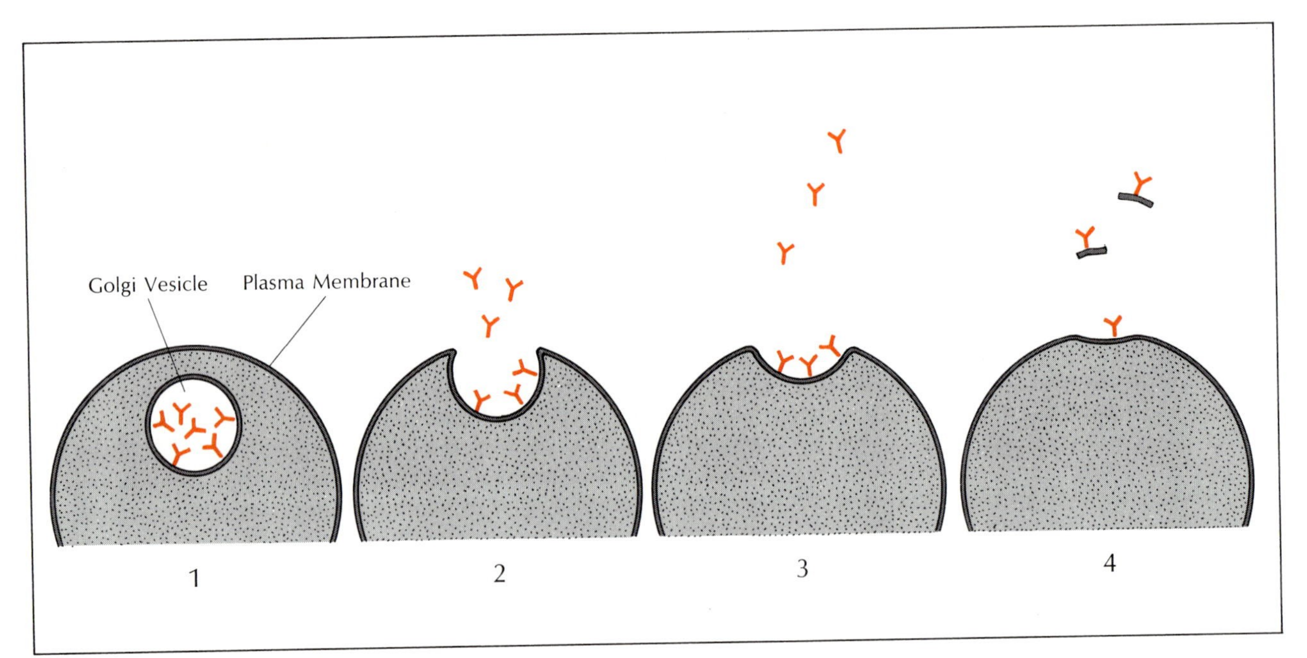

Both Ig destined for cell surface and Ig to be secreted are thought to be transported from synthesis site (the endoplasmic reticulum) in Golgi vesicles, with the former attached to vesicle interior (1). When vesicle merges with plasma membrane (2), free Ig is ejected, bound Ig remains attached to segment of "new" membrane (3). It may, however, be shed later – i.e., when bits of membrane are detached (4). Plasma cells primarily secrete Ig, whereas lymphocytes develop cell surface Ig.

not know the reason for these differences. Unlike the Ig molecule, which (as mentioned earlier) can have only a small hydrophobic portion, the H-2 molecule probably contains a fairly long sequence of hydrophobic amino acid residues. As a result, the molecule could be deeply embedded in the lipids of the cell membrane, quite possibly passing all the way through it, as has been deduced for erythrocyte glycophorin (see Chapter 5, "The Structure and Orientation of a Membrane Protein," by Marchesi). It may well be that the end of the H-2 molecule projecting into the cell interior is attached to some other molecule. This would in fact anchor it.

All this is, of course, speculative. What is not speculative is a relationship we have uncovered between the H-2 alloantigens and another lymphocyte surface molecule, known as the thymus leukemia (TL) antigen, first described by Boyse and his coworkers. As its name implies, this substance is not found on all nucleated cells, as are the H-2 antigens, or even on all lymphocytes. Specifically, TL antigen occurs on normal T cells only in certain strains of mice during an early phase of T cell development (in the thymus), disappearing as the cells become fully mature. However, it shows up again in these and all other strains of mice if the animals become leukemic (nearly all murine leukemias, in contrast with human leukemias, involve T cells), and indeed is pathognomonic of the disease.

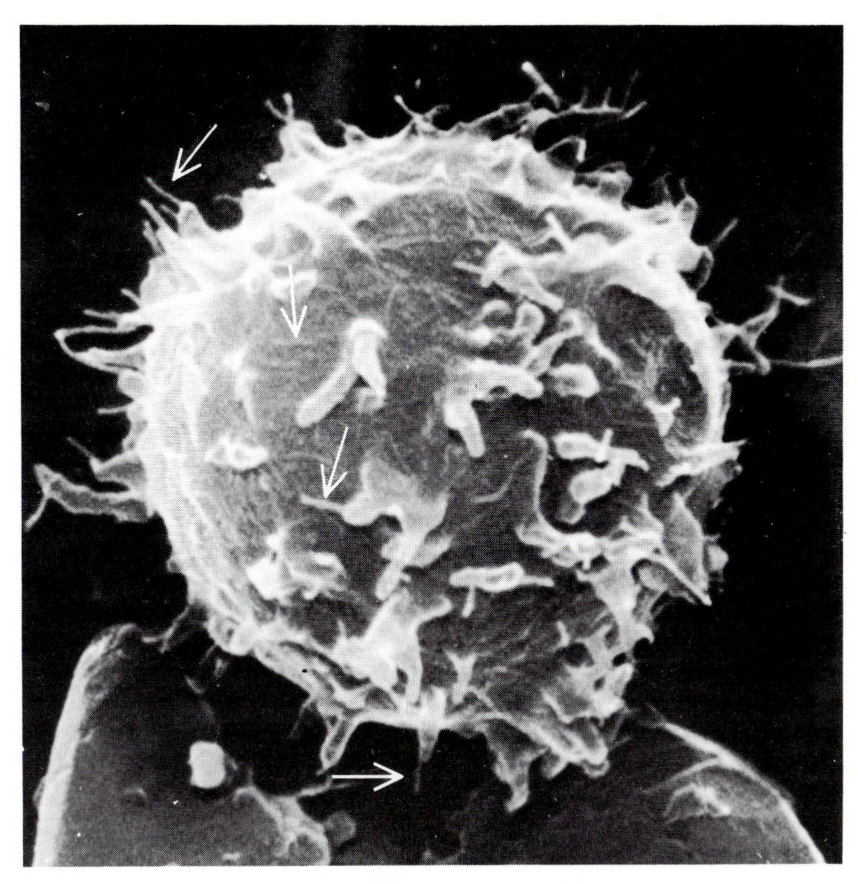

Scanning electron micrograph prepared by Dr. M. Lipscomb shows a murine B lymphocyte with its long finger-like projections called microvilli. Location of Ig on surface is "marked" by slender rods (see arrows) of the tobacco mosaic virus (TMV). This was accomplished by treating cells with "hybrid" antibody, which binds to both Ig and TMV and adding TMV to treated cells. Shedding could be due to pinching off of microvilli.

Thanks to the heavy labeling that the radioiodination technique makes possible, we have been able to put TL antigen through the usual processes of protein analysis. The results indicate that it has a major and minor subunit with the identical apparent molecule weights of the H-2 subunits. Thus, both H-2 and TL have a β-2 microglobulin-like subunit. The obvious possibility is that the TL gene arose from the H-2 genes – and that its appearance (in leukemic cells, at least) is somehow linked with the development of neoplasia. But what the link is (if indeed it exists) remains to be determined.

The reader, as he reaches the end of this chapter, may well feel that it is rather fragmentary, containing a number of facts that, while perhaps interesting in and of themselves, have only an obscure relationship to one another. In fact, this fragmentary quality precisely reflects our present state of knowledge concerning the surfaces of immunocompetent cells. Conceptually, and perhaps physically as well, these surfaces are a jigsaw puzzle, but one in which many of the pieces are missing and in which even the pieces we have are only gradually beginning to reveal their distinctive shapes. The shapes, as we get to know them, are interesting ones indeed – but we are a long way from knowing how they all fit together.

23

Immune Reactions Of Model Membranes

STEPHEN C. KINSKY

Washington University

As is well known, the body possesses a number of mechanisms for dealing with foreign cells (including cancerous ones), many of which involve damage to the intruder's plasma membranes. Some of these processes are cell mediated, requiring the participation of immunocompetent lymphoid cells, most often T-lymphocytes. In this chapter, however, I shall focus on the second category of immunolytic reactions, those that are humoral mediated, involving substances found in the serum. These substances are specific antibodies, directed against antigens localized in the cell membrane of the invasive cell, working in conjunction with the system of nine protein components known collectively as complement.

As pharmacologists, we got into the subject of humoral immunolysis via the back door, through studying a group of antibiotics known as the polyenes. Clinically important members of this group include amphotericin B and nystatin, which are used primarily for treating fungal infections, both systemically and topically. In 1959, when we started work on the polyenes, little was known of their mode of action apart from their remarkable selectivity: though they are exceedingly potent against fungi, they have absolutely no effect on bacteria. Subsequently, it was possible to pinpoint the site of action of the polyenes as the cell membrane. In the presence of these agents, the cell membranes lose their ability to function as restraining barriers so that essential cytoplasmic constituents escape from the cell; conversely, some compounds in the external environment, which are normally excluded, can now enter. At the present time, it is generally accepted that the polyenes exert this effect by reacting with ergosterol, which is the principal sterol present in fungi cell membranes. Their inactivity against bacteria is due to the fact that bacterial membranes contain neither ergosterol nor any other compound of related structure.

Indeed, in the course of these studies, we found that mammalian cells, whose predominant membrane sterol is cholesterol, are also vulnerable to the polyenes. This may partially account for some of the severe toxic side reactions accompanying prolonged therapy with amphotericin B. For example, this antibiotic induces very rapid hemolysis of washed human erythrocytes; the same applies to filipin, which is another polyene antibiotic that is too toxic for clinical use. In 1965, David Zopf, then a medical student at Washington University, made an observation that turned out to be the key that unlocked the back door. He found that filipin treatment produced circular "pits" about 100 Å in diameter in erythrocyte membranes; these could be readily seen in the electron microscope after negative staining. The intriguing feature of these *apparent* lesions was that, shortly before, Humphrey and Dourmashkin in England had described very similar ones in a variety of natural membranes subjected to quite different treatment: lysis by antibody in the presence of guinea pig or human serum as a source of complement. An obvious implication of these findings was that the terminal stages of polyene-induced and immune lysis might possess certain features in common, and it was to explore this possibility that we began studying the latter process.

Many laboratories have been actively engaged in studying complement mechanism for at least 75 years and, at this point, the reader may well wonder what was unique about our experimental approach. Whereas all previous studies had employed natural cells (notably sheep erythrocytes) as targets for immune lysis, we set about to develop a model membrane that might be subject to the action of the complement system. The three types of model membranes (lipid monolayers, black lipid bilayer films, and liposomes) have been thoroughly reviewed by Bangham in Chapter 3. His take-home message, with which we are in complete agreement, is that the structure of model membranes is now well established and, accordingly, these models might provide information about membrane-associated phenomena, such as complement-mediated immune damage, which would be more difficult to obtain otherwise.

This faith in the utility of model membranes was firmly established while our laboratory was still engaged with the polyenes, since all three models behaved as expected when exposed to these antibiotics. To cite one example, consider liposomes – the tiny spheroidal assemblages of concentric, onion-like lipid bilayers, which Dr. Bangham's laboratory first described in 1965. With this model membrane, we were able to demonstrate that glucose trapped in the liposomes at the time of their manufacture was rapidly released by treatment with low concentrations of the potent polyene filipin, provided that the liposomal bilayers also contained a sterol such as cholesterol. Furthermore, Zopf detected in liposomes the same pits that he originally found in mammalian cell membranes after incubation with filipin. This observation facilitated one decision that had to be made at the outset: namely, if any model membrane could be made that might undergo immune damage, our best bet would be liposomes.

Before taking up our experiments in this area, however, I should say a few words about our basic methods. As already indicated, these involved the generation of liposomes (from phospholipids and other ingredients) in a medium containing glucose. The portion of glucose that was not trapped in the aqueous liposomal compartments was then removed by dialysis. The glucose was not the ^{14}C-labeled variety often used in experiments with liposomes but ordinary glucose. To measure its release after incubation with the polyenes or antibody-complement (or, for that matter, a variety of lytic agents), we incorporated into the assay cuvettes certain enzymes and coenzymes (hexokinase, ATP, glucose-6-phosphate dehydrogenase, nicotinamide adenine dinucleotide phosphate [NADP], and magnesium ions), which oxidized any glucose released from the liposomes. The oxidative process rapidly reduced the NADP to NADPH, whose appearance could be easily detected in a spectrophotometer. The advantage of this technique over those involving labeled glucose (or, for that matter, any other radioactive trapped-marker compound) resided in the fact that glucose released did not have to be separated from liposomes – and whatever label they still contained. This is because the enzymatic-spectrophotometric assay responds only to glucose liberated into the medium, since only this glucose is accessible to oxidation with the concomitant production of NADPH. Moreover, with this sensitive method, glucose release could be continuously monitored and, as described below, we could perform several crucial experiments with complement components and synthetic lipids that were available in limited amounts.

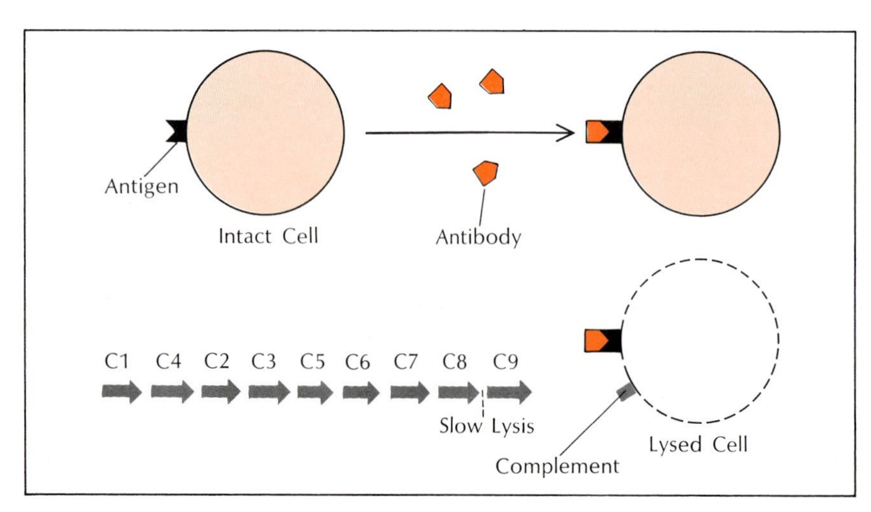

Classic sequence of complement-induced cell lysis is triggered by formation of antigen-antibody complex on the surface of the cell membrane, after which the nine complement components are activated in the order shown.

The initial experiments in 1968 (performed by James Haxby, then a graduate student, ably assisted by my wife) made use of liposomes simply prepared from a lipid extract of sheep erythrocyte membranes. This seemed a logical starting source for two reasons. First, as already mentioned, sheep erythrocytes have long been the favorite test objects for studying complement mechanism, so that a wealth of background information was available in the literature. Second, rabbit antisheep erythrocyte antiserum contains antibodies directed against a compound known as Forssman antigen. Forssman antigen is an amphipathic lipid, like other membrane lipids, having a polar and nonpolar end; specifically, the polar end consists of an oligosaccharide of five sugars. It seemed likely that these amphipathic molecules would mesh neatly with the other lipids that composed the liposomes in such a way that the oligosaccharide end (which carries the antigenic determinant) would project into the aqueous medium, ready to interact with the antibodies and thereby trigger the complement sequence.

To our gratification, these liposomes prepared from the sheep erythrocyte lipids performed as predicted. That is, when the liposomes were incubated with rabbit antisheep erythrocyte antiserum (as a source of anti-Forssman antibodies) in the presence of guinea pig serum (as a source of complement), they released up to 70% of their trapped glucose in about five minutes. If the rabbit antiserum was replaced with normal rabbit serum (from nonimmunized animals), no release occurred, nor did it do so if the guinea pig serum had been heat inactivated (preheated for 30 minutes at 56° C to destroy complement activity). Subsequently, Dr. Carl Alving found that the liposomes did indeed bind the anti-Forssman antibodies and fix (consume) complement activity.

These observations set the stage for the next phase of our investigation: namely, to determine which of the nine complement components were involved in the process of immune damage to liposomes. This was extremely important because, as is

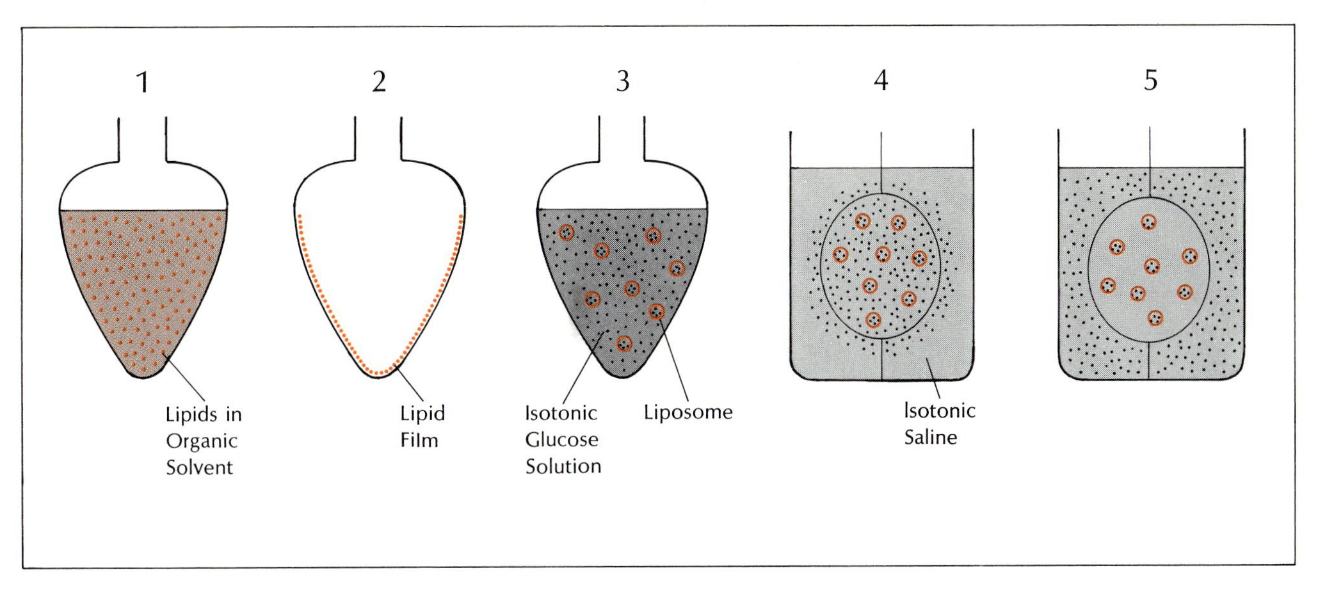

To prepare liposomes with glucose marker so that glucose release can be measured following lysis, appropriate lipids are dissolved in organic solvent (1), which is evaporated to leave lipid film coating flask wall (2). When film is dispersed in isotonic glucose, some glucose is trapped within liposomes (3). Remainder is then removed by dialysis against isotonic saline (4, 5).

well known, many important immunologic phenomena involve some, but not all, of these proteins. For example, phagocytosis, immune adherence, and anaphylatoxin production can occur with just the "early-acting" components C1 through C5; immunolysis, however, requires the "late-acting" components C6 through C9 as well. Fortunately, we were able to do these experiments in the laboratory of Dr. Hans Müller-Eberhard at Scripps Clinic and Research Foundation, La Jolla, California, with his colleague, Dr. Otto Götze; this is one of the few laboratories in the world that has mastered the extremely difficult art of isolating the individual complement components in pure form. Employing these highly purified human complement proteins, we were able to demonstrate that both C2 and C8 were absolutely essential for glucose release from the liposomes. Since activation of C2 requires the prior participation of C1 and C4, and activation of C8 requires the prior participation of C3, C5, C6, and C7, these results indicated the necessary involvement of eight of the nine components. The ninth component was a somewhat special case: though it was not absolutely essential for glucose release, it markedly stimulated the process. This apparently parallels the situation with natural cells – sheep erythrocytes – in that components C1 through C8 can produce some irreversible membrane damage, which is, however, greatly enhanced by the presence of C9.

Needless to say, the preceding experiments convinced us that liposomes were a suitable object for studying complement mechanism and the next step was to investigate how membrane damage occurred. For this purpose, we replaced our original lipid extract of sheep erythrocytes (which was of more or less heterogeneous composition) with various simple combinations of compounds "taken off the shelf." Our basic "recipe" included a phospholipid (e.g., lecithin or sphingomyelin), a sterol (e.g., cholesterol), and a charged compound (dicetyl phosphate for negatively charged liposomes, stearylamine for positively charged ones). Of course, the most important ingredient was the lipid antigen, pure Forssman substance, which was kindly supplied by Drs. S. Handa and T. Yamakawa of the University of Tokyo. With *just these four materials*, we were able to make liposomes that released glucose when incubated with anti-Forssman antiserum and a source of complement!

To explain the implications of

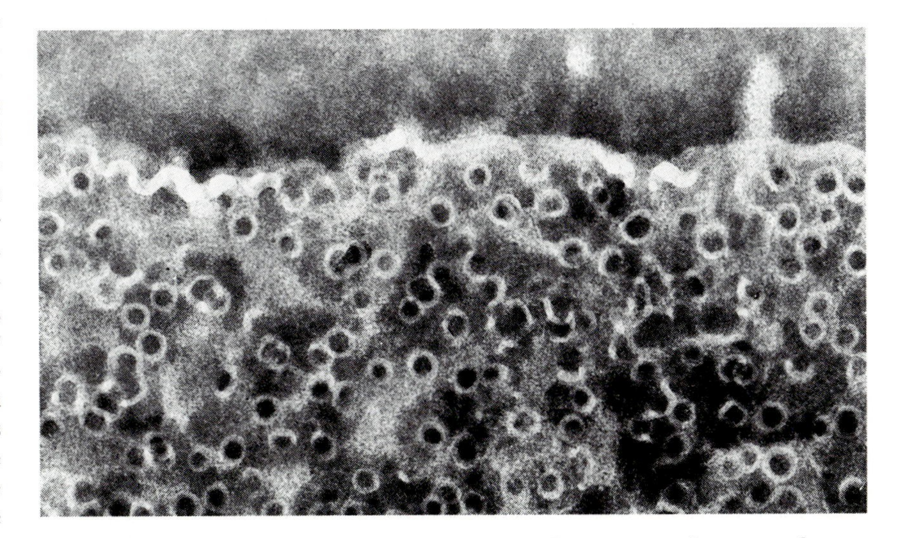

Electron micrograph by Dr. T. W. Tillack shows "pits" appearing in human erythrocyte membrane after lysis by filipin, a polyene antibiotic. About 100 Å in diameter, pits resemble those seen in cell membranes after lysis by appropriate antibodies and complement.

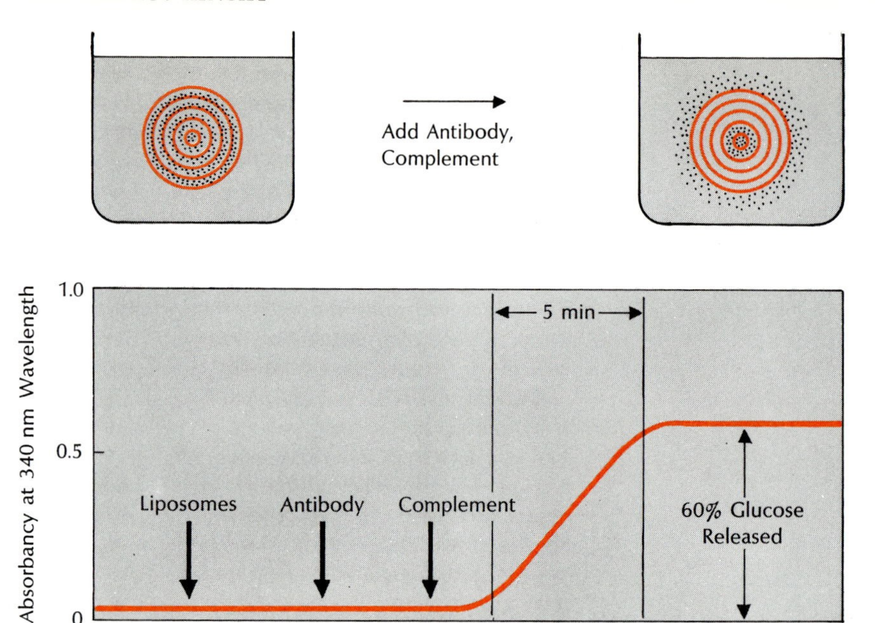

To detect immune damage to liposomes spectrophotometrically, enzymes and coenzymes required to oxidize any released glucose are incorporated into assay cuvets. Concomitant reduction of NADP to NADPH is measurable by increased absorbancy at 340 nm. No absorbancy change occurs until both antibody and complement are added.

these findings, I must cut back for a moment to a theory that had dominated almost all the previous work in this area. For a long time, it had been assumed that the terminal events in the immunolytic process were enzymatic – that is, an enzyme was formed upon activation of either C8 and/or C9 that breaks down some membrane constituent, thereby producing rupture of the membrane. No one, however, had clearly managed to identify the particular membrane constituent that was attacked. Our liposomes of known chemical composition would, we hoped, answer this question because all of the ingredients that go into their preparation could be prepared with radioactive labels.

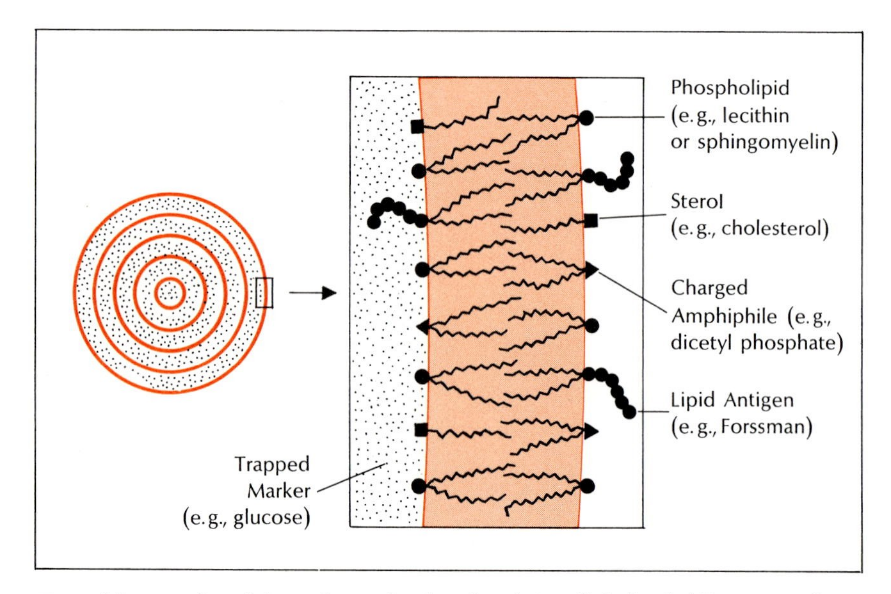

Typical liposomal model membrane developed with four "off-the-shelf" compounds for studies of immune damage is diagrammed; it will release its trapped marker when incubated with antibody against the selected antigen and a source of complement.

The simplified liposomes very literally also simplified the problem, since only one of their constituents seemed at all likely to be the substrate for an enzyme, because it was the only one invariably present in both liposomes and natural membranes. The charged molecules (dicetyl phosphate, stearylamine) were ruled out because they do not occur in *any* natural membranes. The sterols could be excluded because they do not occur in *all* natural membranes – specifically, they are absent in gram-negative bacteria though these organisms are susceptible to the action of complement (the reader should recall that it is their absence that renders bacteria insensitive to the polyene antifungal antibiotics). Finally, the Forssman antigen could be ruled out because a great many cell antigens can trigger immunolysis after combination with antibody, and it seemed unlikely that any single complement-generated enzyme could act on all of them. In fact, throughout this work, we were able to generalize our conclusions by using liposomes that had been sensitized with a variety of naturally occurring lipid antigens. Two of these, globoside I (isolated from human erythrocytes) and galactocerebroside (isolated from beef brain myelin) belonged to the same mammalian "ceramide antigen" category as Forssman but differed from the latter in possessing fewer sugar residues; the other antigens, known as "lipopolysaccharides" were of bacterial origin and had a completely different structure from the ceramides.

By elimination, then, we were left with the phospholipids as the only plausible candidates for enzymatic degradation. Dr. Keizo Inoue thereupon prepared liposomes with radioactive ^{32}P lecithin and sphingomyelin, expecting to find that a breakdown product of these compounds (formed by the presumed enzyme) would show up as a new labeled spot on a thin-layer chromatogram. However, to our surprise – indeed, to our initial disappointment, because these experiments took eight months – we found nothing of the sort. With liposomes that had released 50% to 80% of their trapped glucose, at least 98% of the radioactive label could be recovered in the form in which it had originally been incorporated into the model membranes – as lecithin or sphingomyelin. Thus, the phospholipids were not being broken down to any significant extent, although the membranes that contained them had undergone significant damage.

Though this evidence against the enzymatic theory was persuasive, it was still essentially negative. To buttress our conclusion, we then constructed liposomes that could *not* be enzymatically degraded. Normally, a phospholipid (such as lecithin) is broken down by four enzymes: phospholipases A and B, which hydrolyze the ester (–COO–) bonds that join the fatty acid "tails" to the glycerol "backbone," and phospholipases C and D, which cleave the bonds on either side of the phosphate group. Accordingly, we made use of a unique analogue of lecithin that had been synthesized by Dr. A. F. Rosenthal of the Long Island Jewish Medical Center. In this compound, the ester bonds were replaced by nonhydrolyzable ether (–O–) linkages, while the phosphate bonds were likewise modified. This substance could not be broken down by any of the phospholipases – yet liposomes made with it released just as much glucose as those prepared with lecithin. To our satisfaction at least, this clinched the argument that immunolysis could occur without degradation of phospholipid by any known pathway.

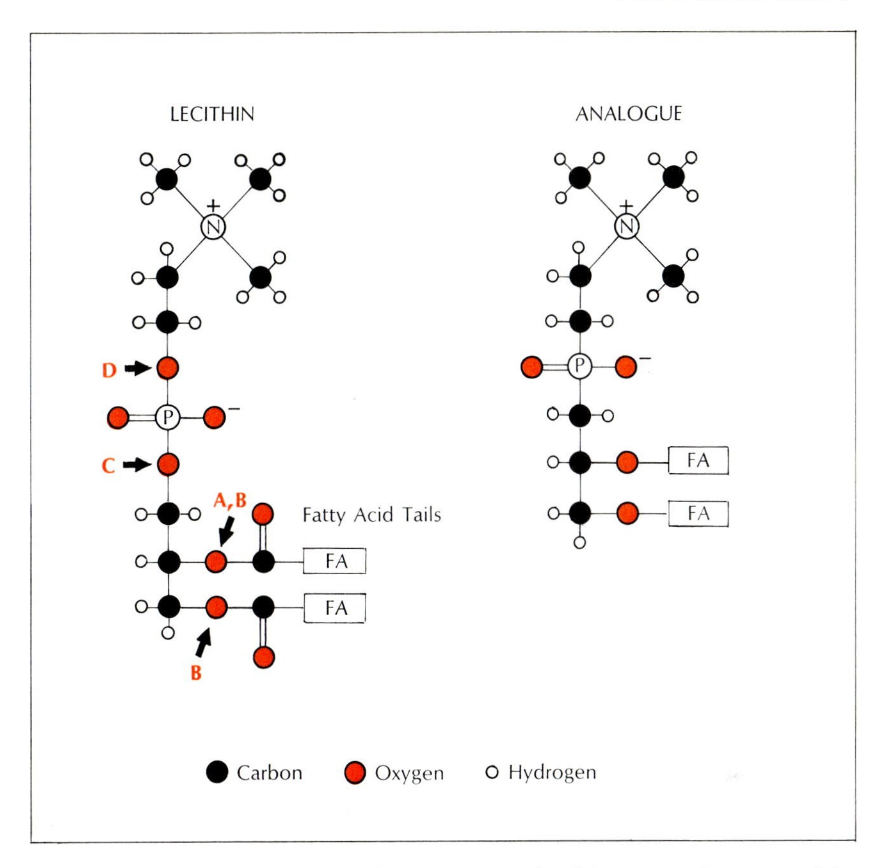

That immunolysis does not proceed via enzymatic breakdown was demonstrated by using the lecithin analogue shown at right to prepare model membranes. Though analogue lacks bonds subject to hydrolysis by phospholipases A, B, C, and D (left), liposomes prepared with it are still subject to immune damage by antibody and complement.

But if immunolysis does not proceed via the enzymatic breakdown of some membrane constituent, how does it in fact occur? Here we must for the moment fall back on speculation. Our present feeling is that the process may be akin to detergency. That is, either the C8 or C9 component of complement (or both), when activated by the preceding complement components, exposes or releases some hydrophobic region in its structure. Normally, this would be enclosed by the protein's hydrophilic portions. Activation, however, might transiently expose a hydrophobic region, which would, by its nature, seek out some other hydrophobic environment where it might feel more "comfortable" – specifically, the hydrophobic interior of the membrane bilayer. Its insertion in the bilayer might well loosen the latter's structure sufficiently to produce lysis.

The concept of complement detergency is currently accepted by most workers in this field owing, in part, to the fact that some genuine immunology laboratories (Humphrey's and Lachmann's in England) performed experiments with liposomes that were compatible with our earlier conclusions. Here, I shall again reveal a personal bias by emphasizing that this hypothesis originally evolved from studies with liposomal model membranes. Of course, its proof (or disproof) can only be undertaken by "complementologists" – namely, those few groups with the ability to isolate complement components in pure form. I should add that if the hypothesis is validated, it is fortunate that the components are proteins (hence biodegradable) because this particular detergent then stands little risk of being banned from the market!

In this regard, mention must be made of one of the major ironies of our work. The reader will recall that we were originally drawn to the study of immunolysis by the similarity of the pits produced by filipin to those produced by complement. In fact, we proceeded on the assumption that these pits represented channels through which cytoplasmic constituents leaked from the cell (or glucose from liposomes). In 1972, Seeman and his collaborators at the University of Toronto looked at freeze-etched (rather than negatively stained) complement-lysed membrane preparations from erythrocytes; they concluded that the pits do not traverse the entire lipid bilayer, as would be expected of functional holes. Subsequently, Dr. Thomas W. Tillack of the Department of Pathology at Washington University and I did analogous experiments with filipin on erythrocytes and liposomes; similar studies were also being done simultaneously in L. L. M. van Deenen's laboratory in the Netherlands. Both groups obtained the same pictures and arrived at the identical conclusion: no convincing evidence that pits formed by the antibiotic could be equated with holes.

In fact, not only is the nature of the pits produced by either filipin or

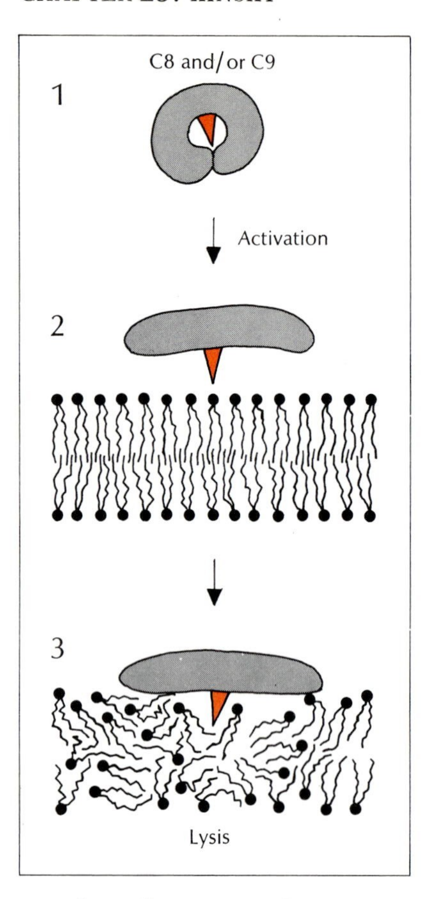

Hypothetic "detergent" mechanism is proposed for complement lysis. Dark wedge represents hydrophobic region normally buried within terminal components that becomes transiently exposed upon activation. Insertion into lipid bilayer produces disorganization that leads to lysis.

complement unclear, but, in the case of complement, there exists considerable controversy as to precisely when in the sequence they first appear. Thus, Müller-Eberhard's laboratory has reported finding pits in erythrocyte membranes as early as the reaction involving C5 – that is, at a stage when the cell membrane is still undamaged. On the other hand, Lachmann and colleagues, working with liposomes, could not detect pits until after the terminal complement components had been added. This situation is made even more puzzling by the fact that we have been able to confirm the observation of D. H. Bing and associates at Michigan State that liposomes that have released 50% or more of their trapped marker in the presence of antibody and guinea pig serum rarely (if at all) display pits when examined by negative staining. Whatever the outcome, I do not wish to disparage the significance of these pits in view of all the research they have stimulated. For our laboratory, they have a special meaning because they obviously encouraged us to work on the right problem although perhaps for the wrong reason.

Now I should like to refer to a tangential issue whose eventual resolution may shed more light on complement mechanism. This initially came about because our work was to some extent handicapped by the difficulties involved in obtaining sufficient quantities of pure, naturally occurring lipid antigens such as the mammalian ceramides and bacterial lipopolysaccharides; also, animals immunized with these substances do not always produce enough antibodies against them. To circumvent these problems, Dr. Kei-ichi Uemura decided to link the dinitrophenyl (DNP) group to the polar residue of phosphatidylethanolamine (PE); PE is an abundant natural membrane phospholipid and can be purchased commercially, and, therefore, large amounts of DNP-PE could be synthesized. We picked the DNP group because it is a potent antigenic determinant when attached to a foreign protein; in addition, the anti-DNP antibodies (made in rabbits) have been well characterized in terms of immunoglobulin class and their affinity for the antigen.

Fortunately, this proved to be the correct choice, in that liposomes, which contained DNP-PE in place of the lipid antigens used previously, released just as much glucose in the presence of anti-DNP antibodies and guinea pig complement. Moreover, by removing one or both of the fatty acid tails from DNP-PE, it was possible to determine precisely what structural features were necessary to render liposomes susceptible to immune damage. We found that the version with one tail removed (DNP-lysoPE) would still sensitize liposomes to immunolysis, while the compound with both tails removed (DNP-GPE) would not. In view of what is known about liposomal structure, the latter finding was perfectly reasonable, since with both tails gone the molecule would be completely water soluble. It therefore would not incorporate itself into the lipid bilayer but would prefer the aqueous compartments of the liposomes, where it would not be accessible to antibody.

So far as triggering the liposomes' response to antibody and complement was concerned, the DNP-lysoPE (one-tailed) molecule possessed a unique property. Up to this point, we had always found it necessary to introduce antigens into liposomes by *active* sensitization – that is, adding them to the lipid components before they were dispersed in the glucose solution. DNP-lysoPE, however, was capable of *passive* sensitization – that is, it could also be incorporated into preformed liposomes. The surprising aspect of these experiments, for reasons that I shall shortly discuss, was that passively sensitized liposomes released the same amount of glucose as the actively sensitized liposomes – between 60% and 70%.

In this connection, I must mention another irony of our work. When these studies began back in 1968, we would have been extremely content with just 10% glucose release, on the assumption that only the outermost lipid bilayer could be destroyed by antibody-complement (it can be calculated that for an "average" liposome of 30 concentric bilayers, about 10% of the trapped marker is contained in the first aqueous compartment). Yet liposomes incubated with excess antibody and complement consistently release much more than this, implying that an appreciable portion must come from the "deeper" aqueous regions as a consequence of damage to internal bilayers (to get even 50% release, about six of the 30 bilayers would have to be affected). For a long time, we have entertained the possibility that this might occur in successive stages, as follows: immune damage to the external bilayer permits the entrance of antibody molecules and the complement components; the antibodies then combine with antigens in the next bilayer; the resulting antigen-antibody complexes initiate the complement sequence, which destroys this second bilayer, resulting in the exposure of the third bilayer, etc. However, we have good reason to believe that in passively sensitized liposomes the antigen is present only in the outermost lipid bilayer. Accordingly, it seems very unlikely that damage to internal bilayers proceeds by the classical complement sequence involving prior formation of an antigen-antibody complex.

In fact, before the phenomenon of passive sensitization was encountered, Dr. Tateshi Kataoka performed an investigation indicating that we might be on the wrong track. In 1970, Weissmann's laboratory at New York University demonstrated that enzymes could be trapped inside liposomes. This prompted us to put into liposomes beta-galactosidase (an enzyme of molecular weight 518,000), which, we found, was released in parallel with glucose (molecular weight 180) when *lecithin* liposomes were incubated with antibody and complement. This observation was consistent with our hypothesis, since if such a large molecule could get out, it seemed equally probable that large molecules (specifically, antibody and complement components) could get in. However, when the same experiment was repeated with *sphingomyelin* liposomes, little (if any) enzyme appeared in the medium though 50% of the glucose was released. This made it highly improbable that molecules such as an IgM anti-Forssman antibody and the first complement component (with molecular weights of 850,000 and about 600,000, respectively) could ever reach the inner bilayers.

To account for these observations, the simplest assumption is that the activated terminal complement component with its exposed hydrophobic region (or the hypothetic "detergent" fragment) is sufficiently small enough to have access to inner bilayers. This hypothesis may also explain why we have never obtained complete (100%) glucose release from multicompartment liposomes, since, by definition, hydrophobic regions have a limited survival time in water and thus may not persist long enough to reach all the liposomal bilayers. This hypothesis is, nevertheless, difficult to reconcile with a recent proposal from Müller-Eberhard's laboratory. Their experiments suggest that erythrocyte lysis is produced not by a single small molecule, but rather by a firm 1,000,000 molecular weight complex consisting of one molecule each of C5, C6, C7, and C8, plus six molecules of C9. In an attempt to resolve this dilemma, William Young has succeeded in attaining one of our cherished goals: to make single-compartment liposomes that would release all of their trapped marker in the presence of antibody-complement. If damage to these tiny model membranes (which have a diameter of 300 Å compared to 3,000-5,000 Å for the multicompartment liposomes) proceeds as suggested for erythrocytes, we should be able to see with the electron microscope the large decamolecular complement complex on their surface.

Finally, I should like to refer briefly to an additional immunologic property of liposomes that eventually may have both theoretical and practical significance. In the course of these studies, Dr. Howard Six synthesized another dinitrophenylated PE derivative abbreviated DNP-CAP-PE (its full name is: dinitrophenylaminocaproylphosphatidylethanolamine). DNP-CAP-PE was originally made because its polar residue has a greater structural resemblance to DNP-lysine residues, which are the predominant antigenic determinants in the proteins used to immunize rabbits. This provided us with a compound that rendered liposomes even more sensitive to anti-DNP antibodies and enabled investigation of such subtle phenomena as the effect of immunoglobulin class and affinity on triggering of the

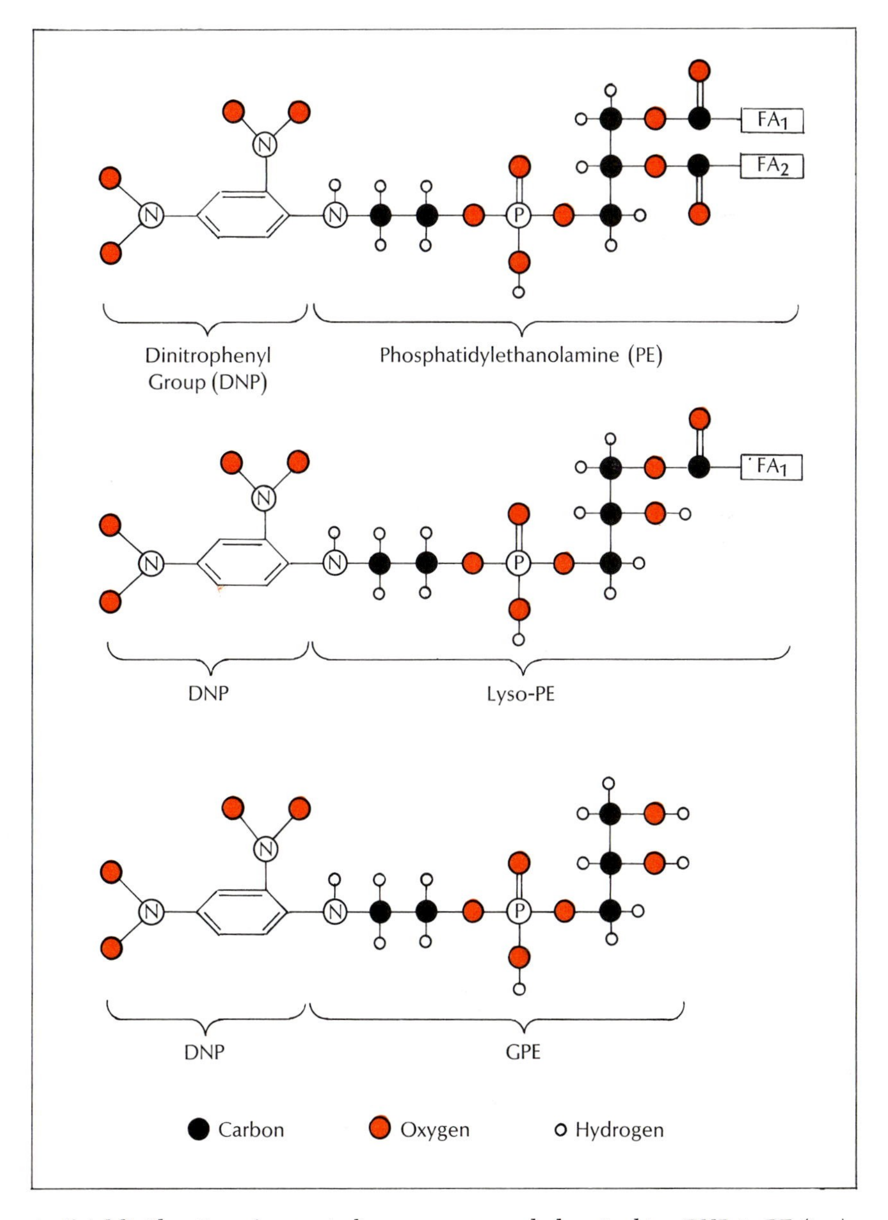

Artificial lipid antigen for use in liposomes was made by attaching DNP to PE (top); when one fatty acid tail is removed (middle), the compound still sensitizes liposomes to immunolysis. With both FA tails removed (bottom), sensitization no longer occurs.

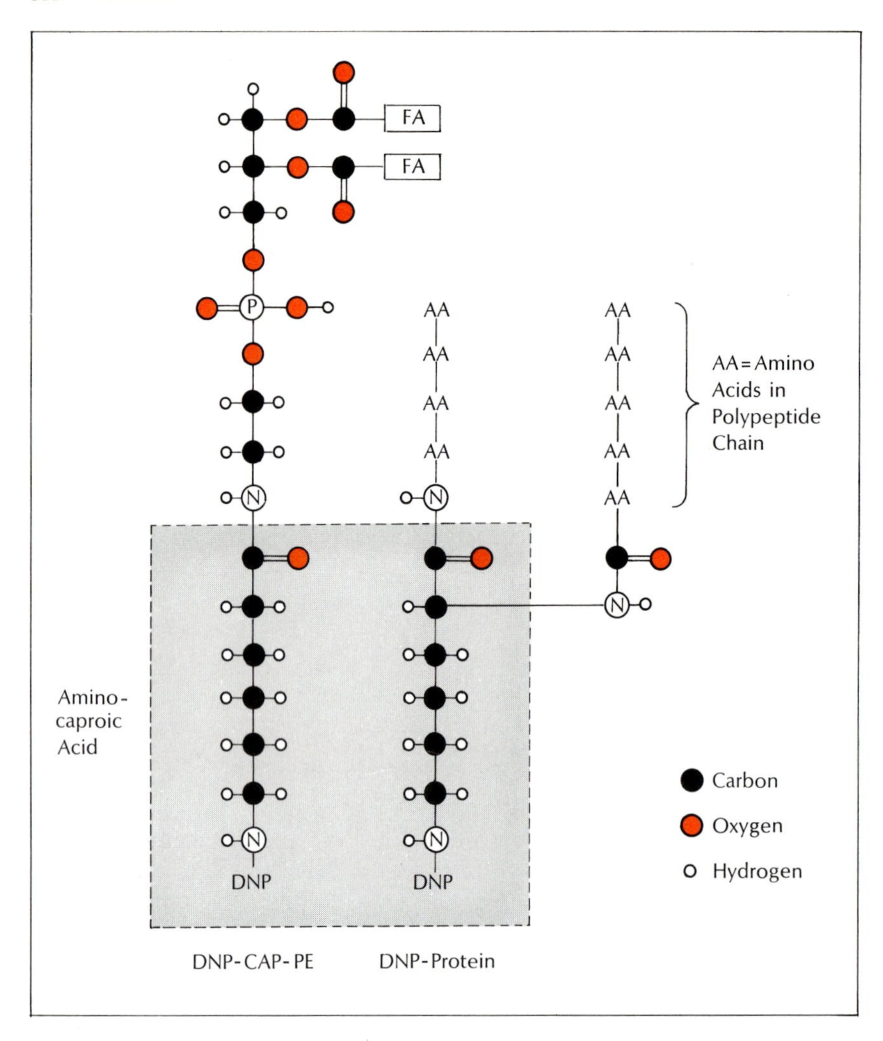

DNP-CAP-PE is compared with DNP-lysine residue in typical protein antigen used to prepare anti-DNP antibodies. Common structure is shown in box. Note that in DNP-CAP-PE, the dinitrophenyl group is farther removed than in DNP-PE (cf figure on preceding page) because of the insertion of an aminocaproic acid "spacer." Anti-DNP antibodies therefore have a greater affinity for DNP-CAP-PE than for DNP-PE.

complement sequence. Fortuitously, one day last summer, we had some "leftover" liposomes containing DNP-CAP-PE; with "nothing better to do," Dr. Uemura decided to inject these into guinea pigs.

To our surprise, the animals responded by producing anti-DNP antibodies. Indeed, three weeks after immunization, the sera of guinea pigs that had been given liposomes actively sensitized with DNP-CAP-PE had an antibody titer almost as great as those that had received a conventional dose of dinitrophenylated protein. Aside from the fact that even naturally occurring lipid antigens are poor immunogens, these results were entirely unexpected for one important reason. Low-molecular-weight compounds such as the dinitrophenyl group generally do not lead to antibody production unless they are first covalently attached to a high-molecular-weight water-soluble carrier such as protein. In the liposomes, however, the dinitrophenyl group is covalently linked to another low-molecular-weight substance, PE, which is itself nonimmunogenic. Furthermore, subsequent experiments by Robert Nicolotti have shown that this phenomenon is not limited to the DNP group; he has attached fluorescein residues to the amino function of PE and found that liposomes actively sensitized with this compound elicit formation of antifluorescein antibodies in guinea pigs. Needless to say, the theoretical basis of liposomal immunogenicity is a problem that currently occupies much of our attention.

Equally intriguing are some of the practical applications, since we believe that liposomes actively sensitized with amino-substituted PE derivatives may represent a convenient alternative to the usual method of preparing antibodies against small molecules – that is, by prior attachment to proteins. To illustrate: consider the many low-molecular-weight drugs that have only a limited solubility in water but, like PE, are soluble in organic solvents. The latter property permits exploitation of countless synthetic routes for making "drug-substituted" PE derivatives, which, when incorporated into liposomes, may elicit "anti-drug" antibodies. Such antibodies would then prove extremely useful in the radioimmunoassay of the drug. Conversely, liposomes containing drug-substituted PE derivatives should release the trapped glucose marker when incubated with a complement source and anti-drug antibodies. This could be developed into a simple procedure for screening patients with suspected drug allergies, whose serum contains such anti-drug antibodies.

I emphasize these possibilities in the hope of convincing skeptics (physicians as well as patients?) that studies of the "immune reactions of model membranes" have some useful purpose. Attainment of these objectives would, of course, provide one final irony: we began this immunologic adventure as pharmacologists and probably shall return in the same capacity!

24

The Actions of Nervous System Drugs on Cell Membranes

PHILIP SEEMAN
University of Toronto

In the light of earlier chapters in this book discussing hormonal and immunologic reactions with the plasma membrane, readers will hardly find it surprising that many drugs also find their site of action on or in this portion of the cell. Many pharmacologically active substances do in fact pass through the plasma membrane and thereafter modify intracellular processes in the cytoplasm or some organelle. Others, however, either do not enter the cell at all or, if they do, still achieve their effects by interacting with one or more components of the cell's membranous envelope.

In discussing the mode of action of membrane-active drugs, an important and central distinction is that between so-called nonspecific and specific action. The former – typified by the effects of nearly all anesthetics and many analgesics – appears to be basically physical, depending entirely on the drug's capacity to enter the membrane structure. The compound must be lipid soluble, but otherwise its chemical structure appears to be essentially irrelevant. In specific action, by contrast, chemical structure is critical. In the opiate narcotics as generally employed, for example, only the levorotatory forms are pharmacologically active; the mirror-image dextrorotatory forms have little or no effect, though their lipid solubility and other physical properties (apart, of course, from their optical activity) are identical. Specific action, though it is not yet well understood, evidently involves the same general mechanisms as hormone action: the triggering by the drug molecule of a distinctive chemical receptor on the membrane surface, which in turn can set off changes within the cell.

There are other distinctions between the two types of action, which I shall discuss later. Meanwhile, however, one should note that specific and nonspecific are not mutually exclusive terms. Certain drugs appear to act in a "semispecific" manner, partaking of both modes, while others combine both specific and nonspecific actions, the dominance of one or the other often depending on the concentration of the drug employed.

The relationship between lipid solubility and anesthetic action was proposed as long ago as the beginning of this century, by Meyer and Overton. Basing themselves on experiments with ordinary, grocery-store olive oil, they even calculated the level of concentration in lipid (they called it "lipoid") that should produce anesthesia with any drug. The actual figure turned out to be only a very rough first approximation – notably because of the difference in solubility characteristics between olive oil and animal lipid – but the basic principle has proved valid. As a second approximation, we can say that the capacity of a drug to block the depolarization of nerve fibers (which is the basis of both anesthesia and many types of analgesia) shows a strong correlation with its solubility, not in "lipoid" per se but in *plasma membrane*. This solubility can be measured quite easily by labeling the drug molecule isotopically and mixing it with a preparation of purified membrane. After the membrane is washed, the amount of label remaining gives an index of the quantity of drug absorbed per gram of membrane. The source of the membrane seems unimportant – preparations from erythrocytes, brain cells, muscle cells, or liver cells give much the same results.

The concentration required for general anesthesia falls in the rather narrow range of 200 to 500 μmole/100 gm of membrane, for a wide variety of drugs. The main variable within this range seems to be the size of the molecule; large molecules (e.g., chlorpromazine) are effective at lower concentrations than small ones (e.g., ethanol, halothane).

Naturally the clinical potency of a given drug does not depend simply on its membrane solubility; no less impor-

The Membrane Concentrations of Anesthetics

A. Nerve-Blocking Concentration *Moles/LH_2O*

B. Volume of Anesthetic in Membrane *ml/kg Membrane*

	A	B
H_2	4.2×10^{-1}	1.8
1. Ne	2.68×10^{-1}	1.2
2. Ar	2.16×10^{-1}	2.1
3. Kr	1.24×10^{-1}	2.2
4. Xe	5.73×10^{-2}	3.5
5. N_2	8.8×10^{-2}	4.1
6. CH_4	1.35×10^{-1}	1.5
7. N_2O	1.28×10^{-1}	1.5
8. C_2H_4	5.2×10^{-2}	1.2
9. c-C_3H_6	1.93×10^{-2}	1.8
10. SF_6	1.01×10^{-2}	4.0
11. CF_2Cl_2	8.2×10^{-3}	3.0
12. Benzylalcohol	2.0×10^{-2}	5.1
13. Methanol	2.4	2.3
14. Ethanol	5.0×10^{-1}	2.2
15. Propanol	2.18×10^{-1}	4.1
16. Butanol	6.8×10^{-2}	5.3
17. Pentanol	2.1×10^{-2}	4.7
18. Hexanol	6.0×10^{-3}	2.9
19. Heptanol	1.75×10^{-3}	2.4
20. Nonanol	6.4×10^{-5}	2.5
21. Isopropanol	3.51×10^{-1}	4.1
22. Tert-amyl-alcohol	8.1×10^{-2}	7.8
23. Menthol	5.8×10^{-4}	12.2
24. Thymol	2.2×10^{-4}	8.5
25. β-naphthol	3.0×10^{-4}	3.3
26. Phenol	7.0×10^{-3}	2.7
27. Chloroform	5.0×10^{-3}	4.6
28. Diethylether	5.0×10^{-2}	3.1
29. Halothane	5.0×10^{-3}	3.4
30. Urethane	2.25×10^{-1}	6.3
31. Pyridine	5.9×10^{-2}	2.3
32. Aniline	2.0×10^{-2}	1.8
33. Nitrobenzene	2.95×10^{-3}	2.6
34. Acetanilide	1.5×10^{-2}	3.4
35. Quinoline	2.0×10^{-3}	3.0
36. Hydroquinone	2.5×10^{-2}	1.2
37. Antipyrine	6.0×10^{-2}	2.6
38. Procaine	4.6×10^{-3}	2.0
39. Cocaine	2.6×10^{-3}	6.7
40. Tropacocaine	2.2×10^{-3}	6.0
41. Chlorpro-mazine	1.0×10^{-5}	2.6
42. RAC 109 I	8.0×10^{-4}	1.3
43. RAC 109 II	2.3×10^{-3}	3.7
44. Morphine	5.5×10^{-2}	2.2
45. Methadone	2.3×10^{-4}	~2.0
46. Pento-barbital	1.7×10^{-3}	2.2
47. Diphenyl-hydantoin	8.3×10^{-4}	4.8
48. Barbital	28.0×10^{-3}	1.9
49. Pheno-barbital	5.7×10^{-3}	4.5

In table at left, left-hand column lists sciatic nerve-blocking concentrations of various drugs. Right-hand column lists the occupying volumes of anesthetic in the membrane phase under conditions of local anesthesia; note that these volumes are stated in units of cm^3 of drug per kg of membrane. An occupying volume of 3 cm^3/kg is thus essentially identical to a fractional occupation of 0.3%. Numbers given to drugs in table are also used to identify them in graphs on opposite page.

tant is the dosage required to achieve the critical membrane concentration. This, in turn, will depend on the drug's solubility ratio between membrane and body fluid (essentially, water) and—for inhaled anesthetics—on its relative solubility in air and water (the air-water partition coefficient), which gives the concentration in inspired air required to produce a given concentration in alveolar blood.

Ethanol, for example, produces anesthesia only in relatively enormous doses, which is explained by the fact that its membrane-serum solubility ratio is less than unity, so that its anesthetic blood concentration must actually be higher than the critical level of membrane concentration. Many tranquilizers, by contrast, have solubility ratios of 5,000 or more, meaning that the critical concentration can be obtained with relatively tiny doses.

In a more-or-less intermediate position is an anesthetic such as halothane. Here the solubility ratio is about 25, meaning that to achieve the critical membrane concentration of 500 μmole/100 gm of membrane will require a serum concentration of only about 20 μmole/100 ml. And since the solubility in air is the same as that in water (i.e., the air-water partition coefficient is unity), the required concentration in inhaled air should be the same 20 μmole/100 ml. By the application of a standard rule of chemistry (1 mole of any gas occupies 25.4 liters at body temperature and atmospheric pressure) one can convert this figure into the volume ratio of 0.5 L of halothane to 100 L of alveolar air, or a concentration of 0.5%. This calculated value agrees closely with the empirically determined concentration of halothane normally used in surgery: 0.5% to 0.6%. (A clinical note: the anesthetist normally *begins* with a considerably higher concentration – 2% to 3% – which speeds induction of anesthesia by loading the body with the drug. As surgical anesthesia is achieved, the concentration is gradually reduced to the lower figure.)

Put this way, it all sounds very simple; if the only variables were membrane solubility and molecular size, then predicting the action of an anesthetic drug would be a matter of rather elementary arithmetic. The reality, of course, is far more complicated, with drug action modified by all sorts of other physiologic factors, some of which are known, others of which can only be guessed at. One of the simplest variables appears to be the size of the nerve fiber, which emerges most clearly in distinguishing between general and local anesthesia. Rather unexpectedly, we find that concentrations required to produce local anesthesia – i.e., by perineural or intraspinal injection – are on the order of 10 times those required to induce general anesthesia. Thus alcohol, for example, produces unconsciousness at a serum concentration of 0.2% to 0.3%, whereas neural blockade requires 4% to 5%.

The reason for this difference seems to be that in local anesthesia we are dealing with much larger neurons than those of the CNS affected in general anesthesia. Exactly why this should so radically alter the required concentration is still uncertain; it may have something to do with the larger neurons' smaller surface to volume ratio.

Other variables emerge when we consider the actual mode of action of the nonspecific drugs on the membrane. In view of everything that has been said about lipid solubility, it might seem obvious that the drugs act predominantly on the lipid rather than the protein moiety of plasma membrane, and for some years precisely this assumption was made. Moreover, it seemed to be supported by certain experimental findings. As has been noted in Chapter 3 of this book, anesthetics applied to the model membrane systems of pure lipid, called liposomes (sometimes, after their creator Alec Bangham, "Bangosomes"), produce an expansion of the membrane and a drop in its electrical resistance. In neurons, it was reasoned, such changes would lower the normal ionic imbalance (Na^+ high outside and low inside, K^+ high inside and low outside) required for neural discharge.

More recent experiments suggest that this view may be the reverse of the truth. Nonspecific anesthesia, it appears, is more likely caused by reaction of the drug, not with the lipid moiety but with the membrane protein; its effect (so far as nerve conduction is concerned) may be not to increase ionic permeability but to decrease it, by blocking the channels through which Na^+ rapidly enters the cell during depolarization.

The evidence for this revised view, some of it fairly complex, is partly indirect, but is nonetheless persuasive if not conclusive. In the first place, while drug molecules undoubtedly enter the lipid portions of the plasma membrane, there is no a priori reason why they cannot also incorporate themselves into the protein regions. Membrane proteins, as we well know, contain many amino acid residues, and even whole sequences of residues, that are hydrophobic, and with which lipid-soluble drugs would be quite as compatible as they are with the hydrophobic "tails" of the membrane lipids.

A second fact is that anesthetics demonstrably combine X17 with purified enzymatic proteins, both soluble and membrane-associated, and alter their functioning (presumably by distorting their structure); depending on the drug, the functional change may be either stimulatory or inhibitory. Enzymes studied in this manner include Na^+-K^+-ATPase, cholinesterase, and adenylate cyclase, all of which are known to be located in or on the membrane. Admittedly, these experiments have been done primarily with enzymes of disrupted cell membranes in vitro; to what extent they would exhibit the same changes in vivo remains to be determined.

Even more persuasive evidence comes from studies of the expansion of membranes induced by anesthetic drugs. Qualitatively, this can be observed in erythrocyte ghosts under the microscope; quantitatively, it can be measured in terms of the density of the membrane (i.e., essentially the reciprocal of the expansion), using a high-precision density meter recently developed by an Austrian group. At

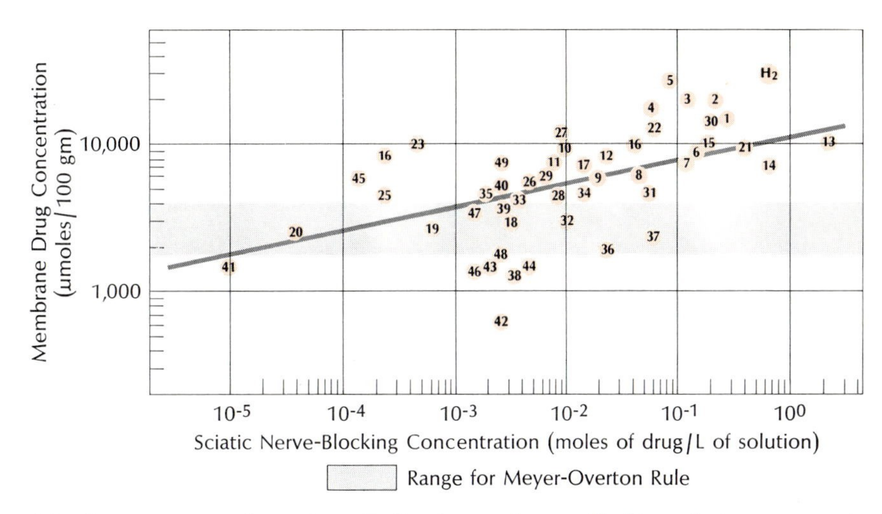

Membrane concentrations of anesthetics for equal nerve-blocking effects are approximately but not exactly predicted by Meyer-Overton rule for local anesthesia – i.e., around 3,000 µmoles of drug/100 gm of membrane. Their rule for general anesthesia predicts much lower concentrations – around 200 to 500 µmoles/100 gm – presumably because nerves affected are more vulnerable. For code to drugs, see table opposite.

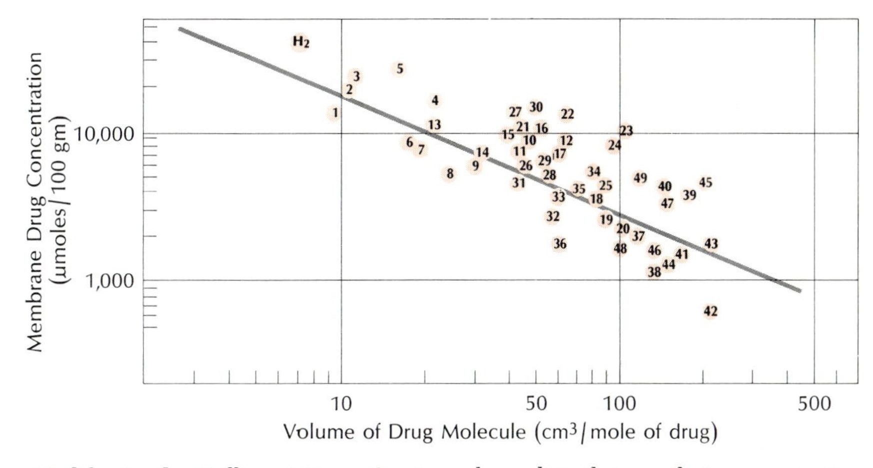

Modification by Mullins of Meyer-Overton rule predicts that anesthetic concentration (for nerve block) in membrane phase (ordinate) is higher for smaller drug molecules.

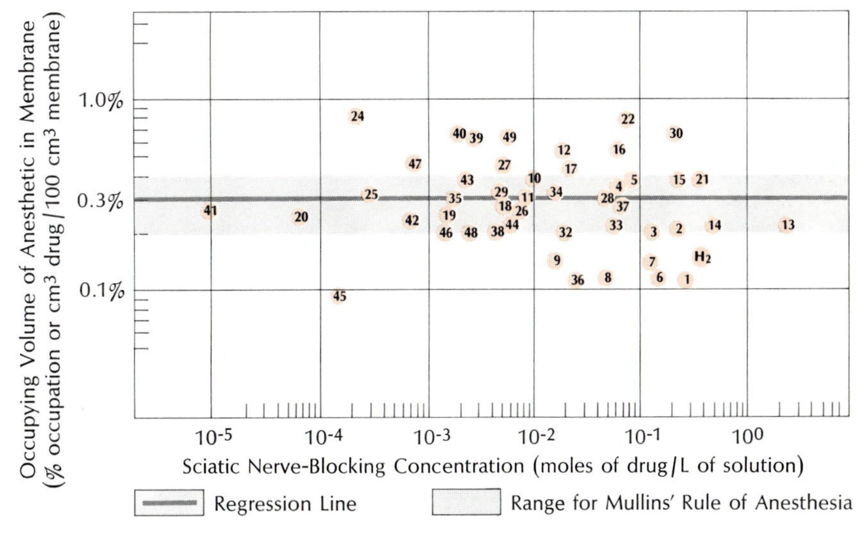

Occupying volume of anesthetic (for nerve block) in membrane phase has a constant value of about 0.3% for all. Although this generalization fits Mullins' rule, observed expansion of membrane volume (or area) is actually about 10 times greater, suggesting that large perturbations occur in membrane's molecular architecture.

concentrations of drugs (such as ethanol) known to block nerve fibers, the membrane expands on the order of 2% to 3% and, significantly, the expansion is something like 10 times what would be predicted from the amount of drug known to enter the membrane. That is, the increase in membrane volume, or decrease in density, is far greater than the volume of the anesthetic molecules incorporated in the membrane; evidently their effect has somehow been amplified by a factor of 10. No less significantly, when the same experiments are done with liposomes, in which protein is not present, no such amplification occurs. The membrane expands, but to a degree that merely reflects the added volume of the drug molecules.

Evidently, then, anesthetics produce changes in membrane proteins that – quantitatively, at least – are far more marked than the changes they induce in membrane lipids. It is possible, of course, that the expansion of the proteins might be due, not to direct interaction of protein and drug, but to a drug-induced perturbation of the lipid matrix, which in turn could perturb the protein structure; no one has yet devised an experiment that can distinguish between these two possibilities. Simply on probability grounds, however, the direct drug-protein interaction seems more likely. We know that for a number of proteins minor spatial changes in one portion of the molecule can be amplified into much larger changes in other regions. A notable example is the hemoglobin molecule, in which amplification of various sorts has been elegantly demonstrated by Max Perutz and associates at Cambridge.

(As a side note, one of the drugs known to produce membrane expansion of this type is tetrahydrocannabinol, the active principle of marijuana and hashish, so that the designation of these substances as "mind expanding" is physiologically, if not psychologically, true. One might also note that if a psychiatrist employs tranquilizers in his practice, it would be the reverse of the truth to describe him as a "shrink"!)

By this reasoning, then, the nonspecific drugs would block the "channels" in the membrane through which sodium enters the cell during depolarization. That such channels exist is generally accepted, as is their proteinacious character. Evidence on the latter point includes studies with the poison tetrodotoxin, obtained from the puffer fish (a delicacy relished by Japanese gourmets who do not mind living dangerously). Tetrodotoxin also blocks the sodium channels, but very specifically; even minor changes in its molecule render it inactive. This high specificity implies that the drug is reacting with a molecule with a no less specific structure, which again suggests a protein rather than a lipid.

So far as the nonspecific drugs are concerned, all this still does not completely dispose of the possibility that they may be altering the protein indirectly, by changing the structures of adjacent lipids, rather than directly. Postive proof one way or the other will require more experiments.

The relationship between membrane expansion and neural blockade implies that if the expansion can be reversed, for example by subjecting the cells to high pressures, the blockade should be removed. In fact, as Bangham mentioned in Chapter 3, this can be done: tadpoles that have been anesthetized with ethanol can be "sobered up" by placing them in a pressure tank. I am told that some people at the U.S. National Agency for Alcoholism and Addiction have been considering the possibility of treating acute alcohol poisoning by placing the patient in a hyperbaric tank. In principle, the approach would probably work, but since the pressures required would be on the order of 70 atmospheres – far higher than those in any current form of hyperbaric apparatus – it does not sound very practical from a clinical standpoint.

Granted that the nonspecific anesthetics and analgesics produce ionic blockade by reacting directly or indirectly with the protein moiety of the plasma membrane, this is by no means to say that they are reacting *only* with the proteins. On the contrary, there is much evidence that they also alter the properties of membrane lipids, and that it is these lipid interactions that probably account for many of the clinically significant differences between various anesthetics, tranquilizers, and so on. From the standpoint of their mode of action, these substances may all be lumped together as nonspecific, but from the clinical standpoint they may be very specific indeed in their individual effects.

A number of experiments have established that the membrane expansion induced by nonspecific drugs is accompanied by an increase in lipid fluidity. This can be shown, for example, by measuring the speed with which the membrane proteins reorganize their spatial relationships under various conditions – the rapidity with which they can move through the lipid "sea" that surrounds them. Likewise, measurements of electron spin resonance and nuclear magnetic resonance show that the various lipid membrane components, including cholesterol and phospholipid, possess greater rotational freedom in the expanded membrane.

Perhaps the most important effect of this increased fluidity, at least so far as the CNS is concerned, is that it apparently facilitates the release of neurohumors. As noted earlier in this book (see Chapter 17, "Membranes in Synaptic Function," by Whittaker), neurotransmitting chemicals – acetylcholine, norepinephrine, dopamine, etc. – seem to be characteristically released at synapses and motor end-plate junctions through the formation of neurohumor-containing vesicles in the neuron's presynaptic region, which by fusing with the plasma membrane expel their contents into the synaptic gap. And there is reason to believe that the more fluid the membrane, the more readily the vesicles fuse with it. It has been shown, for example, that ethanol, which is known to increase membrane fluidity, increases the frequency of the so-called miniature end-plate potentials in synapses, each of which is presumed to correspond to the release of a single vesicle. The result is a relatively massive discharge.

The capacity of the nonspecific drugs to facilitate nerve discharge, by their fluidizing of membrane lipids, as well as to inhibit it, by blocking the protein sodium channels, makes for all sorts of complications. Most of these are still poorly understood. It is quite possible, for example, that these potentially competing effects may explain the different stages of anesthesia observed with many drugs, such as nitrous oxide. In the first stage – cor-

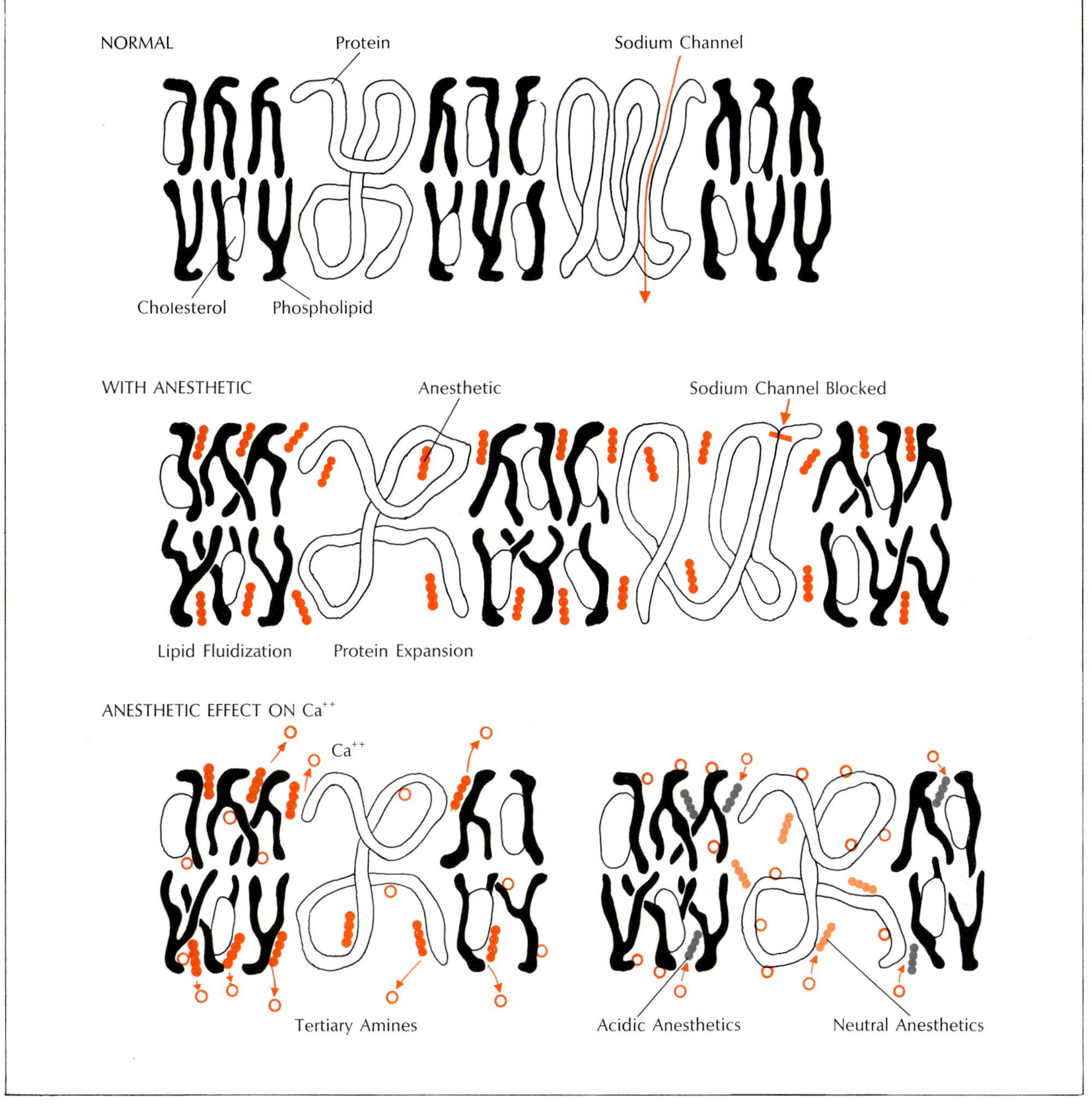

Preanesthesia (normal) state of cell membrane is shown at top; channel through which sodium enters cell during depolarization is open. With nonspecific anesthesia, membrane expands, primarily through distortion of proteins, resulting in blocking of sodium channel; fluidizing of lipid regions, also present, probably participates to lesser extent in membrane expansion. Differential effects of drugs' electrical potential on calcium ion are suggested in bottom panel. Positively charged tertiary amines displace Ca++ from membrane by competing with it for negatively charged binding sites (left). Negatively charged drugs (e.g., acidic anesthetics) increase binding of Ca++ (right); this also occurs with many neutral anesthetics.

responding to 50% to 60% N_2O – the patient experiences analgesia, which could reflect a blockade of the smaller neurons in the spinal cord that mediate pain. In the second stage, corresponding to perhaps 70% N_2O (the balance being oxygen), the predominant effect is excitation and exhilaration; this might reflect the facilitated release of norepinephrine in the CNS. Finally, in the third stage, corresponding to about 80% N_2O, the patient becomes anesthetized, suggesting that the sodium-blockade effect has become dominant again, this time to the point of affecting the larger neurons in the CNS that mediate consciousness.

This explanation is probably oversimplified, but at any rate it provides a relatively straightforward rationale for certain observed drug effects. Other potential complexities in anesthetic-tranquilizer action, however, suggest that the mode of action of at least some of these substances, though it can be categorized as nonspecific, is anything but straightforward. There is the fact, for instance, that some of the neurons whose discharge may be

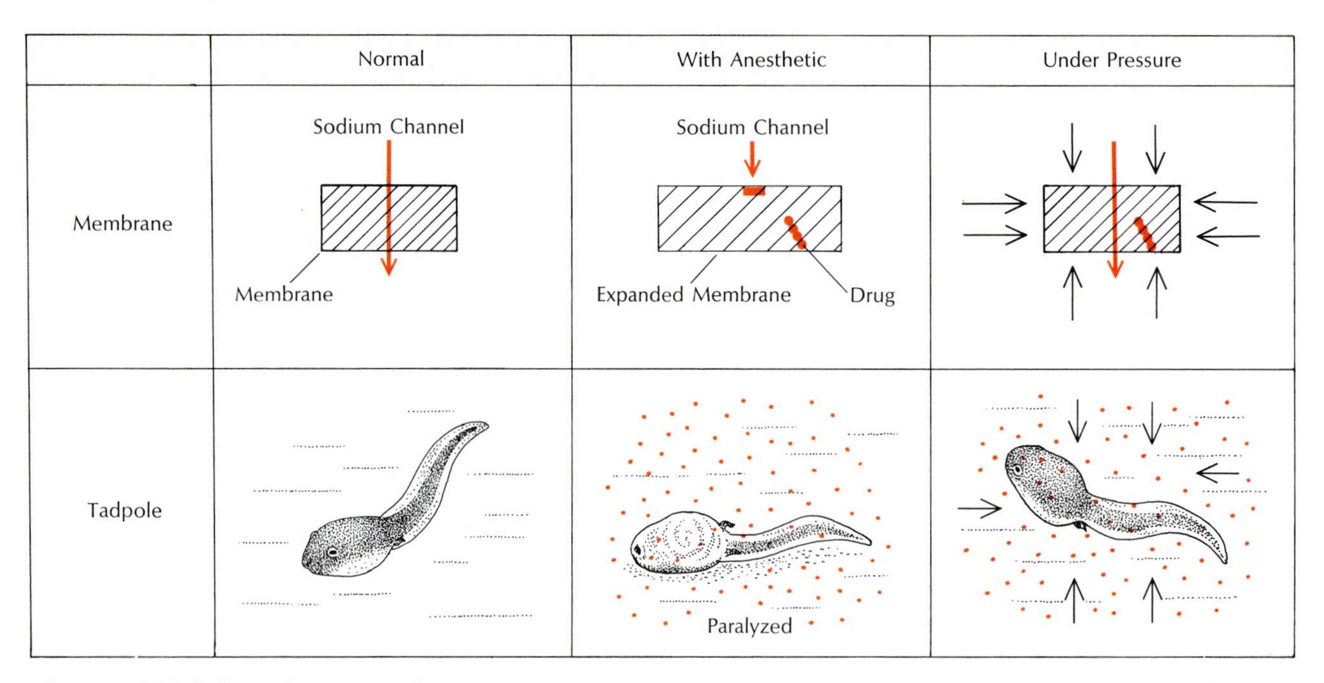

That neural blockade results from membrane expansion is supported by experiments using tadpoles anesthetized with ethanol. Under high pressure membrane expansion is reversed, Na channels reopen, and tadpoles revive despite continued drug presence.

either facilitated or inhibited, depending on the drug and its concentration, are themselves inhibitory rather than stimulatory in relation to the rest of the CNS. This means that inhibiting them becomes equivalent to stimulation and that facilitating their discharge fosters inhibition. To this we can add the virtual certainty that neurons mediated by different neurohumors–for example, dopamine and acetylcholine–differ in the protein moiety of their membranes, and perhaps in the lipid moiety as well. Similar differences may perhaps exist between excitatory and inhibitory neurons. Finally, there is good reason to expect that different drugs will possess different affinities for protein vs lipid, as well as for different species of proteins and lipids.

Yet another complexity of nonspecific drug action concerns membrane calcium. Since Ca^{++} influx contributes to the inward current of the neuron's action potential, changes in concentration of the ion will obviously affect neuronal function. At the same time, however, Ca^{++} concentrations can also affect the neurons' target

	Lytic Concentrations			
	1/50th	1/10th	Sublytic	Lytic
Neutral Drugs	Normal	Membrane-Expanded	Stomatocyte	
Organic Tertiary Amines	Membrane-Expanded	Stomatocyte	Invaginated	
Organic Acidic Drugs	Echinocyte I	Echinocyte II	Echinocyte III	

Although membrane expansion is a common phenomenon in relation to the cytotoxic effects of the different classes of nonspecific anesthetic drugs, from that point on their effects vary, as experiments with erythrocytes have shown. Variations may be associated with the drugs' differential effects on calcium binding (see drawing, page 243) but exact relation has not yet been defined.

cells, since presence of the ion is known to be essential for such processes as secretion and muscle contraction.

In this area, however, we find significant differences among certain classes of nonspecific drugs. The tertiary amines (examples are procaine, chlorpromazine, quinidine) displace Ca^{++} from the membrane, evidently because their positively charged molecules compete with the ion for negatively charged binding sites. This phenomenon would explain the capacity of some of these drugs to act as muscle relaxants, to inhibit secretion by the adrenal medulla (chlorpromazine), and to decrease the irritability of heart muscle (procaine, quinidine).

By contrast, the acidic anesthetics (such as the barbiturates) increase binding of Ca^{++} to membrane lipid. This is what one would expect given their negatively charged molecules; the same thing occurs – for unknown reasons – with many neutral anesthetics (ethanol, ether, chloroform). Calcium-binding of this sort would explain the ability of these drugs to potentiate muscle contraction in response, for example, to caffeine.

Given all these complexities, beginning with the size of the drug molecule, continuing with its relative affinity for various membrane constituents, and concluding with its capacity to bind Ca^{++} to or release it from the membrane, the possible variations in terms of specific clinical action of a given drug become virtually infinite. Even if we limit our field of inquiry to neuroleptic drugs, and further limit it to those drugs that act nonspecifically, it will evidently be a long time before we can set down a chemical formula and predict precisely where the molecule will affect the CNS (and/or its target organs), what it will do there, and in what order. The Meyer-Overton rule of anethesia was a good first approximation, and its refinement in terms of membrane (rather than "lipoid") solubility and molecular size, a good second approximation. But really to comprehend the actions of even this restricted class of drugs we will need to move on to third, fourth, and fifth approximations.

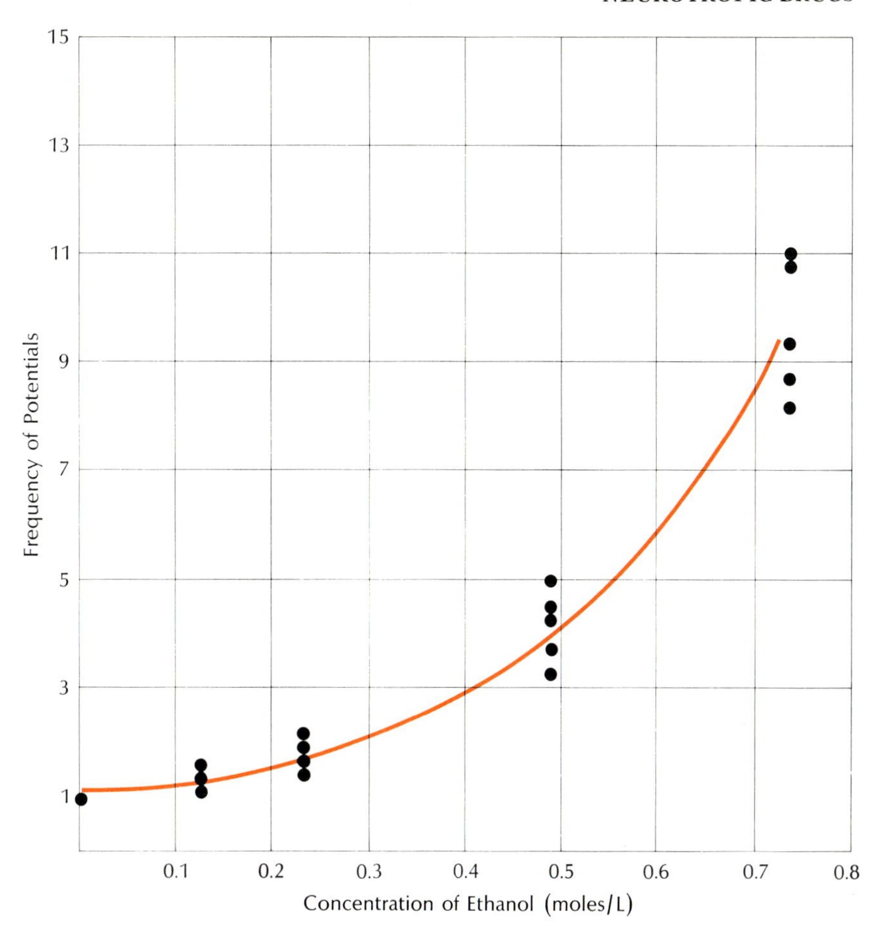

Increased fluidity of all membranes caused by nonspecific anesthetics appears to facilitate fusion of vesicle membranes with the presynaptic nerve terminal membrane; resulting discharge of neurohumor causes a miniature end-plate potential in synapse. As graph shows, with increased ethanol concentration, which is known to increase membrane fluidity, there is a corresponding rise in frequency of such potentials.

It is interesting, though not particularly surprising, that the differences among positive, negative (acidic), and neutral drug molecules show up in relation to their cytotoxic effects. In erythrocytes, for example, the neutral drugs produce simple expansion; at the same time the cells lose their characteristic disc configuration and become cupped (stomatocytes). If carried far enough this process produces membrane breakdown and cell lysis. The positively charged drugs also produce expansion and cupping, but at higher concentration also produce an inward buckling (invagination) of the membrane. The negative drugs, finally, induce an opposite effect: after initial expansion, the cell forms spurs, or pseudopods, and is called an echinocyte (from the Greek word for hedgehog). These differences related to molecular charge suggest that one ultimate cause of these varied membrane distortions may be related to the drugs' differential effect on calcium ion, but the exact relationship remains to be defined.

Turning now to the specific-acting drugs, we find, as noted earlier, that, to the extent that it is understood, their mode of action closely parallels that of hormones. The relationship is expected, since the two classes of substances overlap, with several hormones (such as insulin or ACTH) being employed as drugs. The parallelism between specific-acting drugs and hormones and the contrast between both groups and the nonspecific drugs show up very clearly in the capacity of membrane to bind them, expressed as the number of receptors or binding sites per square micron of cell surface. As measured by various investigators, these figures fall within a relatively narrow range for specific drugs or hormones: $1/\mu^2$ for digitalis, 2 for insulin, 10 for norepinephrine and glucagon, 27 for tetrodotoxin and 600 for angiotensin (adrenal cells).

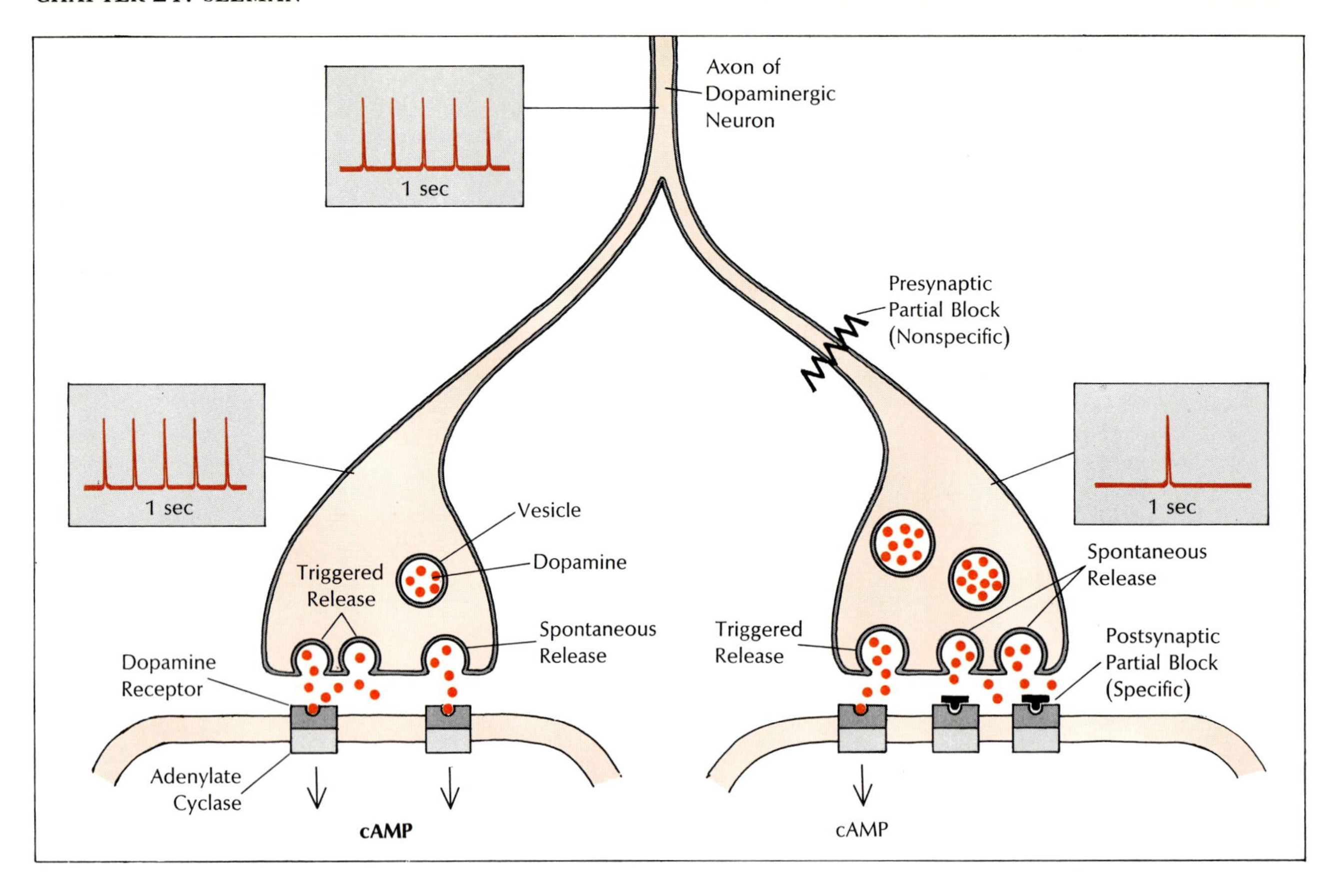

Whether action of chlorpromazine is nonspecific, specific (hormone-like), or both has been a subject of much research. In the diagram of dopaminergic synapse above, normal sequence of events is shown at left, alternative drug mechanisms at right. The nonspecific impulse-modulating action of the neuroleptic blocks the impulse-triggered release of dopamine from the vesicles, but at the same time, the fluidizing action of the neuroleptic enhances spontaneous neurosecretion; with specific action, the adenylate cyclase-cyclic AMP sequence would be blocked by drug competition for the dopamine binding site.

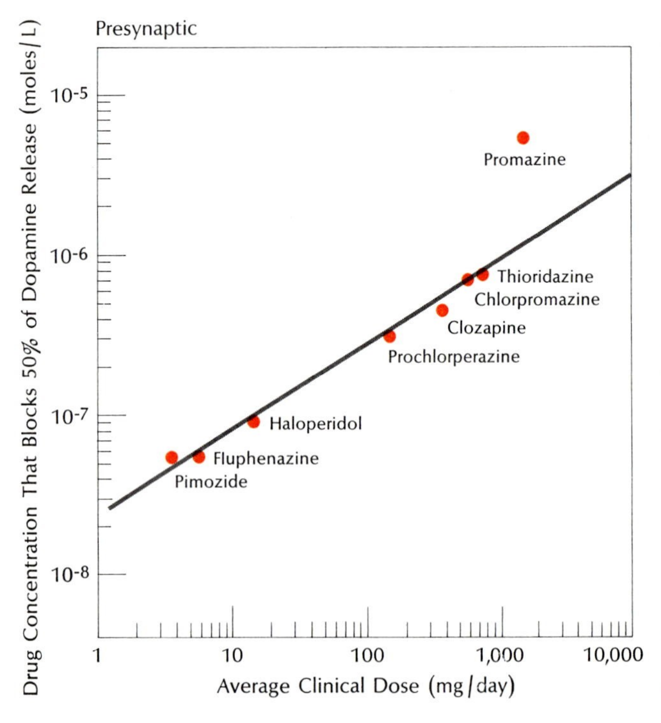

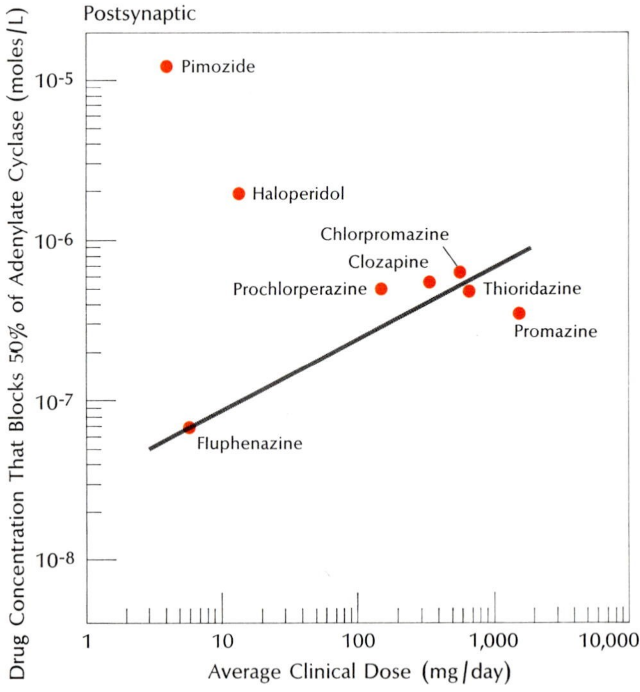

A presynaptic, nonspecific mode of action for major tranquilizers gains support from experiments with rat caudate slices stimulated electrically to produce dopamine release. A much closer correlation is found of the drugs' clinical potencies with presynaptic blockade of dopamine release (left) than with postsynaptic blockade of dopamine-sensitive adenylate cyclase (latter based on Y. C. Clement-Cormier, J. W. Kebabian, G. L. Petzold, P. Greengard: Proc Natl Acad Sci, 71:1113, 1974).

ACTH is a special case, since it apparently binds to two different types of receptors: "high affinity," with a density of only $0.1/\mu^2$; and "low affinity," whose density approximates that for angiotensin receptors. Another special case is acetylcholine, whose receptors in muscle end-plate number on the order of $10{,}000/\mu^2$. Note, however, that this refers to only *part* of the muscle-cell membrane, and that there is, moreover, a good physiologic reason for this very dense concentration: the need for the cell to respond simultaneously to a great many acetylcholine molecules.

When we consider the number of "receptors" for nonspecific drugs, we find ourselves dealing with a wholly different order of magnitude. Using the "critical figure" of 200 μmoles/100 gm of membrane, one can calculate that each square micron of cell surface must be binding a million or more molecules of the drug. Much the same figure obtains for such substances as saponin and the polyene antibiotics such as amphotericin B, which are known to bind preferentially to (and attack) membrane cholesterol. When binding occurs at such densities, however, the term "receptor" has perhaps been stretched beyond the bounds of usefulness, since it implies a stereospecific molecule – a glycoprotein or glycolipid–"keyed" to a particular drug or hormone molecule; for nonspecific binding the term "binding site" is preferable.

A number of drugs appear to combine both specific and nonspecific effects. One example is morphine, which in low concentrations is specific, producing analgesia only in its L-rotatory form, the D-form being far less potent. In much higher concentrations, however, it appears to obey nonspecific rules, with both forms producing anesthesia at concentrations in the critical range of 200 to 500 μmoles/100 gm of membrane. According to a recent report, indeed, a Boston surgical group has been employing massive doses of opiates (around 10 mg/kg) for anesthesia in cardiac surgery. In most forms of surgery this would be unfeasible, since the patient's respiratory centers are totally paralyzed, but in this instance the patient must be artificially oxygenated with a heart-lung machine in any case.

A less certain candidate for inclusion among the dual-acting drugs is chlorpromazine (and also, presumably, its chemical relatives). This compound, widely used as a major tranquilizer, can also serve as a local anesthetic at the characteristic nonspecific concentrations in membrane. It can also, of course, suppress hallucinations and delusions in such conditions as schizophrenia, probably by interfering with dopaminergic transmission in certain parts of the brain such as the limbic system. The question is whether it does so by nonspecific blocking of the dopamine-releasing neurons or by specifically inactivating dopamine receptors. Cited in favor of the latter hypothesis is the recent finding by Greengard and his colleagues at Yale that the dopamine receptor apparently activates or potentiates the enzyme adenylate cyclase, so that blocking the receptor would decrease cyclic AMP levels in the receptor cycle – and tranquilizers, it has been reported by both Greengard's group at Yale and Iversen's group at Cambridge, do indeed block the dopamine-sensitive adenylate cyclase. However, the enzyme-blocking potencies of these major tranquilizers do not correlate with their antipsychotic potencies (see page 246, bottom right). My associates and I have been studying the basis of neuroleptic action, using slices of rat caudate nucleus that we convulse electrically to produce dopamine release. We find that these neuroleptics (or major tranquilizers) block the *impulse-triggered* (i.e., electrically stimulated) release of dopamine from these slices, and that there is a surprisingly good correlation between this impulse-blocking potency of any neuroleptic with its clinical potency in relieving symptoms of schizophrenia (see page 246, bottom left).

Thus far I have been writing as if the term "membrane-active" applied only to drugs affecting the plasma membrane. In fact, there is no reason to believe that some compounds do not produce their characteristic physiologic response in or on the interior membranes of the cell's organelles – though such actions are obviously more difficult to study than those affecting the plasma membrane. Nonetheless, we know, for example, that the antibiotic oligomycin, by specific action, blocks the ATPase "pump" in the mitochondrial membrane, just as digitalis specifically blocks the enzyme Na^+-K^+-ATPase in the plasma membrane. Likewise, we know that certain drugs, in what appears to be one form of nonspecific action, either block or (in other cases) facilitate movement of Ca^{++} across the membranes of both the endoplasmic reticulum and the mitochondria.

Obviously much more must still be learned before we can say that we understand how and when drugs affect these internal membranes, or even how they affect the considerably better understood plasma membrane. Considering the central importance of membranes, both external and internal, to cellular physiology in general, however, it would be surprising if they did not turn out to be no less central in cellular pharmacology.

25

Pathologic States of the Erythrocyte Membrane

HARRY S. JACOB

University of Minnesota

Because red cells are readily obtainable in quantity and can be prepared rather easily in "ghost" form, freed of their cytoplasmic contents, the membranes of erythrocytes have been widely used as experimental prototypes of cell membranes in general. As such, and despite the major differences between erythrocytes and other cells, they have made possible major advances in membrane physiology and biochemistry. The erythrocyte membrane, however, is also of considerable interest in its own right – most especially in connection with diseases of the red cell. Many of these diseases, though of quite diverse origins, are alike in manifesting themselves as membrane abnormalities that beget hemolysis and other pathologic sequelae. It is with this group of pathologic states that the present chapter is concerned.

As is well known, the mature erythrocyte represents a drastically simplified version of the normal mammalian cell pattern. Possessing neither nucleus nor endoplasmic reticulum, it cannot synthesize enzymes or other proteins; for related reasons, it cannot synthesize lipids. Possessing no mitochondria, it cannot carry on oxidative phosphorylation, in which glucose yields energy through conversion to CO_2 and water, but must meet its energy needs through the process of anaerobic glycolysis, in which glucose is converted to lactate. The red cell has, in fact, one and only one physiologic function: to take up oxygen in the lungs and transport it to the tissues.

To perform this function, the erythrocyte must possess two critical properties. First, the hemoglobin it contains must bind oxygen tightly enough for transportation to the tissues, yet not so tightly that it cannot be released there. A number of erythrocyte pathologies, in fact, involve abnormalities of the hemoglobin molecule that alter its oxygen-binding capacity, making the latter either too "loose" for efficient uptake and transport or too "tight" for efficient release. Some of these hemoglobinopathies are also marked by membrane abnormalities, but the latter are distinctly secondary from the pathologic and clinical standpoint, and therefore need not concern us here.

The other essential erythrocyte property is a capacity to move through the body's microvasculature where its freight of oxygen must be delivered. The smallest of these vessels are less than 2μ in diameter, while the red cell itself has a diameter of about 7μ. Clearly, then, its membrane must be highly deformable if the cell is to squeeze through these narrow capillary passages. It is fair to say, I think, that the great majority of erythrocyte pathologies in which oxygen binding is *not* altered manifest themselves in the first instance as a reduction in the deformability, or the plasticity, of the cell membrane.

To obtain a proper perspective on the pathologic erythrocyte, let us first consider some normal properties of the cell's membrane. To begin with, that membrane does not differ greatly in gross composition from other plasma membranes: about 50% protein, 25% cholesterol, and 25% phospholipid. The erythrocyte's special plastic properties are reflected in the protein moiety, much of which consists of fibrous proteins – long chains formed by the polymerization of smaller peptides – which are thought to form a flexible framework for the cell, and probably of contractile proteins which, by "pulling" on the framework, alter the cell's shape. Yet even these are by no means unique to the erythrocyte; one finds similar proteins in, for example, the ameba, which also requires a highly deformable structure in order to function properly (see Chapter 5, "The Structure and Orientation of a Membrane Protein," by Marchesi).

The normal erythrocyte has a lifetime in the circulation of about 120 days, following which it is filtered

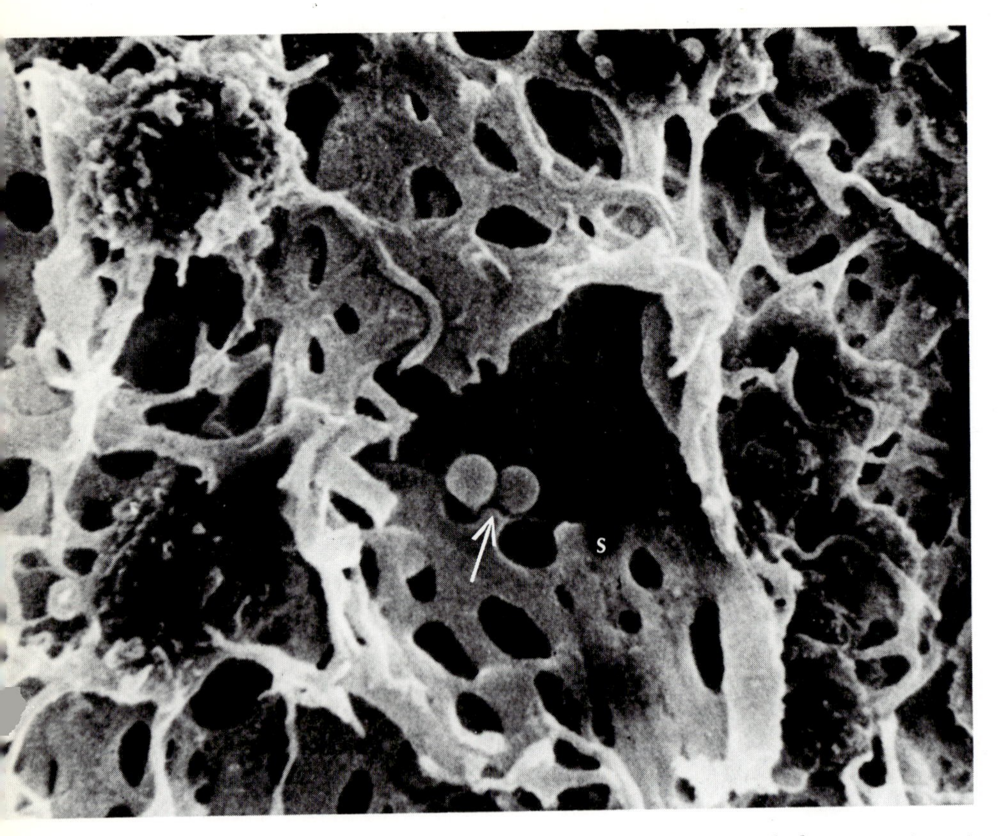

Walls of sinuses (S) in the spleen contain tiny fenestrations, as little as one micron in diameter, through which much larger erythrocytes must squeeze before they return to the circulation. Red cells whose plasma membranes have lost flexibility, as the result of aging or for pathologic reasons, are trapped by this splenic filter and destroyed. Arrow points to two ends of a single erythrocyte attempting to penetrate two fenestrations at the same time (scanning electron micrograph by Miyoshi and Fujita).

out of the blood mainly by the spleen and broken down to its various constituents – lipid, amino acids, iron, bilirubin – which are either eliminated or recycled. Some investigators have suspected that the spleen's recognition of over-age erythrocytes is chemical, comparable to the recognition processes involved in immune mechanisms, but the truth seems to be that the process is purely mechanical. The spleen appears to have been designed by evolution to act as a filter; its capillaries contain myriads of tiny fenestrations, no more than 1μ in diameter, through which the erythrocytes must squeeze to get back into the circulation. The senescent red cell, it appears, has lost so much of its plasticity that it can no longer pass through the filter but remains trapped in the spleen, where it is broken down.

The processes by which the aging erythrocyte loses its plasticity are obviously of key significance in understanding not merely normal red cell aging but the pathologic losses of membrane flexibility already alluded to. In the first place, the erythrocyte's lack of synthetic capability severely limits its ability to replace or repair its membrane. Cholesterol, indeed, can be and is taken up quite rapidly from the serum, with about half the membrane cholesterol being replaced every six hours, and the same process obtains, albeit more slowly, with preformed phospholipids, to the extent these are present in serum. The membrane proteins, however, are irreplaceable – not merely the fibrous and contractile proteins but also (and this seems to be the key point) the enzymes that enable the membrane to carry on its very limited metabolic activities.

These activities appear to be mainly one type: ion transport. Like all plasma membranes, the erythrocyte membrane pumps sodium out and potassium in (see Chapter 10, "Ionic Transport Across the Plasma Membrane," by Hoffman); even more crucial is its ability to expel calcium from the interior of the cell. The concentrations of Ca ion normally within the erythrocyte are kept in the micromole range, at about 1% to 2% of the serum concentration. If calcium concentrations rise above these low levels, the fibrous proteins become denatured, probably by cross-linking with the calcium ion, producing stiffness or even virtual rigidity in the membrane. This process can be demonstrated in the test tube by extracting pure protein from erythrocytes and adding calcium; the protein precipitates almost immediately.

Operation of the calcium pump, like that of the sodium-potassium pump, requires energy in the form of ATP, which, as already noted, is generated (in the cytoplasm) by the breakdown of glucose into lactate. This energy-yielding process, though inefficient, is simple, requiring far fewer steps (and therefore far fewer enzymes) than oxidative phosphorylation; it may be that only two or three different enzymes are directly involved. Nonetheless, these enzymes must be present, and to the extent they disappear – presumably as the result of normal metabolic wear and tear during the cell's lifetime – ATP synthesis will be limited. The calcium pump will be slowed, calcium will accumulate in the cell, proteins will be denatured, and the membrane will lose its plasticity. Eventually the point is reached at which the cell will be unable to pass the spleen's filter. In fact, the deterioration in erythrocyte metabolism, as manifested by a diminished ability to synthesize ATP, can be detected in the test tube as early as 90 days, reaching a critical point during the following month or so.

Let us now consider how the erythrocyte's normal aging processes are altered – in some cases, accelerated – by various pathologic conditions. One of the best understood of these is hereditary spherocytosis, the most common congenital hemolytic anemia in Caucasian populations (its incidence is about 1 in 5,000). Originally, spherocytosis was believed to be caused by an enzyme deficiency, as is the case with so many hereditary diseases. This suspicion was strengthened, in the period around 1960, by studies of another type of hemolytic anemia that showed up among black soldiers given certain antimalarial drugs during service in Korea. The condition occurs in some 15% of American blacks and also, less frequently, in Mediterranean populations. Eventually the defect in these people was traced to a hereditary deficiency in the enzyme glucose-6-phosphate dehydrogenase.

Efforts to demonstrate a similar enzyme abnormality in cells from spherocytosis patients, however, were uniformly unsuccessful. The first clue to the real nature of the disease came from observations that the cells were abnormally permeable to sodium, which suggested a membrane disorder. The membrane was further implicated by the finding that pure membranes – ghosts – of these cells were intrinsically rigid. This can be demonstrated by inserting a micropipette with an orifice of about 3μ into a ghost preparation and, while observing through the microscope, measuring how much negative pressure is needed to suck a ghost into the pipette. For the spherocytosis ghosts, the required pressure was much greater than for normal ghosts. Other observations indicated that the spherocytosis cells, when incubated, tended to bud, with bits of membrane pinching off and falling into the extracellular fluid. This process, by reducing the ratio of membrane area to cell volume, accentuated the cells' spherical form and also rendered them more fragile, rather like overinflated balloons.

At this point, my associates and I focused our attention on the fibrous membrane proteins, which were known to be involved in changing, or fixing, the shapes of other cells such as amebas and platelets. We found that when we made extracts of normal red cell protein in an essentially ion-free medium and then increased the concentration of potassium ion (or calcium ion) toward normal intracellular levels, the protein polymerized. Others demonstrated that these polymers appeared as long fibers visible under the electron microscope. The increase in molecular weight with polymerization was reflected in the sedimentation coefficients in the ultracentrifuge, which increased from 2 to 2.5 S for the ion-free preparation to about 9 S after ionic strength was increased. Significantly, however, protein from spherocytotic red cells behaved quite differently; starting from about the same coefficient in the ion-free state, it changed very little when ions were added, indicating that it was barely polymerizing at all.

Here we had a clear membrane abnormality, and one that evidently involved only the protein moiety, since all lipids had been removed from the preparation. Moreover, it was an abnormality that, in principle, seemed sufficient by itself to explain the observed erythrocyte abnormalities in hereditary spherocytosis. As we shall see, however, the full story has turned out to be much more complicated.

A central fact about this disease is that its expression, even within a single family, can be very variable. It is inherited as an autosomal dominant, meaning that only one parent need possess the defective gene, and cases are on record in which the parent may show almost normal erythrocyte survival, yet one of the children will be severely affected. Some of the disease's variability (though not within a given family) could be rationalized by postulating not a unique or single abnormality in the fibrous protein but several different abnormalities, some having more severe effects on polymerization than others. The particular expression would then depend on the particular amino acid affected by the mutation and on the "substitute" amino acid that replaced it. This situation has long been known to obtain among the hemoglobinopathies, among which at least a hundred different molecular defects are now known; some of them are "silent," producing no functional abnormality, while others engender pathologies as severe as sickle cell disease.

We have, in fact, obtained evidence that at least two distinct types of mutation do exist. Hereditary spherocytosis, one should note, is essentially a disease of heterozygotes; that is, the affected individual possesses one abnormal gene (from the affected parent) and one normal one, which is to say that half his fibrous protein would presumably be normal and half not, at least on the average. Our sedimentation tests, however, showed that in most cases there was no evidence of two types of protein; essentially *none* of it would polymerize. One plausible explanation of this finding is that the abnormal protein is not merely incapable itself of polymerizing but also, by combining with the normal protein, prevents the latter from forming polymerized chains. It is possible to visualize the normal protein as possessing a sort of link at both ends, each capable of joining with a similar link on another molecule, while the abnormal protein has a link on only one end. The abnormal protein could thus combine with a normal molecule but there would be no possibility of the process continuing.

We have turned up a patient, however, in whom some of the protein does polymerize; the sedimentation diagrams show two peaks, one with low coefficient, the abnormal protein, and one with normal sedimentation. Here we have postulated that the ab-

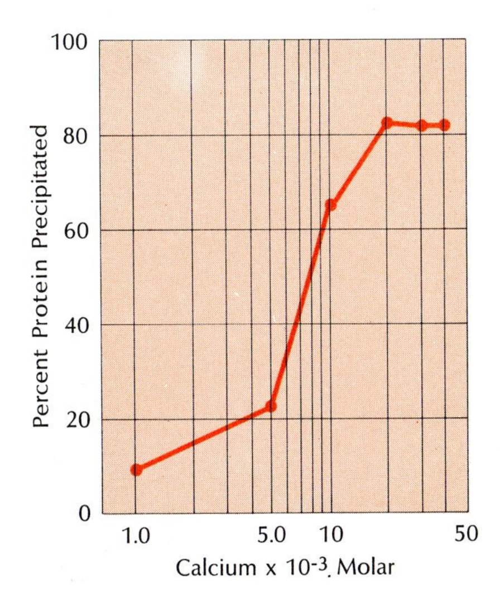

Increases in intracellular calcium ion are believed to play a key role in many situations where erythrocyte membrane loses flexibility. As Wilson et al showed, rapid rise in precipitation of membrane protein occurs as Ca concentration increases.

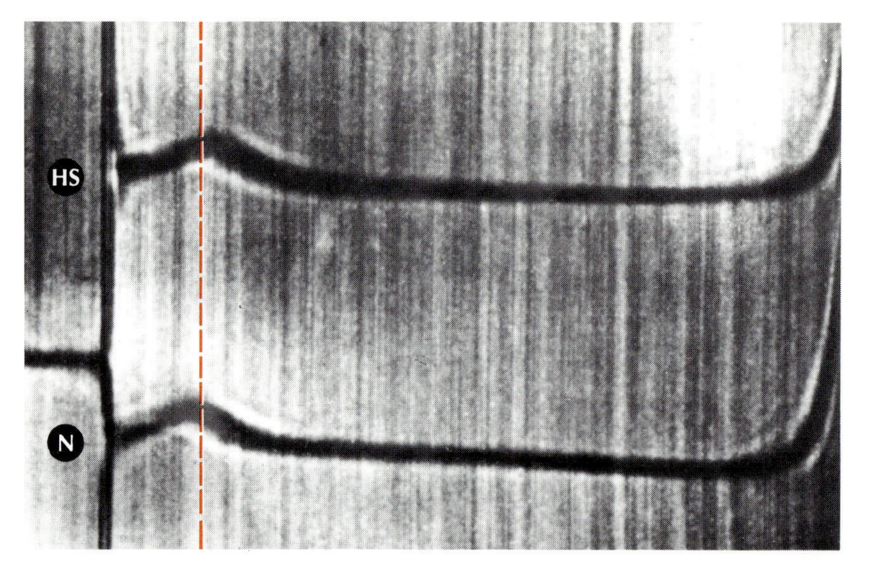

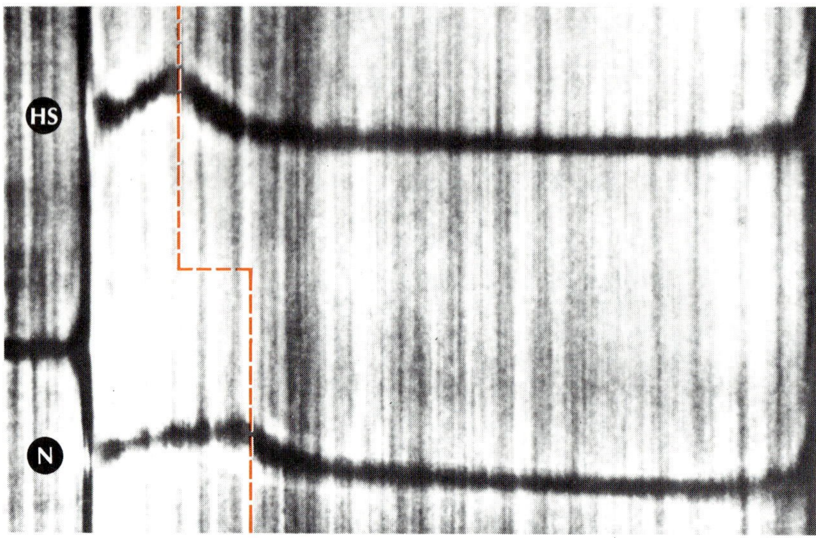

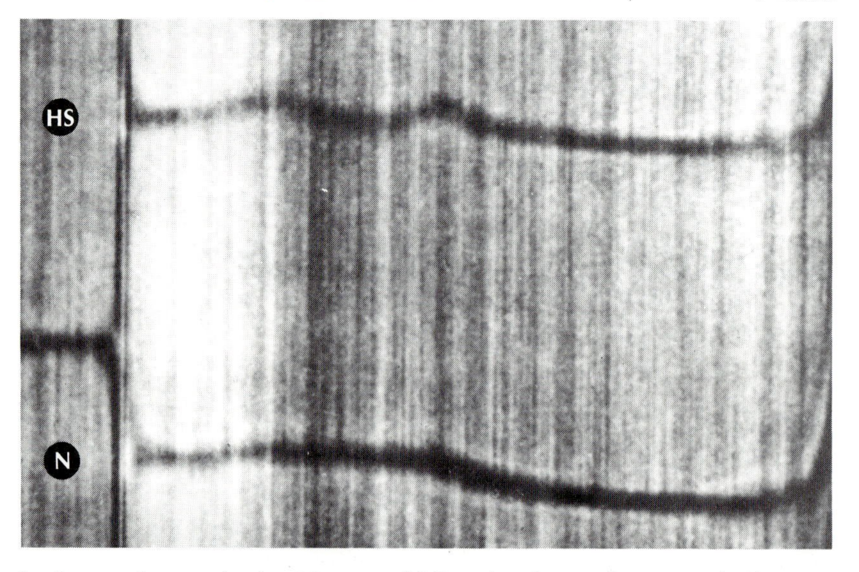

In absence of cations (top), neither normal (N) nor hereditary spherocytosis (HS) protein will aggregate. Cations produce aggregation only in normal protein, as shown by shift to right of sedimentation peak (center). In one form of HS, two peaks appear (bottom), indicating presence of both aggregating and nonaggregating protein.

normal protein has, as it were, no links at all. Not only can it not polymerize but it also cannot combine with and thereby cannot block the polymerization of the normal protein.

Yet another possible source of variability in spherocytosis might involve the particular protein affected by the mutation(s). One can conceive of the sphering as coming about in two ways: defects of the structural protein or of the contractile protein. In the first case, the contractile protein would be normal, but would have a defective structure to "pull against"; in the second, the structure would be normal but the contractile protein would be overly reactive – in a state of "rigor mortis" as it were. In either case, the result would be a cell form determined largely or entirely by surface tension – in other words, a sphere.

With respect to variability *within* families, we have thus far no hard evidence. Analogies from some other hereditary disorders, however, suggest a possible explanation. In enzyme defects such as glucose-6-phosphate dehydrogenase deficiency, for instance, it has been found that the individual heterozygote (a carrier of the trait rather than a sufferer from the disease) does not necessarily possess half the normal enzyme-containing cells and half abnormal, as one might expect. Rather, the proportion of normal to abnormal cells appears to be distributed statistically along the normal distribution curve: for most individuals, in the neighborhood of 50:50, but for some 60:40 or 40:60 and for a few even 70:30 or 30:70. If such a situation obtains in hereditary spherocytosis, and there seems to be no intrinsic reason why it should not, then the different proportions of protein inherited by the individual, plus the difference in the particular abnormality inherited (as between the inhibitory and inert abnormal proteins described above), would be sufficient to explain observed variations in expression within families.

I should stress that a great deal of what I have just said is hypothetical. And even if it should prove true, it still does not exhaust the complexities of pathology in spherocytosis. We and other groups have found that erythrocyte membrane proteins are substrates for phosphorylation mediated by cyclic nucleotides, in particular cyclic AMP. (See Chapter 19 and Chapter 20.)

The phosphorylation appears to be involved in giving the polymer its normal and functionally appropriate shape (i.e., its tertiary structure). It turns out that in spherocytosis, phosphorylation is abnormally sluggish, though as yet there is not enough evidence to say whether this is caused by an abnormality of the fibrous membrane protein, of the enzyme (protein kinase) that directly presides over the phosphorylation, or even, conceivably, of the enzymes that produce ATP, which is of course the phosphate donor in the process.

What we can say quite certainly is that hereditary spherocytosis can no longer be properly described as *a* disease. Rather, it is a syndrome, produced by any one of several different pathologic processes – or even, conceivably, by more than one acting together. The common factor is the result: sphered rigid cells lacking the normal, biconcave-disc shape, which because they cannot fold and otherwise contort themselves, pass the spleen filter with difficulty if at all – leading to rapid hemolysis. Not surprisingly, splenectomy, which enables the abnormal cells to remain in the circulation, is treatment of choice in most spherocytosis cases.

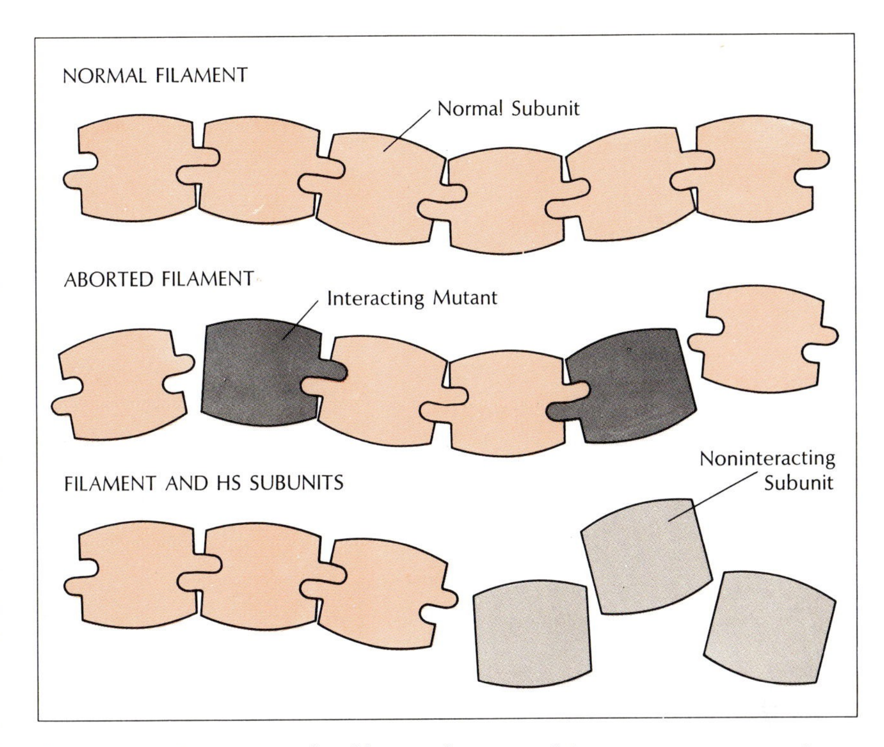

Two forms of HS are rationalized by postulating two different mutations in membrane protein. In one (center), abnormal protein can combine onesidedly with normal, block-formation of filament chains; in the other (bottom), abnormal protein cannot itself aggregate but also cannot combine with normal protein, thus permitting latter to aggregate.

Among the hemolytic diseases, sickle cell anemia shares the limelight with hereditary spherocytosis, and in some populations is far more common. As is well known, the mutation is found mainly among African, and less often Mediterranean, people and their descendants in the U.S. and elsewhere. In some African populations as much as 40% of the population may be heterozygous for the trait – generally a benign condition – while homozygosity and frank disease affect something like one birth in 25. The figures for American blacks are considerably lower – perhaps 10% heterozygous, with homozygosity in about one birth out of 400.

Sickle cell disease occupies a special place in the annals of erythrocyte disease, and indeed of genetic disease generally: it was the first of these conditions to be traced to a specific molecular abnormality. Owing to a particular amino acid substitution at a particular position in the hemoglobin molecule, the abnormal molecules polymerize when deprived of oxygen, distorting the cell into the characteristic, relatively rigid sickle shape. The result is a logjamming of cells in the capillaries, in the spleen, liver, and elsewhere, resulting in hemolysis, ischemia, infarction, and severe pain.

Such, at least, was the theory, but the reality turns out to be a bit more complicated. The polymerization of sickle hemoglobin is fully reversible by oxygenation, but the distorted cell shape frequently is not. If, for example, one examines a blood smear from a sickle cell patient under the microscope, one finds that even under fully oxygenated conditions, with all the hemoglobin fully depolymerized, from 10% to 30% of the cells may not regain their normal biconcave shape. Hemoglobin extracted from these cells is completely fluid, but the cell ghosts retain the sickle shape. Clearly, then, the reversible hemoglobin polymerization has induced a secondary, and irreversible, membrane abnormality.

The cause seems to be the calcium ion, which as we have already noted can itself induce membrane rigidity if its intracellular concentration rises above the almost infinitesimal normal level. The mechanism appears to involve a stretching of the membrane when it is forced into the sickle configuration, which in turn permits an influx of calcium beyond the capacity of the calcium pump to handle. The presence of excess calcium can be demonstrated by means of a new technique known as energy-dispersive x-ray spectroscopy, essentially derived from scanning electron microscopy, in which an electron beam is bounced off the cells. The electron impact on various atoms produces emission of x-rays, with wavelengths characteristic of the element that has been hit. In this manner, one can show that irreversibly sickled cells contain a far higher

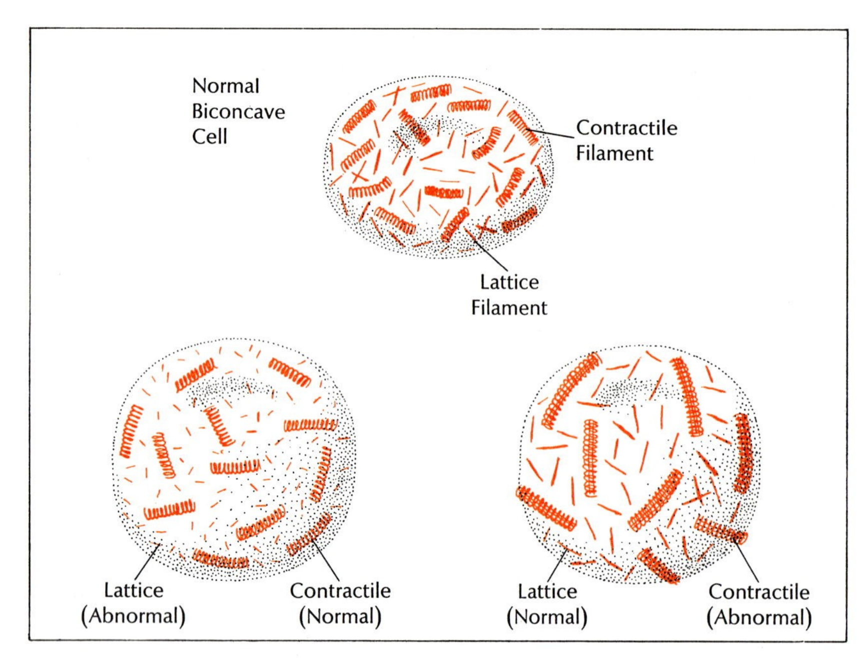

Membrane abnormality in HS can be explained in two ways. Normal cell is assumed to maintain disc shape by contractile protein pulling against lattice "skeleton." In HS, normal contractile protein may be combined with fragmented lattice; alternatively, normal lattice may be combined with abnormal, "frozen" contractile protein.

calcium concentration than their normally shaped neighbors – i.e., those that can still regain the biconcave form when oxygenated.

We believe, though we cannot yet prove, that it is the irreversibly sickled cells that cause a major part of the pathologic developments in sickle cell disease. Not only is their shape permanently changed, but their increased rigidity also makes them more vulnerable to mechanical damage in their passage through the vasculature. (In vitro experiments indicate that deoxygenated sickled blood in movement is more vulnerable to mechanical damage than its stagnant counterpart.) Such damage normally seals itself, even in a relatively rigid cell, but with each lesion a certain amount of membrane is lost. This in itself makes for a still more rigid cell, since the size of the "container" has in effect been reduced, while the volume of the contents has not – or at least not proportionately. Even worse, however, calcium can enter the cells before the lesions seal up, producing or aggravating irreversible protein denaturation.

We find the same process of mechanical damage, resulting in both membrane loss and calcium influx, in erythrocytes from patients with certain types of artificial heart valves; here the torn-off configurations of membrane can actually be observed in blood smears. A similar process occurs in the Coombs-positive or immune hemolytic anemias such as erythroblastosis fetalis. Here damage is done by monocytes and neutrophils, which attach themselves to the antibody coated erythrocyte and phagocytose it by sucking in pieces of membrane. The result, as one would expect, is a more spherical cell – much resembling those in hereditary spherocytosis – and also, it seems very probable, one whose increased rigidity is aggravated by calcium influx.

All these conditions involve either chemical modification of or damage to the protein moiety of the erythrocyte membrane, or mechanical damage to the membrane as a whole and, in some cases, both. Another type of red-cell pathology derives from abnormalities involving the membrane lipids – in particular, cholesterol. This condition is called acanthocytosis, from the spiny protuberances the erythrocytes develop; it is found in two situations. One is a rare hereditary disease in which affected individuals do not produce beta-lipoproteins, the molecules that serve as the major carriers for cholesterol in the serum, which is only minimally soluble by itself. One can also think of beta-lipoprotein as a sort of sponge that, depending on conditions, can either soak up or release cholesterol.

As noted earlier, the erythrocyte membrane normally carries on an active exchange of cholesterol with the serum; labeling experiments show a half-turnover time of around six hours. The same sort of labeling experiments, however, show that cholesterol will not efficiently leave the membrane *unless* the beta-lipoprotein "sponge" is present to absorb it. One result of beta-lipoprotein deficiency, therefore, is a moderate increase in the membrane's cholesterol content, variously estimated at from 10% to 20%. This increase, however, would not be expected to damage the cell's functioning; indeed, from everything we know about membrane chemistry, the effect would be to make the cell more flex-

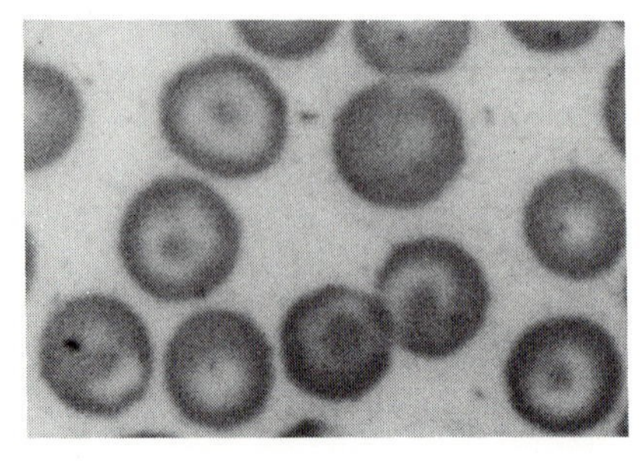

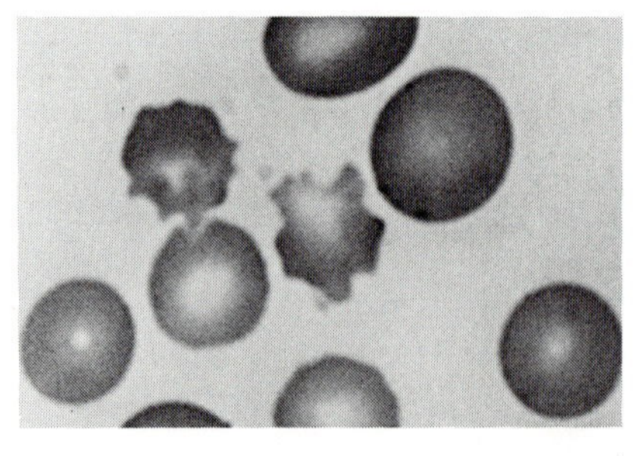

Increased proportion of cholesterol in erythrocyte membrane produces "target" cells (left), in which hemoglobin, unable to fill the expanded membrane, locates in center and around edge. Reduction in cholesterol turnover produces acanthocytes (right), whose spiny protuberances may indicate presence of crystallized, stagnant cholesterol.

ible, not more rigid. Far more significant as a pathologic factor, in my view, is a drastic reduction in cholesterol *turnover*, which leads to its stagnation in the membrane. Exactly what happens at this point we do not yet know; my own belief is that the cholesterol crystallizes – as it does in atherosclerotic plaques – forming needle-like structures that may be the red cell "spines" visible under the light microscope. Whatever the mechanism, there can be no doubt that acanthocytic cells become more rigid, with the usual sequelae of accelerated cell destruction and hemolysis.

Acanthocytosis also occurs in patients with severe, often terminal, liver disease such as cirrhosis. The spiny cells appear identical with those in beta-lipoprotein deficiency, and the causes are similar. We know that beta-lipoprotein is manufactured in the liver, and with severe liver disease it is evidently either being made in inadequate quantities or made incorrectly. Antibody experiments show that in at least some acanthocytosis patients of this type, the beta-lipoproteins are immunologically different from those of normal individuals, and it seems not unlikely that they are functionally abnormal as well – i.e., unable to accept cholesterol.

So long as cholesterol turnover remains normal, a hypernormal concentration of this lipid in the erythrocyte membrane seems to have no pathologic result and may, indeed, be somewhat beneficial in certain conditions. One finds such high-cholesterol cells especially in patients with obstructive jaundice, whose blood shows numbers of so-called target erythrocytes. These, in effect, have too much membrane for their hemoglobin content, which therefore locates partly as a bull's-eye in the middle of the cell and partly in a ring around the edge – hence the "target" designation. The chemical rationale for high cholesterol in jaundice involves the balance between two forms of cholesterol in the serum. Normally, about 70% of serum cholesterol is esterified with fatty acids, with only 30% in "free" form (actually, as already noted, "free" cholesterol is loosely bound to beta-lipoprotein). It has been determined, however, that bile salts, which are of course at high levels in the serum of obstructive jaundice patients, act as potent inhibitors of the enzymes that preside over esterification, with the result that the normal 70:30 ratio is shifted to something like 30:70. The additional nonesterified cholesterol, by the law of mass action, will tend to enter the erythrocyte membrane and expand it. The expanded membranes seem to function normally; far from being pathologically rigid, they tend to be superflexible.

This phenomenon explains a clinical observation of many years standing: that hereditary spherocytosis patients who developed obstructive jaundice, say from a common duct stone, showed markedly less anemia; the added cholesterol was restoring much of the lost flexibility in their erythrocytes, leading to reduced trapping and cell destruction in the spleen. Pigment gallstones are in fact not uncommon in such individuals, since the increased bilirubin produced by their excessive hemoglobin breakdown tends to precipitate in the biliary tract. The amelioration of spherocytosis under these conditions suggests a possible therapeutic approach to the disease less drastic than the usual splenectomy. One could not, indeed, induce jaundice in these patients in the interest of controlling their hemolysis. But it seems conceivable that pharmacologists might be able to develop a drug that would inhibit esterification of cholesterol as does an excess of bile salts. Such a drug might also prove beneficial in sickle cell patients, where, as we have seen, an important problem is also a loss of flexibility in the erythrocyte membrane.

In a few rare conditions, the erythrocyte membrane abnormality involves an abnormal permeability to ions – notably, sodium. We find this situation even in hereditary spherocytosis, in which sodium permeability is increased two- or threefold. In a few patients with stomatocytosis (cup-shaped erythrocytes), the sodium flux may be as much as six times normal. In all these cases, however, the sodium pump appears to increase its activities in step with the added sodium influx, which therefore has little if any effect on cell function or survival. In one very rare (possibly hereditary) form of stomatocytosis, however, the sodium flux increases 20 to 30 times, to the point where the pump cannot keep up. The result is accumulation of both sodium and water in the erythrocytes, which therefore swell, with the probable result that they are more easily trapped by and destroyed in the spleen.

The discovery that virtually all forms of hemolytic anemia, despite their very diverse origins, apparently derive from a single proximate cause – alterations (usually involving increased rigidity) in the erythrocyte membrane – is rather unusual in the history of pathology; for the most part, the trend has long been toward "splitting" clinically similar syndromes rather than "lumping" them. It is perhaps reassuring to find that progress in physiology, which so often seems to complicate our understanding of the body's functioning, may occasionally simplify it.

26

The Molecular Basis Of Acute Gout

GERALD WEISSMANN

New York University

"*28. Eunuchs do not take the gout, or become bald. 29. A woman does not take the gout unless her menses be stopped. 30. A young man does not take the gout until he indulges in coition.*"

As these aphorisms of Hippocrates indicate, gout is one of the most venerable of man's diseases. Always prominent among the inborn errors of metabolism, it now has the distinction of being the first disease of this type in which the mechanisms responsible for the acute clinical manifestations have been fully demonstrated. We know the offending agent, the urate crystal. We believe we have at last identified the chemical and physical events involved in the crystal's interaction with intracellular components and in the subsequent release of lytic contents that produces acute gouty inflammation and tissue damage. We are close to pinpointing the molecular basis for the greater proclivity of some individuals than others to the development of the acute state. And we can with greater assurance venture the suggestion that the mechanisms involved in acute gout represent a pathway common to many inflammatory disorders.

This chapter will propose a rationale for the affirmations stated above, but the details will be better appreciated if presented in their historical context. For one, our work is deeply rooted in both the recent and distant past. For another, the story of gout, illuminating the kind of insights and pitfalls that are inherent in all metabolic research, is interesting in its own right and may furnish instructive analogies to other fields of inquiry.

The modern scientific era in gout research can be said to have begun in the late 18th century when Scheele (1776), Wollaston (1797), and Pearson (1798) identified uric acid in the urine and tophi of patients with gouty arthritis. Here was the first instance in which a specific chemical substance was implicated in a metabolic disease. Some 50 years later, Garrod performed the experiments that established that high levels of serum uric acid are the prerequisite for a diagnosis of gout. He devised the famous thread test by which uric acid crystals could be shown to adhere to fine linen thread that had been immersed in the serum of gouty subjects and then dried. Most importantly for what transpired (and failed to transpire) later, he projected the unitary view that acute and chronic gout were only different manifestations of a common disorder related to the hyperuricemia, the nephropathy, the tophi, and crystal deposition.

At the turn of the century the great Swiss anatomist His and his student Freudweiler demonstrated that acute arthritis could be provoked in animals by injections of sodium urate microcrystals and that white cells took up this substance. In addition to confirming the theories of Garrod and restating his unitary concept, His and Freudweiler went so far as to incorporate the ideas then being proposed by Metchnikoff that one consequence of phagocytosis was inflammation. This, they suggested, was a possible mechanism to explain gouty inflammation, but they assumed that phagocytosis was required to remove the offending particle.

For reasons easier to explain than support, a wrong turn was then taken and for the next six decades the relationship between hyperuricemia and acute (but not chronic) gout was cast into question. Interestingly, this occurred when metabolic chemists preempted the subject and began to suspect variations between that which they could measure and direct clinical or experimental observation. Thus, Minkowski claimed that the histologic changes induced by His and Freudweiler failed to implicate uric acid in acute gout because no inflammatory or toxic effect could consistently be attributed to soluble urate carried in the circulation. This statement was perfectly true but proved to be irrelevant since the proposition was that crystalline, not soluble, urate was the culprit. The lack of correlation between serum uric acid levels and acute gouty attacks furnished the basis for what came to be a central dogma of rheumatology, namely, that uric acid disturbances were the result, not the cause, of the acute disease.

The detour ended over a decade ago when two groups of investigators returned to the path blazed by His and Freudweiler. McCarty and Hollander demonstrated the presence of urate crystals in cells of the synovial fluid of patients with acute gout and, following concepts pioneered by Metchnikoff, proposed that the uptake of these crystals might be related to the inflammation. Seegmiller and Howell and Faires and McCarty, at about the same time, were able to induce acute arthritis in animals and humans with injections of monosodium urate (MSU) crystals. McCarty and Phelps, also producing an inflammatory response with MSU in dogs, then showed that leukocytes were required for the reaction. When animals were rendered leukopenic by anti-white cell antibody or vinblastine, they demonstrated, the crystals were unable to produce inflammation. When leukocytes were reinfused, arthritis promptly ensued. A search of the literature then resulted in the rediscovery of His' and Freudweiler's work.

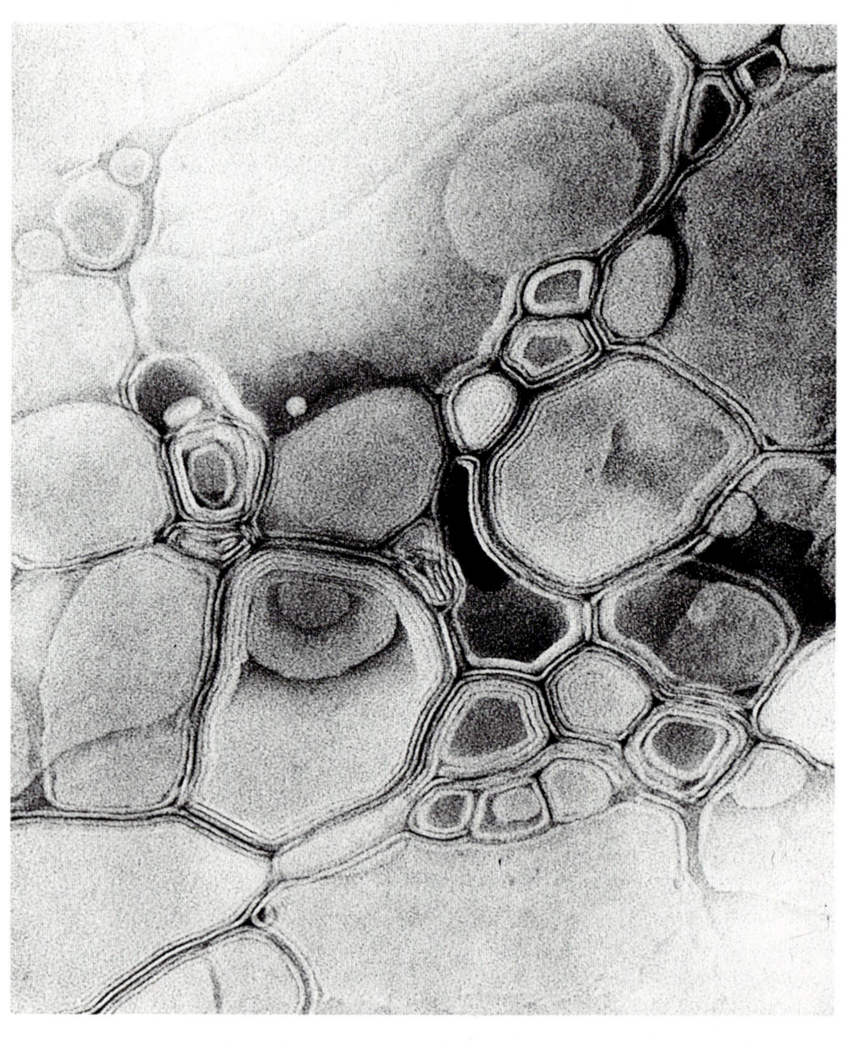

Electron micrograph shows liposomes magnified 80,000x (courtesy of Dr. John Freer of New York University). The artificial organelle consists of lipid membranes formed in concentric layers approximating in lipid composition the natural lysosomal membrane (see Chapter 3 in which A.D. Bangham describes their properties). Liposomes were exposed to crystals of monosodium urate in order to test whether the crystals perturbed lipid membranes.

It was not clear at this point, however, whether urate crystals formed de novo within the cell or were taken up from the outside. Schumacher and Phelps in very recent studies showed that leukocytes do indeed ingest urate crystals. With electron microscopy, they were able to follow the sequence of events after urate crystals were incubated with human polymorphonucleocytes. The crystals were rapidly phagocytized and within eight minutes could be seen in lysosomal structures, with degranulation and dissolution of the membranes of secondary lysosomes already beginning. By 30 minutes the death of many cells and the release of ingested crystals and cellular contents could be observed. The site of the critical encounter was now narrowed down to the lysosome.

This led to our own involvement in the problem, for we were investigating the role of lysosomes in acute and chronic arthritis. For a little over a decade now, lysosomes have been identified as primary mediators of phagocytosis and tissue injury and have been implicated in the pathogenesis of a number of diseases.

A discussion of these organelles and their role in inflammation, tissue injury, etc., can be found in the Selected Readings for this chapter at the back of the book. Suffice it to say here that in phagocytic cells these enzyme-containing granules (primary lysosomes) join with an invaginated phagosome to form a phagolysosome, or secondary lysosome, in which digestion of the foreign body occurs. Under normal circumstances the undigested residues, encapsulated in an intact membrane, do no harm to the phagocyte, but when, under exceptional circumstances, the membrane is violated, enzymes can be discharged inside the cells and cause lytic damage. This latter feature caused Christian de Duve, who first postulated the existence of the lysosome, to nickname it "the suicide sac." It must be emphasized that this is a rare event; it has been documented only in a few pathologic situations.

After the study by Schumacher and Phelps it became feasible to invoke this process to explain the tissue damage occasioned by cell-ingested urate

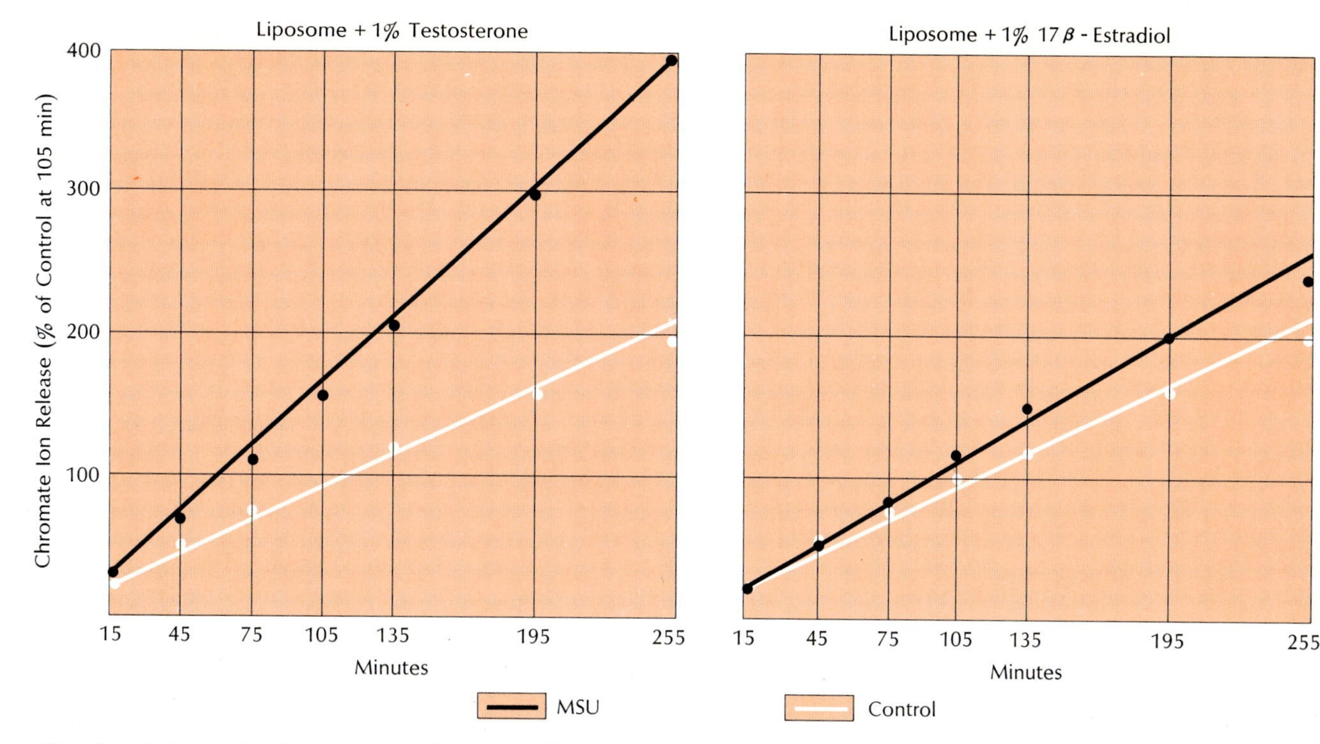

The clinical observation that women are less susceptible to gout than men finds documentation in studies in which one liposomal preparation is made "male" by preincorporating testosterone, the other "female" by preincorporating 17 β-estradiol. Exposure of MSU crystals to the "male" liposome causes chromate ions to escape at a rate far in excess of the control rate (left), suggesting either greatly enhanced diffusion or, more likely, rupture. In the "female" liposome, the crystals do little to increase diffusion of chromate above control levels. It is suggested that the "male" sterol either aids or at least does not impede hydrogen bonding between the crystal and membrane, while the "female" sterol does interfere with bonding, thus preventing membrane rupture.

crystals. One could postulate that the crystals, phagocytized by polymorphonuclear leukocytes and synovial cells, somehow caused the "perforation from within" of lysosomal membranes through which the lytic contents were dumped and caused death or injury of the cell.

But why should urate crystals have this effect? Any number of foreign bodies can be phagocytized without such consequences. It seemed reasonable at one time to speculate that the critical factor was the size and shape of the crystals, that phagolysosomal membranes were mechanically punctured by these needle-like bodies. However, other sharply faceted foreign bodies, diamond dust for example, could be shown to be phagocytized without perforating lysosomes, and ingested particles like silica, not at all acicular, were shown to rupture such membranes.

It was necessary to explore other than mechanical explanations. Perhaps the difference was that between chemically inert and active substances. We considered the strong clue furnished by Allison, who found that silica caused lysis of phagocytic cells by forming hydrogen bonds with the inside of the phagolysosomal membrane, thus disrupting the membrane and causing perforation from within. The proof of this was that when agents that inhibit hydrogen bonding were added, they effectively blocked the damage. Allison pointed out that silica crystals, like a number of other particles – including those of sodium urate – are not inert but are hydrogen donors. Wallingford and McCarty were then able to demonstrate, both with silica and with urate, a similar hydrogen bond-mediated membranolysis of red cells, and it was this observation that directly led to our experiments.

In seeking a demonstration of the nature of the urate-lysosome encounter, we needed to sort out at least three possibilities. The first was suggested by Metchnikoff, who, unaware of intimate details of lysosomal function, proposed that death of the white cell was the result of ingesting foreign bodies, even inert particles, and the release into the surrounding medium of all the cellular enzymes, which he termed cytases. If cell death were truly involved, we reasoned, one would expect to find cytoplasmic enzymes as well as lysosomal markers in the fluid bathing phagocytes exposed to inert or chemically active particles.

The second possibility was suggested by the electron microscope studies, from our own and other laboratories, which clearly showed that white cells "regurgitate during feeding" a portion of their acid hydrolases into the surrounding medium, for while the invaginating phagocytic vacuole is open to engulf the external particle, its internal portion is in contact with the lysosome now merging with it and donating its enzymes. Under these circumstances one can see leakage of lysosomal enzymes without escape of cytoplasmic markers and without death of the cell.

A third possibility was suggested by experiments in which cells were exposed to nonphagocytosable particles, or treated with the agent, cytochalasin B. By "reverse endocytosis"

Crystals and Media

Uric Acid

Silica (Silicic Acid)

Tris

Sucrose

Structural formulas of urate and silica crystals (above) show that both have hydroxyls able to donate hydrogen and thus to participate in hydrogen bonding with lysosomal membranes. Ability of tris and sucrose media to inhibit bonding (below) is apparently related to the fact that they have OH ions and can donate hydrogen. Sucrose also has hydrogen acceptors in the form of ring oxygens. When either sucrose or tris media are used, the release of β-glucuronidase from rabbit liver lysosomes is not enhanced by monosodium urate crystals. In phosphate/saline, release of the enzyme is far more rapid, indicating that the MSU crystal has bonded to the lysosomes and induced rupture.

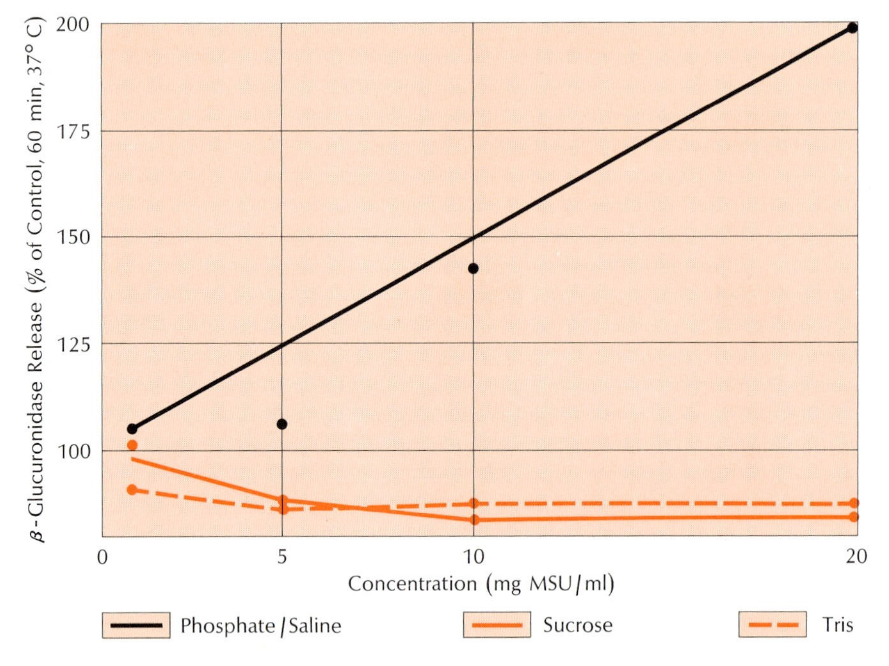

lysosomes near phagocytic vacuoles are perturbed and move directly to merge with the cell membrane where they extrude their contents by evagination, or exocytosis. One would expect, if this were so, to find lysosomal but not cytoplasmic enzymes in experimental preparations.

It seemed to us that the question first to be resolved was whether or not there was a difference between inert particles and MSU crystals in their encounter with white cells. We exposed human white cells to two kinds of inert particles, either zymosan or immune complexes of rheumatoid factor with aggregated IgG. The result of this was that for the first six hours there was release almost exclusively of beta-glucuronidase, a lysosomal enzyme, but not of lactic dehydrogenase, a cytoplasmic enzyme. We could conclude, then, that inert particles cause selective release of lysosomal contents either by regurgitation or exocytosis but without rupture of the lysosome or cell death. That the cells ingesting inert particles were very much alive could be demonstrated by their continued C-1 oxidation and capacity for extruding eosin-Y, as well as by microscopic inspection of their morphology. On the other hand, when white cells were exposed to MSU, and provided the appropriate buffer was used (of which more later), within a couple of hours there was massive release of both lactic dehydrogenase and beta-glucuronidase, indicating a degradation of the cell corresponding to what Metchnikoff had proposed.

Did cell death result because the crystals acted on the cell membrane or on the lysosomal membrane, either from within or without? The sequence of ultrastructural events provided by the electron microscopy of Schumacher and Phelps, cited earlier, practically compelled the working hypothesis that phagolysosomes ingested urate particles, with consequent rupture, and Allison's studies placed high on the agenda for exploration the proposition that hydrogen bonding was the mechanism by which the sequestered crystal perturbed lysosomal membranes. Verification demanded study of the action of urate crystals on isolated lysosomes. While such studies had been conducted, results had been inconclusive.

Our own early experiments with isolated lysosomes were equally inconclusive and provided little hard evidence that urate crystals ruptured lysosomes by either hydrogen bonding or any other means. This rather dampened enthusiasm for the hypothesis until we uncovered the error that explained our disappointing results and that may have been responsible for the ambiguous findings of others. Now, the bonding for which we were looking would take place between the hydroxyl groups in uric acid and the phosphate esters in the phospholipids of lysosomal membranes. Was there anything about our solutions, we wondered, that might inhibit hydrogen bonding? Our lysosomes were prepared, as were those of other workers, in sucrose or in tris buffer.

It occurred to us that tris and sucrose, both of which can form hydrogen bonds, might interfere with or compete for hydrogen bonding between crystals and membranes. Hydrogen bonds are weak bonds at best and are dependent for effectiveness on cooperative abutments between the repeating acceptors on the lysosomal membrane and the long surfaces provided by the acicular crystal. If this were indeed the reason for the discouraging results, a change to a buffer medium without such potential for preempting hydrogen bonding would show it up, demonstrating a dose-dependent lysis of isolated lysosomes by MSU crystals.

And that is exactly what occurred when we prepared lysosomes from rabbit liver in phosphate-buffered saline, accepting the risk of background breakage inherent in saline solutions. In that medium, MSU was demonstrably as lytic for lysosomes as was the proved membranolytic hydrogen-bonding agent, silica, which we used as the control in the system. Confirming results were obtained with human polymorphonuclear leukocytes. Within a hour or two, cells exposed to MSU in saline were dead by all morphologic and metabolic criteria and had leaked cytoplasmic as well as lysosomal enzymes. They had died from within, validating de Duve's "suicide sac" hypothesis at least insofar as MSU crystals were concerned. The reaction was sharply attenuated in tris or sucrose buffers and, of course, was nonexistent with inert particles regardless of preparation. Incidentally, it is useful to record here that the prepared MSU crystals we used were identical, as determined by x-ray crystallography, to those derived from tophi.

We then considered the possibility that the membrane lysis was due not to the crystals but to the soluble uric acid always present when urate crystals are dispersed in an aqueous medium and especially in a saline one. To test this we placed the lysosomes in the same medium but added dialysis sacs that kept them from contact with the crystals but not the solute. No lysis took place, dispelling the possibility of a role for soluble uric acid.

We had shown lysosomal disruption that was occasioned by the presence of

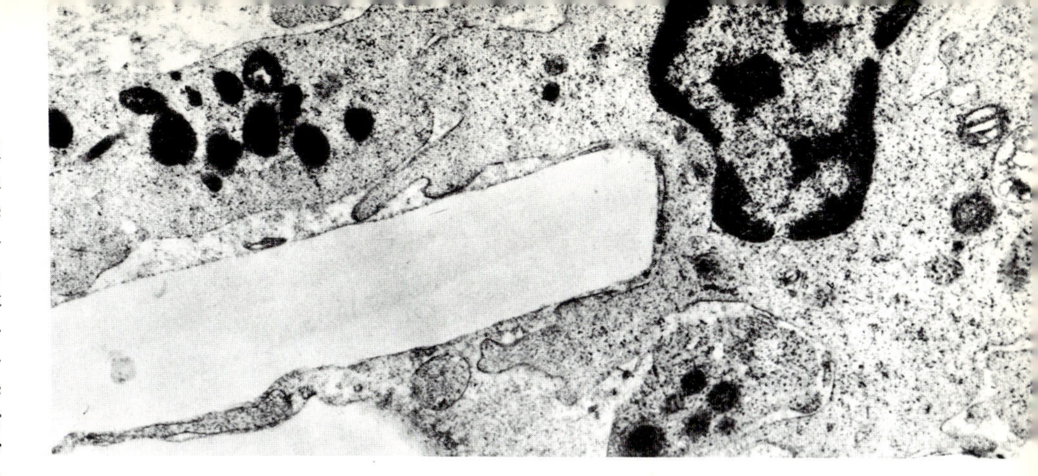

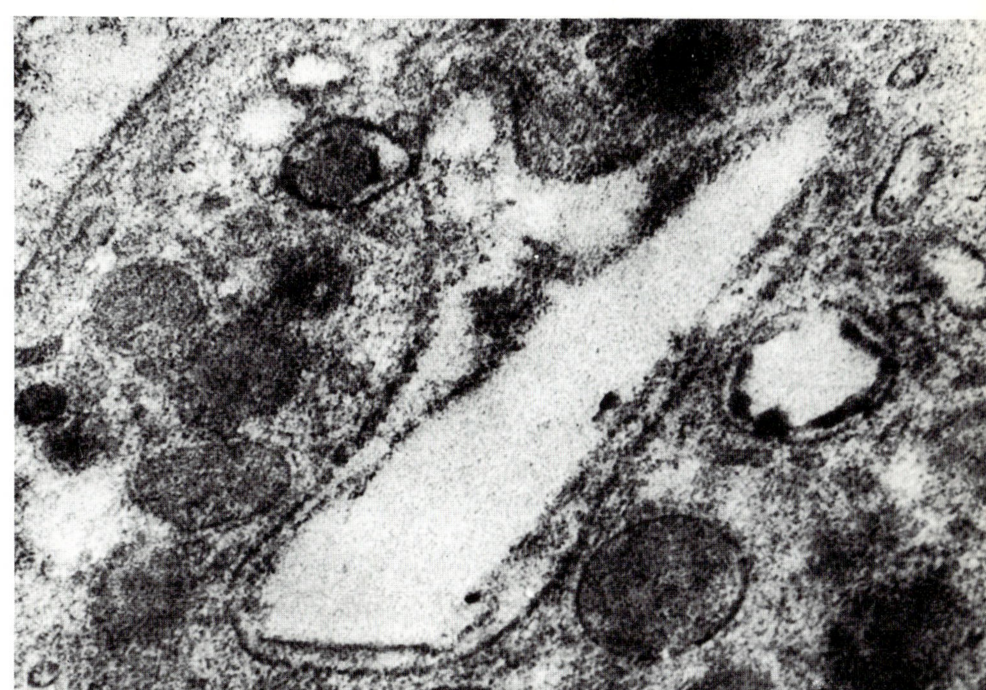

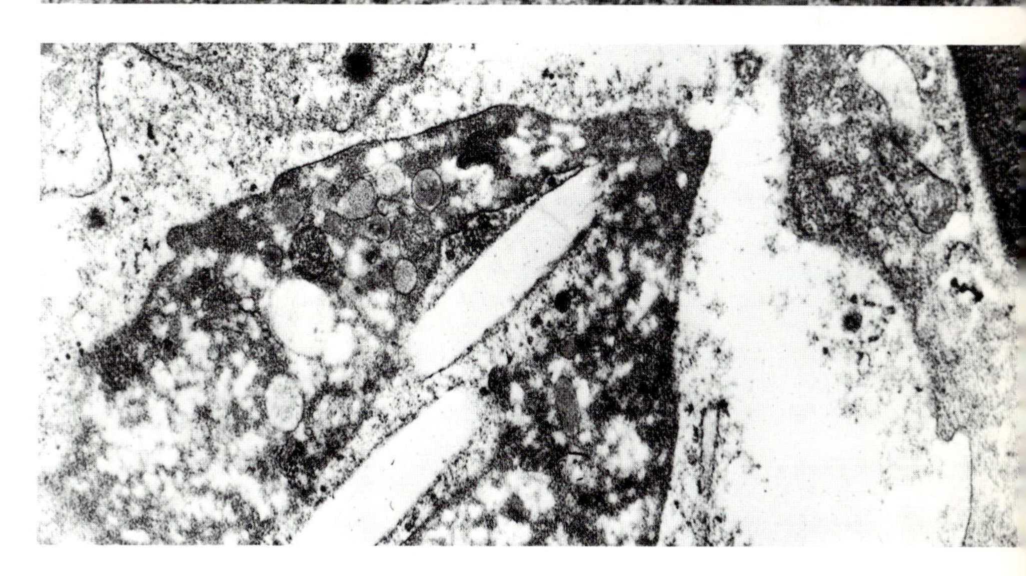

Electron micrographs show three phases in the engulfment of a monosodium urate crystal by a leukocyte. First the crystal is engulfed by the phagosome (top). Then the crystal is seen to lie completely within the phagosome, with lysosomes nearby (center). Finally, 30 minutes after engulfment, the crystals are free in the cytoplasm, the phagolysosomal membranes are no longer intact, and the entire cell shows autolytic changes. (Courtesy Drs. H. Ralph Schumacher and Paulding Phelps, University of Pennsylvania, from Arthritis and Rheumatism, *14:513, 1971.)*

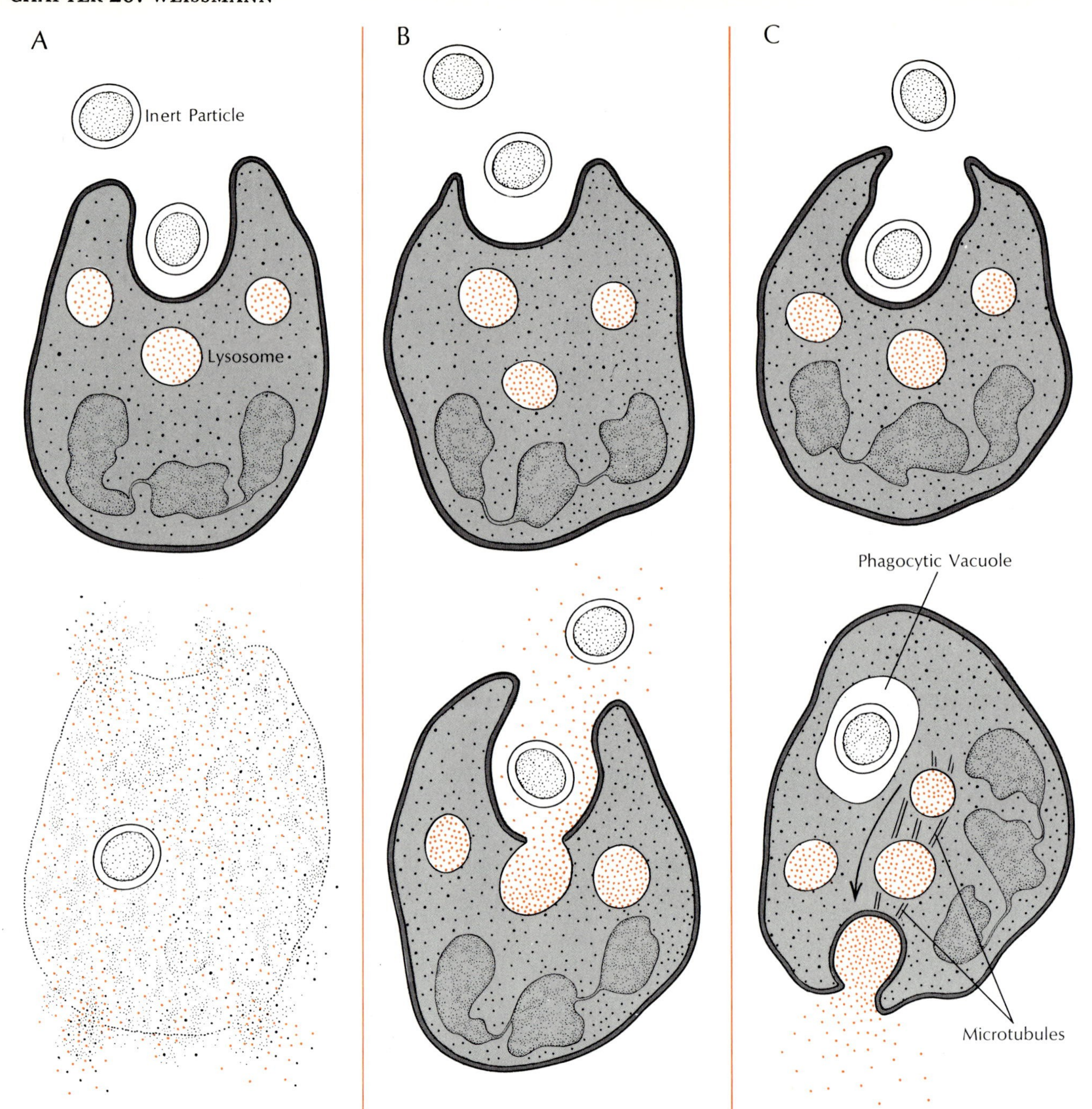

In the past, three hypotheses have been advanced to explain the mechanisms of enzyme release subsequent to phagocytosis of particles. The first, suggested by Metchnikoff, was that as a result of ingesting inert particles, the leukocyte is killed and its enzymes accordingly released into the surrounding medium (A). A second possibility is that cells (B) release lysosomal enzymes by "regurgitation during feeding" as the phagocytic vacuole remains open to the outside after merger with the lysosome. Another possibility is that cells, stimulated by a membrane perturbing agent (e.g. complement component $C5_a$, lysolecithin) directly merge their lysosomes with the plasma membrane when phagocytosis is completely or partially blocked (as with cytochalasin B) as in (C).

MSU crystals, but had we truly established that this occurred from within? Why could it not be the case that the crystals acted on the outside of the lysosomal membrane or even on the outside of the whole cell? A simple answer exists: 10% serum was present in all media and even 0.1% serum completely inhibits interaction of crystals with membranes, as repeatedly shown by experiments we conducted. The crystals had to gain entry into the serum-free interior of the cell, even as in vivo, and had to be stripped there of plasma protein if direct contact with the membrane and hydrogen bonding were to take place.

To test further whether the rupture of membrane was attributable to hydrogen bonding or to hydrophobic insertions of the crystal in some way analogous to the action of vitamin A on membranes, we compared silica and MSU actions with those of known hydrophobic substances, namely, vitamin A, etiocholanolone, and nystatin. The latter agents would not be expected to have their action limited by buffer solutions that inhibited hydrogen bonding. Experiments showed that the hydrophobic agents ruptured lysosomes as readily in sucrose as in saline, but the action of silica and MSU was restricted to saline. We could only conclude, therefore, that the mode of action of silica and MSU was different and associated with the effect of these solutions on hydrogen bonding, and that MSU shared the hydrogen-bonding property with silica. Indeed, the direct action of silicic acid with phospholipids, with which membranes are heavily endowed, is one of the mainstays of silicic acid chromatography.

If it were true that hydrogen bonding between biomembranes and urate crystals causes lysosomal rupture from within, we should be able to demonstrate this action and learn more about its character by exposing the crystals to recognized models for biomembranes. Such models (described in Chapter 3 and Chapter 23) had been developed by us in collaboration with A. D. Bangham of Cambridge, and had proved their worth as research tools. These models, called liposomes, are artificial lipid spherules prepared in aqueous dispersions.

They can trap in aqueous spaces between their lipid lamellae substances such as sodium, potassium, chromate, glucose, lysine, etc. The release of these markers from liposomes can be studied in much the same way as release of enzymes from lysosomes. Indeed, the recent incorporation of lysozyme into these artificial organelles provides a quite adequate model of lysosomes, in which phenomena of enzyme-substrate interaction, enzyme release, and properties of biomembranes can be observed under conditions closely mimicking the natural. Moreover, the lipid composition of the structures can be varied at will; they can be made to incorporate cholesterol or omit it, they can be charged positively or negatively, and they can even be made "male" or "female" by including estrogens or testosterone.

In our first experiments we exposed cholesterol-containing liposomes to silica particles and MSU crystals. As predicted, release of markers was abundantly enhanced in the presence of these substances, duplicating our experiments with natural lysosomes. However, a surprise was in store when we repeated the experiment with liposomes that contained no cholesterol – in this circumstance the lysis did not take place.

Now, the formation of a hydrogen bond with the lipid membrane might be dependent on the presence of sterol. One of the effects of cholesterol on biomembranes is to condense the phospholipid layer and thus present a more concentrated and favorable long-surface run for multiple, cooperative hydrogen bonding. This suggested to us the possibility that hormonal or nutritional differences involving trace sterol components might influence the response of membranes to MSU crystals. If this could be shown, it would furnish some insight into the known hormonal involvement in gout. Indeed, the experiments appear to confirm this reasoning, for artificial membranes containing 0.1% 17-beta-estradiol are far less susceptible to lysis by MSU crystals than those containing testosterone. It will be most interesting to see if this sex difference can be duplicated in experiments with human white cells. Now that we understand the importance of the correct buffer solutions, such experiments can be expected to provide more straightforward results than were formerly achieved.

We can now with some confidence propose the mechanism for acute gouty inflammation. It ensues, we believe, when white cells in the joint or extracellular spaces engulf acicular crystals of monosodium urate. When this happens the crystals, still filmed and protected by plasma proteins, find themselves in the matrix of a phagocytic vacuole. This vacuole, or phagosome, then merges with a primary lysosome that contributes its lytic enzymes, the union now being termed the secondary lysosome, or phagolysosome. Dehydration and digestion of plasma proteins then occur, placing the naked crystal in direct apposition with the interior surface of the phagolysosomal membrane and creating the conditions for hydrogen bonding between them. The phagolysosomal membrane, derived as it is from plasma membrane, is particularly cholesterol-rich and therefore susceptible to hydrogen bonding, which perturbs the membrane from within. The reaction may be expected to occur more readily in men than in women, indeed, perhaps in gouty patients compared with normals, because of differences in membrane sterol composition. Upon rupture of

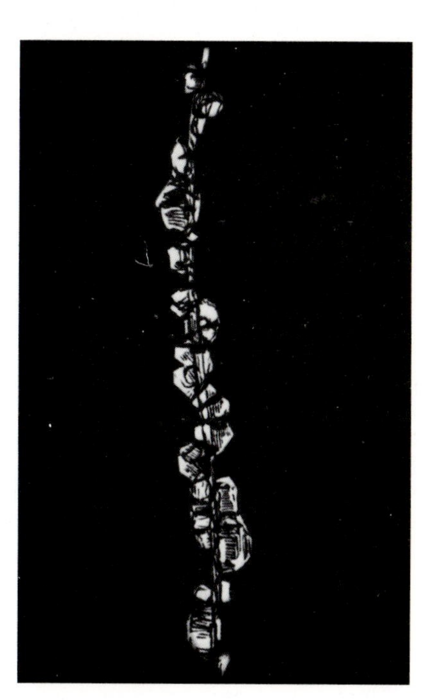

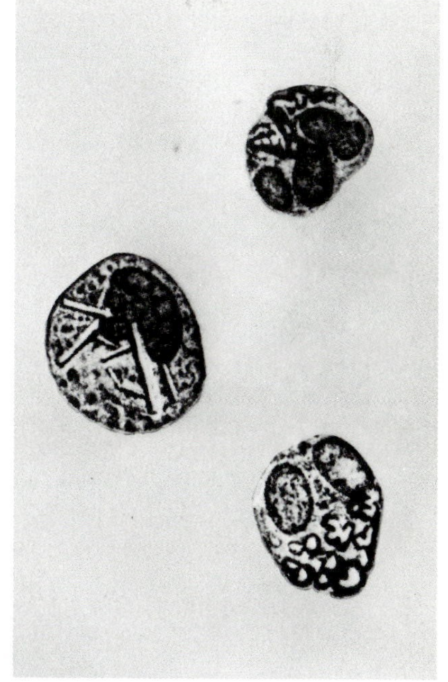

Two historical views show there was early recognition of the role of urate crystals in acute gout. The Garrod string test (left) was first published in 1859. Urate crystals can be seen adhering to thread immersed in acidified serum that was allowed to evaporate. Sketches by the Swiss anatomist His, made around 1900, show phagocytosis of synthetic microcrystalline sodium urate by leukocytes. His interpreted the phenomenon as reflecting a means of removing crystals from serum.

the membrane, lysosomal enzymes escape into the cell and wreak damage on cytoplasmic structures and the cell wall, with final cell death, dissolution, and release into the surrounding medium of toxic cytoplasmic and lysosomal enzymes.

Readers will note that this construction contains no important role for kinins. Phelps and McCarty have cast doubts on the role of kinins in gouty inflammation since they showed that urate-induced arthritis could not be modified by carboxypeptidase, the enzyme that breaks down bradykinin. While kinins may be important in the rubor, dolor, and tumor of early inflammation, they do not go on to produce the tissue damage that is central to the definition of arthritis. For that one must look elsewhere, to hydrolytic enzymes released, as we believe is now solidly demonstrated, following interaction between urate crystals and phagocytic cells.

This construction leads to some suggestions concerning the action of a few of the drugs shown to be effective in the treatment of gout. Malawista has suggested that colchicine, by disrupting the microtubules of white cells, prevents the merger of primary lysosomes with the phagocytic vacuole. Thus, although the membrane may consequently be perforated, there need not be inflammation because no lytic enzymes are present in the ruptured sac. Two other possibilities, for which some evidence also exists, are that colchicine interferes with the motility of white cells or with their capacity for phagocytosis. It is of interest that the interaction of colchicine with mammalian cells is completely inhibited by tris buffer, as our current studies show.

Ignarro has demonstrated that phenylbutazone and its derivatives have a directly stabilizing effect on lysosomal membranes under certain conditions and at certain pH's. We are testing these agents as well as colchicine to see if they a) inhibit white cell motility and access to crystals, b) inhibit phagocytosis, or c) inhibit intracellular merger of primary lysosomes with ingested crystals. Cortisone has many effects on joints, one of which appears to be the direct stabilization of lysosomal membrane, as shown by ourselves and other investigators in the macrophages and

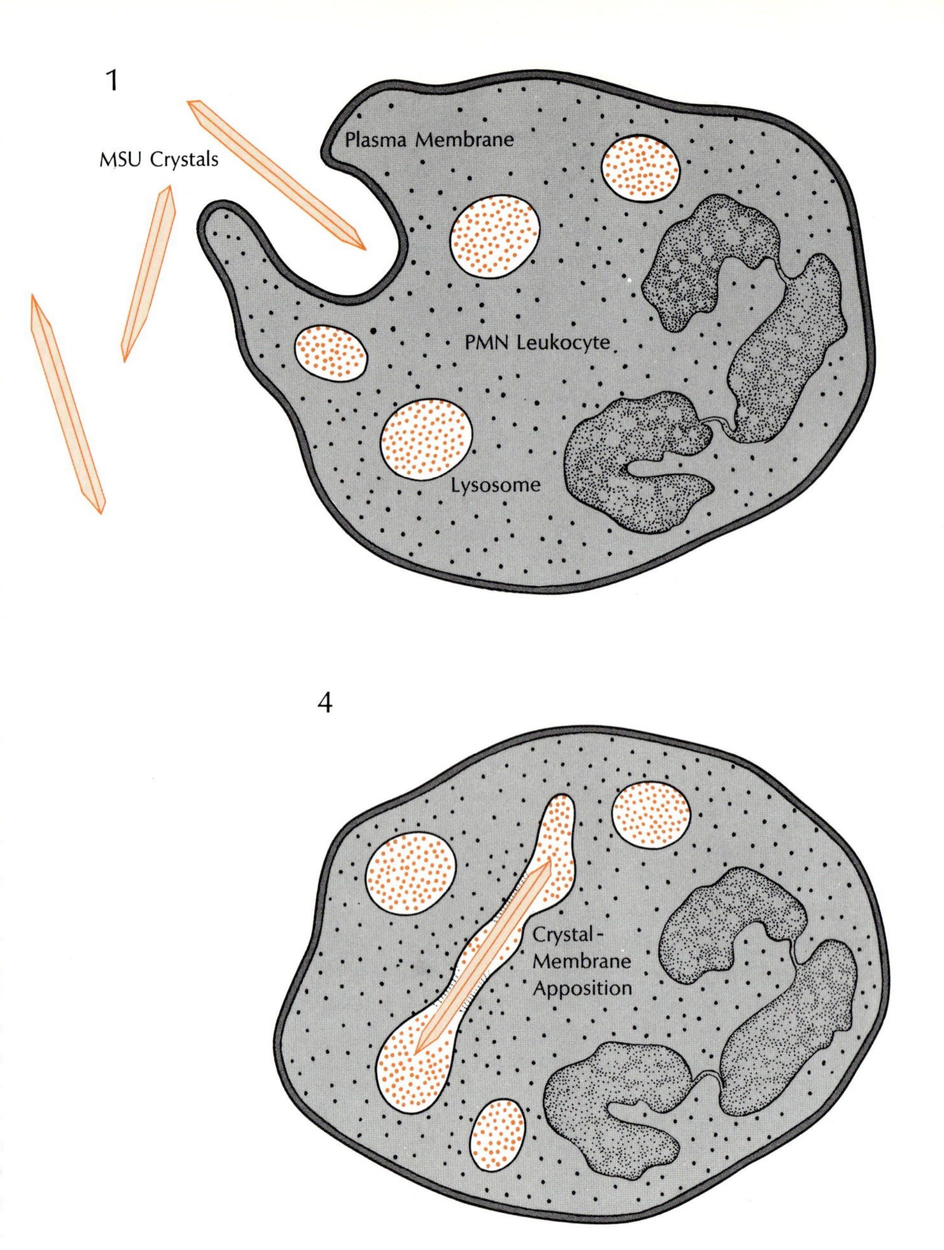

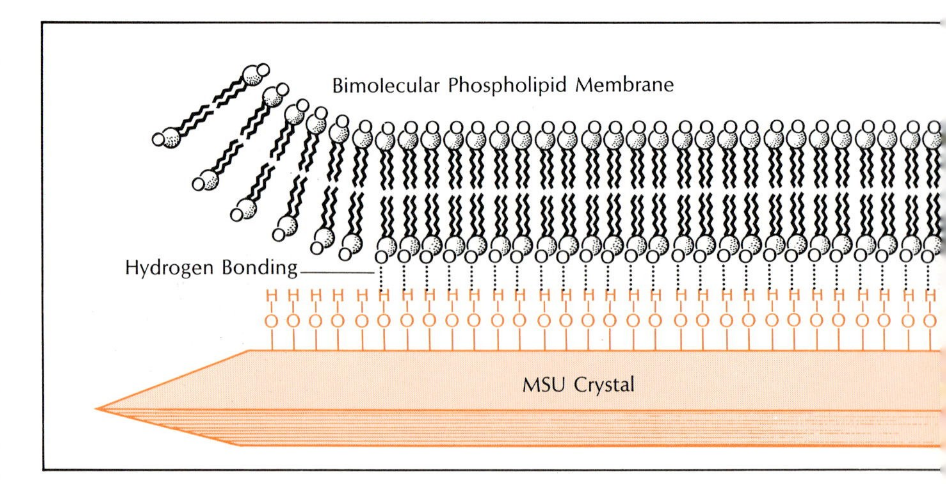

The pathogenesis of acute gout is initiated by the phagocytosis of monosodium urate crystals from the serum (1). After the MSU crystal has been ingested, a phagosomal vacuole is formed (2); it merges with the primary lysosomes to form a phagolysosome with the MSU crystal inside (3). Hydrogen bonding then causes the crystal to become

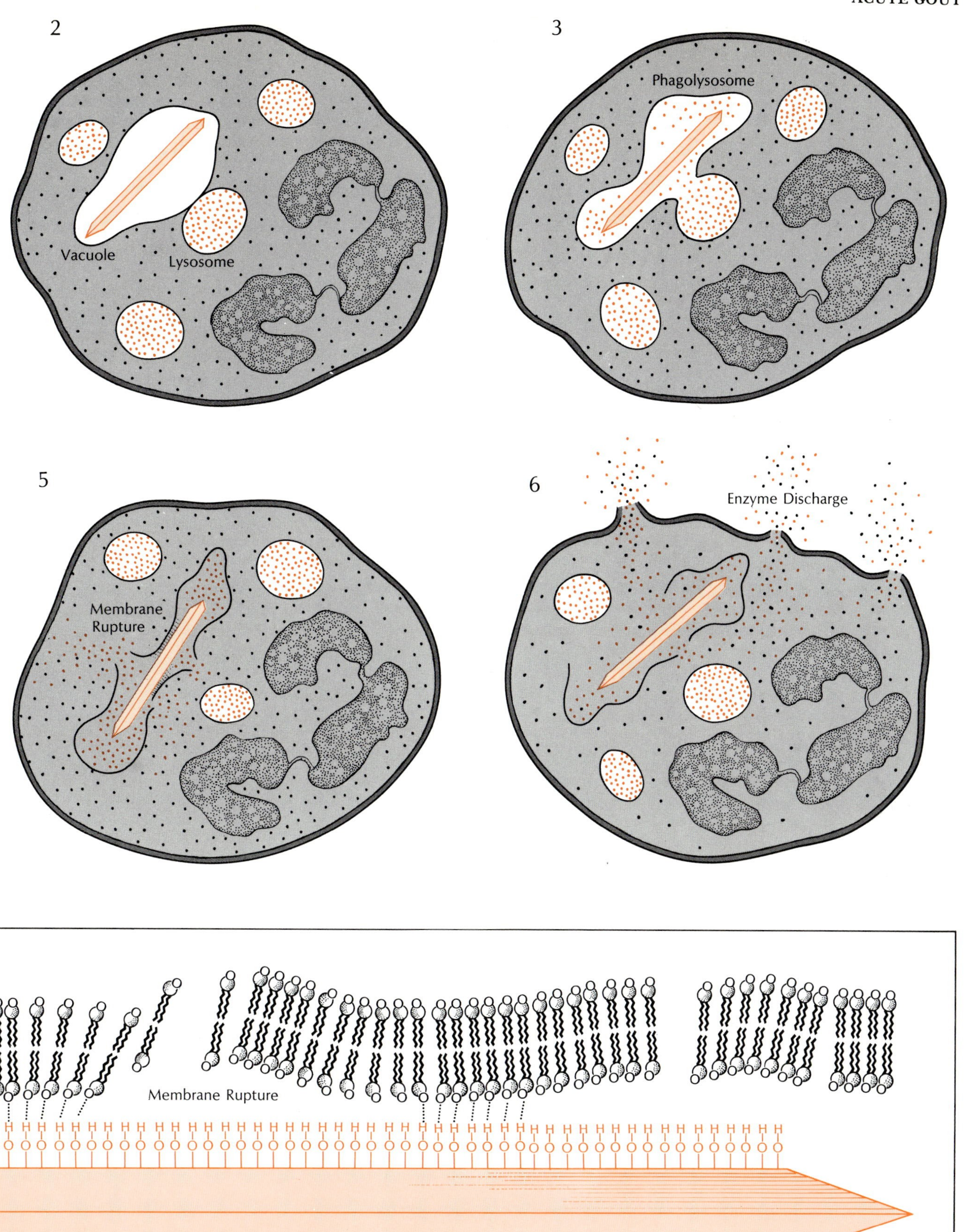

"stuck" to the phagolysosomal membrane (4), which is in motion. The membrane is torn (5) and the lysosomal enzymes are discharged into the cytoplasm, producing autolysis, cell rupture, and enzyme loss into joint fluid (6). In the inset, the lysosomal membrane is seen as a bimolecular phospholipid structure with oxygens at the polar heads of the phospholipids, permitting hydrogen bonding with the crystal's OH groups. Rupture occurs as membrane moves with segments adhering to the crystal.

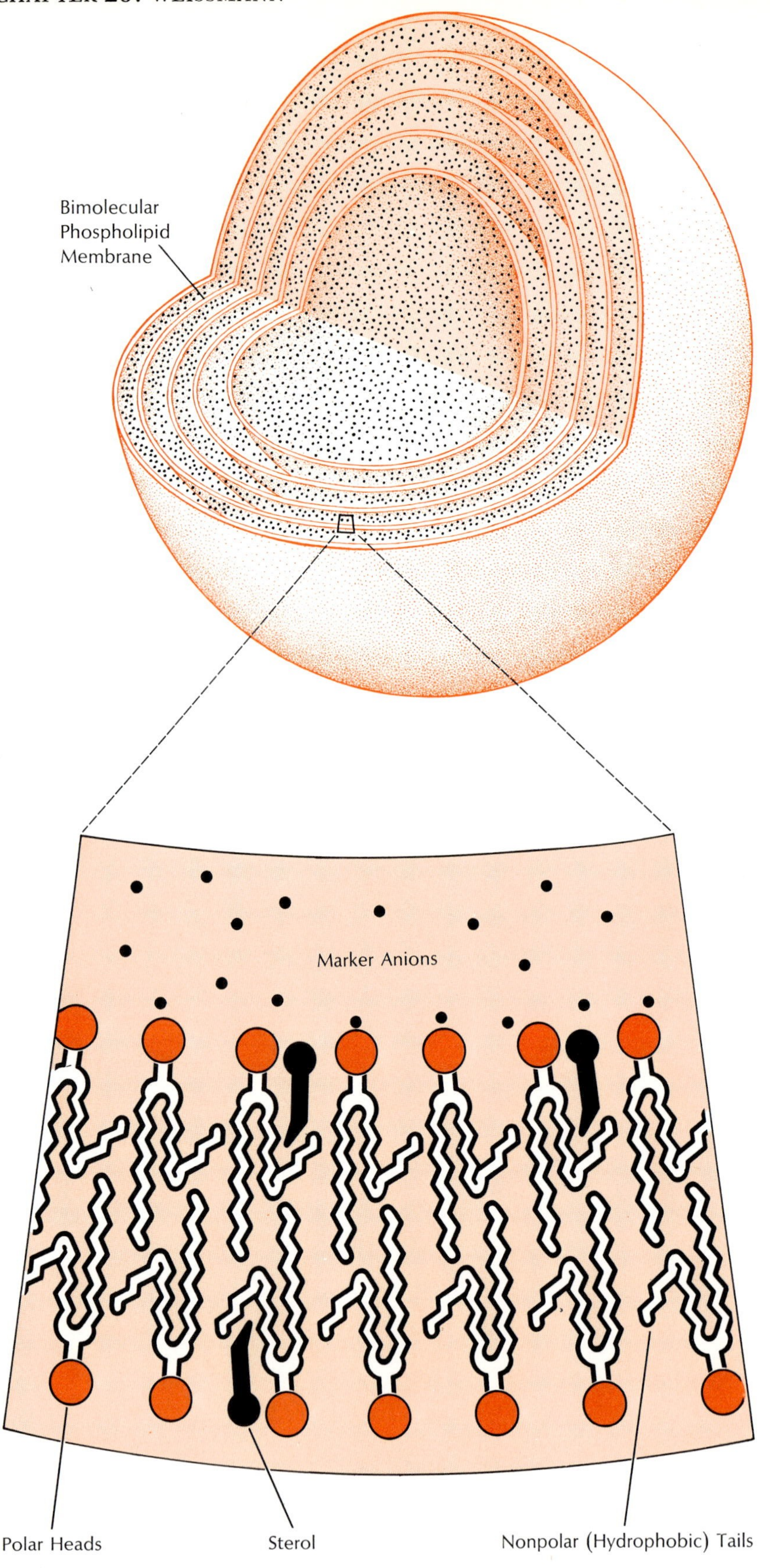

The artificial lysosome, or liposome, is shown schematically, with the concentrically arranged bimolecular phospholipid membranes containing marker anions (top). In the detail of the membrane structure, one can note the introduction of sterol molecules, such as testosterone or 17 β-estradiol, into the membrane to permit studies designed to evaluate the effects of sex differences on membrane behavior.

liver cells. It may very well turn out that part of this stabilization consists of an inhibition of the proper merger of intracellular organelles. A final possibility for all agents, and one we are also now examining, is that they interfere with the crystal-membrane interaction. Here our artificial lysosomes may provide useful findings.

There may even be a role for prostaglandins and cyclic AMP. In our studies of phagocytosis of inert particles we found that selective release of lysosomal enzymes could be blocked by agents that elevate cAMP levels within cells. These agents included prostaglandin E_1, which stimulates adenocylase to form cAMP, and such compounds as theophylline or 2-chloroadenosine, which inhibit the phosphodiesterase required to break down cAMP. We presumed that the blockage was an effect on a translational step required to carry the primary lysosome to the secondary or to the plasma membrane. This step may well involve the site at which colchicine may act, namely, the microtubules. Indeed, there is some patchy evidence that the aggregation of microtubules is sensitive to the level within cells of cyclic AMP. Consequently, we added prostaglandin E_1 to white cells taking up MSU and found that this effectively blocked the expected consequences of that uptake. Whether or not this observation will prove to have clinical interest remains to be seen.

Should the preceding explanation of the mechanism of acute gouty inflammation bear up, it will fortify a concept we suggested several years ago – that there is a final common pathway, namely the release of lysosomal enzymes, in all acute arthritis. Admittedly, the explanation arrives somewhat late on the therapeutic scene of gout, an ancient clinical entity for which relatively satisfactory therapies have been empirically found. But, with better understanding of their action, the therapies can probably be improved. Even more effective agents may be found as a result of searching for more selective inhibitors of intracellular hydrogen bonding and enhancers of lysosomal integrity, and these, if the concept of a final common pathway be sound, may help mount a more effective attack against the full range of arthritic disorders.

Section Four

Selected References

Selected References

Chapter 1

Danielli JF, Davson H: A contribution to the theory of permeability of thin films. J Cell Comp Physiol 5:4:495, 1935

Danielli JF: Some properties of lipoid films in relation to the structure of the plasma membrane. J Cell Comp Physiol 7:3:393, 1936

Danielli JF: The present position in the field of facilitated diffusion and selective active transport. Vol 7, Colston Papers: Proceedings of the Seventh Symposium of the Colston Research Society, March 29, 1954. Butterworths Scientific Publications, London

Hendler RW: Biological membrane ultrastructure. Physiol Rev 51:1:66, 1971

Vandenheuvel FA: Structure of membranes and role of lipids therein. J Lipid Res 9:161, 1968

Chapter 2

Biological Membranes: Physical Fact and Function, Vol 1, Chapman D, Ed. Academic Press, London, New York, 1968

Biological Membranes, Vol 2, Chapman D, Wallach DFH, Eds. Academic Press, London, New York, 1973

Chapman D: Biological membranes. Science Journal 3:55, 1968

Chapman D: The chemical and physical characteristics of biological membranes, page 23, Vol 1, Membranes and Ion Transport, Bittar EE, Ed. Wiley-Interscience, New York, 1970

Chapman D: An Introduction to Lipids. McGraw Hill, New York, 1968

Chapman D: Lipids, malnutrition and the developing brain. CIBA Foundation Symposium, October 1971

Oldfield E, Chapman D: Dynamics of lipids in membranes: Heterogeneity and the role of cholesterol. FEBS Letter 23:3:285, 1972

Chapter 3

Bangham AD: Lipid bilayers and biomembranes. Annu Rev Biochem 41:753, 1972

Bangham AD: Membrane models with phospholipids. Prog Biophys Mol Biol 18:29, 1968

Bangham AD, Hill MW, Miller NG: Method of Membrane Biology, Vol 1, pp 1-68, Korn ED, Ed. Plenum Publishing Corp., New York, 1974

Bangham AD, Standish MM, Watkins JC: Diffusion of univalent ions across the lamellae of swollen phospholipids. J Mol Biol 13:238, Aug 1965

Haydon DA, Hladky SB: Ion transport across thin lipid membranes: a critical discussion of mechanisms in selected systems. Q Rev Biophys 5:187, May 1972

Henn FA, Thompson TE: Synthetic lipid bilayer membranes. Annu Rev Biochem 38:241, 1969

Mueller P, Rudin DO: Action potentials induced in bimolecular lipid membranes. Nature (Lond) 217:713, 24 Feb 1968

Mueller P, Rudin DO, Ti Tien H, Wescott WC: Reconstitution of excitable membrane structure in vitro. Circulation 26:1167, Nov 1962

Mysels KJ: Dynamic processes in soap films. J Gen Physiol 52:1:Part 2:113s, Jul 1968

Pressman B, Haynes DH: in The Molecular Basis of Membrane Function. Tosteson D, Ed. Prentice-Hall, Inc., Englewood Cliffs, N.J., 1969

Shemyakin MM, Antonov VK, Bergelson LD, et al. *Ibid.*

Ti Tien H: Bilayer Lipid Membranes (BLM) Theory and Practice. Marcel Dekker, Inc., New York, 1974

Chapter 4

Frye CD, Edidin M: The rapid intermixing of cell surface antigens after formation of mouse-human heterokaryons. J Cell Sci 7:319, 1970

Guidotti G: Membrane proteins. Annu Rev Biochem 41:731, 1972

Pinto da Silva P, Branton D: Membrane splitting in freeze-etching. J Cell Biol 45:598, 1970

Singer SJ: The molecular organization of membranes. Annu Rev Biochem 43:805, 1974

Singer SJ, Nicolson GL: The fluid mosaic model of the structure of cell membranes. Science 175:720, 1972

Singer SJ: The molecular organization of biological membranes. Structure and Function of Biological Membranes, Rothfield LI, Ed. Academic Press, New York, 1971

Strittmatter P, Rogers MJ, Spatz L: The binding of cytochrome b_5 to liver microsomes. J Biol Chem 247:7188, 1972

Taylor RB, Duffus PH, Raff MC, de Petris S: Redistribution and pinocytosis of lymphocyte surface immunoglobulin molecules induced by anti-immunoglobulin antibody. Nature (New Biol) 233:225, 1971

Chapter 5

Fairbanks G, Steck TL, Wallach DFH: Electrophoretic analysis of the major polypeptides of the human erythrocyte membrane. Biochemistry 10:2606, 22 Jun 1971

Hughes RC: Glycoproteins as components of cellular membranes. Prog Biophys Mol Biol 26:189, 1973

Marchesi VT, Tillack TW, Jackson RL, Segrest JP, Scott RE: Chemical characterization and surface orientation of the major glycoprotein of the human erythrocyte membrane. Proc Natl Acad Sci USA 69:1445, Jun 1972

Segrest JP, Kahane I, Jackson RL, Marchesi VT: Major glycoprotein of the human erythrocyte membrane: Evidence for an amphipathic molecular structure. Arch Biochem Biophys 155:167, Mar 1973

Tillack TW, Scott RE, Marchesi VT: The structure of erythrocyte membranes studied by freeze-etching. II. Localization of receptors for phytohemagglutinin and influenza virus to the intramembranous particles. J Exp Med 135:1209, Jun 1972

Chapter 6

Curtis ASG: Cell contact and adhesion. Biol Rev 37:82, 1962

Manly RS: Adhesion in Biological Systems. Academic Press, New York, 1970

Moscona AA: Rotation-mediated histogenic aggregation of dissociated cells. Exp Cell Res 22:455, Jan 1961

Roseman S: The biosynthesis of cell-surface components and their potential role in intercellular adhesion. The Neurosciences. Third Study Program. Schmitt FO, Worden FG, Eds. The MIT Press, Cambridge, Massachusetts, 1974

Roseman S: Complex carbohydrates and intercellular adhesion. Biology and Chemistry of Eukaryotic Cell Surfaces. Lee EYC, Smith EE, Eds. Academic Press, New York, 1974

Roseman S: The synthesis of complex carbohydrates by multiglycosyltransferase systems and their potential function in intercellular adhesion. Chem Phys Lipids 5:270, Oct 1970

Townes PL, Holtfreter J: Directed movements and selective adhesion of embryonic amphibian cells. J Exp Zool 128:1:53, Feb 1955

Trinkaus JP: Cells into Organs: The Forces that Shape the Embryo. Prentice-Hall, Englewood Cliffs, New Jersey, 1969

Chapter 7

Buck CA, Glick MC, Warren L: A comparative study of glycoproteins from the surface of control and Rous sarcoma virus transformed hamster cells. Biochemistry 9:4567, 10 Nov 1970

Moscona AA: Analysis of cell recombinations in experimental synthesis of tissues in vitro. J Cell and Compar Physiol 60:65, Supplement 1, Oct 1962

Rapp F: Viruses and neoplasia. Hospital Practice 6:5:49, 1971

Schoenheimer R, Rittenberg D: The study of intermediary metabolism of animals with the aid of isotopes. Physiol Rev 20:1:218, 1940

Warren L, Critchley D, Macpherson I: Surface glycoproteins and glycolipids of chicken embryo cells transformed by a temperature-sensitive mutant of Rous sarcoma virus. Nature (Lond) 235:275, 4 Feb 1972

Warren L, Fuhrer JP, Buck CA: Surface glycoproteins of normal and transformed cells: a difference determined by sialic acid and a growth-dependent sialyl transferase. Proc Natl Acad Sci USA 69:1838, Jul 1972

Warren L, Glick MC: Membranes of animal cells. II. The metabolism and turnover of the surface membrane. J Cell Biol 37:729, Jun 1968

Chapter 8

Ahkong QF et al: Mechanisms of cell fusion. Nature 253 (5488):194, 17 Jan 1975

Ahkong QF et al: Studies on chemically induced cell fusion. J Cell Sci 19:769, 1972

Bennett HS: The cell surface: movements and recombinations. Handbook of Molecular Cytology. Lime-de-Faria A, Ed. North-Holland, Amsterdam, 1969

Harris H: Cell Fusion. Clarendon Press, Oxford, 1970

Harris H: Cell fusion and the analysis of malignancy. Proc R Soc Lond (Biol) 179:1, 12 Oct 1971

Lucy JA: The fusion of biological membranes. Nature (Lond) 227:815, 22 Aug 1970

Lucy JA: Lipids and membranes. FEBS Letter 40:(Suppl):105s, 23 Mar 1974

Marx JL: Somatic cell hybrids: impact on mammalian genetics. Science 179 (4075):785, 23 Feb 1973

Okada Y: Factors in fusion of cells by HVJ. Curr Con Microbiol Immunol 48:102, 1969

Poste G, Allison AC: Membrane fusion reaction: a theory. J Theor Biol 32:165, Jul 1971

Poste G, Allison AC: Membrane fusion. Biochim Biophys Acta 300:421, Dec 1973

Rapp F: Viruses and neoplasm. Hospital Practice 6:5:49, 1971

Stormorken H: The release reaction of secretion: a general phenomenon related to phagocytosis/pinocytosis. Scand J Haematol, Suppl 9:1, 1969

Watkins, JF: Fusion of cells for virus studies and production of cell hybrids. Methods in Virology. Vol. 5. Maramorosch K, Koprowski H, Eds. Academic Press, New York, 1971

Chapter 9

Bennett MVL: Function of electrotonic junctions in embryonic and adult tissues. Fed Proc 32:1:65, Jan 1973

Fawcett DW: The Cell: Its Organelles and Inclusions (An Atlas of Fine Structure). W. B. Saunders Company, Philadelphia, 1966

Gilula NB: Junction Between Cells. Page 1, Cell Communication, Cox RP, Ed. John Wiley and Sons, Inc., New York, 1974

Keeter JS, Pappas GD: Gap junctions in embryonic skeletal muscle. Anat Rec 175:355, Feb 1973

Pappas GD, Asada Y, Bennett MVL: Morphological correlates of increased coupling resistance at an electrotonic synapse. J Cell Biol 49:1:173, Apr 1971

Pappas GD, Waxman SG: Synaptic fine structure—morphological correlates of chemical and electrotonic transmission. Pages 1-43, Structure and Function of Synapses, Pappas GD, Purpura DP, Eds. Raven Press, New York, 1972

Chapter 10

Dunham PB, Hoffman JF: Active cation transport and ouabain binding in high potassium and low potassium red blood cells of sheep. J Gen Physiol 58:94, Jul 1971

Glynn IM, Hoffman JF, Lew VL: Some "partial reactions" of the sodium pump. Philos Trans R Soc Lond (Biol Sci) 262:91, 20 Aug 1971

Glynn IM: Membrane adenosine triphosphatase and cation transport. Br Med Bull 24:165, May 1968

Hoffman JF: Cation transport and structure of the red cell plasma membrane. Circulation 26:1201, Nov 1962

Hoffman JF: The red cell membrane and the transport of sodium and potassium. Am J Med 41:666, Nov 1966

Hoffman JF: The interaction between tritiated ouabain and the Na-K pump in red blood cells. J Gen Physiol 54:1:Part 2:343s, Jul 1969

Hoffman JF: Molecular aspects of the Na+-K+ pump in red blood cells. Organization of Energy Transducing Membranes. Nakao M, Parker L, Eds. University of Tokyo Press, 1973; also University Park Press, Baltimore

Hoffman PG, Tosteson DC: Active sodium and potassium transport in high potassium and low potassium sheep red cells. J Gen Physiol 58:438, Oct 1971

Sachs JR et al: Antibody-induced alterations in the kinetic characteristics of the Na-K pump in goat red blood cells. J Gen Physiol 63:4:389, Apr 1974

Chapter 11

Azarnia R, Larsen WJ, Loewenstein WR: The membrane junctions in communicating and noncommunicating cells, their hybrids and segregants. Proc Natl Acad Sci USA 71:880, Mar 1974

Furshpan EJ, Potter DD: Low-resistance junctions between cells in embryos and tissue culture. Pages 95-127, Current Topics in Developmental Biology, Vol 3, Moscona AA, Monroy A, Eds. Academic Press, New York, 1968

Loewenstein WR: Communication through cell junctions. Implications in growth control and differentiation. Devel Biol (27th Symp Soc Dev Biol) 19 Suppl 2:151, 1968

Loewenstein WR: Intercellular communication through membrane junctions and cancer etiology. Pages 103-125, Membrane Transformations in Neoplasia. Schultz J, Block RE, Eds., Academic Press, Inc., New York, 1974

Loewenstein WR: Permeability of membrane junctions. Conference on Biological Membranes. Ann NY Acad Sci 137:441, Jul 1966

Loewenstein WR: Some reflections on growth and differentiation. Perspect Biol Medicine 11:2:260, winter 1968

Rose B, Loewenstein WR: Permeability of cell junctions depends on local cytoplasmic calcium activity. Nature 254:(5497):250, 20 Mar 1975

Chapter 12

Ernster L, Orrenius S: Substrate-induced synthesis of the hydroxylating enzyme system of liver microsomes. Fed Proc 24:1190, 1965

Goldberg AL, Dice JF: Intracellular protein degradation in mammalian and bacterial cells. Annu Rev Biochem 43:835, 1974

Siekevitz P: Biological membranes: The dynamics of their organization. Annu Rev Physiol 34:117, 1972

Siekevitz P: On membranes. Page 181, Horizons of Bioenergetics. San Pietro A, Gest H, Eds. Academic Press, New York, 1972

Wirtz KWA: Transfer of phospholipids between membranes. Biochim Biophys Acta 344:95, 1974

Chapter 13

de Duve C: The lysosome in retrospect. Pages 3-40, Lysosomes in Biology and Pathology, Vol 1, Dingle JT, Fell HB, Eds. North Holland Publishing Co., Amsterdam and London, 1969

Glauert AM: The high voltage electron microscope in biology. J Cell Biol 63:(3):717, Dec 1974

Neufeld EF: Mucopolysaccharidoses: The biochemical approach. Hospital Practice 7:2:107, 1972

Smuckler EA, Arcasoy M: Structural and functional changes of the endoplasmic reticulum of hepatic parenchymal cells. Int Rev Exper Pathol 7:305, 1969

Trump BF, Arstila AU: Cell injury and cell death. Pages 9-95, Principles of Pathobiology, LaVia MF, Hill RB, Eds. Oxford University Press, New York, 1974

Trump BF, Laiho KA, Mergner WJ, Arstila AU: Studies on the subcellular pathophysiology of acute lethal cell injury. Beitr Pathol 152:(3):243, 1974

Trump BF, Mergner WJ: Cell injury. Pages 115-257, The Inflammatory Process, Zweifach BW, Grant L, McCluskey RT, Eds. Vol 1, 2nd Ed. Academic Press, New York, 1974

Whaley WG, Dauwalder M, Kliphart JE: Golgi apparatus: Influence on cell surfaces. Science 175:596, 11 Feb 1972

Chapter 14

Racker E: Pages 269-281, Dynamics of Energy-Transducing Membranes. Ernster, Estabrook, Slater, Eds. Elsevier Scientific Publishing Company, Amsterdam, 1974

Suolinna EM, Lang DR, Racker E: Quercetin, an artificial regulator of the high aerobic glycolysis of tumor cells. J Natl Cancer Inst 53:(5):1515, Nov 1974

Racker E, Ed: Mechanisms in Bioenergetics. Academic Press, Inc., New York, 1965

Racker E: Function and structure of the inner membrane of mitochondria and chloroplasts. Page 127, Membranes of Mitochondria and Chloroplasts. Racker E, Ed. Van Nostrand Reinhold Company, New York, 1970

Mitchell P: Chemiosmotic coupling in oxidative and photosynthetic phosphorylation. Biol Rev 41:445, Aug 1966

Racker E: Bioenergetics and the problem of tumor growth. Am Sci 60:56, Jan-Feb 1972

Luft R et al: A case of severe hypermetabolism of nonthyroid origin with a defect in the maintenance of mitochondrial respiratory control: A correlated clinical, biochemical and morphological study. J Clin Invest 41:9:1776, Sep 1962

Chapter 15

Farquhar MG: Processing of secretory products by cells of the anterior pituitary gland. Page 79, Memoirs of the Society for Endocrinology, No 19. Heller H, Lederis K, Eds. Cambridge University Press, New York, 1971

Jamieson JD: Transport and discharge of exportable proteins in pancreatic exocrine cells: In vitro studies. Current Topics in Membranes and Transport. Bronner F, Kleinzeller A, Eds. Academic Press, New York, 1971

Palade GE: Intracellular aspects of the process of protein secretion. 1974 series Nobel Lectures. Elsevier Publishing Co., New York

Palade GE, Siekevitz P, Caro G: Structure, chemistry and function of the pancreatic exocrine cell. CIBA Foundation Symposium on the Exocrine Pancreas. de Reuch AVS, Cameron MP, Eds. J and A Churchill Ltd., London, 1962

Tartakoff AM, Greene LJ, Jamieson JD, and Palade GE: Parallelism in the processing of pancreatic proteins. Page 177, Advances in Cytopharmacology, Vol. 2. Ceccarelli B, Clementi F, Meldolesi J, Eds. Raven Press, New York, 1974

Chapter 16

Eytan G, Jennings RC, Forti G, Ohad I: Biogenesis of chloroplast membranes. J Biol Chem 249: 738, 1974

Goldberg AL, Dice JF: Intracellular protein degradation in mammalian and bacterial cells. Annu Rev Biochem 43: 835, 1974

Goodenough U, Levine RP: The genetic activity of mitochondria and chloroplasts. Sci Am 223:5:22, 1970

Kellenberger E: The genetic control of the shape of a virus. Sci Am 215:6:32, 1966

Masur SK, Holtzman E, Walter R: Hormone-stimulated exocytosis in the toad urinary bladder; possible implications for turnover of surface membranes. J Cell Biol 52:211, 1972

Miller PL, Ed: Control of Organelle Development (Symp Soc Exp Biol 24). Academic Press, New York, 1970

Nomura M: Assembly of bacterial ribosomes. Science 179:864, 1973

Novikoff AB, Holtzman E: Cells and Organelles. Holt, Rinehart, and Winston, New York, 1970

Racker E, Ed: Symposium on Assembly of Intracellular Structures. Fed Proc 31:10, 1972

Reinert J, Ursprung H, Eds: Origin and Continuity of Cell Organelles (Results and Problems in Cell Differentiation, Vol 2). Springer-Verlag, New York, 1971

Rosenbaum JL, Moulder JE, Ringo DL: Flagellar elongation and shortening in *Chlamydomonas*. The use of cycloheximide and colchicine to study the synthesis and assembly of flagellar proteins. J Cell Biol 41:600, 1969

Rubin MS and Tzagoloff A: Assembly of the mitochondrial membrane system X. Mitochondrial synthesis of three of the subunit proteins of yeast cytochrome oxidase. J Biol Chem 248:4275, 1973

Weisenberg RC, Rosenfeld AC: In vitro polymerization of microtubules into asters and spindles in homogenates of surf clam eggs. J Cell Biol 64:146, 1975

Wood WB, Edgar RS: Building a bacterial virus. Sci Am 217:1:60, Jul 1967

Young RW: The renewal of rod and cone outer segments in the rhesus monkey. J Cell Biol 49:303, 1971

Chapter 17

Gray EG, Whittaker VP: The isolation of nerve endings from brain: An electron-microscopic study of cell fragments derived by homogenization and centrifugation. J Anat (Lond) 96:79, 1962

Whittaker VP, Michaelson IA, Kirkland RJA: The separation of synaptic vesicles from nerve-ending particles ('synaptosomes'). Biochem J (Lond) 90:293, 1964

Whittaker VP, Sheridan MN: The morphology and acetylocholine content of isolated cerebral cortical synaptic vesicles. J Neurochem (Lond) 12:363, 1965

Whittaker VP, Essman WB, Dowe GHC: The isolation of pure cholinergic synaptic vesicles from the electric organs of elasmobranch fish of the family *Torpedinidae*. Biochem J (Lond) 128:833, 1972

Whittaker VP: The synaptosome. Pages 327-364, Handbook of Neurochemistry. Structural Neurochemistry. Lajtha LA, Ed., Vol 2, Plenum Press, New York, 1969

Whittaker VP, Dowdall MJ, Boyne AF: The storage and release of acetylcholine by cholinergic nerve terminals: Recent results with non-mammalian preparations. Biochem Soc Symp 36:49, 1972

Whittaker VP: The biochemistry of synaptic transmission. Naturwissenschaften (Berlin) 60:281, 1973

Whittaker VP: The current state of research on cholinergic synapses. Cholinergic Mechanisms, Waser P, Ed. Raven Press, New York, 1975 (in press)

Zimmermann H, Whittaker VP: Effect of electrical stimulation on the yield and composition of synaptic vesicles from the cholinergic synapses of the electric organ of Torpedo: A combined biochemical, electrophysiological and morphological study. J Neurochem (Lond) 22:435, 1974

Zimmerman H, Whittaker VP: Different recovery rates of the electrophysiological, biochemical and morphological parameters in the cholinergic synapses of the Torpedo electric organ after stimulation. J Neurochem (Lond) 22:1109, 1974

Chapter 18

Bennett V, O'Keefe E, Cuatrecasas P: The mechanism of action of cholera toxin and the mobile receptor theory of hormone receptor-adenylate cyclase interactions. Proc Natl Acad Sci USA, 1975 (in press)

Birnbaumer L: Hormone-sensitive adenylyl cyclases. Useful models for studying hormone receptor functions in cell-free systems. Biochim Biophys Acta 300:129, 10 Sep 1973

Cautrecasas P: Insulin receptor of liver and fat cell membranes. Symposium on Membranes and Mechanisms of Hormone Action. Fed Proc 32: 1838, Aug 1973

Cautrecasas P: Membrane receptors. Annu Rev Biochem 43:169, 1974

Cautrecasas P: Affinity chromatography of macromolecules. Pages 29-89, Advances in Enzymology, Vol 36, Meister A, Ed. John Wiley & Sons, Inc., New York, 1972

Cautrecasas P, Hollenberg MD, Chang KJ, Bennett V: Hormone receptor complexes and their modulation of membrane function. Recent Progress in Hormone Research, Vol 31, Academic Press, New York, 1975 (in press)

Illiano G, Tell GPE, Siegel ME, Cautrecasas P: Cyclic guanosine 3′:5′ monophosphate and the action of insulin and acetylcholine. Proc Natl Acad Sci USA 70:2443, Aug 1973

Lockwood DH, Livingston JN, Amatruda JM: The relation of insulin receptors to insulin resistance. Fed Proc, 1975 (in press)

Roth J: Peptide hormone binding to receptors: A review of direct studies in vitro. Metabolism 22: 1059, Aug 1973

Chapter 19

Robison GA, Butcher RW, Sutherland EW, Eds: Cyclic AMP. Academic Press, New York, 1971

Goldberg ND, Dietz SB, O'Toole AG: Cyclic guanosine 3′:5′ monophosphate in mammalian tissues and urine. J Biol Chem 244:4458, 25 Aug 1969

George WJ, Polson JB, O'Toole AG, Goldberg ND: Elevation of cyclic guanosine 3′:5′ monophosphate in rat heart after perfusion with acetylcholine. Proc Natl Acad Sci USA 66:398, Jun 1970

Grund VR, Goldberg ND, Hunninghake DB: Effect of histamine on lipolysis and adenosine 3′:5′ monophosphate level in canine adipose tissue. Biochem Pharmacology 22:769, 15 Mar 1973

Hadden JW, Hadden EM, Haddox MK, Goldberg ND: Cyclic guanosine 3′:5′ monophosphate: A possible intracellular mediator of mitogenic influences in lymphocytes. Proc Natl Acad Sci USA 69:3024, Oct 1972

Goldberg ND, Haddox MK, Hartle DK, Hadden JW: The biological role of cyclic guanosine 3′:5′ monophosphate. Proc Fifth International Congress on Pharmacology. San Francisco, 1972; Vol 5, pp 146-169, Cellular Mechanisms. Karger, Basel, 1973

Voorhees JJ, Colbrun NH, Duell EA, Haddox MK, Goldberg ND: Imbalanced cyclic AMP and cyclic GMP levels in the rapidly dividing, incompletely differentiated epidermis of psoriasis. Vol I, pp 635-648, The Cold Spring Harbor Symposium on Regulation of Proliferation in Animal Cells. Clarkson B, Baserga R, Eds. Academic Press, New York, 1974

Goldberg ND, Haddox MK, Estensen R, White JG, Lopez C, Hadden JW: Evidence for a dualism between cyclic GMP and cyclic AMP in the regulation of cell proliferation and other cellular processes. Pages 247-262, Cyclic AMP in Immune Response and Tumor Growth. Lichtenstein L, Parker C, Eds. Springer-Verlag, New York, 1974

Goldberg ND, O'Dea RF, Haddox MK: Cyclic GMP. Vol 3, pp. 155-223, Recent Advances in Cyclic Nucleotide Research. Greengard P, Robison GA, Eds. Raven Press, New York, 1973

Goldberg ND, Haddox MK, Dunham E, Lopez C, Hadden JW: The yin yang hypothesis of biological control: Opposing influences of cyclic GMP and cyclic AMP in the regulation of cell proliferation and other biological processes. Pages 609-626, The Cold Spring Harbor Symposium on the Regulation of Proliferation in Animal Cells. Clarkson B, Baserga R, Eds. Cold Spring Harbor Laboratory, New York, 1974

Estensen R, Hill HR, Quie PG, Hogan N, Goldberg ND: Cyclic GMP and cell movement. Nature 245:458, 26 Oct 1973

Dunham EW, Haddox MK, Goldberg ND: Alteration of vein cyclic 3′:5′ nucleotide concentrations during changes in contractility. Proc Natl Acad Sci USA 71:815, Mar 1974

Kuehl FA, Jr., Ham EA, Zanetti ME, Sanford C, Nicol SE, Goldberg ND: Estrogen-related increases in uterine cyclic guanosine 3′:5′ monophosphate levels. Proc Natl Acad Sci USA 71:1866, May 1974

Bernlohr RW, Haddox MK, and Goldberg ND: Cyclic guanosine 3′:5′ monophosphate in *Escherichia coli* and *Bacillus licheniformis*. J Biol Chem 249:4329, 10 Jul 1974

Chapter 20

Anderson RGG: Cyclic AMP and calcium ions in mechanical and metabolic responses of smooth muscles; influence of some hormones and drugs. Acta Physiol Scand (Suppl) 382:1, 1972

Borlé AB: Calcium metabolism at the cellular level. Fed Proc 32:1944, Sept 1973

Brostow CO et al: The regulation of skeletal muscle phosphorylase kinase by Ca++. J Biol Chem 246:(7):1961, 10 Apr 1971

Luckasen JR, White JC, Kersey JH: Mitogenic properties of a calcium ionophore, A23187. Proc Natl Acad Sci USA 71:5088, 1974

Maino VC, Green NM, Crumpton MJ: The role of calcium ions in initiating transformation of lymphocytes. Nature 251 (5473):324, 27 Sep 1974

Malaisse WJ: Insulin secretion: Multifactorial regulation for a single process of release. Diabetologia 9:167, Jun 1973

Prince WT, Berridge MJ, Rasmussen H: Role of calcium and cyclic adenosine 3′:5′ monophosphate in controlling fly salivary gland secretion. Proc Natl Acad Sci USA 69:553, Mar 1972

Rasmussen H: Cell communication, calcium ion and cyclic adenosine monophosphate. Science 170:(3956) 404, 23 Oct 1970

Rasmussen H, Goodman DB, Tenenhouse A: The role of cyclic AMP and calcium in cell activation. CRC Crit Rev Biochem 1:95, Feb 1972

Schultz G, Hardman JG, Schultz K: The importance of calcium ions for the regulation of cyclic guanosine 3′:5′ monophosphate levels. Proc Natl Acad Sci USA 70:3889, Dec 1973

Chapter 21

Rapin AM, Burger MM: Tumor cell surfaces: General alterations detected by agglutinins. Pages 1-91, Advances in Cancer Research, Vol 20. Klein G, Weinhouse S, Haddow A, Eds. Academic Press, New York, 1974

Burger MM: Surface changes in transformed cells detected by lectins. Fed Proc 32:91, Jan 1973

Sharon N, Lis H: Lectins: cell agglutinating and sugar-specific proteins. Science 177:949, 15 Sep 1972

Stoker M: Contact and short-range interactions affecting growth of animal cells in culture. Pages 107-128, Current Topics in Developmental Biology. Vol II. Monroy A, Moscona AA, Eds., Academic Press, Inc., New York, 1967

Dulbecco R: Topoinhibition and serum requirement of transformed and untransformed cells. Nature (Lond) 227:802, 22 Aug 1970

Fox TO, Sheppard JR, Burger MM: Cyclic membrane changes in animal cells: transformed cells permanently display a surface architecture detected in normal cells only during mitosis. Proc Natl Acad Sci USA 68: 244, Jan 1971

Chapter 22

Boyse EA, Bennett D: Differentiation and the cell surface: Illustrations from work with T cells and sperm. Page 155, Cellular Selection and Regulation in the Immune Response. Edelman GM, Ed. Raven Press, 1974

Katz DH, Benacerraf B: The regulatory influence of activated T cells on B cell responses to antigen. Adv Immunol 15:1, 1972

Vitetta ES, Uhr JW: Synthesis, transport, dynamics and fate of cell surface Ig and alloantigens in murine lymphocytes. Transplant Rev 14:50, 1973

β-2-microglobulin and HL-A antigen. Göran Möller, Ed. Transplant Rev 21: entire volume, 1974

Vitetta ES et al: Cell surface immunoglobulin XI. J Exp Med 141:(1):206, Jan 1975

Chapter 23

Good RA, Fisher DW: Immunobiology. Sinauer Associates, Inc., Stamford, Conn., 1971

Kinsky SC: Antibiotic interaction with model membranes. Annu Rev Pharmacol 10:119, 1970

Müller-Eberhard HJ: The molecular basis of the biological activities of complement. Harvey Lectures 66:75, 1972

Kinsky SC: Antibody-complement interaction with lipid model membranes. Biochim Biophys Acta 265:1, 1972

Uemura K, Kinsky SC: Active vs. passive sensitization of liposomes toward antibody and complement by dinitrophenylated derivatives of phosphatidylethanolamine. Biochemistry 11:22:4085, 1972

Six HR et al: Effect of immunoglobulin class and affinity on the initiation of complement dependent damage to liposomal model membranes sensitized with dinitrophenylated phospholipids. Biochemistry 12:20:4003, 1973

Six HR et al: Effect of antibody-complement on multiple vs. single compartment liposomes. Application of a fluorometric assay for following changes in liposomal permeability. Biochemistry 13:4050, 1974

Uemura K et al: Antibody formation in response to liposomal model membranes sensitized with *N*-substituted phosphatidylethanolamine derivatives. Biochemistry 13:8:1572, 1974

Chapter 24

Eyring H: Untangling biological reactions. Science 154:3757:1609, Dec 1966

Goodman LS, Gilman A: The Pharmacological Basis of Therapeutics. 4th Ed., The Macmillan Company, New York, 1970

Kwant WO, Seeman P: The membrane concentration of a local anesthetic (chlorpromazine). Biochim Biophys Acta 183:530, 1969

Miller KW, Paton WDM, Smith RA, Smith EB: The pressure reversal of general anesthesia and the critical volume hypothesis. Mol Pharmacol 9:131, Mar 1973

Seeman P: The membrane expansion theory of anesthesia: direct evidence using ethanol and a high-precision density meter. Experientia 30:750, 15 Jul, 1974

Seeman P: The membrane actions of anesthetics and tranquilizers. Pharmacol Rev 24:583, Dec 1972

Seeman P, Roth S: General anesthetics expand cell membranes at surgical concentrations. Biochim Biophys Acta 255:171, 17 Jan 1972

Chapter 25

Cooper RA, Arner EC, Wiley JS, Shattil SJ: Modification of red cell membrane structure by cholesterol-rich lipid dispersions. J Clin Invest 55:1:115, 1975

Eaton JW, Skelton TD, Swofford HS, Kolpin CE, Jacob HS: Elevated erythrocyte calcium in sickle cell disease. Nature 246:105, Nov 1973

Jacob HS: Dysfunctions of the red cell membrane. Page 269, The Red Blood Cell, 2nd Ed. Surgenor DM, Ed. Academic Press, New York, 1974

Jacob HS: The abnormal red-cell membrane in hereditary spherocytosis: evidence for the causal role of mutant microfilaments. Br J Haematol 23: Suppl 35, Sep 1972

McBride JA, Jacob HS: Abnormal kinetics of red cell membrane cholesterol in acanthocytes: studies in genetic and experimental abetalipoproteinaemia and in spur cell anaemia. Br J Haematol 18:383, Apr 1970

Nathan DG, Shohet SB: Erythrocyte ion transport defects and hemolytic anemia: "hydrocytosis" and "desiccocytosis." Semin Hematol 7:381, Oct 1970

Weed RI, LaCelle PL, Merrill EW: Metabolic dependence of red cell deformability. J Clin Invest 48:795, May 1969

Chapter 26

Wallingford WR, McCarty DJ: Differential membranolytic effect of microcrystalline sodium urate and calcium pyrophosphate dihydrate. J Exp Med 133:100, Jan 1971

Weissmann G, Zurier RB, Spieler PJ, Goldstein IM: Mechanisms of lysosomal enzyme release from leukocytes exposed to immune complexes and other particles. J Exp Med 134:149s, Sep 1971

Weissmann G, Rita GA: The molecular basis of gouty inflammation: Interaction of monosodium urate crystals with lysosomes and liposomes. Nature New Biol 240:167, 6 Dec 1972

Zurier RB, Hoffstein S, Weissmann G: Mechanisms of lysosomal enzyme release from human leukocytes. I. Effect of cyclic nucleotides and colchicine. J Cell Biol 58:27, June 1973

Weissmann G: Crystals, lysosomes, and gout. Advances in Internal Medicine 19:239, 1974

Hoffstein S, Weissmann G: Mechanisms of lysosomal enzyme release from leukocytes IV. Interaction of monosodium urate crystals with dogfish and human leukocytes. Arthritis & Rheumatism, 18:153, Mar-Apr 1975

Section Five

Index and Credits

Index

A

B

C

Page number in italics denotes illustration or table

Page number in italics denotes illustration or table

H

I

J

K

L

Page number in italics denotes illustration or table

Illustration Credits

CHAPTER 1: 4-6, Bunji Tagawa; 7, Albert Miller; 8, 9, Bunji Tagawa

CHAPTER 2: 14-17, 19-21, Albert Miller; 22, Stephanie Marcus, adapted from Caspar and Kirschner, Nature, 1971

CHAPTER 3: 26, 27, Albert Miller; 28, Robin Ingle; 29-31, Albert Miller; 32, 33, Robin Ingle

CHAPTER 4: 36, Albert Miller, data from G. Guidotti Annu Rev Biochem 41: 732, 1972; 37 (top), Bunji Tagawa; 37 (bottom), Bunji Tagawa based on Science 175: 720, 1972; 38 (top), Bunji Tagawa; 39, Albert Miller; 40, Bunji Tagawa based on Science 175:720, 1972; 41-44, Bunji Tagawa

CHAPTER 5: 46 (top), 49 (top), 50, 51, 53, Bunji Tagawa; 49 and 52 photos, © 1972 J Exp Med 135:6:1218, 1221

CHAPTER 6: 56 (bottom), 57, Albert Miller; 58, Bunji Tagawa; 59, Albert Miller; 60, 61, Bunji Tagawa; 62, Albert Miller; 63, Bunji Tagawa

CHAPTER 7: 67-73, Albert Miller

CHAPTER 8: 78-81, Alan Iselin

CHAPTER 9: 88-90, 92 (top), 94, Bunji Tagawa

CHAPTER 10: 96, 97, 99, Bunji Tagawa; 100, Bunji Tagawa with Alan Iselin; 101, 102 (top) Albert Miller

CHAPTER 11: 106, 107 (bottom), 112, Bunji Tagawa

CHAPTER 12: 117-121, Bunji Tagawa

CHAPTER 13: 126-129, 131, Bunji Tagawa

CHAPTER 14: 136-138, 140, 141, Bunji Tagawa

CHAPTER 15: 149-151, Bunji Tagawa

CHAPTER 16: 154-161 (drawings), 162, 165, Bunji Tagawa

CHAPTER 17: 169 Nancy Lou Gahan; 170, 172, Bunji Tagawa

CHAPTER 18: 178-181, 182 (bottom), 183, 184, Bunji Tagawa

CHAPTER 19: 187, 189, 191, 192, 194, 195, Bunji Tagawa; 196, 199, Albert Miller

CHAPTER 20: 204 (top), Bunji Tagawa; 204 (bottom), Albert Miller; 205-207, 208 (top), Bunji Tagawa; 208 (bottom), Albert Miller; 209, 210, Bunji Tagawa

CHAPTER 21: 217-221, Bunji Tagawa

CHAPTER 22: 224, 225, 226 (top), Bunji Tagawa; 226 (bottom), Albert Miller; 227 (top), Bunji Tagawa; 227 (bottom), Albert Miller; 228, Bunji Tagawa

CHAPTER 23: 232, 233 (top), 234-238, Bunji Tagawa

CHAPTER 24: 241, Albert Miller; 243, 244, Bunji Tagawa; 245, Albert Miller; 246 (top), Bunji Tagawa; 246 (bottom), Albert Miller

CHAPTER 25: 253, Albert Miller; 254 (top), Bunji Tagawa

CHAPTER 26: 259, 260, Albert Miller; 262, 264-266, Irving Geis

Data and Photo Sources

Chapter 6, 22 © 1975

Chapter 10, 11, 14, 16-20, 23-25 © 1974

Chapters 1-5, 7-9, 12, 13, 15, 21 © 1973

Chapter 26 © 1971

Sources of data used by authors are frequently mentioned in captions. Following are acknowledgements of sources not made in captions.

CHAPTER 8: 76 (top), photograph copyright Proc Natl Acad Sci USA 69:5:1134, 1972; 76 (bottom), photograph copyright Nature 227:5260:813, 1970; 82, photographs copyright Nature (New Biology) 242:215, April 18, 1973

CHAPTER 13: 125 (bottom left and right), courtesy of Histochemie (Springer-Verlag) 34:281:1973; 126 (top right), photograph used through the courtesy of Am J Pathol 72:295:1973

CHAPTER 15: 144, copyright Current Topics in Membrane and Transport Vol 3, p. 276, Academic Press, New York, London, 1972; 146 (left), Ibid. p. 282; (right), ibid. p. 284; 147 (left), ibid. p. 286

CHAPTER 16: 157, Fig. 4, copyright Cells & Organelles, Holt, Rinehart & Winston, Inc., 1970, p. 164; Fig. 5, ibid. p. 165; 160, Fig. 8, ibid. p. 116; 161 (top), ibid. p. 149; Fig. 9 (lower left), copyright J Cell Biol 59:273, 1973; 163, copyright J Cell Biol 41:606, 1969; 164, ibid. 57:359, 1973; 165, ibid. 42:392, 1969

CHAPTER 25: 250, copyright Arch Histol Jap 33:225, 1971; 251, adapted from Wilson et al, Proc Natl Acad Sci USA 66:807, 1970; 252, copyright Br J Haematol (suppl) 23:35, 1972; 254 (bottom), copyright Br J Haematol 18:383, 1970

CHAPTER 26: 263 (left), Gerald P. Rodin, M. D., University of Pittsburgh School of Medicine